Herbalism, Phytochemistry and Ethnopharmacology

Herbalism, Phytochemistry and Ethnopharmacology

AMRITPAL SINGH SAROYA
Herbal Consultant
Punjab
India

CRC Press
Taylor & Francis Group
an **informa** business
www.crcpress.com

6000 Broken Sound Parkway, NW
Suite 300, Boca Raton, FL 33487
270 Madison Avenue
New York, NY 10016
2 Park Square, Milton Park
Abingdon, Oxon OX 14 4RN, UK

Science Publishers
Enfield, New Hampshire

Published by Science Publishers, P.O. Box 699, Enfield, NH 03748, USA
An imprint of Edenbridge Ltd., British Channel Islands

E-mail: ***info@scipub.net*** **Website:** ***www.scipub.net***

Marketed and distributed by:

CRC Press
Taylor & Francis Group
an **informa** business
www.crcpress.com

6000 Broken Sound Parkway, NW
Suite 300, Boca Raton, FL 33487
270 Madison Avenue
New York, NY 10016
2 Park Square, Milton Park
Abingdon, Oxon OX 14 4RN, UK

ISBN 978-1-57808-697-9

CIP data will be provided on request.

The views expressed in this book are those of the author(s) and the publisher does not assume responsibility for the authenticity of the findings/conclusions drawn by the author(s). Also no responsibility is assumed by the publishers for any damage to the property or persons as a result of operation or use of this publication and/or the information contained herein.

Printed in Great Britain

Foreword

It is indeed an honour to be invited to write the Foreword for book titled 'Herbalism, Phytochemistry and Ethnopharmacology' with its twin lofty motto both to educate and to provide much needed scientific validation to herbal medicine.

The author has achieved in this book the multi-faceted objectives of emerging herbal science which is a blend of ancient knowledge coupled with scientific proofs. When I received an invitation from the author to write the Foreword to the book I was somewhat puzzled by the invitation. After all, during the past one decade, my work has focused upon the pharmacological aspects of Ayurvedic and herbal medicine. I have coauthored a number of research and review papers along with the author detailing herbal medicine and ethnopharmacology. And prior to that, my original scientific training and qualifications were as pharmacologist, rather than Herbal Medicine.

The scope of this book is truly impressive, reviewing the phytochemical and pharmacological aspects of herbal drugs. The book is thoroughly referenced and illustrates the scientific approach to herbalism. The book is not limited to one class of Complementary and Alternative Medicine but also highlights medicinal applications of lesser known systems like Amchi. The book clearly warrants application of algae, fungi and mosses in drug discovery. This book was need of the hour to provide clarifications regarding issues related to herbalism.

Herbalism and traditional medicine are backbone of the modern pharmaceutical industry. In China and Western countries, herbal medicine has come into the limelight because of advancement in research and development. Taxol (*Taxus brevifolia*), Silymarin (*Silybum marianum*), Artemisinin (*Artemisia annua*) are some reputed drugs in synthetic system of medicine. All the three drugs owe their origin to plants. We need to explore other systems of Complementary and Alternative Medicine such as Ayurveda, Siddha, Unani, Homeopathy and Amchi for more potent and life saving drugs. I believe that this book should be of great help to people concerned about the value of herbalism in future healthcare.

Dr. Samir Malhotra
Associate Professor, Dept of Pharmacology
Postgraduate Institute of Medical Sciences and Research
Chandigarh

Preface

This book has been written with the primary objective of providing scientific footage to medical herbalism. With advancement of our knowledge of the subject, a large number of texts have appeared which are useful in their own way but do not fully meet the interdisciplinary nature of herbalism. Herbalism coupled with phytochemistry and ethnopharmacology is a powerful tool for drug discovery.

The present treatise has been presented in a simple and lucid style so that a student who has even a limited knowledge of herbalism can understand the phytochemical and pharmacological aspects of herbalism. A simple, clear, well illustrated and a concise account of phytochemicals and experimental pharmacology has been presented. An introductory chapter in the beginning of the text is meant to acquaint the students with a general outline of various issues related to herbalism. Another objective has been to present the entire subject matter in light of new advancements in herbal drug discovery.

The book is primarily meant for students of phytochemistry, ethnopharmacology, phytotherapy and medicinal plants. For the benefit of the readers, concise information in form of tables has been included. Each chapter contains a complete bibliography which may particularly benefit postgraduate and research-oriented students and teachers.

It is certainly not possible to consider all aspects of herbalism in a single book, especially when there are genuine limitations such as lack of standardization and awareness. No originality is claimed in the preparation of this book. It is simply a compilation of work done in a manner so as to meet the needs of the students of herbalism and related subjects. A large of standard books on the subject and research journals have been consulted. I am also greatly indebted to all eminent herbal scientists of the world for sending me copies of original papers.

During the three-year period that this book passed through the different stages of preparation and production, I perforce ignored my family. For all kinds of personal help rendered to me by my parents and wife, I have no words to express my feelings. Special thanks is due to my pretty daughter Seerat Kaur for tolerating my whims and moods while working for the project. My heartiest thanks to Dr. Bhupinder Singh Bhoop,

Professor (Pharmaceutics & Pharmacokinetics), University Institute of Pharmaceutical Sciences, Punjab University, Chandigarh and Dr. Samir Malhotra, Associate Professor (Pharmacology), Postgraduate Institute of Medical Sciences and Research, Chandigarh for providing the stimulus to achieve this ambitious project.

Finally, I wish to express my thanks to my publishers for providing me the facilities in the publication of the book. Constructive suggestions from readers, for the improvement of possible future editions of the book, shall be gratefully acknowledged.

September 2010 **Amritpal Singh**

Contents

Chapter 1

Herbal Drug Industry

1.1 Introduction

A considerable number of definitions have been proposed for the term 'medicinal plant'. According to the World Health Organization, "a medicinal plant is any plant which, in one or more of its organs, contains substances that can be used for therapeutic purposes, or which are precursors for chemopharmaceutical semi synthesis".

This definition distinguishes between the already known medicinal plants whose therapeutic properties or characters are precursors of certain molecules which have been established scientifically, with that of other plants used in traditional medicine which are regarded as medicinal, but have not yet been subjected to a thorough scientific study.

Medicinal plants are a significant source of synthetic and herbal drugs. Patterns of herbal utilization are depicted in Fig. 1.1 Medicinal plants have been used for the treatment of diseases since antiquity. According to Alves and Rosa (2007), 20,000 plant species are used for medicinal purposes. India and China have been on the forefront when one refers to the history of herbal drugs. The traditional systems of medicines viz. Ayurveda, Siddha, Unani, Western Herbal Medicine, Traditional Chinese Medicine and Homeopathy have roots in medicinal herbs. Herbal medicines have been produced by a number of renowned researchers and due to its accessibility to traditions it is still practiced even by lay practitioners.

Ayurveda, the ancient healing system flourished in India in the Vedic era. The classical texts of Ayurveda, Charaka Samhita and Sushruta Samhita were written around 1000 B.C. The Ayurvedic Materia Medica includes 600 medicinal plants along with therapeutics. Herbs like turmeric, fenugreek, ginger, garlic and holy basil are integral parts of Ayurvedic formulations. The formulations incorporate a single herb or more than two herbs (polyherbal formulations).

The history of traditional Chinese medicine is renowned and the herbal system, is very well preserved. It originated about 3000 years ago

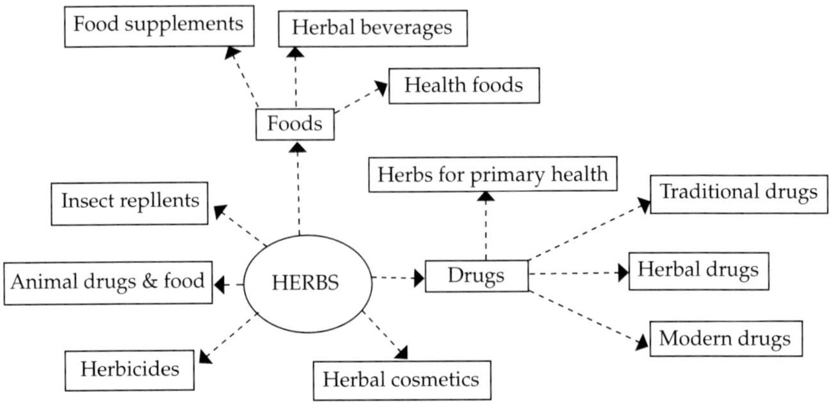

Fig. 1.1 Patterns of herbal utilization.
Source: Riewpaiboon (2003).

and is a popular science in western countries. Some of the medicinal herbs mentioned in Chinese medicine are common to Ayurveda. Traditional Chinese medicine favors the use of medicinal herbs in their natural form rather than by extraction. Herbal drugs have a different history in Europe and America and have produced healers like Culpeper. The use of tinctures in homeopathy is based on medicinal herbs.

Before the availability of synthetic drugs, man was completely dependent on medicinal herbs for prevention and treatment of diseases. The use of the medicinal herbs for curing diseases has been documented in the history of all civilizations. The drugs were used in crude forms like expressed juice, powder, decoction or infusion. Although the formulations mentioned in ancient texts are difficult to understand in terms of scientific parameters, some of them are reputed for their curative values. The Napralert database at the University of Illinois establishes ethno medicinal uses for about 9200 of the 33,000 species of monocotyledons, dicotyledons, gymnosperms, lichens, pteridophytes, and bryophytes.

Ancient healers, who developed formulations based on medicinal herbs, were probably not aware of the chemical composition of the herbs. However the advances they made despite the non-availability of scientific procedures is astonishing. The work on *Terminalia chebula* (chebulic myrobalan) mentioned in Charaka Samhita is quite authentic and modern studies have revealed that the purgative activity mentioned in Ayurveda is justified by the isolation of chebulic acid, the active constituent of chebulic myrobalan.

Initially, the term Materia Medica was coined for the study of natural products. Materia Medica is defined as the knowledge of natural history, physical characteristics, and chemical properties of drugs. It includes the study of herbs, minerals and drugs from the animal kingdom. The Ayurvedic equivalent for Materia Medica is Dravyaguna, which is the study of medicinal herbs in Ayurvedic terms. Now days the term 'Materia Medica' is known as pharmacognosy.

1.2 Phytomedicines and Herbal Extracts

A medicinal herb is considered to be a chemical factory as it contains a multitude of chemical compounds like alkaloids, glycosides, saponins, resins, oleoresins, sesquiterpene lactones and oils (essential and fixed). Some rare compounds like furanocoumarins, hydroxycoumarins, napthoquinones, acylphloroglucinols and sterones are also distributed among the plant kingdom. The active constituents are usually secondary metabolites, derived from biosynthetic pathways present within the plant tissue. In 1985 it wa s recorded that 74% of the 119 plant derived drugs were discovered as a result of chemical studies to isolate the active substances responsible for their traditional use.

Plants are used in different forms varying from powders to extracts. Powder represents the drug in the ground form and these types of preparations are considered to be crude. The Pharmacopoeia mentions standardized vegetable powders for therapeutic application. Herbal systems of medicine have become increasingly popular in recent years.

In light of the growing demand of herbal drugs, quality control and assurance is primarily important. The standardized herbal extracts are considered to be more scientific than crude drugs. The commonly employed technique for removal of the active substance from the crude drug is called ***extraction***. Selection of the solvent is very critical in preparing the extracts, because the active constituent of the plants have an affinity for solvents.

1. Water and petroleum ether are used for extraction of fixed and essential oils and sterones.
2. Chloroform and ether are used for extraction of alkaloids.
3. Water and alcohol are used for extraction of glycosides.
4. Tannins and phenols are extracted with alcohol and ethyl acetate.

Extracts are prepared by separating the soluble matter from vegetable tissues by application of a suitable solvent like alcohol, water or ether. The resultant liquid is concentrated by evaporation to obtain a liquid extract or concentrated almost to dryness to obtain the solid extract. Depending on the solvent used, the extracts are classified as alcoholic, etheral or aqueous.

The ***standardized herbal extract*** is a preparation, which contains a certain fixed proportion of the ***active constituent*** (Table 1.1). Although the most obvious aspect of standardization is the guaranteed content of one or more active constituents or marker compounds, standardization involves much more than guaranteed levels of constituents. For example, a standardized extract of *Papaver somniferum* contains not less than 9.5% of morphine. Standardization has a great impact on the quality of herbal products. Standardization helps in adjusting the herbal drug formulation to a defined content of a constituent or constituents with therapeutic activity.

Table 1.1 Constituents with known activity.

Extract	Constituent
Aesculus hippocastanum	Escin (s)
Atropa belladona	L-Hyoscyamine
Cassia angustifolia/acutifolia	Sennosides
Piper methysticum	Kava-pyrones (mixture of six)
Silybum marianum	Silymarin (mixture of six)

Source: Keller (2001).

The biological source of a drug has great impact on finished products in herbal drug preparation. Proper identification of the drug is significant for phytochemical screening, which further exerts importance on therapeutic activity of the medicinal herb. Thus presence of the identification standard is necessary in finished products of an herbal drug preparation. A constituent of a medicinal herb, which is used for quality control and assurance of the herbal product, is known as ***marker compound*** (Table 1.2). A marker compound may or may not have therapeutic activity.

Table 1.2 Constituents with known activity.

Extract	Constituent
Ginkgo biloba	Ginkgolides and/or Flavonoid(s) *and/or* Bilobalide
Hypericum perforatum	Flavonoids, e.g. Quercetin and/or hypericin and/or Hyperforin
Matricaria chamomilla	Flavonoid (s), e.g. Apigenine *and/or* Chamazulene *and/or* *Levomenol (-(α) Bisabolol)*
Valeriana officinalis	Valerenic acid and/or Hydroxy-valerenic acid and/or Acetoxy-valerenic acid

Source: Keller (2001).

Advantages of standardized extracts

Standardized extracts retain the chemical complexity typical of the natural plant, but offer the added advantage of guaranteed levels of certain key constituents. An increasing number of botanical medicines have had their clinical efficacy confirmed in clinical trials. The vast majority of clinical trials involving botanical medicines have used standardized extracts. The reason is simple: standardized extracts offer consistent and reproducible therapeutic effects and the highest degree of safety. Accordingly, standardized extracts produce the best clinical results.

Why standardization?

As botanical extracts are made directly from crude plant material, they can show very substantial variation in composition, quality, and therapeutic effects. The variation and diversity of life is enormous, even within a species. In other words, two medicinal plants of the same species may look similar, yet be substantially different in the levels of active constituents that they contain. Botanical medicines made from plants that differ markedly in their chemical constituents cannot produce the same therapeutic effects. Since the practitioner or consumer will be unable to assess the difference, they cannot compensate for it.

The consequence will be inconsistent clinical results.

Standardized extracts are:

- High quality extracts containing consistent levels of specified compounds
- Broad spectrum extracts containing recognized active constituents as well as a variety of other plant constituents (some of which may contribute to the overall therapeutic quality of the extract)
- Extracts subjected to rigorous quality controls during all phases of the growing, harvesting and manufacturing processes.

Clinical advantages:

- High quality extracts with consistent activity
- Consistent activity allows for more accurate prescribing
- Consistent activity allows for consistent clinical results
- Extensive quality control ensures the quality and safety of standardized extracts

Quality control ensures:

- That the correct botanical species is used
- That only high quality raw materials are used
- That no other plant material has been used
- That the plant material is not contaminated with pesticides, heavy metals, or other noxious agents, that the final extract complies with international limits for microbial content and that the final product is of a consistent high standard preparation.

Commonly used herbal extracts are listed in the Table 1.3.

Table 1.3 List of standardized herbal extracts prepared from medicinal plants.

S. No	Medicinal Herb	Standard	Percentage
1	*Achellia millefolium*	Essential oil	0.04%
2	*Adhatoda vasica*	Vasicine	0.5%
3	*Allium sativum*	Allicin	0.6%
4	*Andrographis paniculata*	Andrographolide	10%
5	*Arctostaphylos uva-ursi*	Arbutin	20%
6	*Asparagus racemosus*	Saponin	30%
7	*Azadirachita indica*	Azadiractin	2%
8	*Bacopa monneri*	Bacoside	20%
9	*Betula alba*	Flavonoids	1.6%
10	*Boswellia seratta*	Boswellic Acid.	40% & 70%
11	*Camelia sinensis*	Epigallocatechin Gallate	0.2%
12	*Capsicum frutescens*	Capsaicinoids	0.62%
13	*Centella asiatica*	Asiaticoside	3%
14	*Cholorella emersoni*	Chlorophyll	1%
15	*Cimicifuga racemosa*	Triterpenoid glycosides	2.5%
16	*Coleus forskohlii*	Forskohlin	10%–20%
17	*Commiphora mukal*	Guggulsterones	5%
18	*Cratageus oxycanthus*	Vitexin	5%
19	*Curcuma longa*	Curcumin	95%
20	*Cynara scolymus*	Cynarin	1%
21	*Echinacea angustifolia*	Echinacosides	4%
22	*Echinacea purpurea*	Cichoric acid	4%
23	*Embelia ribes*	Embellin	8%
24	*Ephedra sinica*	Ephedrine	6%
25	*Equisetum arvense*	Silicon dioxide	7%
26	*Eschscholtzia californica*	Protopine	0.20–0.24%
27	*Filipendula ulmaria*	Salicylic acid	0.20%
28	*Garcinia cambogia*	Hydroxy Cirtic Acid	50%
29	*Ginkgo biloba*	Flavonoglycosides	24%
30	*Glycyrrihiza glabra*	Glycyrrhizin	20%
31	*Gymnema sylvestre*	Gymnemic Acid	75%
32	*Hamamelis virginiana*	Tannins	8–12%
33	*Harpagophytum procumbens*	Harpagoside	2.5%
34	*Huperzia serrata*	Huperzine	5%
35	*Hydrastis canadensis*	Alkaloids	3%
36	*Hypericum perforatum*	Hypericin	0.3%
37	*Malpighia glabra*	Vitamin C	2.5–25%

Table 1.3 contd...

Table 1.3 contd...

S. No	Medicinal Herb	Standard	Percentage
38	*Matricaria chamomilla*	Apigenin	1%
39	*Matricaria recutita*	Apigenin	0.6%
40	*Melissa officinalis*	Rosmarinic acid	4 %
41	*Momordica Charantia*	Bitters	3%
42	*Myrciaria dubis*	Vitamin C	5–10%
43	*Ocimum Sanactum*	Ursolic Acid	8%
44	*Olea europaea*	Oleuropein	18%
45	*Orthosiphon staminens*	Sinensetin	0.20%
46	*Passiflora Incarnata*	Vitexin	4%
47	*Paullinia cupana*	Caffeine	5–22%
48	*Perilla frutescens*	Polyphenols	3%
49	*Peumus boldus*	Boldine	0.05%
50	*Phylanthus niruri*	Bitters	2%
51	*Picrorrhiza kurroa*	Kutkosides	10%
52	*Piper methysticum*	Kavalactones	30%
53	*Pueraria lobata*	Daidzein	15%
54	*Pueraria tuberosa*	Disogenin	7%
55	*Rhamnus frangula*	Frangulin	10–20%
56	*Rhamnus purshiana*	Cascaroside A	18–22%
57	*Ruscus aculeatus*	Saponins (Ruscogenins)	10%
58	*Salix tetrasperma*	Salicin	10%
59	*Sambucus nigra*	Flavonoids	3%
60	*Saraca indica*	Tannins	8%
61	*Sereno repens*	Fatty acids	20–25%
63	*Sida cordifolia*	Phenylpropanolamine	0.02%
64	*Silybum marianum*	Silymarin	70%
65	*Spirulina maxima*	Phycocyanin	2.5%
66	*Terminalia arjuna*	Tannins	8%
67	*Terminalia belerica*	Tannins	40%
68	*Terminalia chebula*	Tannins	60%
69	*Tribulus terrestris*	Saponin	20% & 40%
70	*Trigonella foenum graecum*	Saponin	10%
71	*Uncaria tomentosa*	Saponin	2%
72	*Urtica dioica*	ß-sitosterol	0.8%
73	*Vaccinium myrtillus*	Anthocyanosides	25%
74	*Valeriana officinalis*	Valerenic acid	0.8%
75	*Vitex agnus castus*	Aucubin	0.6%
76	*Vitis vinifera*	Proanthocyanidins	95%
77	*Withania somnifera*	Withanolides	1.5%
78	*Zingiber officinale*	Gingerols	5%

Source: Singh and Sandhu (2005).

1.3 Markets and Marketing Issues

Herbal medicine and natural pharmaceuticals are moving from the fringe to mainstream, with a larger number of people seeking remedies and health approaches free of the side-effects caused by synthesized chemicals (Fig. 1.2). This was considered one of the most vital and high growth industries of the 90s and is set to expand even further into the next century.

The increasing acceptance of herbal medicines in Australia is well supported by trends around the world. In Germany and France which together represent 39% of the $14 billion global retail market, herbal remedies known as phytotherapeutics are well established, and the cost for therapeutic use is covered by health insurance systems, and the quality criteria applied to regulation and manufacturing are comparable to those for chemical drugs.

The crude botanical raw materials for this industry have been grown for long and traded in many countries around the world. As the Australian market for herbal medicine develops, opportunities are arising for raw materials to be grown in Australia, both for local and export markets. Access to export markets may be facilitated by the `clean green' image that Australian agriculture presents to the world.

Botanical raw materials are comprised of dried plant materials in the form of roots, barks, herbs, flowers, fruits, seeds, and resins. These materials are traded in a whole form or, more commonly, are cut and sifted to a consistently even particle size. Market prices for raw botanical

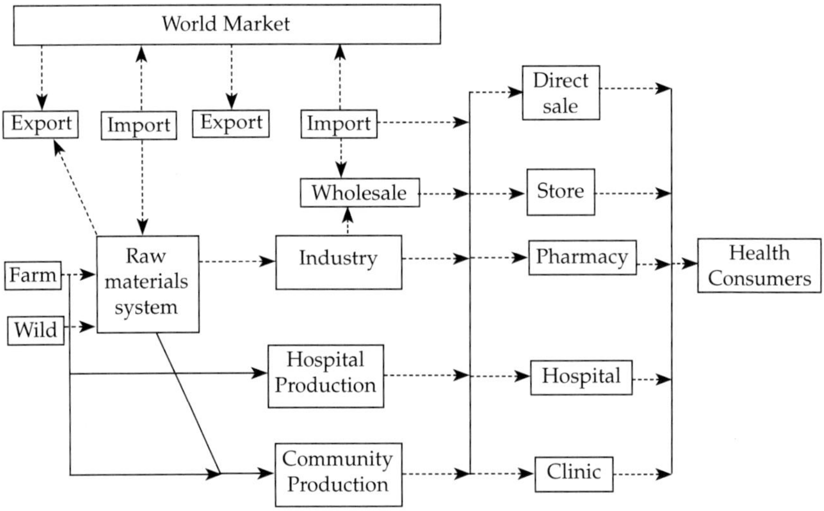

Fig. 1.2 Herbal drug market.
Source: Riewpaiboon (2003).

materials are usually determined by supply and demand, but generally tend to be stable.

At least 25% of medical prescriptions in recent times contain one active compound from plant species, Duke (1990) estimated that the value of drugs containing compounds from plant species is about US$ 10 billion in the United States alone (see Table 1.1). Most traded European materials are priced at source in the range of $2.00 to $6.00/kg. Prices for certified organic produce can be anywhere between $10.00 and $30.00/kg. This supply market is very limited, hence the high prices. Prices for difficult-to-grow, wild-harvested, or certified organic materials, usually North American botanicals, can range in price from $20.00 to $120.00/kg (Table 1.4).

Table 1.4 World market for flavor and fragrances (US$ m in 1990).

Flavours	2300
Fragrances	2400
Essential oils & other natural extracts	1050
Synthetic aroma chemicals	1100

Source: Pearce (1988).

The principal primary market for these raw materials is to industries that manufacture:

- Essential oils (Table 1.5)
- Liquid extracts and tinctures
- Herbal teas
- Concentrated soft extracts (for further industrial application)
- Concentrated dry extracts (for further industrial application)
- Plant-derived pure pharmaceutical drugs.

The pharmaceutical industry is both large and highly successful. Sales of plant derived drugs reached $30 billion worldwide in 2002. At present about 50% of the total plant-derived drug sales come from single entities, while the remaining 50% come from herbal remedies. Although the latter have greater volumes of consumption, the relatively low volumes of single entities, which are mostly prescription products, are more than compensated by their higher prices.

Single entity plant drugs, which mostly treat serious medical ills, include atropine, digoxin, morphine, paclitaxel, pilocarpine, reserpine, scopolamine, topotecan and vincristine, among many others. Several of the compounds have outlived their usefulness in light of better alternatives, and are exhibiting a decline in sales. On the other hand, as a consequence of new drug developments, single entities overall are projected to increase their market share of the combined total future dollar sales.

Table 1.5 Some major essential oils and their compounds.

Source	Compounds
Angelica	Alpha-Pinene, Limonene
Clove leaf	Eugenol, Caryophylene
Citronella	Geraniol, Citronellol
Eucalyptus	Cineol
Lavender	Linalylacetate
Lemon grass	Citral
Lemon	Limonene
Peppermint	Menthol
Spearmint	Carvone

Source: Haq (1993).

Sales of these plant-based drugs in the U.S. amounted to some US$4.5 billion in 1980 and an estimated US$15.5 billion in 1990. Other drugs are derived from animals and microorganisms. Development of drugs from plants is a long and arduous process which involves many disciplines. It has been estimated that only 5 to 15% of the approximately 250,000 species of higher plants have been systematically investigated for the presence of bioactive compounds. In industrialized countries, substances in everyday use derived from plants are -digitalin, ephedrine, morphine, quinine and many more. While the ones used less often like reserpine, guggulipid and artemisinin are equally well known.

All 119 plant-derived drugs used worldwide in 1991 came from fewer than 90 of the 250,000 plant species that have been identified. "Each such plant is a unique chemical factory", according to Norman R. Farnsworth of the University of Illinois at Chicago, "capable of synthesizing unlimited numbers of highly complex and unusual chemical substances whose structures could (otherwise) escape the imagination... forever".

In other words, scientists may be able to synthesize these plant compounds in the laboratory. Commercially, these plant derived medicines are worth about US$14 billion a year in the United States and US$40 billion worldwide.

In 1990s, the U.S. National Institute earmarked US$8 million to screen 50,000 natural substances for activity against 100 cancer cell lines and the AIDS virus. China, Germany, India, and Japan, among others, are also screening wild species for new drugs.

The European market for herbal supplements is estimated at over US$ 2.7 billion and for herbal remedies, a further US$ 0.9 billion. Germany is by far the largest market. The market is growing rapidly at over 4% per annum for herbal remedies and considerably faster for herbal supplements. The US herbal market is nearing saturation and is expected to peak at US$ 6–8 billion in the next few years.

Table 1.6 Synthetic vs. phytomedicines.

S.No.	Parameter	Synthetic	Phytomedicine
1.	Cost	High	Low
2.	Chemistry	Usually simple	Usually complex
3.	Target	High	Usually complex
4.	Affinity	High	low
5.	Potency	High	low
6.	Incidence of side effects	Higher, often unpredictable	Lower, usually predictable
7.	Action	Drastic changes in physiological events	Restore physiological balance
8.	In vitro test	Often adequate	Inadequate
9.	Patents	Easy	Difficult

Source: Rajasekharan (2000).

Their dietary herbal supplement market is estimated at US$4 billion and has been growing at 6–8% per annum. The main producers are manufacturers based in the developed countries, including the large multinational pharmaceutical companies. There are also smaller companies that specialize in herbal products and some have emerged to challenge the multinationals for market leadership in this field.

The main products sold are based on plants such as Echinacea and St. John's Wort that were known for their medicinal properties in the consuming countries. Recent research has helped propel the knowledge of other plants from around the world and this has helped accelerate the development of new supplements and medicines. The market share of herbal products made in developing countries remains comparatively low.

Chinese products, mainly in herbal supplements, have achieved major successes. The EU and the US regulations have special provision for herbal medicines that do not use mixtures of herbs. In this respect their regulations are, comparatively, relaxed. But if the exported products contain herbal mixtures and claim curative properties, the rules become much stricter. For medicines, product trials need to be carried out that cost several millions of dollars.

Scientific knowledge of the products produced in the developing countries, and of their systems of traditional medicine, is limited and this also restricts the market for their herbal products. As markets grow, the search for a wider variety of ingredients is increasing. Phytomedicines have already started to link traditional medicines with modern (allopathic) medicine, with research and development primarily funded by large pharmaceutical manufacturers.

Demand for medicinal plants is expected to continue to expand rapidly, fuelled by the growth of sales of herbal supplements and remedies. Their basic uses in medicine will continue in the future, as a source of therapeutic agents, and as a raw material base for the extraction of semi-synthetic chemical compounds such as cosmetics, perfumes and food industries.

1.4 Medicinal plant analysis: HPLC

High performance liquid chromatography *(HPLC)* is a highly selective chemical analytical technique, which is suitable for the analysis of a range of plant materials and botanical extracts. HPLC provides both qualitative information about the composition of the sample and quantitative data about the amount present of each constituent.

Although HPLC is a sophisticated analytical technique requiring expensive technology, the method is based on the same simple principles as other types of chromatography. A high performance liquid chromatograph contains a narrow stainless steel column. This column is packed with a material *(usually fine silica particles)* that selectively absorbs molecules on the basis of some difference in their chemical structure.

Due to the differential retention of individual molecules traveling through the column, compounds in a sample are separated. Compounds that are absorbed weakly or not at all have the shortest retention time and will exit the column first.

Compounds that are absorbed more strongly have longer retention times and will exit the column more slowly.

As the individual constituents of the sample exit the column, they are monitored by a detection device, which is linked to a printer. Each peak in the chromatogram represents a chemical compound, and the height of the peak is proportional to the concentration of that particular compound in the sample.

High performance liquid chromatography is carried out under high pressure and is therefore much faster than most other chromatography techniques. By way of comparing HPLC 'fingerprints' with reference standards, plant material and botanical extracts can be identified unequivocally. However, HPLC also provides detailed information about the composition of the sample.

HPLC is therefore widely used in the quality control of standardized botanical extracts, to ensure the presence of key constituents in specified amounts. The figure below shows a three-dimensional HPLC chromatogram of the silymarin extract from milk thistle (*Silybum marianum*).

The vertical axis shows the concentration, the horizontal axis is the retention time in minutes, and the third axis shows the wavelengths of detection. The group of peaks visible between 12.5 and 17.5 min *(the peaks in the right-hand part of the chromatogram)* represent the silybin-type compounds: silybin A, silybin B, and isosilybin A/B. Silymarin is standardized to specified levels of these compounds, and the peaks representing them are validated using reference standards.

The peaks that appear to the left of the silybin peaks represent other constituents of the extract. These are not quantified or validated as part of the standardization process, but they contribute to the total 'fingerprint' of silymarin.

The overall HPLC profile is used to monitor individual batches of extract in order to ensure consistent constituent profiles and consistency in therapeutic activity.

1.5 Test Procedures for Herbal Drug Preparations and Products

A specification is defined as a list of tests, references to analytical and biological procedures, and appropriate acceptance criteria, which are numerical limits, ranges, or other criteria for the tests described. It establishes the set of criteria to which an herbal drug preparation (herbal drug) or herbal medicinal product should conform to be considered acceptable for its intended use.

The applicant shall not be required to provide the results of trials if he can demonstrate:

Article 4 No. 8 a) ii

By detailed reference to published scientific literature that the constituent or constituents of the proprietary medicinal product have a well established medicinal use, with recognized efficacy and an acceptable level of safety.

Commission directive 1999/83/EC of 8 September 1999

A. Whereas it is in particular necessary to clarify that 'bibliographic reference' to other sources of evidence (post marketing studies, epidemiological studies, studies conducted with similar products). and not just tests and trials may serve as a valid proof of safety and efficacy of a product if an applicant explains and justifies the use of these sources of information satisfactorily.

B. The Expert report must explain the relevance of any data submitted which concern a product different from the product intended for marketing. A judgment must be made whether the product studied can be considered as similar to the product which will be granted a marketing authorization in spite of the existing differences.

Well established use

The Expert report must explain the relevance of any data submitted which concern a product different from the product intended for marketing. A judgment must be made whether the product studied can be considered as similar to the product which will be granted a marketing authorization in spite of the existing differences.

In the case of 'classical' herbal drug preparations such as tinctures and extracts described in pharmacopoeias and used for long time, a 'comprehensive' specification will not be available from published literature in most cases. For these preparations the starting material and the extraction solvent must be identical. If there are reasons to expect a different pharmacological or toxicological profile, additional data and an update of the specification and/or appropriate data on bioavailability may be necessary.

Aspects of a specification for an extract

The example of Commission E monograph Ginkgo extract (Fig 1.3) is cited here.

Requirements

- Extract (acetone 60%) from dried Ginkgo leaves
- Drug: extract ratio = 35–67: 1 (mean 50: 1)
- 22–27% flavone-glycosides 5–7% terpene lactones (2.8–3.4% Ginkgolides and 2.6–3.2% Bilobalide)
- < 5 ppm Ginkgolic acids

Ginkgolic acids are alkylphenols related to compounds found in poison ivy (*Rhus toxicodendron*). They can cause contact allergy, The German Institute for Drugs and Medicinal Plants has set a maximum limit of 5 parts per dermatitis. Ginkgolic acids are primarily found in the seed cover of *Ginkgo biloba* but also occur in small amounts in the leaf. In Germany the two best-selling ginkgo products meet the standard for ginkgolic acid content, as these compounds are eliminated or reduced during the extraction or manufacturing processes. In the United States there is no regulation of ginkgolic acid content of ginkgo preparations at present.

In the case of rapid release herbal medicinal products without constituents with known therapeutic activity, the test for *in-vitro* active ingredient release can be omitted (Fig. 1.4, 1.5).

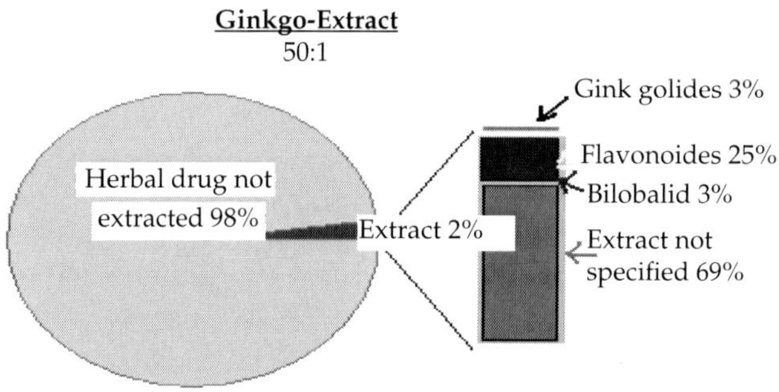

Fig. 1.3 Specification for Ginkgo extract.

Color image of this figure appears in the color plate section at the end of the book.

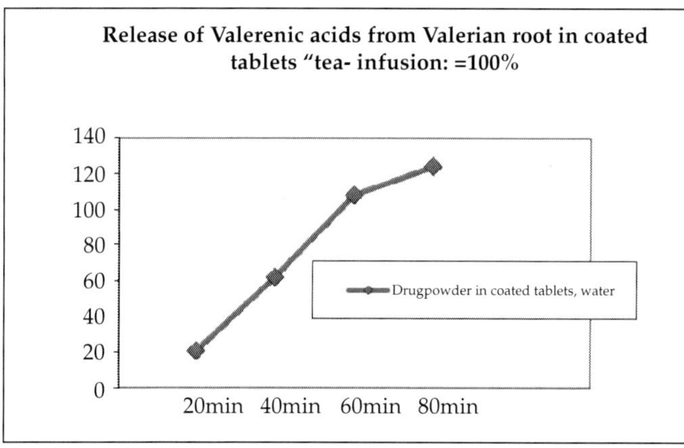

Fig. 1.4

Color image of this figure appears in the color plate section at the end of the book.

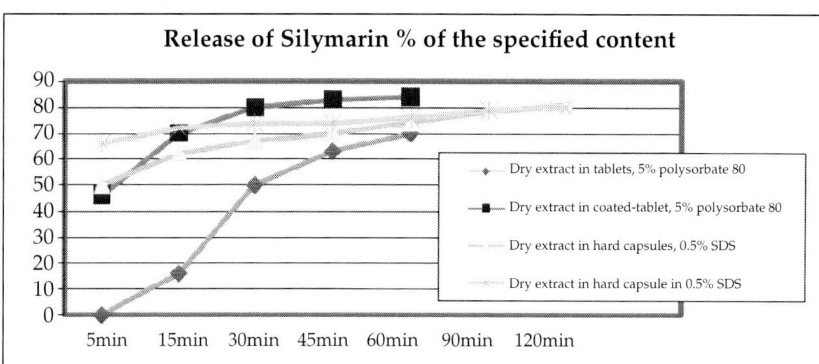

Fig. 1.5

Source: Keller (2001).

Color image of this figure appears in the color plate section at the end of the book.

Bioavailability/bioequivalence

Pharmacokinetic data for complex mixtures, where 'constituents with known therapeutic activity' are not established, should not be required (Fig. 1.6). Requirements for pharmacokinetic data should always relate to the Public Health and safety risks associated with the product. When there are safety concerns (e.g. photosensitization in the case of *Hypericum perforatum*) pharmacokinetic data are useful to provide safety margins. If there is a constituent with known therapeutic activity and a narrow therapeutic range, pharmacokinetic data will be required.

The same principles as defined for pharmacokinetic data shall apply for bioavailability/bioequivalence data. Need to provide bioavailability data for galenic preparations with modified release in view of the difficulty to use bibliographical data in such case (Source: European Medicines Evaluation Agency, London, 2001).

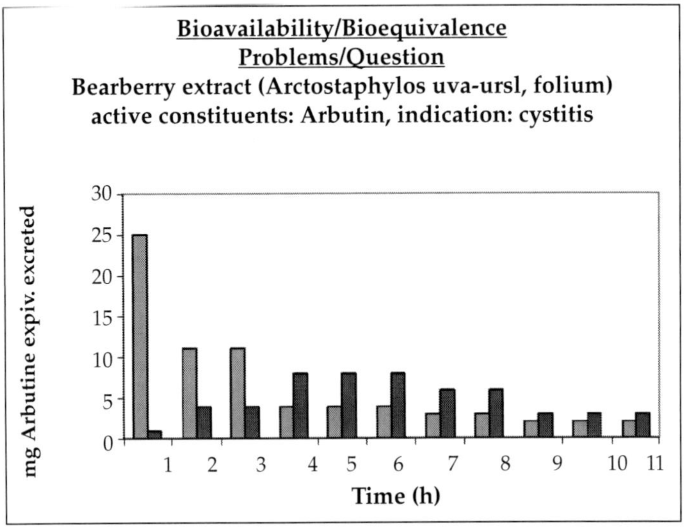

Fig. 1.6

Source: Keller (2001).

Color image of this figure appears in the color plate section at the end of the book.

References and further reading

Alves RRN and Rosa IL. 2007. Biodiversity, Traditional Medicine and Public Health: where do they meet? J. Ethnobiol. Ethnomed.: 3, 1–9

Anonymous. 1998. Indian Herbal Pharmacopoeia Vol. 1. Indian Drug Manufacturers Association. Worli, Mumbai.

Chung B. 2001. Natural Plant Extracts Export Market Opportunities in the USA. Rural Industries Research and Development Corporation, Kingston ACT, Australia.

Eisenberg DM, Kessler RC, Foster C, Norlock FE, Calkins DR, and Delbanco TL. 1993. Unconventional medicine in the United States—Prevalence, costs and patterns of use. NEJM. 328, 246–252.

Farnsworth NR. 1980. The development of pharmacological and chemical research for application to traditional medicine in developing countries. J. Ethnopharmacol. 2, 173–181.

Farnsworth Norman R, Olayiwola A, Bingel AS, Soejarto DO and Zhengang G. 1985. Medicinal Plants in Therapy. Bull. World Health Organ. 63, 965–981.

Flynn R, and Roest M. 1995. Your guide to standardized herbal products. One World Press, Prescott, AZ.

Haq, N. 1993. Breeding and improvement of medicinal and aromatic plants. In: N. Chomchalow and H. V. Henle (eds) Medicinal and Aromatic Plants in Asia. RAPA Pub. No: 19/93.

Keller K. 2001 Herbal Medicinal Products in the EU Biopharmaceutical/Quality Issues. European Medicines Evaluation Agency, London, UK.

Rajasekharan PE. 2000. Medicinal plants and the pharmaceutical industry. Indian Institute of Horticultural Research, Bangalore, India.

Riewpaiboon A. 2004. Increasing herbal product consumption in Thailand during the period 2001–2003. Second International Conference in Improving Use of Medicines. Chiang Mai, Thailand.

Singh AP and Sandhu AS. 2005. A Dictionary of Medicinal Plants. Sundeep Publishers, New Delhi, India, pp. 237–239.

Tyler V. 1993. The Honest Herbal: A Sensible Guide to the Use of Herbs and Related Remedies. 3rd Ed. Pharmaceutical Products Press, New York, NY. pp. 237–238.

Tyler V. 1994. Herbs of Choice: The Therapeutic Use of Phytomedicinals. Pharmaceutical Products Press, Binghamton, NY p. 119.

Wasik J. 1999. The truth about herbal supplements. Consumer's Digest., Deerfield, Illinois, USA July/August: 75–79.

World Health Organisation. 1978. The promotion and development of traditional medicine. World Health Organization, Geneva. (Technical reports series no. 622).

Chapter 2

Reverse Pharmacology

2.1 Introduction

Reverse pharmacology is the science of integrating documented clinical/experiential hits, into leads by transdisciplinary exploratory studies and further developing these into drug candidates by experimental and clinical research (Figs. 2.1, 2.2).

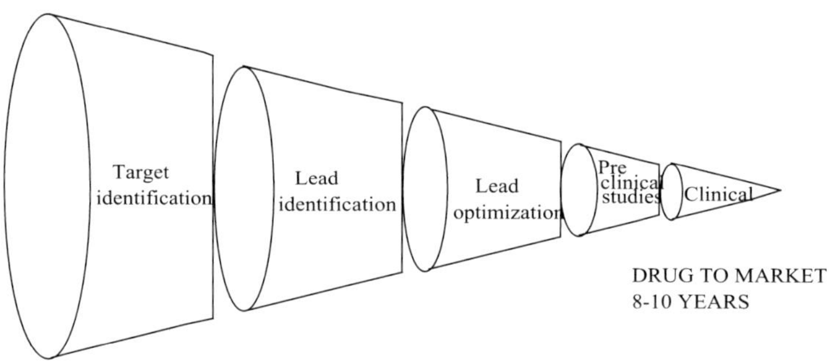

Fig. 2.1 Drug development with conventional pharmacology.

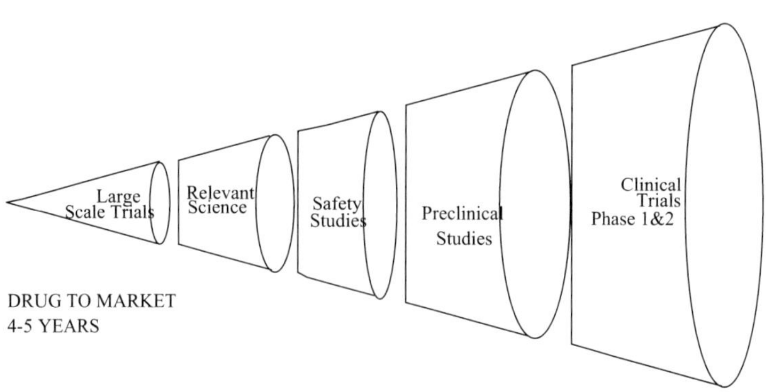

Fig 2.2 Drug development with reverse pharmacology.

2.2 Scope

The scope of reverse pharmacology is to understand the mechanisms of action at multiple levels of biological organization and to optimize safety, efficacy and acceptability of the leads in natural products, based on the relevant science.

2.3 Dimensions

* *Experiential documentation*: Pharmacoepidemiology-standardized formulation with HPLC pattern standardized formulation with HPLC pattern.
* Exploratory human/animal studies exploratory human/animal studies—Relevant models of activity relevant models of activity—Human dose determination human dose determination.
* Experimental programs
  - Levels of biological organization.
  - Rapid drug development path.
  - Leads: Comb. Leads: Combinatorial Chemistry & HTPS.

2.4 Examples

* *Mucuna pruriens* for Parkinson's disease
* *Zingiber officinale* for nausea/vomiting
* *Picrorhiza kurroa* for Hepatitis
* *Curcuma longa* for Oral Cancer
* *Panchavalkal* for burns and wounds
* *Azadirachta indica* for Malaria

Mucuna pruriens for Parkinson's disease

L-dopa content: 2.5g%–12.0g%. Other phytochemicals: mucunain, pruriendine, 5-HTP, aminoacids, dopaquinones, melanin, alanine, linolenic, methionine, and niacin etc. *M. pruriens* seeds and HP-200 standardized for L-dopa content (3.9–4.2% and 3.0–3.7%, respectively). A study by Vaidya et al., (1978) reported the role of *M. pruriens* in the treatment of Parkinson's disease (Fig. 2.3).

Zingiber officinale for nausea/vomiting

Modern-day uses for ginger in Eastern medicine include the use of the herb to treat nausea (including motion sickness and morning sickness of pregnancy). The main components of ginger are the aromatic essential

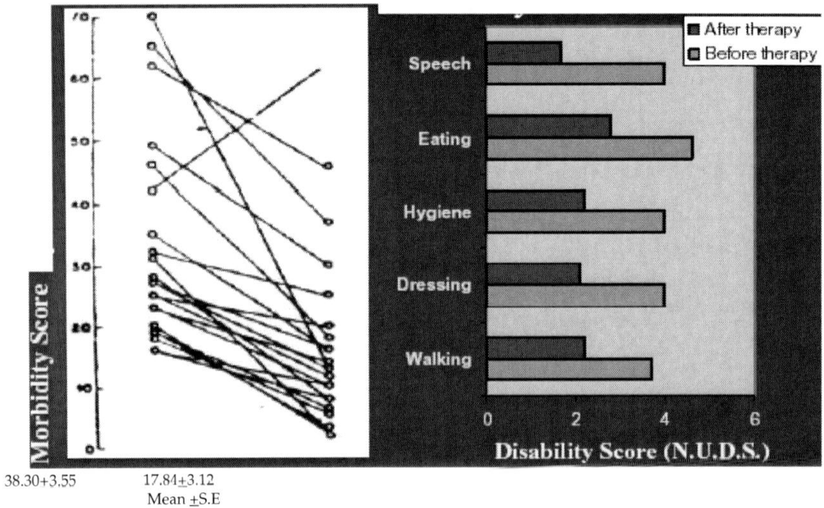

Fig. 2.3 Clinical activity of *M. pruriens* in Parkinson's disease.

Color image of this figure appears in the color plate section at the end of the book.

oils, antioxidants, and the pungent oleo-resin. These aromatic or pungent compounds have been identified as the phenylalkylketones, known as gingerols, shogaols, and zingerone.

Zingerone, diterpenoid constituent of ginger has been shown to have an activity similar to a 5HT3 antagonist, similar to ondesteron and other anti-emetic drugs used as adjuncts to chemotherapy. 5HT3 receptors are found in both the chemoreceptor trigger zone and on the vagal nerve terminals in the intestine. The anti-emetic effects of ginger are due to its local effect on the vagal receptors in the stomach.

A double-blind randomized placebo-controlled trial was carried to investigate the effect of a ginger extract on the symptoms of morning sickness. The participants included 120 women less than 20 wk pregnant, who had experienced morning sickness daily for at least 1 wk and had had no relief of symptoms through dietary changes. Random allocation of 125 mg ginger extract equivalent to 1.5 g of dried ginger or placebo given 4 times per day for 4 d. The nausea experience score was significantly less for the ginger extract group relative to the placebo group after the first day of treatment and this difference was present for each treatment day.

Azadirachta indica for Malaria

There are a number of references in classical Ayurvedic texts for neem as an antimalarial agent. Classical formulations for Vishamajwara include Nimbadi Kwath, Amruta Nimba Kwath, Panchatiktaka Ghrita, and Jwarhara Kwath. Neem is still widely used in India. *In vitro* studies with limonoids isolated from neem extract have shown efficacy against Plasmodium species (Table 2.1). Nimbolide (Fig. 1.2) was identified as an antiplasmodial compound (IC50=0.95 ng ml–1, P. falciparum K1). The derivatives nimbinin and geduine (Fig. 2.4) and its dihydro-derivative were also found to be active *in vitro* against Plasmodium parasites with EC50 values from 0.72 to 1.74 lg ml–1.

Table 2.1 Efficacy of limonoids of neem in Plasmodium species.

Limonoid	**IC50 (μmol/l)**	**Falciparum clone**	**Reference**
Gedunin	0.200	W2(CQ-9)	MacKinnon et al., 1997
Gedunin	0.039	D6(CQ-S)	MacKinnon et al., 1997
Nimbinin	0.77	K1(MDR)	Bray et al., 1990
Nimbolide	2.0	K1(MDR)	Rochanakij et al., 1985
Nimbin	50	K1(MDR)	Bray et al., 1990

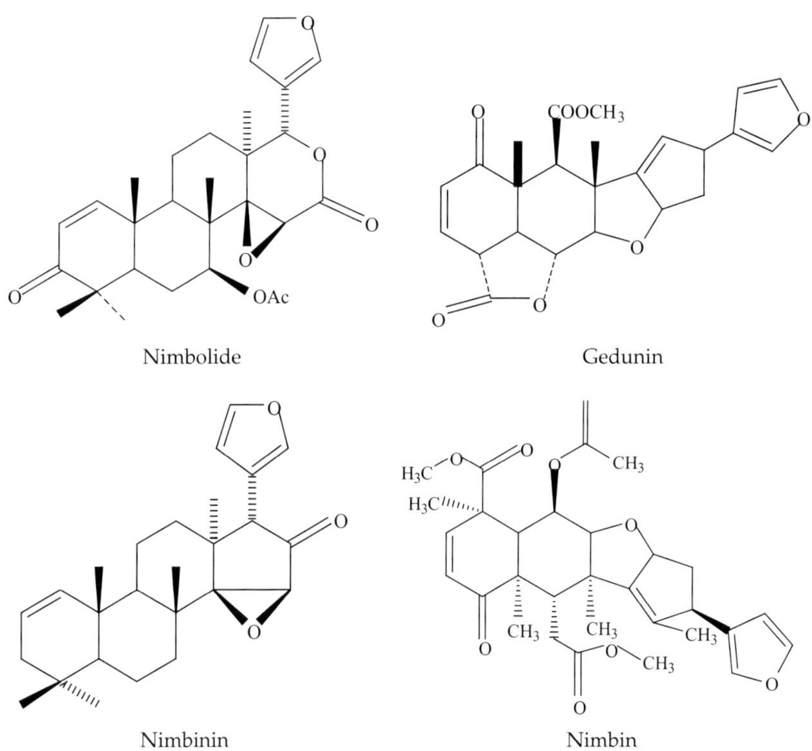

Fig. 2.4 Antimalarial limonoids of neem.

2.5 Reverse pharmacology and Ayurveda

Reverse pharmacology correlates with Ayurvedic drug action. The approach has attracted considerable attention, nationally and internationally. Central Scientific Instrumental Research (CSIR) and Indian Council of Medical Research (ICMR) have done clinical trials with natural products. Moreover, Central Council for Research in Ayurveda and Siddha (CCRAS) has recently adopted the golden triangle approach for some new indications of old drugs, as well as for Ayurveda (Fig. 2.5). The golden triangle approach is a combination of Dravyagunavignyan, systems biology, and reverse pharmacology for the discovery of potent and cost-effective remedies. Dravyaguna deals with the Ayurvedic study of drugs derived from natural (plant, animal or marine) origin. Dravya refers to constituent of the universe and guna signifies property. In the recent past, the study of Dravyaguna has become more important because of global acceptance of the Ayurvedic system of medicine. Ayurveda has its own concept as far as drug formulation is concerned. Ayurveda has documented Dravya, Guna, Rasa, Virya, Vipaka, Prabhava and Karma for all medicinal agents and these represent the pharmacological aspects of drug usage in Ayurveda.

In Ayurveda, drugs have been classified in a number of ways but the classification based on action of the drug is widely accepted. For instance a drug used for alleviating worm infection is known as anthelmintic and in Ayurvedic language, as krimighana. In modern medicine, drugs have been classified according to pharmacological actions. Medicinal plants like Ashwagandha, Brahami, Tulsi, Guggul, Kutki, Kalmegha, Gokshura and Shatavari have been targeted for their application in modern science. Active constituents of the plants have been identified and highly purified extracts are being marketed.

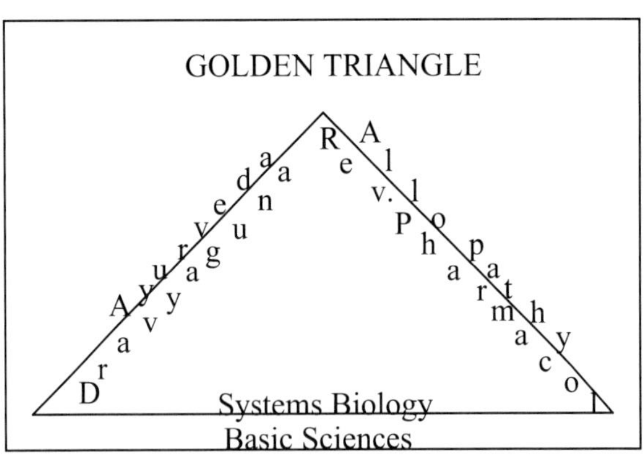

Fig. 2.5 Golden triangle approach.

Studying the Ayurvedic drugs at bimolecular levels may reveal the mechanism of action which has eluded scientists for long time. As far as the disease segment is concerned, hepatology and rheumatology are two areas where Ayurvedic remedies are even prescribed by allopathic physicians. *Silybum marianum* is a well-documented western herbal remedy used for liver diseases. In India, *Picrorhiza kurroa* (kutki), is considered a valued drug for liver diseases. When *P. kurroa* was compared with *S. marianum*, the hepatoprotective effect of *P. kurroa* was found to be similar, or in many cases, superior to the effect of *S. marianum*. However it can be seen that *S. marianum* is more popular than *P. kurroa*. Silymarin the active constituent of *S. marianum* has been isolated and purified and above all pharmacological and pharmacokinetic data is available for the drug.

The reverse approach in pharmacology has been quite successfully applied in the past. The drawback was the long time frame from the observational therapeutics to a new drug. For example, *Rauwolfia serpentina* (sarpagandha) was convincingly demonstrated to be anti-hypertensive by Sen and Bose in 1931. But a drug reserprine, emerged only after 20 yr of work by Vakil, Bein, Muller and Schlitter. This occurred because the path of reverse pharmacology was quite discontinuous.

The paradigm of reverse pharmacology is actually a rediscovery of the path, which founded modern pharmacology. Table 2.2 lists the names of plants, clinical effects, and experimental correlates. The list illustrates how novel clinical bio-dynamic effects can lead to the development of the basic disciplines in pharmacology and biology.

Table 2.2 Rediscovery of the paradigm of reverse pharmacology (see Fig. 2.6 also).

Medicinal Plant	Clinical Effect	Experimental Correlate
Chondrodendron tomentosum	Paralysis and death	Neuromuscular block
Cinchona officinalis	Fever	Antimalarial
Digitalis purpurea	Dropsy	Na+-K+ ATPase
Papaver somniferum	Analgesia	Opioid receptors
Physostigma venenosum	Ordeal poison	Anticholinesterase
Salix alba	Fever and pain	Prostaglandins
Strychnos nux-vomica	Stimulant and convulsant	Glycinergic receptors

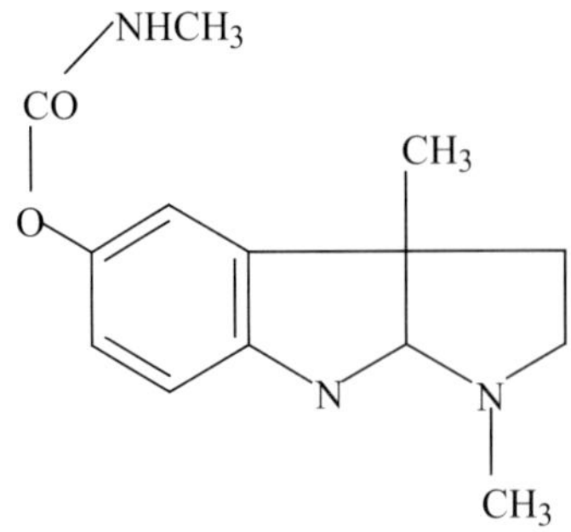

Physostigmine

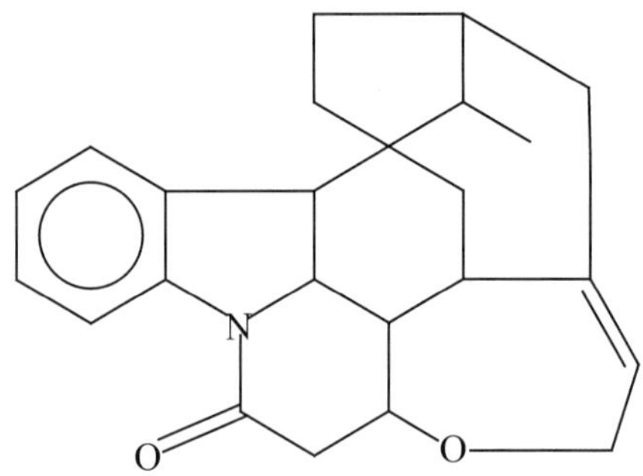

Strychnine

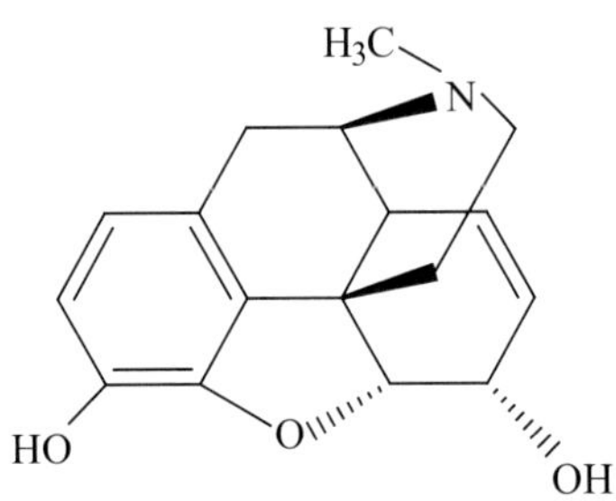

Morphine

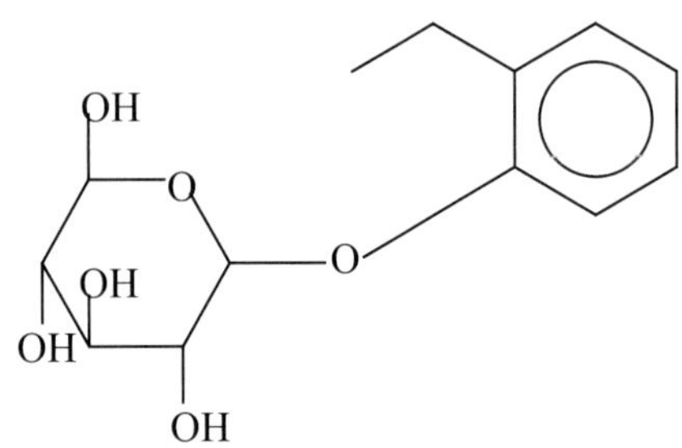

Salicin

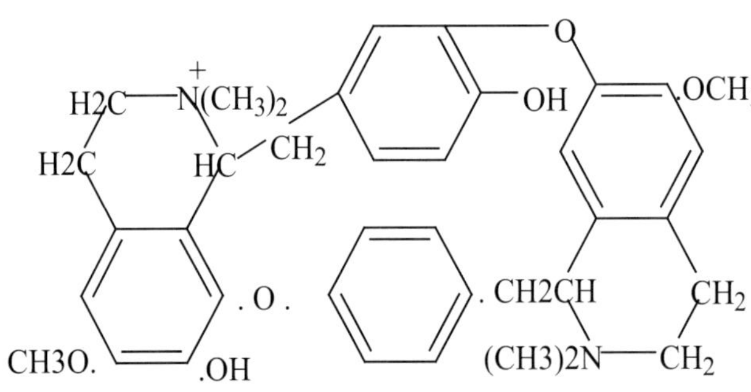

D-tubocurarine

Fig. 2.6 Active constituents of plants listed in Table 1.6.

There has been a renaissance in Ayurvedic research that western and Indian pharma companies have just started to notice. Reverse pharmacology was only sporadically applied to new drug development. It is the need of the hour to document unknown, unintended and desirable novel prophylactic and therapeutic effects in observational therapeutics. Several new classes of drugs have accidentally emerged by this path (Fig. 2.7).

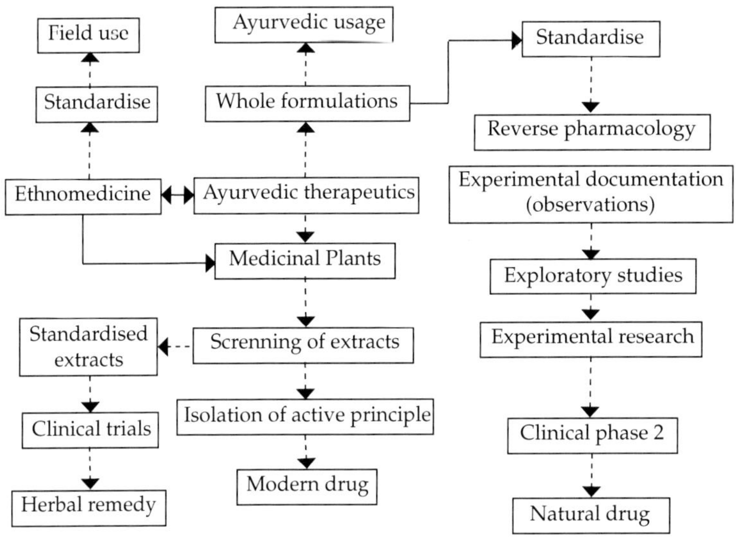

Fig. 2.7 Scheme for drug discovery from plants used in Ayurveda.

References and further reading

Arfeen Z, Owen H, Plummer JL, Ilsley AH, Sorby-Adams RA, and Doecke, CJ. 1995. A double-blind randomized controlled trial of ginger for the prevention of postoperative nausea and vomiting. Anaes. Inten. Care. 23, 449–452.

Bray DH, Warhurst DC, Connolly JD, O'Neill MJ and Phillipson JD. 1990. Plants as sources of antimalarial drugs. Part 7. Activity of some species of Meliaceae plants and their constituent limonoids. Phytother. Res. 4, 29–35.

Damodaran M, and Ramaswamy R. 1937. Isolation of L-Dopa from the seeds of *Mucuna pruriens*. Biochemistry. 31, 2149.

MacKinnon S, Durst T, Arnason JT, Angerhofer C, Pezzuto J, Sanchez-Vindas PE, Poveda LJ, and Gbeassor M. 1997. Antimalarial activity of tropical Meliaceae extracts and gedunin derivatives. J.Nat. Prod. 60, 336–341.

Rochanakij S, Thebtaranonth Y, Yenjai C and Yuthavong Y. 1985. Nimbolide, a constituent of *Azadirachta indica*, inhibits Plasmodium falciparum in culture. Southeast. Asian. J. Trop. Med. Public Health. 16, 66–72.

Singh AP. 2005. Dravyaguna Vijnana. Chaukhambha Orientalia, New Delhi, India.

Vaidya AB Rajagopalan TG, Mankodi NA, Antarkar DS, Tathed PS and Purshit AV. 1978 Treatment of Parkinson's diseases with cowhage plant—*Mucuna pruriens Bak.* Neurol. India. 4, 171–176.

Vaidya ADB. 2006. Reverse pharmacological correlates of Ayurvedic drug action. Ind. J. Pharmacol. 38, 311–315.

Vaidya ADB, Raut AK, and Amonkar AJ. 2007. Active Principle Identification and Determination. USP-IP 6th Ann. Meeting, Hyderabad.

Chapter 3

Ethnopharmacology

3.1 Introduction

With the isolation of quinine from Cinchona in 1820, an ancient herbal cure was transformed into a chemical drug. This was the inspiration for a new scientific discipline—ethnopharmacology—as Western scientists began to reinvent traditional herbal cures by extracting their active principles to make new and profitable drugs.

According to preliminary definition, ethnopharmacology may be defined as a multidisciplinary study of biologically active agents used in traditional medicine.

Or

Ethnopharmacology is the scientific study correlating ethnic groups, their health, and how it relates to their physical habits and methodology in creating and using medicines.

Traditional medicine is a term applied to pre-scientific medical systems that possess bodies of medical knowledge, including pharmacopoeia, that are passed through generations from healer to healer.

Differentiation between ethno pharmacology and pharmacognosy

According to Wardwell, ethno pharmacology and pharmacognosy overlap considerably, but can be differentiated by several distinct features as explained below:

(1) Pharmacognosy considers a broad range of natural products used as therapeutic agents, for medical purposes (such as cotton), as pharmaceutical aids, biologics, and as poisons; ethnopharmacology is restricted to natural products used in a traditional context.

(2) Pharmacognosy is concerned with the history, economics, and commercial processing of natural substances that affect human health; ethno pharmacology seeks to provide a comprehensive view of the human use of crudely processed drugs that includes ethnographic information.

(3) Pharmacognosy has tended to move toward specialized fields such as biosynthesis and fermentation microbiology; ethnopharmacology aims to support interdisciplinary collaboration.

3.2 Ethnopharmacology and Traditional Medicine

Numerous drugs have entered the international pharmacopoeia via the study of ethnopharmacology and traditional medicine. For traditional medicines, newer guidelines of standardization, manufacture and quality control and scientifically rigorous research on the scientific basis for traditional treatments will be required. Traditional medical traditions can offer a more holistic approach to drug design and myriad possible targets for scientific analysis.

Powerful new technologies such as automated separation techniques, high-throughput screening and combinatorial chemistry are revolutionizing drug discovery. Traditional knowledge can serve as a powerful search engine, which will greatly facilitate and rediscover intentional, focused and safe natural product drug discovery. By looking at the historical trends in drug and medical developments, it is possible to understand how current drug development will benefit from this partnership.

Ayurvedic and traditional Chinese systems are great living traditions. These traditions have relatively organized databases, and more exhaustive descriptions of botanical material that are available and can be tested using modern scientific methods. Both systems of medicine thus have an important role in bioprospecting of new medicines. Good botanical practices which can improve the quality control procedures of monitoring impurities, heavy metals and other toxins in the raw material can make ethnopharmacology research more meaningful.

Drug discovery in the current scenario has become unproductive to the point where the economic future of the industry is questionable. The research and development thrust in the pharmaceutical sector needs to focus on development of new drugs, innovative processes for known drugs and development of plant-based drugs through investigation of leads from the traditional systems of medicine. Traditional medicine can provide novel inputs into the drug development process. However, bioprospecting—the search for economically valuable natural resources—by pharmaceutical companies, or on their behalf, has not been conspicuously successful in recent years.

3.3 Role for Physicians in Ethnopharmacology and Drug Discovery

Ethnopharmacology investigations classically involved traditional healers, botanists, anthropologists, chemists and pharmacologists. The role of some groups of researchers but not of physicians has been highlighted and well defined in ethno pharmacological investigations. Historical data has shown that the discovery of several important modern drugs of herbal origin owe to the medical knowledge and clinical expertise of physicians. Current trends indicate a negligible role of physicians in ethnopharmacological studies.

The rising cost of modern drug development is attributed to the lack of using a classical ethnopharmacological approach. Physicians can play multiple roles in the ethnopharmacological studies to facilitate drug discovery as well as to rescue the authentic traditional knowledge of use of medicinal plants. These include:

(1) Ethnopharmacological field work which involves: interviewing healers; interpreting traditional terminologies into their modern counterparts; examining patients who are consuming herbal remedies; and identifying the disease for which an herbal remedy is used.
(2) Interpretation of signs and symptoms mentioned in ancient texts and suggesting the proper use of old traditional remedies in the light of modern medicine.
(3) Clinical studies on herbs and their interaction with modern medicines.
(4) Advising pharmacologists to carry out laboratory studies on herbs that have been observed during field studies.
(5) Work in collaboration with local healers to strengthen the traditional system of medicine in a community.

A physician's involvement in ethnopharmacological studies will lead to more reliable information on traditional use of medicinal plants both from field and ancient texts, more focused and cheaper natural product based drug discovery, and will bridge the gap between traditional and modern medicine.

3.4 Ethnopharmacology and Cultural Relativism

It is commonly accepted that people differ culturally. In the Giger and Davidhizar Transcultural Assessment Model, cultural differences are evident in communication, spatial relationships and needs, social organizations, time orientation, the ability or desire to control the environment, and biological variations. While many individuals appreciate that there are differences between cultures, what is less well recognized is that people also vary according to biological variations depending on their racial and ethnic group. In the last 15 years, information about biological variations has rapidly expanded and that knowledge is essential in order to understand and provide care to individuals from another culture or another racial and ethnic group.

Attention to biological variations related to race and ethnicity, the last component of the Giger and Davidhizar Transcultural Nursing Assessment Model, is a critical phenomenon that needs to be assessed in order to develop and implement a culturally sensitive plan of care in an effort to understand ethnopharmacolgy.

The Giger and Davidhizar Transcultural Assessment Model was developed in 1988 in response to the need for nursing students in an undergraduate program to assess and provide care for culturally diverse patients . The model included six cultural phenomena: communication, time, space, social organization, environmental control, and biological variations. These provide a framework for patient assessment and from which culturally sensitive care can be designed.

3.5 Ethnobotany/Ethnopharmacology and Mass Bioprospecting

Ethno botany/ethnopharmacology has contributed to the discovery of many important plant-derived drugs. Field explorations, to seek and document indigenous and traditional medical knowledge (IMK/TMK), and/or the biodiversity, with which the IMK/TMK is attached, and its conversion into a commercialized product is known as bioprospecting or biodiversity prospecting. When performed in a large-scale operation, the effort is referred to as mass bioprospecting.

Experiences from the mass bioprospecting efforts undertaken by the United States National Cancer Institute, the National Cooperative Drug Discovery Groups (NCDDG) and the International Cooperative Biodiversity Groups (ICBG) programs demonstrate that mass bioprospecting is a complex process, involving expertise from diverse areas of human endeavors, but central to it is the Memorandum of Agreement (MOA) that recognizes issues on genetic access, prior informed consent, intellectual property and the sharing of benefits that may arise as a result of the effort. Future mass bioprospecting endeavors must take heed of the lessons learned from past and present experiences in the planning for a successful mass bioprospecting venture.

Traditional medicines play an important role in the provision of health care in many developing countries. Their use is also significant in developed countries, increasing their commercial value. Several 'high-profile' cases of patenting of traditional medicines, without consent from or compensation to their holders, have further focused attention on their importance. Traditional medicine usually involves biological resources and the knowledge of local and indigenous peoples and/or healers regarding their medicinal use; thus, it is interlinked with biodiversity conservation and indigenous peoples' rights over their knowledge and resources.

At this multi-faceted interface, complex ethical questions arise. The key issues involve dilemmas and challenges in utilizing and protecting traditional medicines. Are there ways to modify and devise new forms of intellectual property ownership that may better suit the needs of those who seek to protect traditional medicine? There also is the question whether such protection, which may restrict access, is the preferred option. While intellectual property protection for traditional medicines has multiple and diverse objectives, the priorities are often not clear and the strategies which could be deployed may interfere with each other, as well as with the prioritization of objectives. This is further aggravated by differences in stakeholders' concepts on ownership of knowledge and by uncertain or paradoxical effects of some potentially useful strategies. Thus, policymakers should address the multiple, multi-layered issues and questions, and try to develop a range of solutions in order to address and balance the various objectives and interests.

When new plant-derived therapeutics based on indigenous knowledge are being explored, it is important that the pharmaceutical companies return benefits to the native populations and the local governments from which the research material was obtained. When a potentially marketable plant product is being developed, it is essential that equitable agreements have already been established between the pharmaceutical companies and the people and/or countries from which this indigenous knowledge was acquired.

Equally important is the commitment to provide immediate reciprocity that will enhance the welfare, the biocultural diversity and the well-being of the forest peoples. These measures should commence when a research project begins and continue during its duration. The development of these measures must be based upon the expressed needs of the indigenous communities. There is a relationship between the stability of the rain forest biocultural diversity, the creation and development of agro-forest resources and long-term benefits to the forest people.

3.6 Ethnopharmacology: Novel Applications Vaccines

The combined use of vaccines and immunostimulants is emerging as one of the innovative approaches in adjuvant development. The role of herbal drugs as immunomodulator agents are well-documented and their importance in bioprospecting is obvious. Recently researchers have suggested using ethno pharmacology in contrast to the random screening approach since it is more cost and time effective.

Antiviral

Herbal medicinal products have been used as a source of putative candidate drugs in many diseases. However, in case of viral diseases, the development of antivirals from natural sources is less explored, probably because within the virus there are few specific targets where the small molecules can interact to inhibit or kill the virus. The currently available antiherpes drugs are nucleoside analogs that did not cure the lifelong or recurrent infections and the use of these drugs often leads to the development of viral resistance coupled with the problem of side effects, recurrence and viral latency. However a wide array of herbal products, used by diverse medicinal systems throughout the world, showed a high level of antiherpes virus activities and many of them have complementary and overlapping mechanisms of action, either by inhibiting viral replication, or by viral genome synthesis.

Further reading

Burna WR. 2008. East meets West: how China almost cured malaria. Endeavour. 32, 101–6.

Chattopadhyay D, and Khan MT. 2008. Ethnomedicines and ethnomedicinal phytophores against herpesviruses. Biotech. Annu. Rev. 14, 297–348.

Cordell GA, and Colvard MD. 2005. Some thoughts on the future of ethnopharmacology. J. Ethnopharmacol. 100, 5–14.

Davidhizar R, and Giger JN. 2002. The Giger and Davidhizar transcultural assessment model. J. Transcult. Nurs. 13, 185–188.

Davidhizar R, and Giger JN. 2008. Understanding ethnopharmacolgy: implications for cultural relativism. J. Natl. Black. Nurses. Assoc. 19, 63–68.

Gautam M, Gairola S, Jadhav S and Patwardhan B. 2008. Ethnopharmacology in vaccine adjuvant discovery. Vaccine. 26, 5239–5240.

Gilani AH, and Rahman A. 2005. Trends in ethnopharmacology. J. Ethnopharmacol. 100, 237–243.

King SR, Carlson TJ, and Moran K. 1996. Biological diversity, indigenous knowledge, drug discovery and intellectual property rights: creating reciprocity and maintaining relationships. J. Ethnopharmacol. 51, 45–57.

MacKinnon S, Durst T, Arnason JT, Angerhofer C, Pezzuto J, Sanchez-Vindas PE, Poveda LJ, and Gbeassor M. 1997. Antimalarial activity of tropical Meliaceae extracts and gedunin derivatives. J. Nat. Prod. 60, 336–341.

Patwardhan B. 2007. Drug Discovery and Development: Traditional Medicine and Ethnopharmacology. New India Publishing Agency, Pitam Pura, New Delhi, India.

Raza M. 2006. A role for physicians in ethnopharmacology and drug discovery. J. Ethnopharmacol. 104, 297–301.

Singh AP. 2007. Herbal Medicine–Dream Unresolved. Ethnobot. Leaf. 11, 195–198.

Soejarto DD, Fong HH, Tan GT, Zhang HJ, Ma CY, Franzblau SG, Gyllenhaal C, Riley MC, Kadushin MR, Pezzuto JM, Xuan LT, Hiep NT, Hung NV, Vu BM, Loc PK, Dac LX, Binh LT. Chien NQ, Hai NV, Bich TQ, Cuong NM, Southayong B, Sydara K, Bouamanivong S, Ly HM, Thuy TV, Rose WC, and Dietzman GR. 2005. Ethnobotany/ethnopharmacology and mass bioprospecting: issues on intellectual property and benefit-sharing. J. Ethnopharmacol.100, 15–22.

Timmermans K. 2003. Intellectual property rights and traditional medicine: policy dilemmas at the interface. Soc. Sci. Med. 57, 745–756.

Wardwell WI.1994.Alternative medicine in the United States. Sue. Sci.Med. 38, 1061-1068.

Chapter 4

Medicinal Phytochemistry

4.1 Introduction

Phytochemistry is defined as the study of chemical composition of medicinal plants or phyto-drugs. Phytochemistry is a significant subject of the pharmacognosy curriculum. Today, one is witnessing an explosive growth of the herbal drug industry. Standardized herbal extracts and phytochemicals are in high demand for applied research as well as commercial use. The phytochemistry course taught in pharmacy schools deals with the process of isolation, purification and pre-clinical studies of herbal extracts and phytochemicals.

Before the availability of synthetic drugs, phytodrugs or herbal drugs were the mainstay of treatment. Several factors including the lack of emergency medicines, surgery, research and standardization, led to the decline of phytomedicine. With the discovery of opium alkaloid, morphine, chemists started targeting herbal drugs for bioactive compounds (referred to as phytochemicals). Phytochemicals may affect health, but are not essential nutrients. Most of the phytochemicals used in the modern pharmaceutical industry are secondary metabolites. Emerging factors like drug-resistance, cost-effectiveness and side-effects of synthetic drugs have led to the resurgence of phytomedicine.

World Health Organization (WHO) has identified the growing popularity of phytomedicine. Several articles and documents have been published highlighting the future of phytomedicine. Preclinical or non-pharmacological studies published in index journals have demonstrated the utility of standardized herbal extracts and phytochemicals. Clinical-studies done on phytodrugs like garlic, St. John's wort, purple cone-flower, ginseng, black-cohosh and Ginkgo have proved ancient theories to be accurate. More so, these phytodrugs have been documented in reputed medical text-books.

The herbal drug industry is in the transition stage. The time is ripe for initiating more research activities in this unexplored field. Non-standardized herbal preparations like infusions, decoctions, tinctures and powders have been largely replaced by standardized herbal preparations like extracts. Extracts are standardized to marker-compounds or active-constituents. Marker-compounds are used in identification of plant material. It is of more industrial importance. Active-constituents are responsible for pharmacological or medicinal activity of the extract.

4.2 Classification of phytochemistry

In our view, phytochemistry can be divided into two classes:

1. Conventional or traditional phytochemistry.
2. Medicinal or applied phytochemistry.

Conventional phytochemistry, adopts procedures for isolation and purification of active constituents of medicinal plants. The study of this discipline is of more relevance to the pharmaceutical industry. A detailed discussion of conventional phytochemical techniques is beyond the scope of this book. Medicinal phytochemistry is emerging as a new subject keeping in mind recent trends in phytotherapy. This subject is of interest to medical and herbal professionals. Recently, some institutes have introduced courses or modules on herbal medicine or phytotherapy in the regular medical syllabus.

4.3 Scope of medicinal phytochemistry

Medicinal phytochemistry seems to be an interdisciplinary subject. For a well-trained medicinal phytochemist, knowledge of subjects like medicinal or pharmaceutical botany, anatomy, physiology, pathology, pharmacognosy, ethnopharmacology, chemical ecology, conventional phytochemistry, toxicology, traditional systems of medicine (Ayurveda, Siddha, Homeopathy, Traditional Chinese Medicine, and Unani), clinical research and biostatistics are prerequisites. While several complementary and alternative systems of medicine are in practice, the Traditional Chinese System (TCM) has made tremendous progress regarding the application of phytochemistry in medicine.

A typical medicinal phytochemistry course can be postgraduate and of a 2-yr duration. In case of the Ayurvedic curriculum, medicinal phytochemistry can combine well with Dravyaguna. Dravyaguna is defined as a study of Ayurvedic perspective of medicinal plants. A comparative curriculum blending ancient and modern concepts of ancient therapy will be helpful for students of Ayurveda. A curriculum of herbal medicine taught in Australia, Europe and the United States have some courses of medicinal phytochemistry but it needs significant expansion.

4.4 Career prospects

A well-trained medicinal phytochemist can join academic institutes conducting research on natural medicinal products or as an expert in the herbal drug industry. There is a cut-throat competition in the herbal drug industry and well-trained medicinal phytochemists are highly sort after.

4.5 Challenges in the future

Herbal medicine is expanding rapidly. There is a need to create specialty subjects in the existing curriculum of herbal medicine. Many allopathic practitioners are interested in understanding the mechanism of actions of phyto-drugs. Most of the CAM practitioners have little interest towards integrating the modern and traditional systems of medicine. However, some CAM practitioners have begun to understand the importance of the integrative approach. In order to sustain medicinal phytochemistry as a subject in the near future, more funds for research and development are needed. The course is essential for creating super-specialty subjects like phytopharmacology and phytopharmacotherapy.

Further reading

Gillespie SG. 1995. Ethno pharmacology: Design of an Undergraduate Course. Am. J. Pharmaceut. Edu. 59, 34–37.

Gottlieb OR, Borin MR, De MB Brito, and de Brito NRS. 2002. Integration of ethno botany and phytochemistry: Dream or reality? Phytochemistry. 60, 145.

Singh AP. 2007. A Short Note on Designing Curriculum for Medicinal Phytochemistry. Ethnobot, Leaflets. 11, 204–205.

Singh AP. 2008. Reviving the Dravyaguna Curriculum. Light on Ayur. 7, 33–34.

Chapter 5

Phytochemicals

5.1 Introduction

'Phyto' is the Greek word for plants. Phytochemicals literally means chemicals produced by plants. The term is reserved for those plant chemicals that have a beneficial effect on human-health but are not essential from the point of view nutrition (see phytonutrients). If the term is analyzed minutely, phytochemical means 'fight-o-chemicals'. Twenty five thousand phytyochemicals are known to exist in the plant flora. It is estimated that approximately 10,000 alkaloids and 4,000 flavonoids are distributed in the plant kingdom.

Medicinal plants rich in phytochemicals have been used for centuries in the treatment and prevention of diseases. Some phytochemicals can be dangerous and some have no effect on human health. Sometimes some of the compounds in plants with potent medicinal properties may not necessarily be chemicals, but may be elements like selenium or chromium.

Thousands of phytochemicals have been isolated and characterized from plants, including fruits and vegetables. The isolated phytochemicals are grouped into distinctive classes by the number and kind of constitutive atoms and the structure of the basic skeleton. Phytochemicals derived from the plants remain the basis for a large proportion of the commercial medications used today for the treatment of a wide range of diseases.

5.2 Definition of phytochemical

Definition 1

A phytochemical is a natural bioactive compound found in plant foods that works with nutrients and dietary fiber to protect against diseases.

Definition 2

Phytochemicals are chemicals produced by plants. They may affect health, but are not-essential nutrients as our diet does not require them to sustain life in the same way as vitamins and minerals.

Or

Phytochemicals are non-nutritive plant chemicals that contain protective, disease-preventing compounds.

Or

Phytochemicals are defined as non-nutritive biologically active compounds.

5.3 Evolution process of phytochemicals

Gottlieb (1993) proposed the redox theory for evolution of phytochemicals. According to this theory, when the evolution of plants took place, a negligible amount of oxygen was present in the atmosphere. With an increase in atmospheric oxygen, the environment was polluted as a direct result of plant metabolism. With the passage of time, plants kept on acquiring antioxidants, which neutralized the ill-effects of free-radicals. Small plants like algae, fungi, liverworts and mosses survived the increase in atmospheric oxygen. Over a period of time, higher plants which were oxygen resistant, like ferns, gymnosperms and angiosperms evolved.

The plants then evolved phytochemicals possessing a fighting capacity (fight-o-chemicals) against microorganisms and offering a protective effect to the genetic material. Phytochemicals also exert protective effects against solar-radiation and insect invasion. When the evolution of animal species took place, it made dependent on the plants for food and shelter. The plants consumed by animals, including human-beings, provided them with their own chemical defense mechanisms. Consumption of phytochemicals by humans has shown to keep cells in the body healthy and stable in many ways.

Phytochemicals accumulated by plants play a significant role in interaction with the surroundings. Recently, it has been postulated that the underlying mechanism in the evolution of phytochemicals by plants includes biosynthetic enzymes. However the role of transcription factors in coordinating expression of all the genes in the evolution pathways is poorly understood. According to Grotewold (2005), duplication of regulatory genes results in control of new pathways. New pathways are evolved as a result of mutation, which impairs the routine function, resulting in inactivating one or more steps in a metabolic pathway. As a result of which, intermediates start accumulating and are converted into new compounds by specific enzymes.

5.4 Classification of phytochemicals

A brief account of the type of phytochemicals distributed in the plant flora is given below.

5.4.1 Alkaloids

Alkaloids are basically nitrogen bases. The amino acids act as building blocks for the biosynthesis of alkaloids. Alkaloids are responsible for physiological effects in man or in animal. Alkaloids have basic properties and are alkaline in reaction, turning red litmus paper blue.

Alkaloids are basically compound ammonias, where various radicals replace one or more atoms of hydrogen. Alkaloids combine with acids to form crystalline salts without the production of water. The majority of alkaloids exist in solid form like atropine and contain oxygen. Some alkaloids like lobeline or nicotine occur in a liquid form and contain carbon, hydrogen, and nitrogen.

Alkaloids have one peculiarity regarding solubility in organic solvents. They are readily soluble in alcohol and sparingly soluble in water. The salts of alkaloids are usually soluble in water. In nature, the alkaloids exist in many plants: in larger proportion in the seeds and roots often in combination with vegetable acids. The solutions of alkaloids are intensely bitter.

The name of alkaloids, end in -ine suffix. The salts of alkaloids are official in various pharmacopiea. Codeine, atropine, morphine, ergotamine and ephedrine are common examples. As discussed earlier alkaloids are responsible for physiological effects in man or animals. The physiological effects are due to secondary metabolites arising from biochemical pathways operating in the plant cell. Alkaloids constitute the largest group of secondary chemical constituents. Alkaloids are classified as shown in Fig. 5.1.

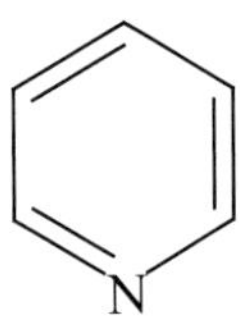

Pyridine

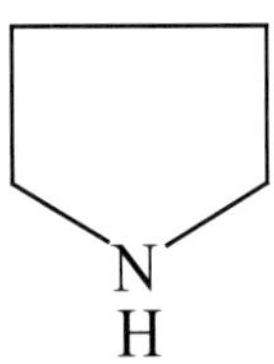

Piperidine

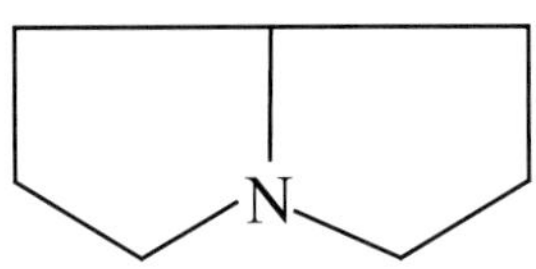

Pyrrolidine

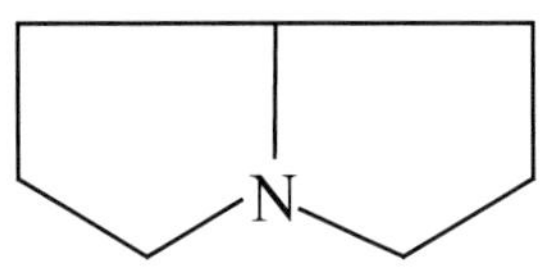

Pyrrolizidine

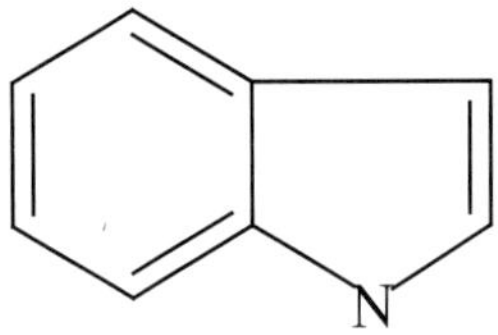

Indole

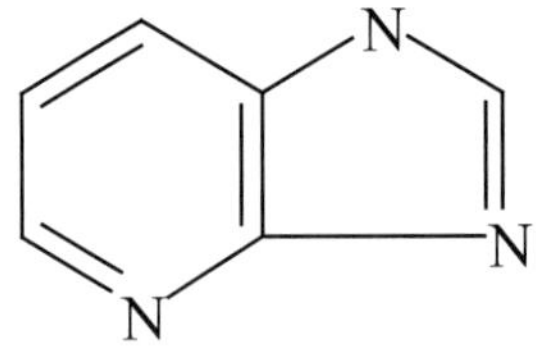

Purine

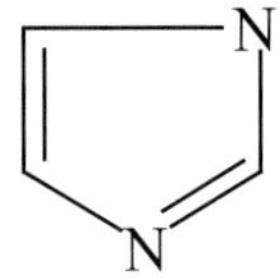

Imidazole

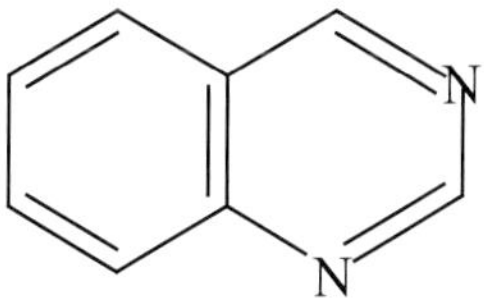

Quinoline

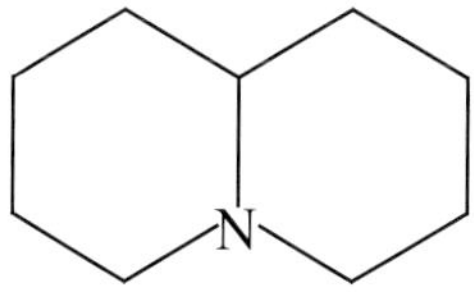

Quinolizidine

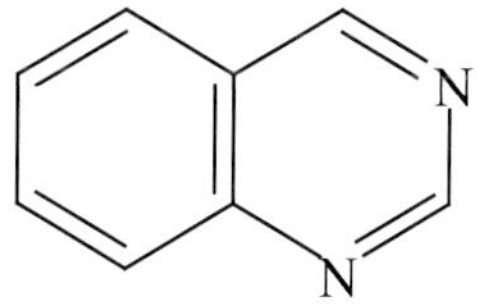

Quinazoline

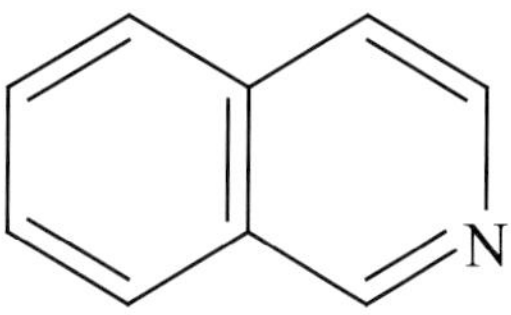

Isoquinoline

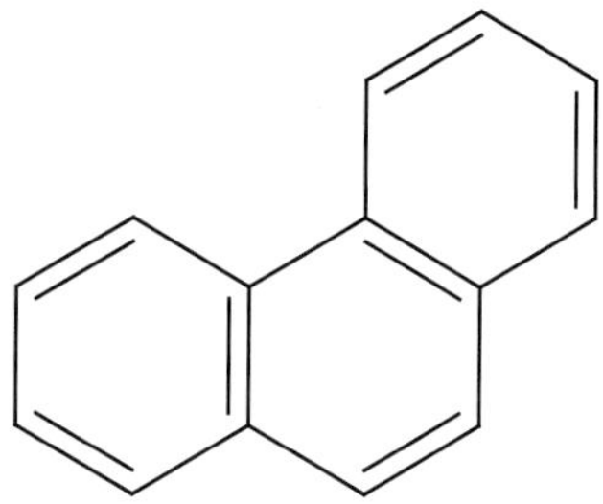

Phenanthrene

Fig. 5.1 Classification of alkaloids.

Amino acids act as a precursor for biosynthesis of alkaloids. Ornithine and lysine are common amino acids used as starting material for alkaloid biosynthesis. Cocaine and nicotine are classical examples from this series (Fig. 5.2).

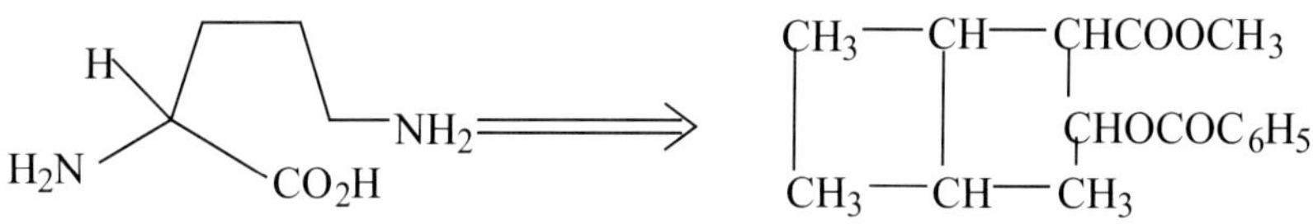

Ornithine Cocaine

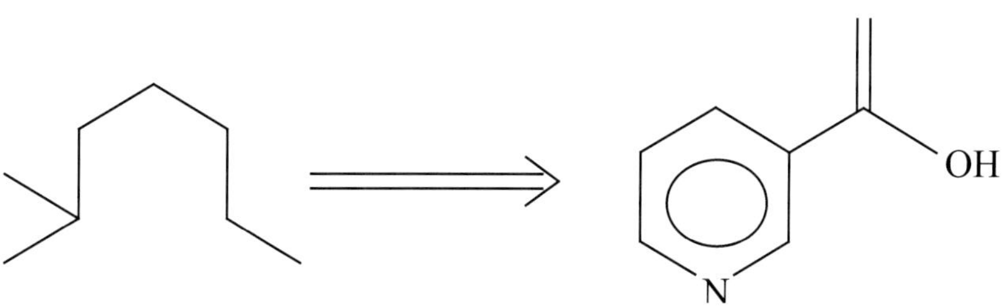

Lysine Nicotine

Fig. 5.2 Diagrammatic representation of alkaloid biosynthesis.

Phenylalanine and tyrosine act as precursors for opium alkaloid biosynthesis. Tryptophan is a significant source of Vinca alkaloids. Alkaloids are derived from anthranilic acid (Fig. 5.3), which is an intermediate in biosynthesis of tryptophan. Some alkaloids are derived from acetate, terpene or shikimic acid. Shikimic acid is a significant metabolite as most of the aromatic constituents are derived from shikimic acid pathway (Fig. 5.3).

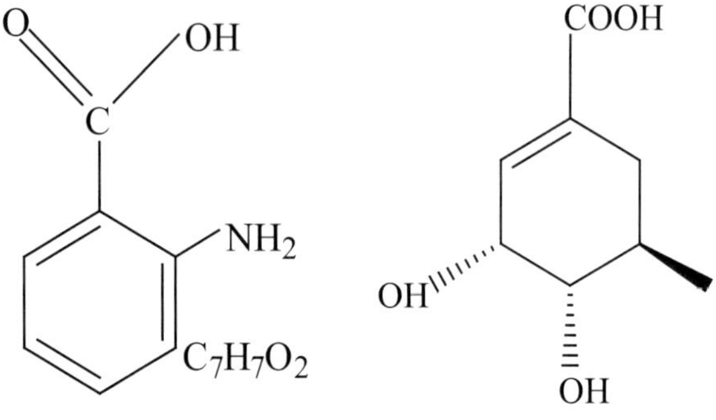

Fig. 5.3 Structure of anthranilic acid and shikimic acid.

Steroidal alkaloids

They are also known as azosteroids. Depending upon the parent carbon skeleton, they are classified in four types (Fig. 5.4). They are found in Apocynaceae, Buxaceae, Solanaceae, and Liliaceae.

Alkaloids are significant sources of pharmaceutical drugs. More than 12,000-alkaloids are known to exist in green flora and only few have been exploited for medicinal purposes. Structures of important alkaloids are presented in Figs. 5.5 and 5.6.

5.4.2 More about alkaloids

Cystisus scoaprius is known to contain a volatile alkaloid, ***sparteine*** and a non-volatile alkaloid, ***sarothamnine***. *Abies webbiana* contains a crystalline alkaloid, ***taxine*** (Fig. 5.7).

Asimina triloba contains benzyl isoquinoline alkaloid ***anolobine***. *Galipea officinalis* is reported to contain quinoline alkaloids including (cusparine, galipine, galipoline, quinaldine, cuspareine, galipoidine, and l-methyl-2-quinolone). *Vaccinium myrtillus* contains quinolizidine alkaloids: ***myrtine*** and ***epimyrtine.***

Alkaloids **(*atidine, atisine, hetidine, hetisine, heteratisine, heterophylline, heterophyllisine, heterophyllidine, isoatisine*),** are present in *Aconitum heterophyllum* (Fig. 5.8).

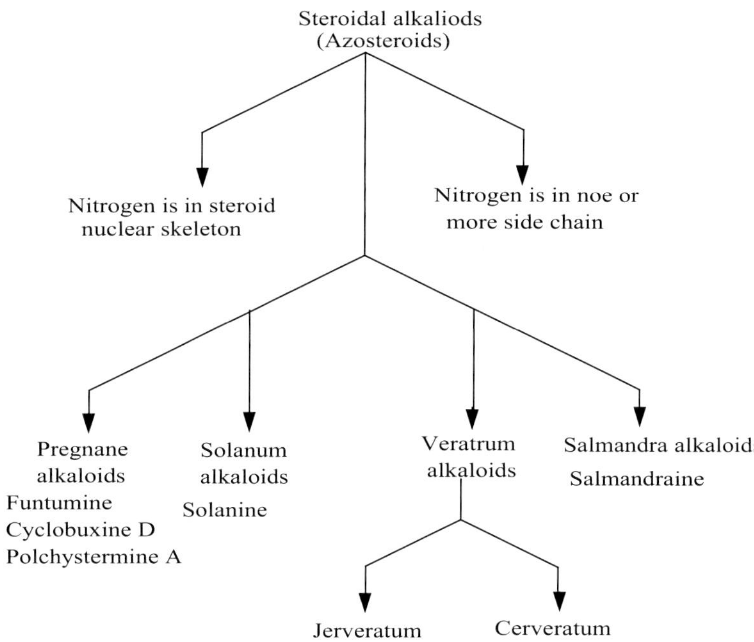

Fig. 5.4 Classification of steroidal alkaloids.

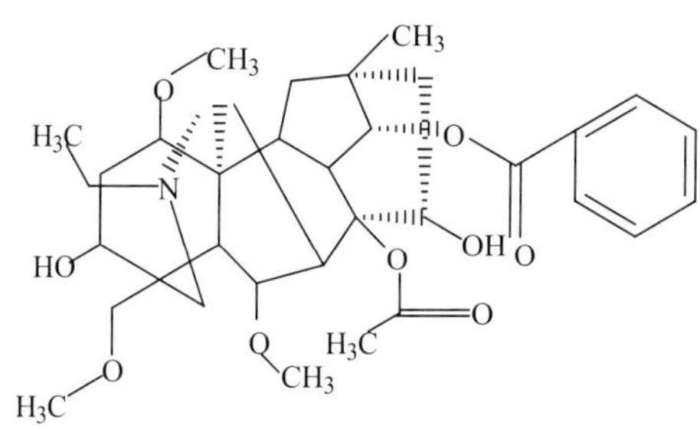

Aconitine

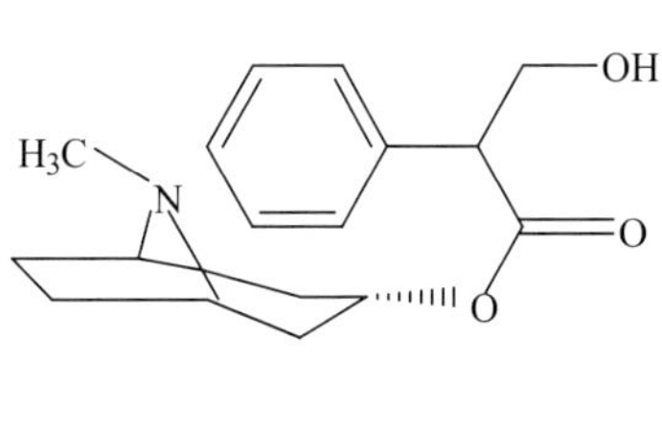

Atropine

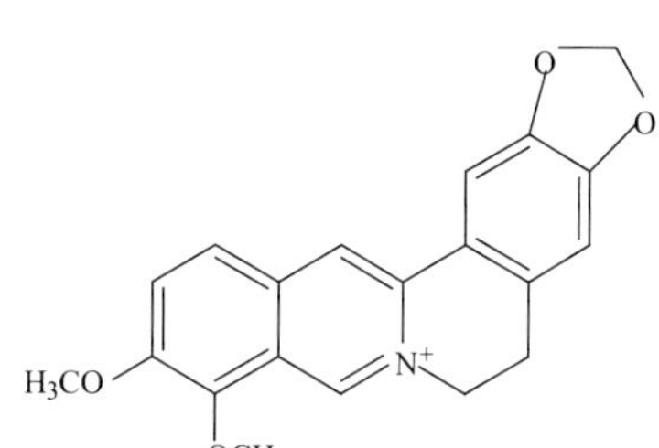

Berberine

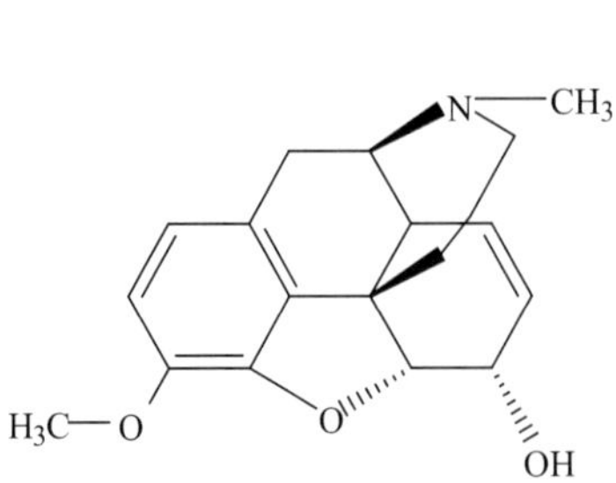

Codeine

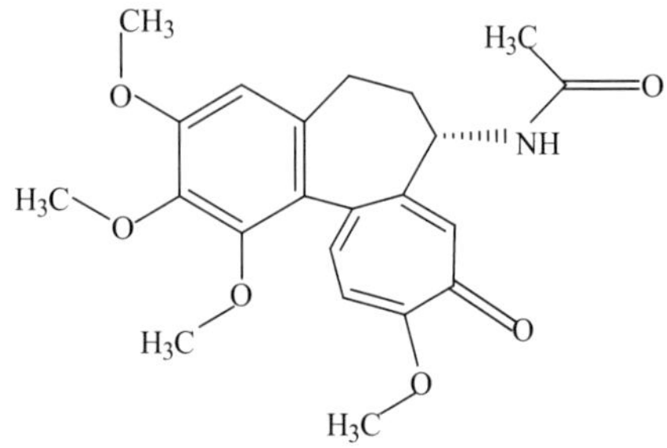

Colchicine

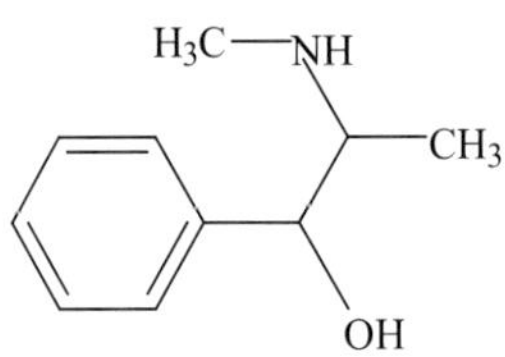

Ephedrine

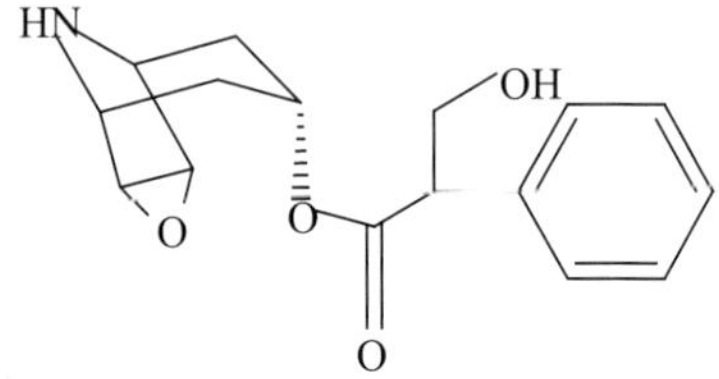

Hyoscine

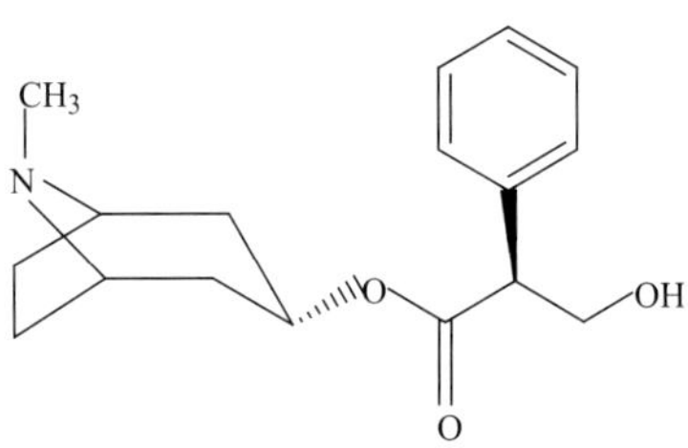

Hyoscyamine

Fig. 5.5 Structure of important alkaloids.

Pelosine or cissampeline is an amorphous alkaloid present in *Cissampelos pareira*. It is related to berberine. Protoberberine alkaloids, ***bebeerine, d-isochondrodensrine, menismine, hayatine*** (*dl*-bebeerine), ***hayatidine*** and ***cissamine*** or ***cyclanoline*** (Fig. 5.9), ***pareirine,*** are reported from the roots of *Cissampelos pareira*. ***Cissampareine*** has an inhibitory action against human carcinoma cells of the nasopharynx in cell culture. ***dl-curine*** is used as a muscle relaxant.

Cyclea peltata contains alkaloids including ***d-tetrandrine, dl-tetrandrine, isotetrandrine, limacine, berbamine, bomoarmoline, cycleapeltine, cyleadrine, cycleacurine, cycleanorine*** and ***cycleahomine***. Tetrandrine is an irritant to mucosa, produces hypoglycemia in rats and emesis in pigeons. It is also a cardiac depressant. *Coscinium fenestratum* contains alkaloids including ***berberine, jatorrhizine, berbaruine, N,N-di-Me-lindacarpine, thalifendine*** and ***palmitine***.

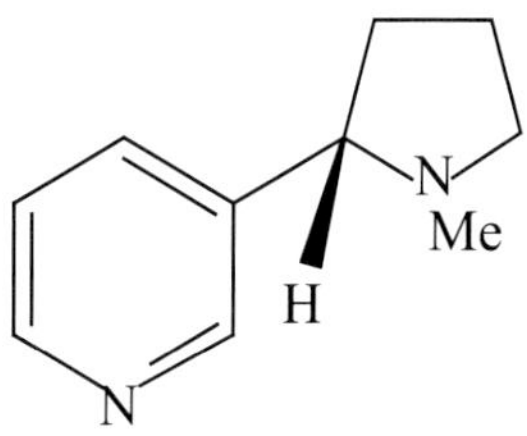

Nicotine

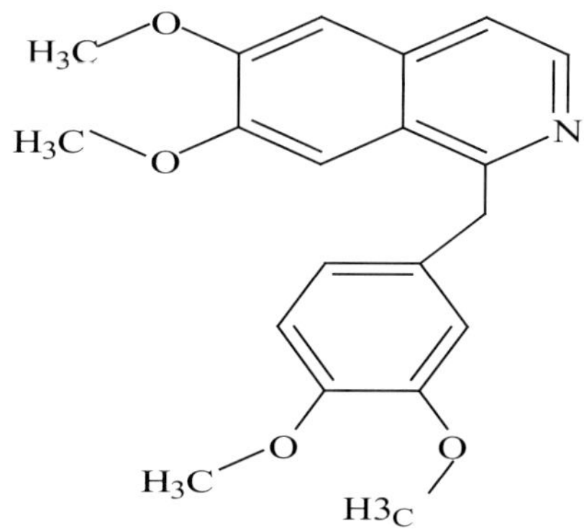

Papaverine

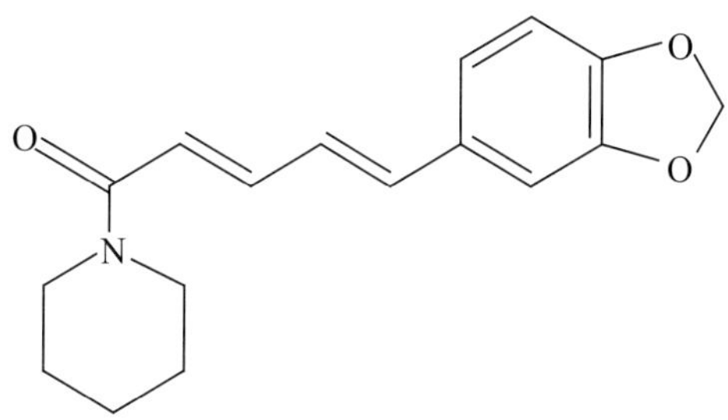

Piperine

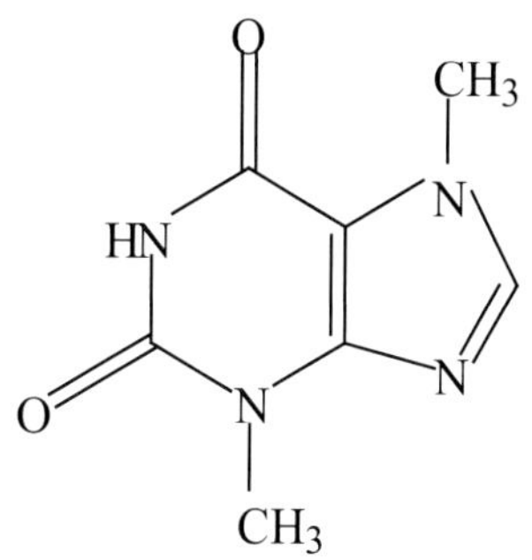

Theobromine

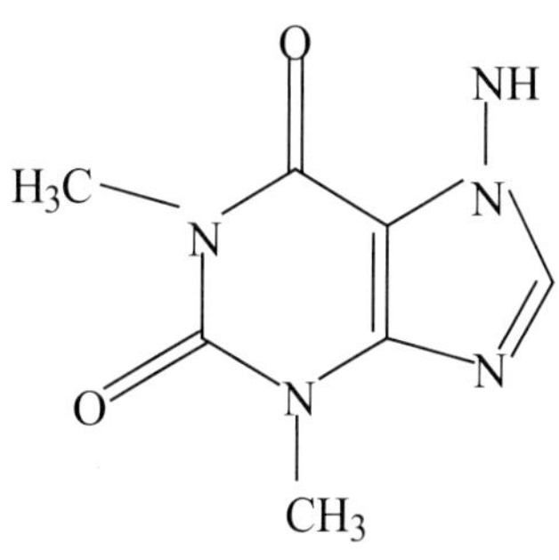

Theophylline

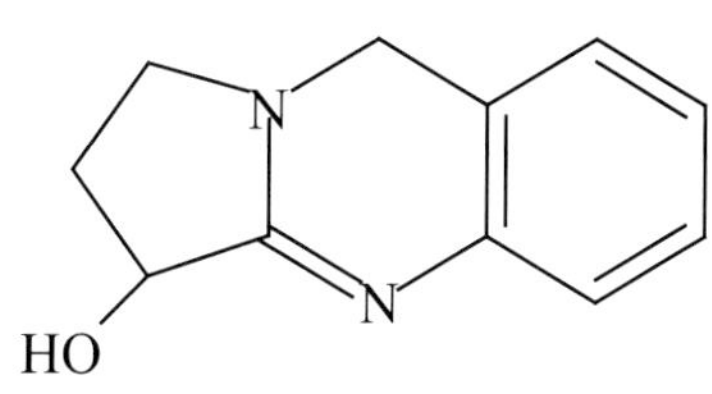

Vasicine

Fig. 5.6 Structure of important alkaloids continued.

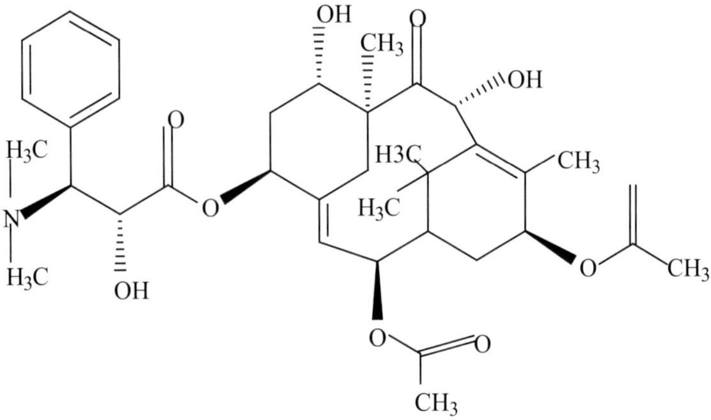

Fig. 5.7 Structure of Taxine.

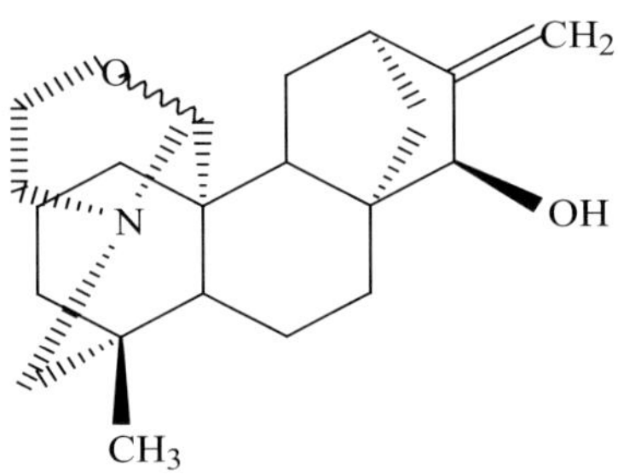

Atisine

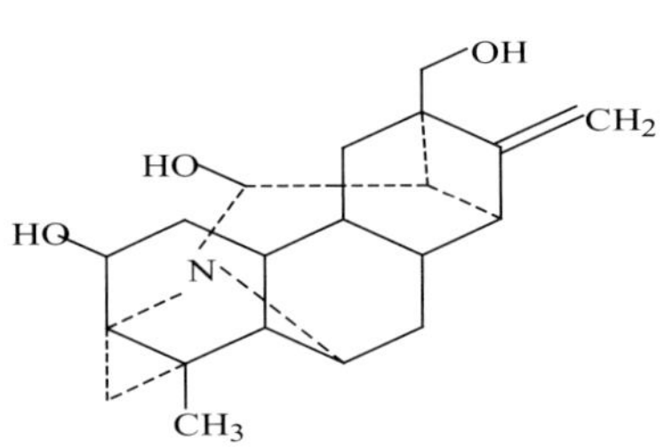

Hetisine

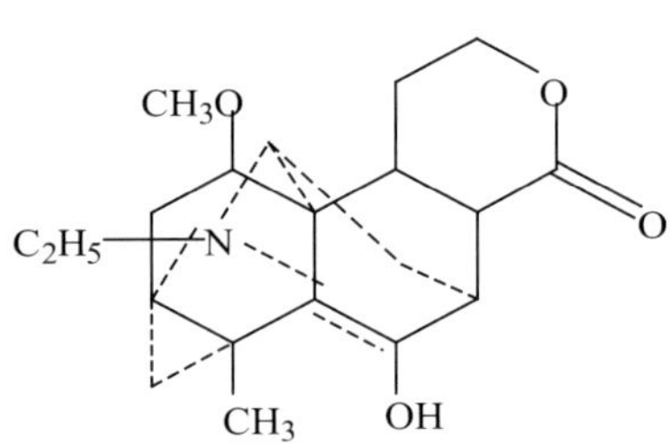

Heteratisine

Fig. 5.8 Structure of major alkaloids of *Aconitum heterophyllum*.

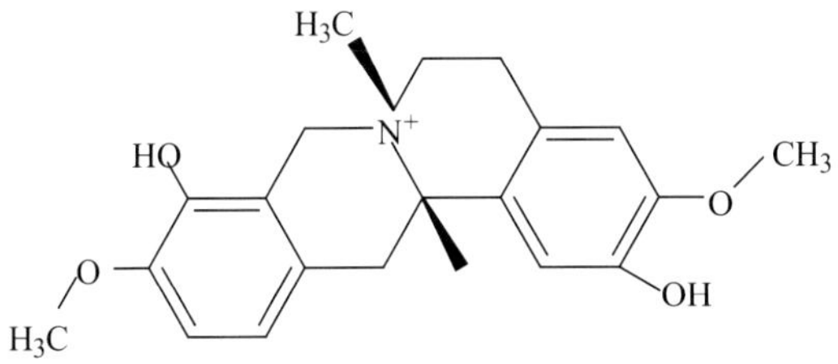

Fig. 5.9 Structure of Cissamine.

Ervatamia heyneana contains alkaloids including, ***coronaridine, heyneanine, voacangine, ψ-indoxy voacangine, and ibogamine. Isovoacrisitine*** is anticholinergic and antihistaminic. Coronaridine, on oral administration, prevents pregnancy in rats. It is cytotoxic against P 388 lymphocytic leukemia. It also contains ***camptothecine***. ***Rutaecarpine*** and ***dehydroevodiamine*** from *Euodia rutaecarpa* show, *in vitro*, uterotonic effect on the rat uterus. It also contains a sympathomimetic alkaloid, ***synephrine***. Alkaloidal fraction of *Fritillaria imperialis* containing ***imperialine*** which reduces blood pressure and shows relaxant and antispasmodic activity. Bulb of *Fritillaria royeli* contains alkaloid, ***kashmirine***.

Celatrus paniculatus, popularly known as *Jyotishmati* in Ayurveda, is reported to contain two alkaloids, ***celastrine*** (Fig. 5.9) ***and paniculatine***. The former is recorded to be powerful stimulant and pyrogenic. Some authors report the presence of ***celopagine, celophanigine,*** **and** ***cepapanine***. *Eclipta alba* is reported to contain ***ecliptine*** and ***nicotine*** (0.078%).

Loturine (Fig. 5.11), ***loturidine*** and ***colloturine***, present in *Symplocos racemosa*, are chemically related to harmine found in *Peganum harmala*. ***Gentianine*** present in medicinal plants of Gentianaceae has been reported to have antipsychotic, antimalarial and antiamoebic activities (Fig. 5.11). Other alkaloids distributed in medicinal plants of gentianaceae are ***gentianaine, gentianadine, gentialutine, gentiocrucine and gentianamine*** (Fig. 5.10). ***Erthyrocentaurine*** is a monoterpene alkaloid reported from *Swertia laurii* and *Enicostemma hyssopifolium*.

Phyllantidine and ***phyllantine*** (Fig. 5.11) are reported from leaves, fruits, and *in vitro* tissue cultures of *Emblica officinalis*.

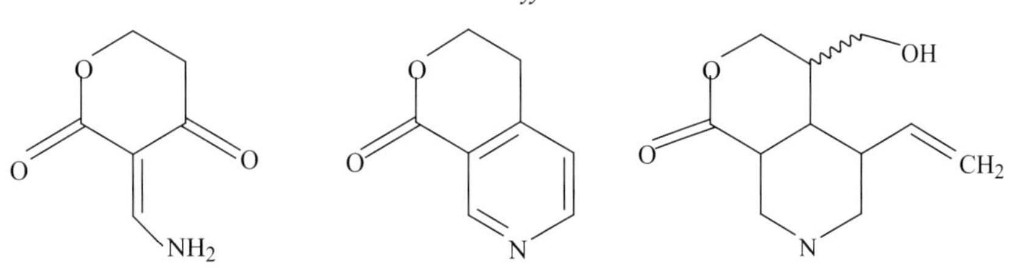

Gentianaine Gentianadine Gentianamine

Fig. 5.10 Structure of alkaloids reported from Gentianaceae.

Ryania speciosa, indigenous to South America, contains an insecticidal alkaloid, ***ryanodine***. *Alstonia scholaris* is known to contain antimalarial

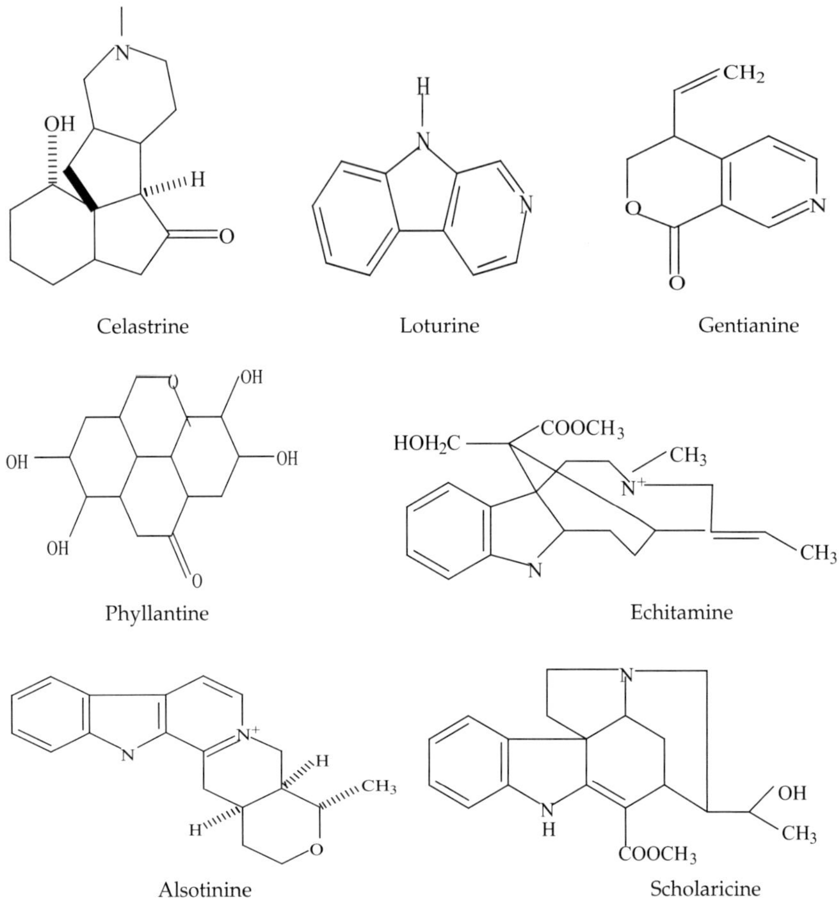

Fig. 5.11 Structure of alkaloids reported from Indian medicinal plants.

alkaloids including ***alsotinine, ditamine, echitamine*** and ***echitanine*** (Fig. 5.11). Ditamine possesses anti-periodic properties equal to the best quinine sulfate. *A.scholaris* also contains an alkaloid called ***scholaricine*** (Fig. 5.11). *Alstonia constricta contains* indole alkaloids including ***reserpine, deserpidine, alstonine, tetrahydroalstonine, alstonidine,*** and ***yohimbine***. ***N-isobutyl 2, 4 decadieamide***, a waxy alkaloid from *Piper longum* has antitubercular activity (Fig. 5.12).

$CH_3(CH_2)_4\ CH{=}CH{-}CH{=}CH{-}CO{-}NHCH_2\ CH\ (CH_3)_2$

Fig. 5.12 Structure of N-isobutyl 2, 4 decadieamide.

Actinidine is present in *Nardostachys jatamansi*. ***Coptine, copsine, coptisine*** and ***palmatine*** are reported from *Coptis teeta*. ***Mimosine*** is present in *Mimosa pudica* and is a protoplasmic poison. ***Moringine*** reported from *Moringa olifera* acts as a cardiac stimulant, bronchodilator, and has a depressant action on smooth muscle fibers and an inhibitory effect on the tone and movements of the intestine in rabbits and guinea pigs (Fig. 5.13).

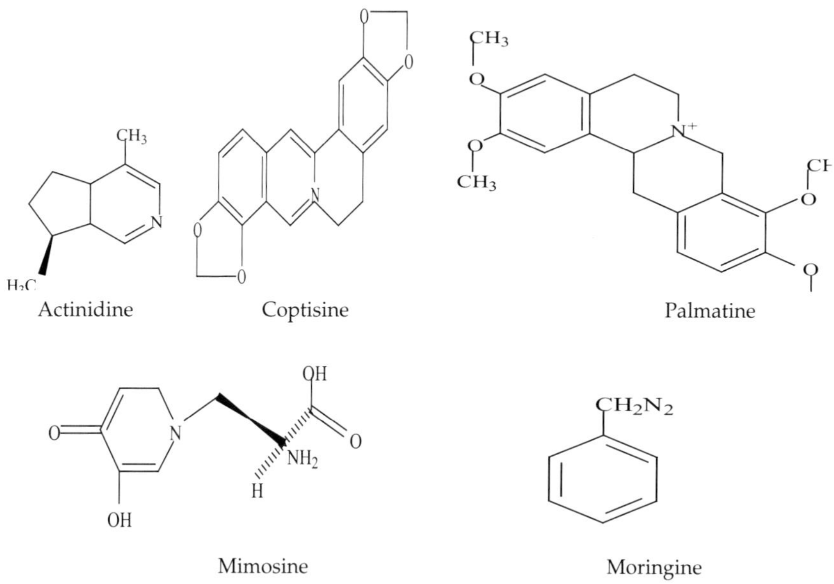

Fig. 5.13 Structure of alkaloids reported from medicinal plants.

Strychnine, Brucine, icajine and ***vomicine*** (*Strychnos nux vomica*), ***carpaine*** (*Carica papaya*), ***colchicine*** (*Colchium luteum*), ***diaboline*** and ***novacine*** (*Strychnos potatorum*), ***lycorine*** (*Crinum latifolium*), ***noroxyhydrastinine*** (*Coscinium fenestratum*), ***piperlongumine, piperlonguminine and piperlatine*** (*Piper longum*), ***precatorine*** (*Abrus precatorius*) and ***asparagamine*** (a polycyclic alkaloid from *Asparagus racemosus*) are other examples of alkaloids isolated from medicinal plants (Fig. 5.14).

Experiments conducted with ***tylpohorine***, a phenanthroindalizidine alkaloid, present in *Tylophora asthmatica* in various animal models, has shown significant anti-inflammatory, anti-anaphylactic and anti-spasmodic activities (Fig. 5.15).

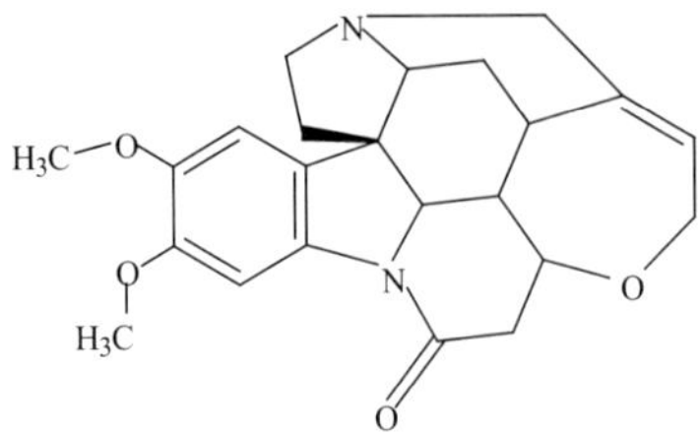

Brucine

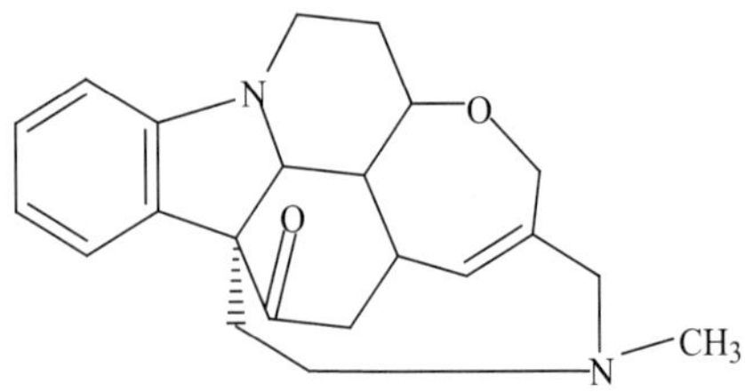

Icajine

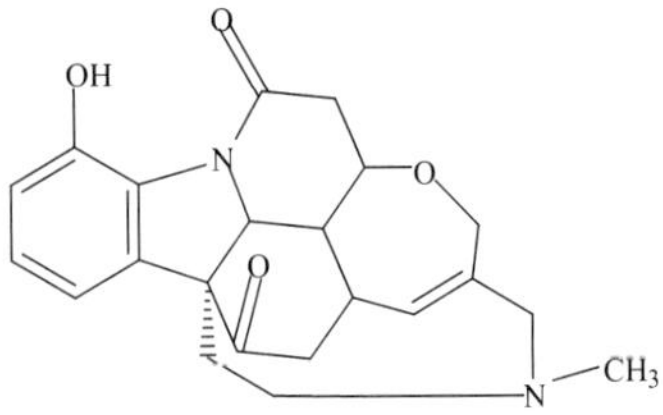

Vomicine

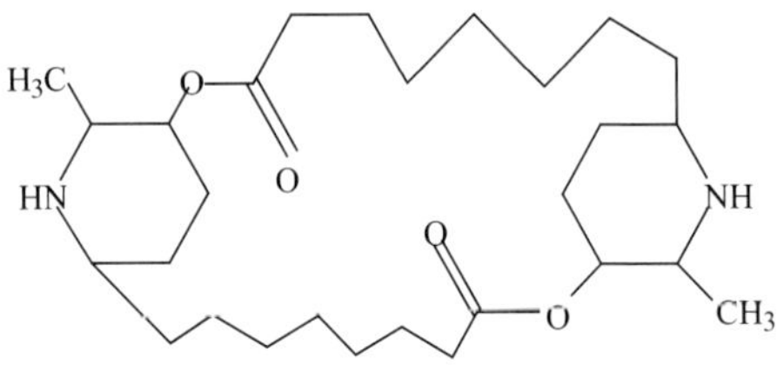

Carpaine

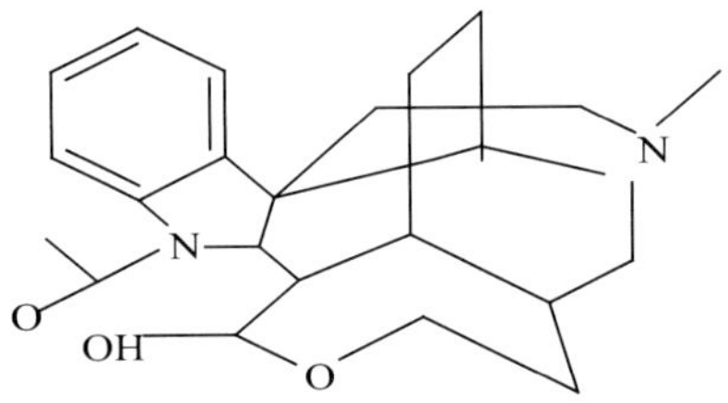

Diaboline

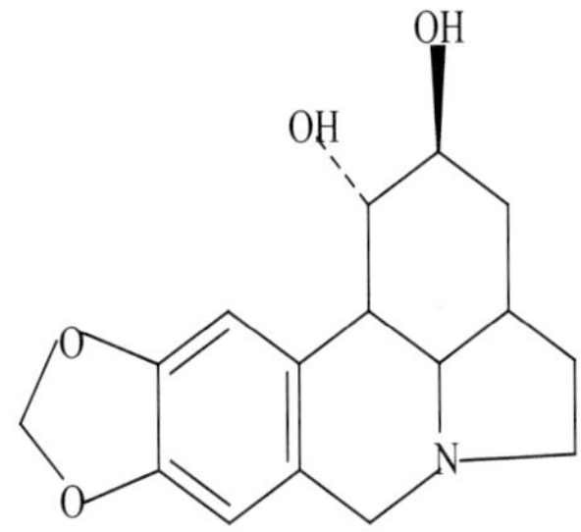

Lycorine

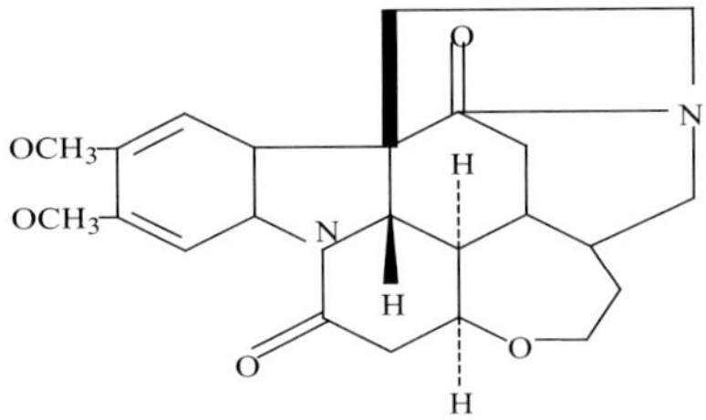

Novacine

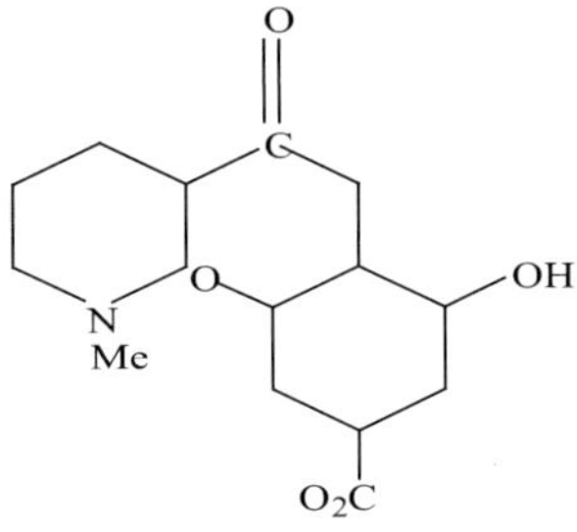

Precatorine

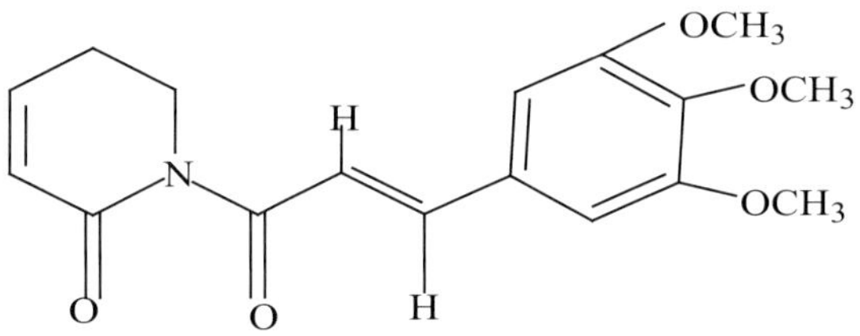

Piperlongumine

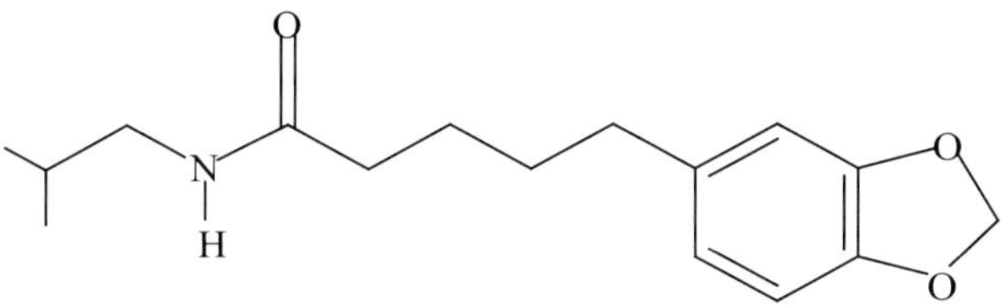

Piperlonguminine

Fig. 5.14 Structures of alkaloids reported from medicinal plants.

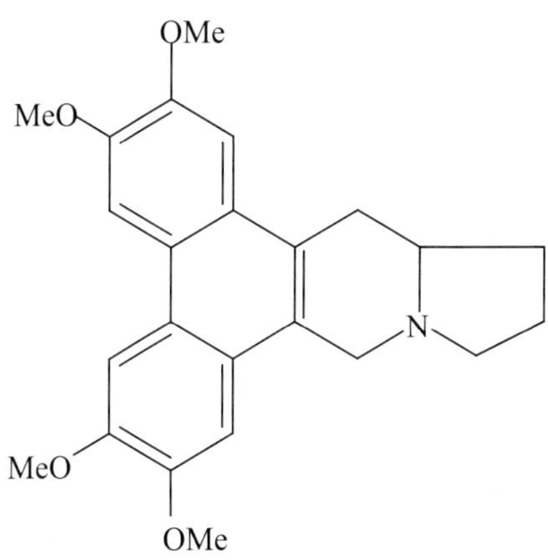

Fig. 5.15 Structure of Tylpohorine.

Imidazole alkaloid, ***chaksine*** isolated from *Cassia absus* has antibacterial activity. In a test for antifungal activity, chaksine iodide at 0.5% inhibited all fungi tested (Fig. 5.16). ***Isochaksine*** has similar activities to chaksine, but usually at higher doses.

A new carbazole alkaloid, ***clausenol,*** isolated from an alcoholic extract of the stem bark, of *Clausena anisata* was found to be active against Gram-positive and Gram-negative bacteria and fungi (Fig. 5.17). Alkaloids; ***3-formylcarbazole, mukonal*** (Fig. 5.18), ***3-methoxycarbonylcarbazole,***

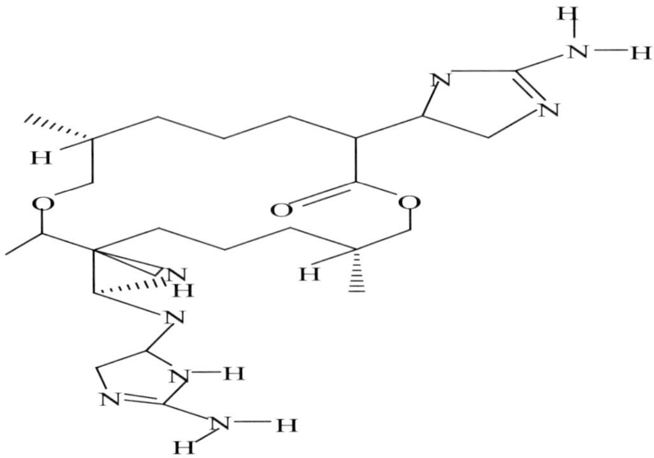

Fig. 5.16 Structure of Chaksine.

2-hydroxy-3-formyl-7-methoxycarbazole, and ***clauszoline*** J from *Clausena excavata* eliminates *Mycobacteria* with MIC values ranging from 50–200µg/mL. 3-Formylcarbazole, mukonal, 3-methoxycarbonylcarbazole, and 2-hydroxy-3-formyl-7-methoxycarbazole show antifungal activity with IC50 values of 13.6, 29.3, 9.5, and 2.8µg/mL, respectively. Carbazole alkaloids of the *Clausena* species and Rutaceae in general are related in structure to the antitumor alkaloid ellipticine and hold some potential as cytotoxic and antiviral agents.

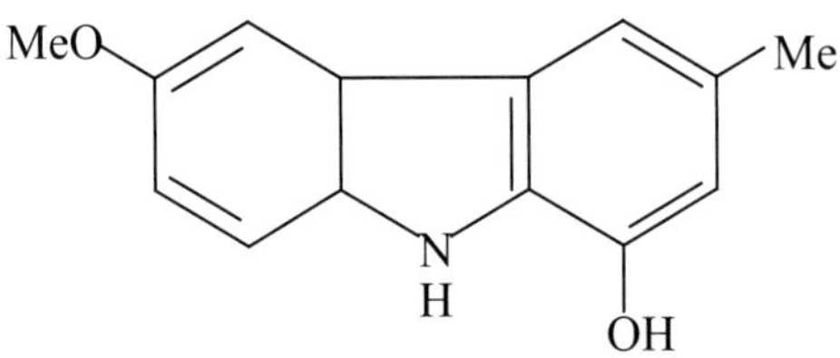

Fig. 5.17 Structure of Clausenol.

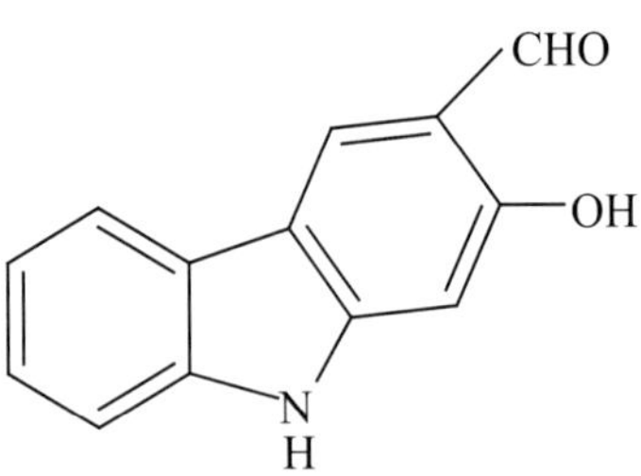

Fig. 5.18 Structure of Mukonal.

Mahanine (Fig. 5.19), a carbazole alkaloid, isolated from *Micromelum minutum*, induces apoptosis in the human myeloid cancer cell HL-60 at 10μM via activation of caspase-3 through a mitochondrial dependent pathway.

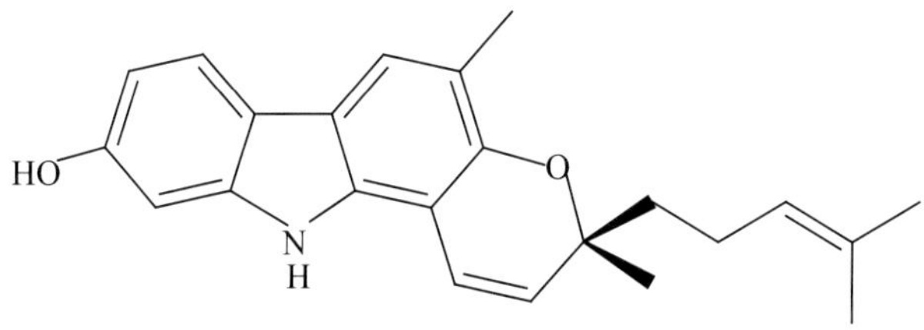

Fig. 5.19 Structure of Mahanine.

Bicuculline acts as an antagonsit at GABA receptor. It has been shown to compete with GABA for specific sites on synaptosomes (Fig. 5.20). The alkaloid is reported from *Corydalis govniana*. The trans-conformation of GABA fits the structure of (+)–bicuculline.

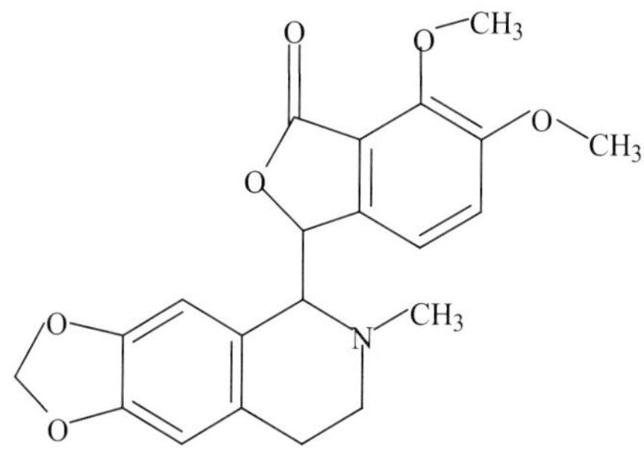

Fig. 5.20 Structure of (+) – Bicuculline.

Hydrastine from *Hydrastis canadensis* is a potent $GABA_A$ competitive antagonist, twice as potent as biccuculline (Fig. 5.21).

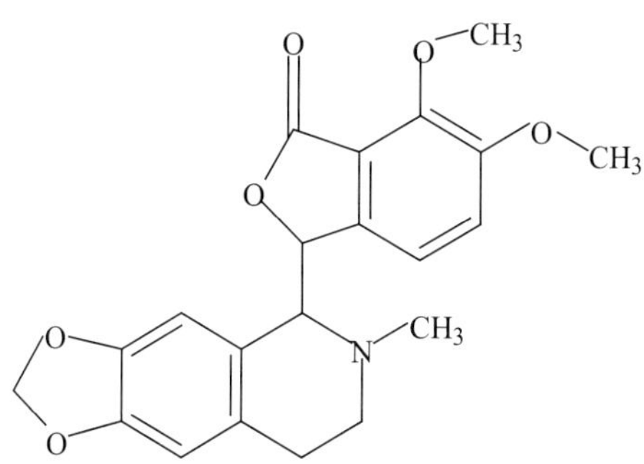

Fig. 5.21 Structure of Hydrastine.

Isoquinoline alkaloid, ***liriodenine***, isolated from *Fissistigma glaucescens* is known to block the muscarinic receptors and therefore impede the secretion of gastric juices and the contraction of gastric smooth muscles. Isoquinoline alkaloid, ***cryptofolione*** (Fig. 5.22), isolated from *Cryptocarya tomentosa* seems to have anxiolytic potential as it resembles Kawain, the active principle of *Piper methysticum*.

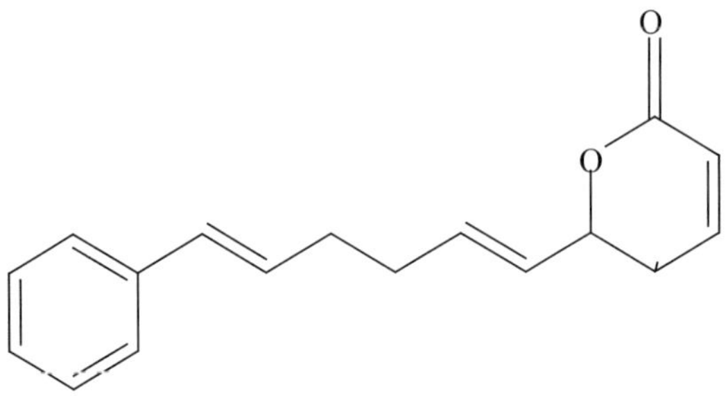

Fig. 5.22 Structure of Cryptofolione.

Chelerythrine from *Sanguinaria canadensis* is a narcotic (Fig. 5.23). ***Corycavine*** isolated from *Corydalis tuberosa* is antihelmintic (Fig. 5.24). ***Corydaline*** isolated from *Corydalis govaniana* is antiperiodic (Fig. 5.25).

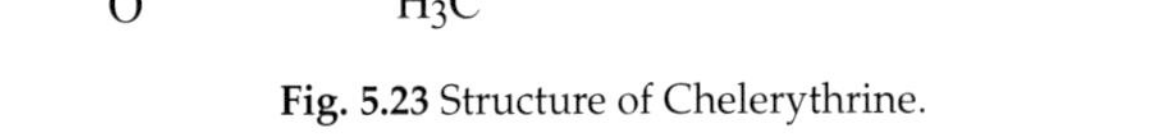

Fig. 5.23 Structure of Chelerythrine.

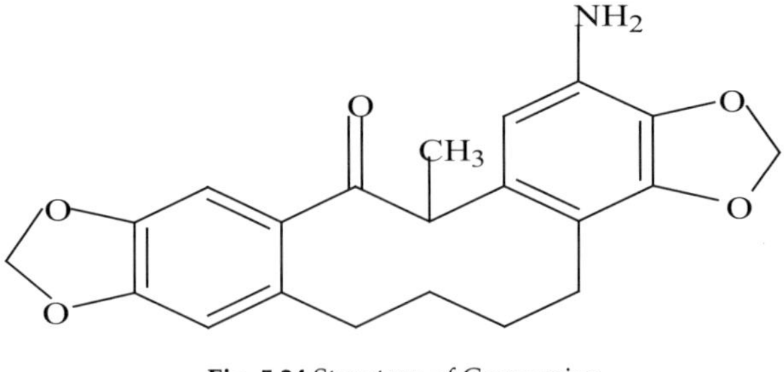

Fig. 5.24 Structure of Corycavine.

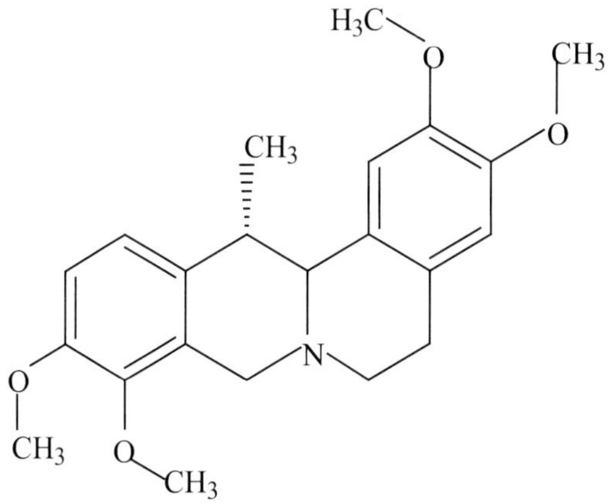

Fig. 5.25 Structure of Corydaline.

Berberine (Fig. 5.26) is the chief alkaloid from roots and stem-bark of *Berberis aristata, B. asiatica, B.lycium, B. vulgaris* and *B. umbellata.* Berberine has diverse pharmacological activites including antibacterial, antihypertensive, and cholagouge.

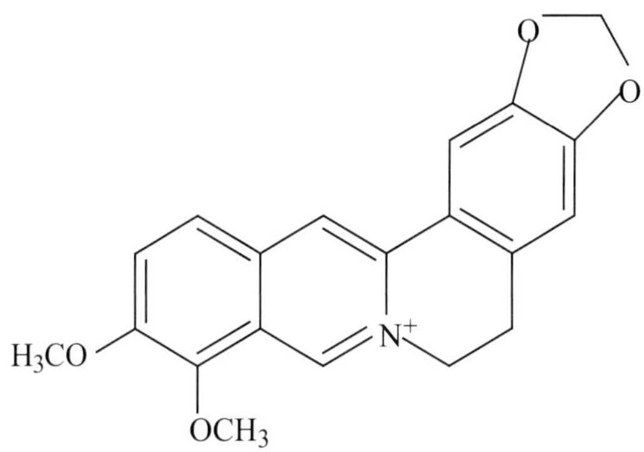

Fig. 5.26 Structure of Berberine.

Berberine sulfate has been shown to inhibit the growth of *Entamoeba histolytica, Giardia lamblia* and *Trichomonas vaginalis, in vitro.* The parasites all exhibited morphological changes after exposure to berberine sulfate. In one experiment, berberine hydrochloride reduced the cholera toxin-induced secretion of water, sodium and chloride in perfused rat ileum. Berberine was also found to inhibit the intestinal secretory response of *Vibrio cholerae* and *Escherichia coli* enterotoxins without causing histological damage to the intestinal mucosa.

Berberine is also active against other intestinal infections that cause acute diarrhea such as *Shigella dysenteriae,Salmonella paratyphi* and various *Klebsiella species.*Berberine sulfate has been shown to block the adherence of *Streptococcus pyrogenes* and *E. coli* to host cells, possibly explaining it's mechanism of action against numerous pathogens.

Berberine was found to be the active constituent in an extract of *Hydrastis canadensis* root that demonstrated activity against a multiple drug-resistant strain of *Mycobacterium tuberculosi.* Berberine also inhibits *Helicobacter pylori.* Berberine has a long history of use for eye infections. In one study that looked at the effectiveness in treating trachoma, berberine was better than sulfacetamide in eradicating *Chlamydia trachomatis* from the eye and preventing a relapse of symptoms.

Berberine has a choleretic (bile-stimulating) effect and has been shown to lower bilirubin levels. Berberine inhibited the effects of tumor promotors on the skin using a mouse model. There is evidence that berberine also has a direct tumor killing effect and has the ability to stimulate production of white blood cells.

Berberine has hypotensive, antisecretory and sedative effects. The mechanism for these effects may be explained by the fact that berberine has platelet α 2 adrenoceptor agonist activity that is similar to that of clonidine.

Berberine inhibits the growth of HepG2 cells by direct interaction with DNA in which it intercalates. Berberine, extracted from *Arcangelisia flava,* inhibits the enzymatic activity of *Plasmodium falciparum* telomerase dose-dependently at doses ranging from 30–300mM. ***Palmatine*** (Fig. 5.87), ***berberine, jatrorrhizine*** (Fig. 5.87), and dihydroberberine (Fig. 5.27) inhibit the growth of *Babesia gibsoni* cultured *in vitro* at very small doses.

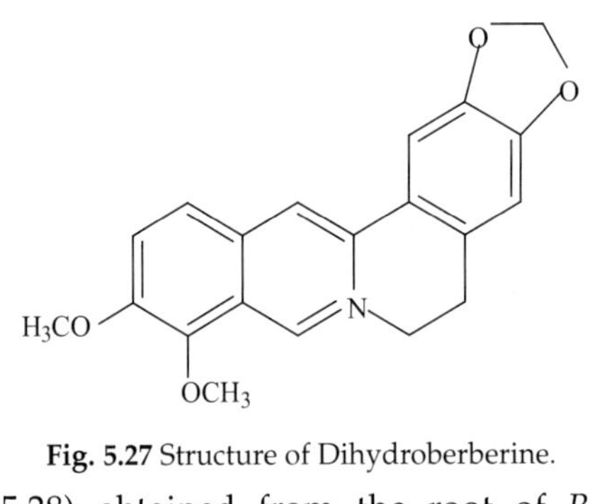

Fig. 5.27 Structure of Dihydroberberine.

Berbamine (Fig. 5.28) obtained from the root of *B. poiretii* increases leucocytosis. ***Oxycanthine*** (Fig. 5.29) is considered to be less potent as compared to berberine. Other alkaloids distributed among Berberis species include ***oxyberberine, berbinine, berbericine*** and ***berbericinine***. Berbericine is identical with berberine-acetone complex.

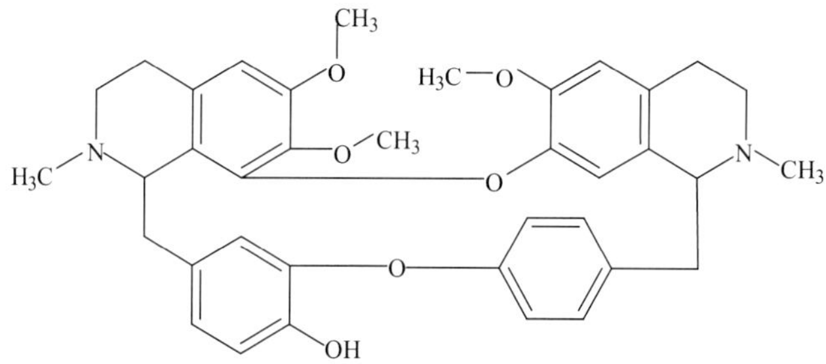

Fig. 5.28 Structure of Berbamine.

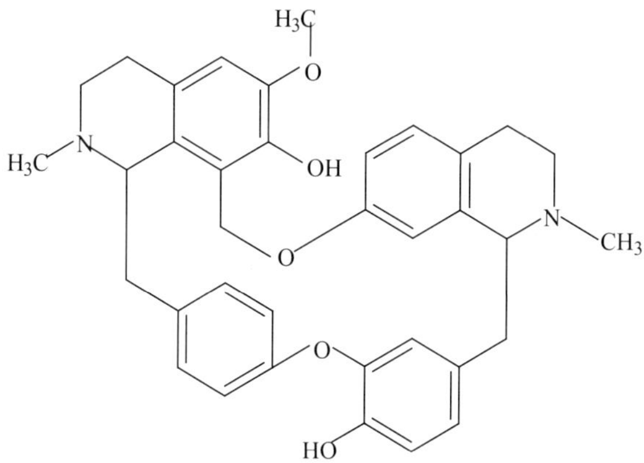

Fig. 5.29 Structure of Oxycanthine.

Alkaloids of *Uncaria tomentosa* include ***rhynchophylline*** (antihypertensive), ***hirsutine*** (local anesthetic) and ***mitraphylline*** (diuretic) (Fig. 5.30). The plant also contains ***5-α-carboxystrictosidine, isopteropodine, isomitraphyllin*** and ***isorynchophylline***.

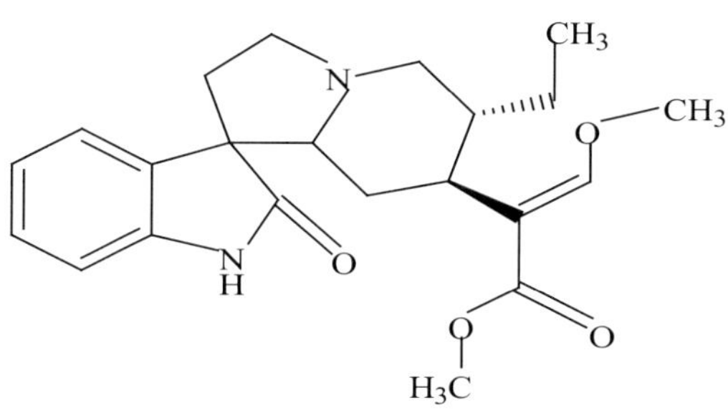

Rhynchophylline

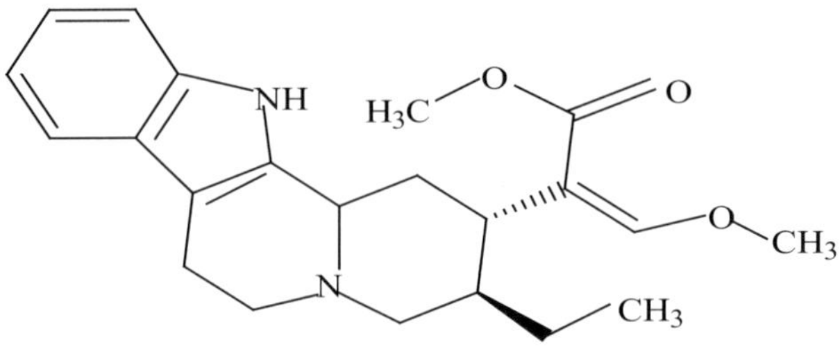

Hirsutine

Fig. 5.30 Structure of alkaloids of *Uncaria tomentosa.*

Pausinystalia yohimbe contains stimulant alkaloid, ***yohimbine or quebrachine*** (Fig. 5.31). Yohimbine causes an increase in epinephrine release from the adrenal gland and results in a dose-dependent increase in plasma epinephrine. Yohimbine, given in moderate doses; increases systolic blood pressure in patients with orthostatic hypotension due to primary autonomic failure.

Yohimbine significantly enhanced the overall analgesic effect of morphine with post operative dental pain. One study demonstrated yohimbine reversed sedation and shortened the duration of analgesia associated with clonidine administered post operatively.The alkaloid has been investigated clinically in erectile dysfunction, post traumatic stress disorder and Parkinson's disease.

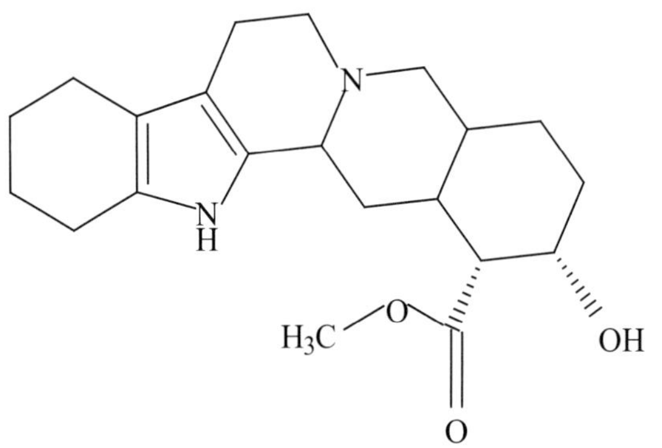

Fig. 5.31 Structure of Yohimbine.

Sanguinaria canadensis contain isoquinoline alkaloids, ***sanguinarine, chelerythrine and oxysanguinarine***: protoberberine-type: ***berberine, coptisine***:protopine-type:***protopine.α-andβ-allocryptopine***.Sanguinarine is antimicrobial and anti-inflammatory. Its use as an antiplaque agent and for gingivitis is plausible and has been documented in diverse studies.

The alkaloids initially act as a narcotic, causing severe cramps followed by local paralysis of sensitive nerve endings. *Caulophyllum thalictroides* contain isoquinoline alkaloid, ***magnoflorine***.

Brahmine (0.01%) is a highly toxic alkaloid obtained from *Hydrocotyle asiatica*. In therpaeutic dosage, it resembles strychnine. It does not produce reflex irritation. ***Herpestine*** is another alkaloid from same plant. ***Adifoline***, alkaloid obtained from acetone fraction of wood of *Adina cordifolia* is a central nervous sytem depressant and hypotensive in experimental animals.

Punarnavine, an alkaloid present in *Boerrhavia diffusa* (punarnava) is bitter in taste. Punarnavine hydrochloride is the salt in crystalline form. Punarnavine sulfate is a needle shaped crystalline form brownish-white in color. Yeild of punarnavine sulfate from the plant is 0.053%. Punarnavine is a diuretic and its acts chiefly on renal epithelium. It is non-toxic, however the nature of the alkaloid is not known.

Himbacine, obtained from *Gabulimina baccata* is a muscarinic antagonist (Fig. 5.32). ***Aegeline*** from *Aegle marlemols* is reported to be a cardiac tonic and useful in asthma (Fig. 5.33). ***Aegelenine*** is identical with fagarine isolated from *Zanthoxylum acanthopodium*. The plant, in addition, contains alkaloids including; ***skimmianine*** (Fig. 5.34), ***shahidine***, ***O-methylhalfordinol*** and ***isopentylhalfordinol***. ***α-pederine*** and ***β-pederine*** are obtained from *Paederia foetida*.

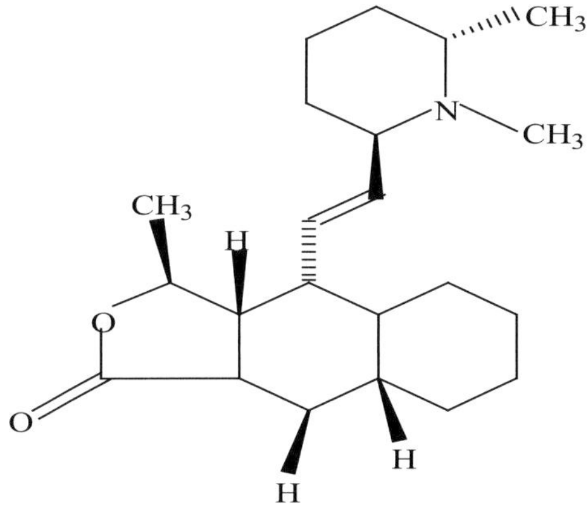

Fig. 5.32 Structure of Himbacine.

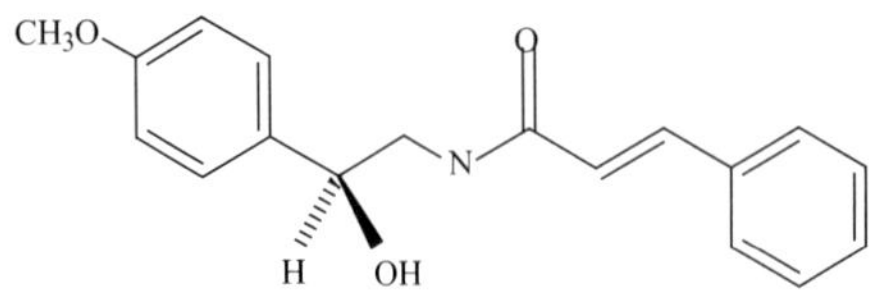

Fig. 5.33 Structure of Aegeline.

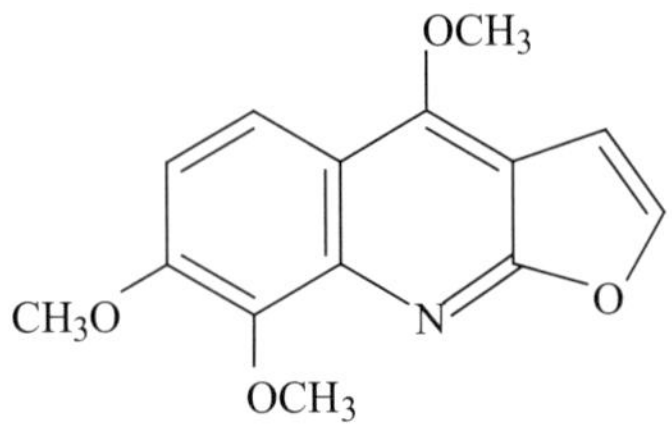

Fig. 5.34 Structure of Skimmianine.

Christembine is obtained from *Embelia ribes*. ***Hymenodictyoline*** is one of the few alkaloids which do not contain oxygen and is obtained from *Hymenodictyon excelsum*. ***Fagamide***, an alkaloid present in *Zanthoxylum acanthopodium* is the possible irritating principle. ***Nishindine*** is extracted from *Vitex negundo*. *Viola odorata* yields ***violine***, which is emetic. ***Anistrocladinine*** isolated from *Ancistrocladus heyneanus* has antispasmodic acitivity on isolated guinea pig ileum comparable with papaverine.

Achilleine, an alkaloid of *Achillea millefolium* is considered to be an active haemostatic. It reduces blood-clot in rabbits. The total alkaloidal content (***mucuadine,mucuadinine,mucucuadinine,pruriendine,mucunine,mucunadine and nicotine***) of *Mucuna prurita* demonstrated increased spermatogenesis in albino rats. ***O-methylsolanocapsine*** isolated from *Solanum pseudocapsicum* leaves possess strong cytotoxic properties (Fig. 5.35).

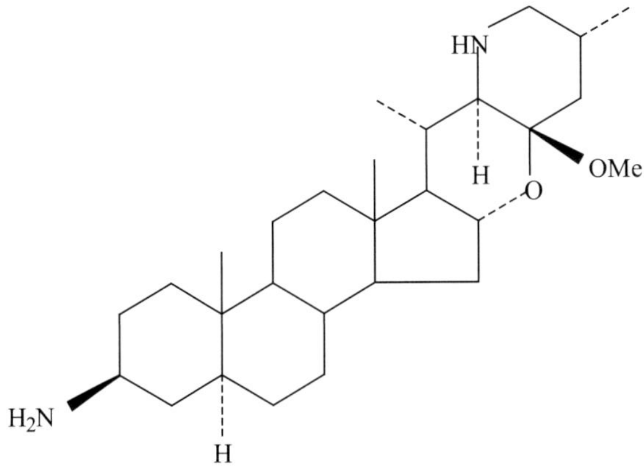

Fig. 5.35 Structure of O-methylsolanocapsine.

Saussurine, an alkaloid isolated from *Saussurea lappa* is a potent bronchodilator. It has depressant action on the involuntary muscle fibers of the bronchioles and gastrointestinal tract. Saussurine tartrate has been shown to produce a definite relaxation of the bronchioles in the same way as adrenaline does. Saussurine has a positive inotropic action on the heart muscle.

The hydrochloride of alkaloid, ***evolvine*** present in *Evolvulus alsinoides* is reported to exhibit lobeline-like action on the cardiovascular system. In cats, the drug has significant sympathomimetic activity. The blood pressure remains elevated for a longer duration as compared to adrenaline. Increase in peripheral pressure is observed on local injection of the drug. *Phaeanthus ebracteolatus* contains bis-benzylisoquinoline known as ***phaeantharine*** which has exhibited some levels of antibacterial activity.

Lycoriside, at lower concentrations (1–20 μg/ml), *in vitro*, produces statistically significant protection against Tween 80-induced degranulation, as also to sensitized mast cells challenged with an antigen (horse serum). It also provided protection against compound 48/80-induced degranulation of mast cells when administered *in vivo* in a dose of 1–5 mg/kg, po.

Cerpegin, a novel furopyridine alkaloid isolated from *Cerpegia juncea* demonstrates dose-related analgesic effect in mice. Cerpegin do not produce any autonomic or behavioral changes up to a dose of 200 mg/kg but doses of more than 400 mg/kg produced excitation and later convulsions in mice (Fig. 5.36).

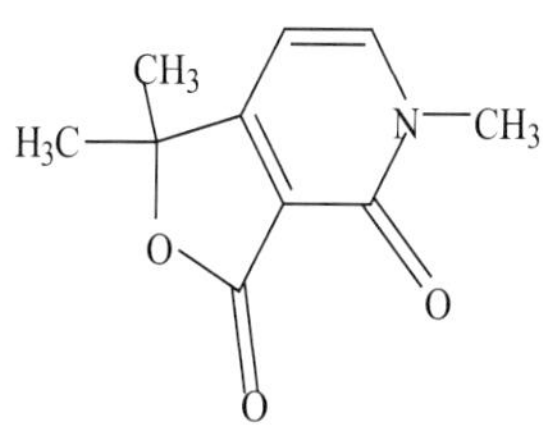

Fig. 5.36 Structure of Cerpegin.

Premnazole, an isoxazole alkaloid isolated from *Premna integrifolia* α. and *Gmelina arborea* has significant anti-inflammatory activity in reducing cotton pellet-induced granuloma formation in rats (Fig. 5.37). The anti-inflammatory activity is comparable to that of phenylbutazone.

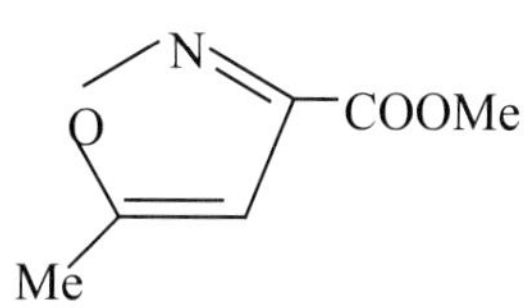

Fig. 5.37 Structure of Premnazole.

The water-soluble alkaloid, ***achyranthine***, isolated from *Achyranthes aspera* has been screened for its anti inflammatory and antiarthritic activity against carrageenin-induced foot oedema, granuloma pouch, formalin induced arthritis and adjuvant arthritis in rats. It shows significant anti-inflammatory activity in all the four models employed but is less active than phenylbutazone and betamethasone. Incidence of gastric ulcers is maximum with betamethasone and minimum with achyranthine.

Anti-inflammatory activity of ***crotalaburnine*** (anacrotine) isolated from *Crotalbria laburnifolia* has been investigated against several models of inflammation (Fig. 5.38). The effect has been compared with the activity of hydrocortisone, phenylbutazone, sodium salicylate and cyproheptadine against different types of inflammation.

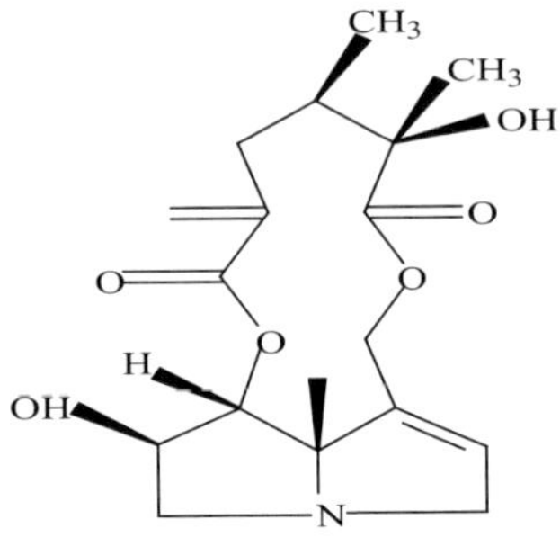

Fig. 5.38 Structure of Crotalaburnine.

In normal rats crotalaburnine (10 mg/kg s.c.), phenylbutazone (100 mg/kg oral) and sodium salicylate (500 mg/kg i.p.) inhibits hyaluronidase-induced oedema. However, in adrenalectomized rats, there is a reduction of the inhibitory effect of sodium salicylate but not of phenylbutazone or crotalaburnine.

Solkhasianine is a new glycoalkaloid from *Solanum khasianum*. ***Solasodine,*** an alkaloid of *Solanum xanthocarpum* demonstrates antifertility activity in males and dogs (Fig. 5.39). The alkaloid on oral administration results in inhibition of spermatogenesis and sperm motility. The mechanism of action of solasodine may be attributed to inhibition of testosterone release. ***Yuehchukene,*** an alkaloid from *Murraya paniculata* is used to regulate fertility in China.

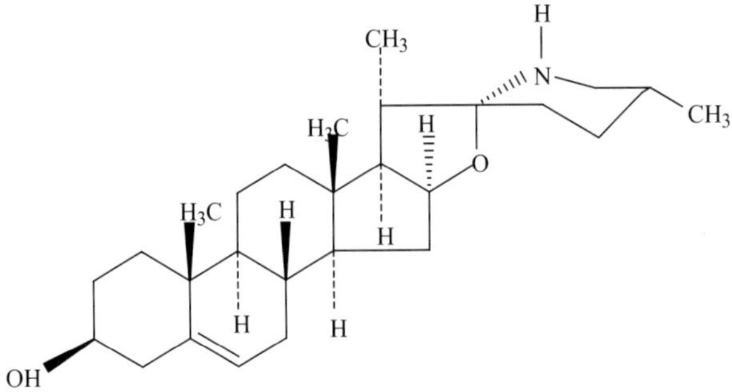

Fig. 5.39 Structure of Solasodine.

Pergularinine, tylophorinidine and deoxytubulosine isolated from the Indian medicinal plants *Pergularia pallida* and *Alangium lamarckii* respectively, inhibits (IC50 = 50 microM) the elevated TS activity of leukocytes in cancer patients with clinically diagnozed chronic myelocytic leukemia (n = 10), acute lymphocytic leukemia and metastatic solid tumors (Figs. 5.40 and 5.41).

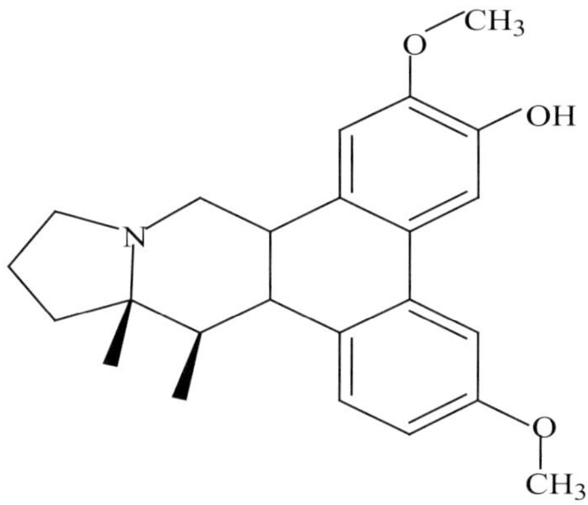

Fig. 5.40 Structure of Tylophorinidine.

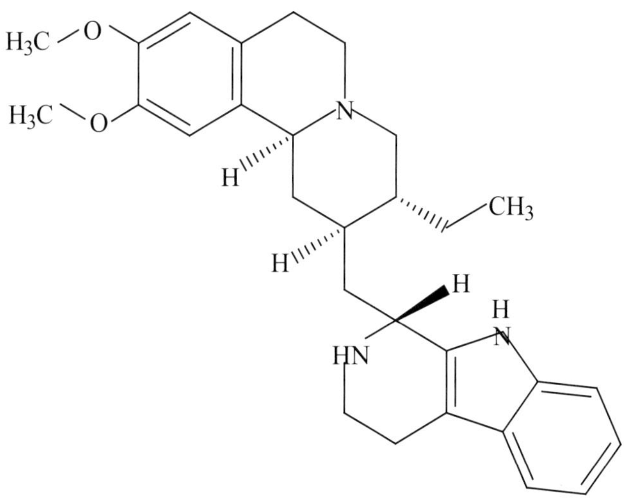

Fig. 5.41 Structure of Deoxytubulosine.

Cryptolepine (Fig. 5.42) is an indolisoquinoline antimalarial alkaloid with IC50 value approximately half of chloroquine. It was shown that the alkaloid might interact with DNA, and it appeared that two nitrogen atoms N and N-CH3 of cryptolepine interact with adenine–thymine base pair. There is also a possibility of formation of *p–p* charge transfer complex between purine–pyrimidine bases and cryptolepine and electron environment.

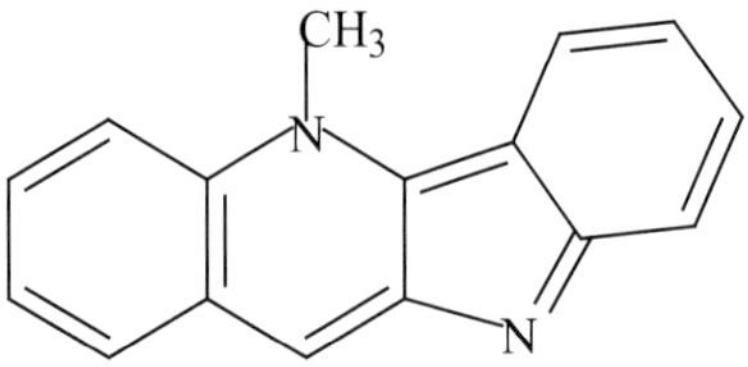

Fig. 5.42 Structure of Cryptolepine.

Another interesting antimalarial compound, ***tubulosine*** was found active *in vitro* against both sensitive and resistant strains of *P. falciparum*. The indole moiety in tubulosine enhances the affinity for protozoan receptor, when compared with psychotrine and cephaeline. Extracts from *Triphophyllum peltatum* led to the isolation of ***dioncophylline B and dioncophylline*** C both exhibiting high antiplasmodial activity (Fig. 5.43). Dioncophylline C cured malaria-infected mice completely after a 4-d oral treatment with 50 mg kg–1 d–1 without noticeable toxic effects.

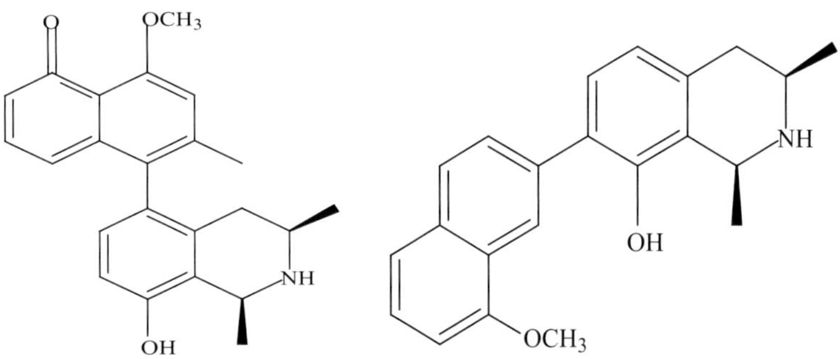

Fig. 5.43 Structure of dioncophylline B and dioncophylline C.

A bisbenzylisoquinoline alkaloid ***dehatrine*** isolated from the wood of *Beilschmiedia madang*, exhibited potent inhibitory activity (IC50 0.017 μ M) against the proliferation of malaria pathogen *P. falciparum*, which was comparable to quinine. Recently, a novel dimeric naphthylisoquinoline alkaloid heterodimer with antiplasmodial activity, ***korundamine A*** has been isolated from *Ancistrocladus korupensis* (Fig. 5.44).

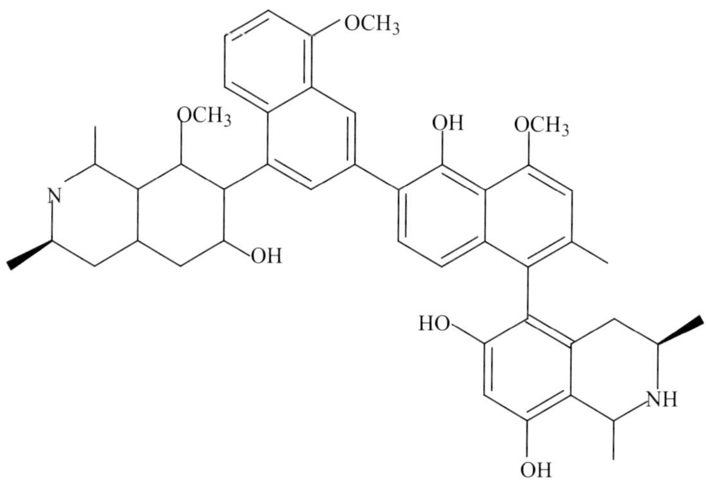

Fig. 5.44 Structure of Korundamine A.

Topotecan, an analog of a plant alkaloid, ***camptothecin*** discovered in the Chinese tree species *Camptotheca acuminata*, for the treatment of ovarian and small cell lung cancers (Fig. 5.45). Irinotecan, another chemical analog which has been developed from another plant alkaloid discovered in the same tree, *C. acuminata*, is used for the treatment of metastatic colorectal cancer.

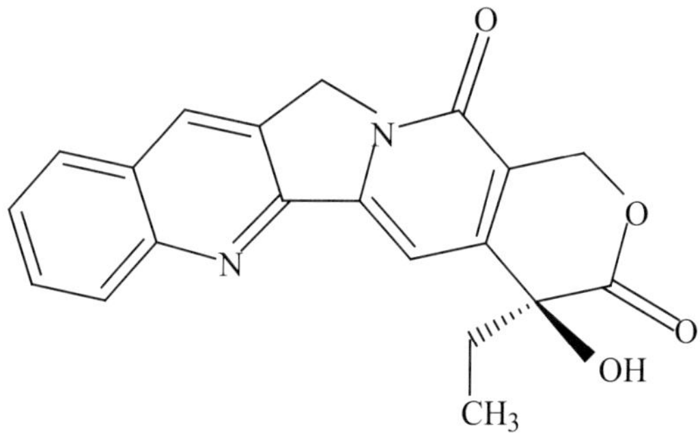

Fig. 5.45 Structure of Camptothecin.

(+)-Calanolide A and (-)-Calanolide B (costatolide) isolated from *Calophyllum lanigerum* and *Calophyllum teysmanii*, respectively. (+)-Calanolide A is currently in early clinical trials in the United States. ***Maytansine*** from *Maytenus ilicifolia* exhibits significant cytotoxic and antitumoral efficacy. Additionally, an ulcer-preventing effect has been demonstrated in both animal and human studies.

Michellamine B, from the leaves of *Ancistrocladus korupensis*, a vine found in the Korup rainforest region of southwest Cameroon, has undergone extensive preclinical study, but is considered too toxic for advancement to clinical trials (Fig. 5.46).

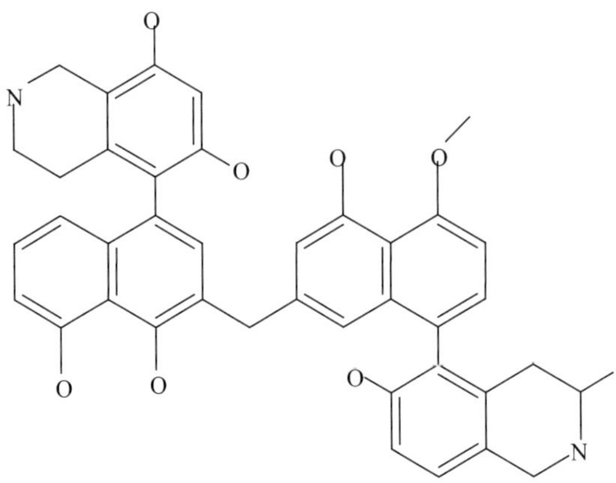

Fig. 5.46 Structure of Michellamine B.

Homoharringtonine from *Cephalotaxus harringtonia* is used in acute nonlymphoblastic leukemia and chronic myelogenus leukemia (Fig. 5.47).

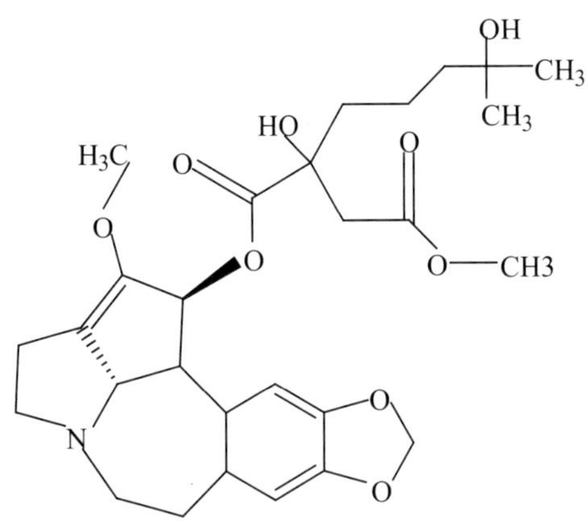

Fig. 5.47 Structure of Homoharringtonine.

Ellipticine (Fig. 5.48) and related alkaloids, e.g. ***9-methoxyellipticine*** are found in the bark of *Ochrosia elliptica*. Clinical trials with these alkaloids and a number of synthetic analogues showed them to be potent inhibitors of several cancerous disorders, but pre-clinical toxicology indicated a number of side-effects, including haemolysis and cardiovascular effects.

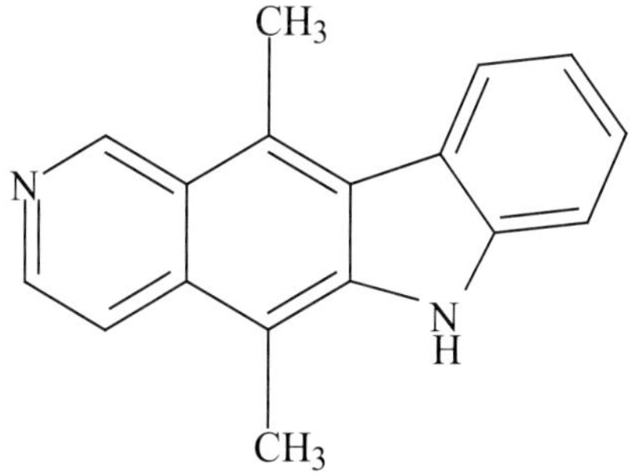

Fig. 5.48 Structure of Ellipticine.

Evodiamine (Fig. 5.49) extracted from *Evodia elleryana,* inhibits cell proliferation and migration of several types of cancer cell lines, including leukemia human caucasian acute lymphoblastic leukaemia cell-line (CCRF-CEM) cells in a concentration-dependent manner with an IC50 of 0.57μM via apoptosis following microtubular cytoskeleton abrogation. Methanol extracts of various parts of *Euodia elleryana* inhibits the growth of a broad spectrum of bacteria. While the antibacterial principle here is unknown, one could reasonably expect that evodiamine and ***rutaecarpine*** (Fig. 5.50)-like quinazoline alkaloids are involved.

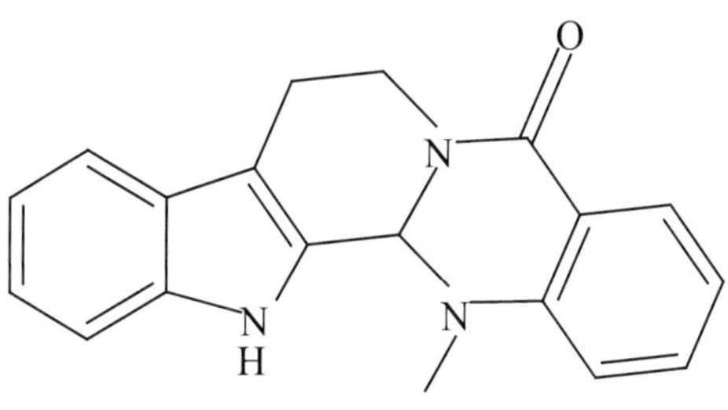

Fig. 5.49 Structure of Evodiamine.

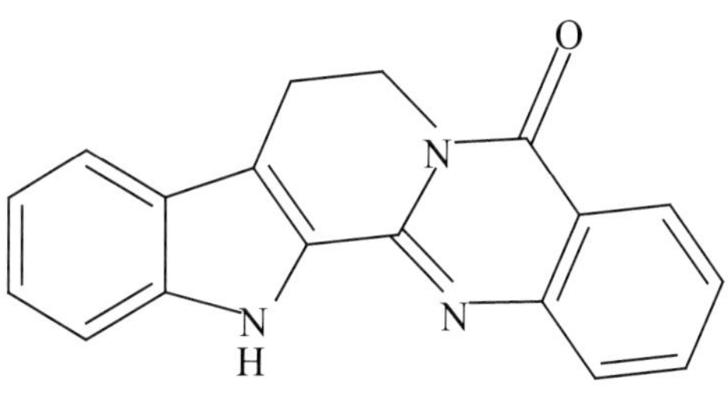

Fig. 5.50 Structure of Rutaecarpine.

Rhizome of *Nuphar pumilum* contains dimeric sesquiterpene thioalkaloids, such as ***6-hydroxythiobinupharidine, 6, 6″-dihydroxythiobinupharidine, and 6-hydroxythionuphlutine B*** (Figs. 5.51, 5.52), which inhibited the invasion of B16 melanoma cells across collagen-coated filters *in vitro* with IC50 0.029, 0.087, and 0.36μ M, respectively, indicating a clear antimetastatic potential.

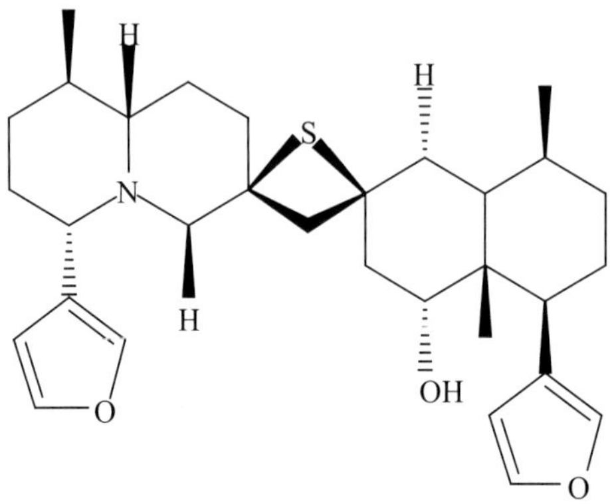

Fig. 5.51 Structure of Thiobinupharidine.

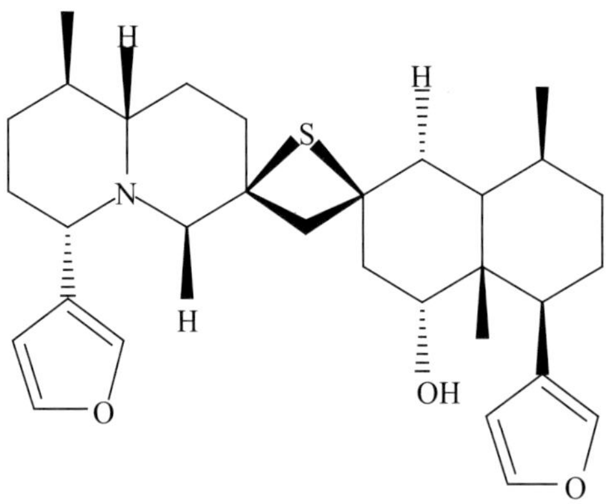

Fig. 5.52 Structure of 6-hydroxythionuphlutine B.

Nelumbo nucifera elaborates antiviral isoquinolines ***(+)-1(R)-coclaurine*** (Fig. 5.53) ***and 1(S)- norcoclaurine***, which inhibits the replication of HIV *in vitro* with EC 50 values of 0.8 and <0.8g/ μL, and therapeutic index values of >125 and >25, respectively. ***Liensinine*** (Fig. 5.54) and ***isoliensinine*** showed potent anti-Human Immunodeficiency Virus (HIV) activities with EC50 values of <0.8g/μ L and Therapeutic Index values of >9.9 and >6.5g/μ L.

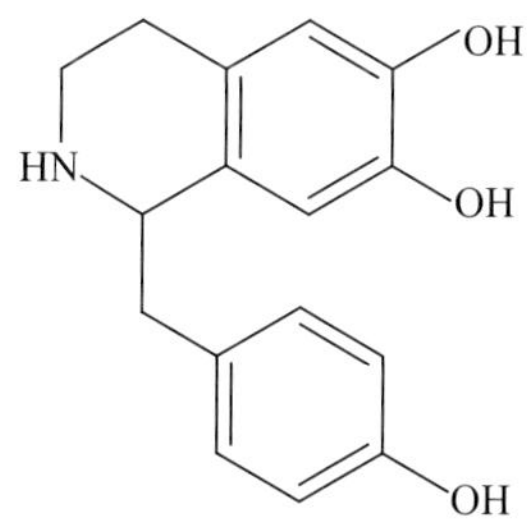

Fig. 5.53 Structure of Coclaurine.

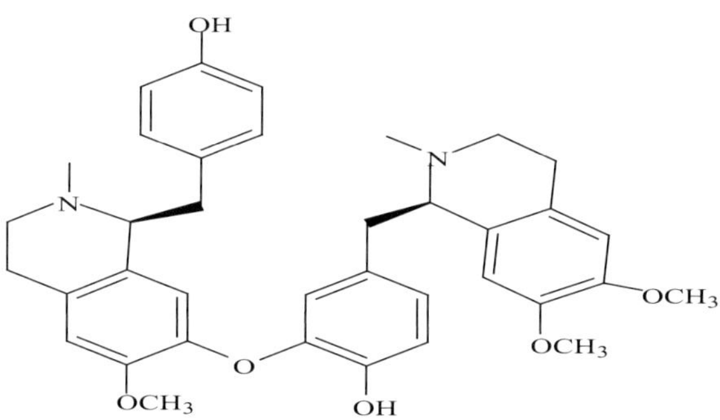

Fig. 5.54 Structure of Liensinine.

Nuciferine (Fig. 5.55), an aporphine alkaloid, had an EC50 value of 0.8g/ μ L and a Therapeutic Index value of 36. Isoliensinine exhibited a significant inhibitory effect on bleomycin-induced pulmonary fibrosis in male mice. The alkaloid lowered the hydroxyproline content and limited the lung histological injury induced by bleomycin, and also inhibited the overexpression of TNF-α and TGF-β induced by bleomycin, and showed potent activity against bleomycin-induced pulmonary fibrosis.

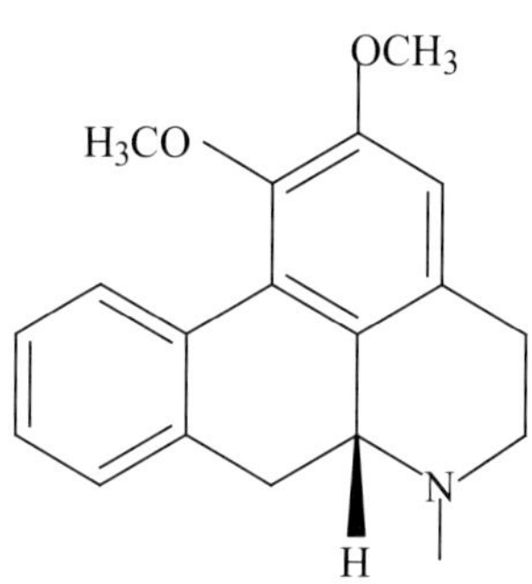

Fig. 5.55 Structure of Nuciferine.

Cappamansin A (Fig. 5.56), from the roots of *Capparis sikkimensis* subsp. *formosana* has encouraged potently the survival of ovarian (1A9), lung (A549), ileocecal (HCT-8), breast (MCF-7), nasopharyngeal (KB), and vincristine resistant (KB-VIN) human tumor cell lines cultured *in vitro.*

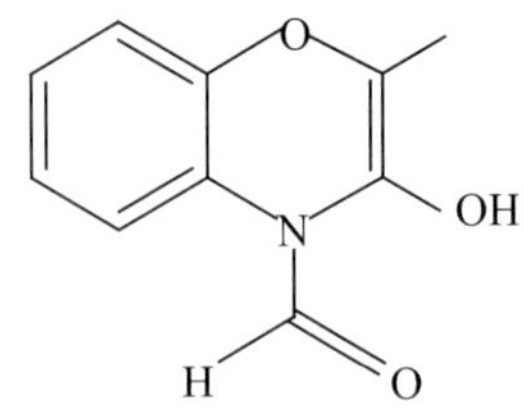

Fig. 5.56 Structure of Cappamansin A.

Tetrandrine (Fig. 5.57), a bis-benzyl isoquinoline alkaloid, isolated from the roots of *Stephania tetrandra*, showed that it not only inhibits Hep G2 growth but also induced apoptosis and blocked cell cycle progression in the GI phase.

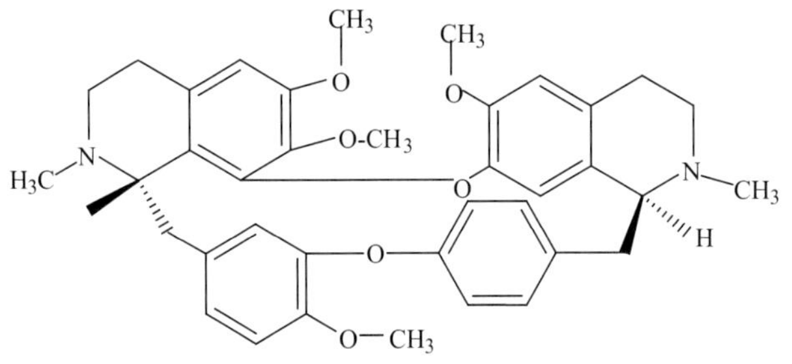

Fig. 5.57 Structure of Tetrandrine.

Cassaine is a cardiotoxic alkaloid in *Erythhrophloeum guinnense* (Fig. 5.58). It causes anorexia, defective vision, increases heart sounds and dyspnoea.

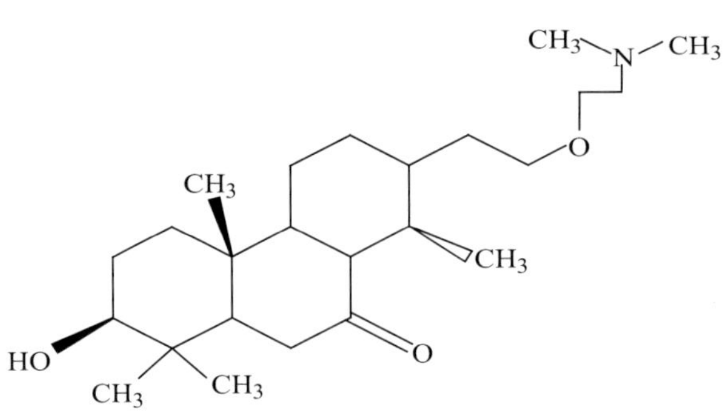

Fig. 5.58 Structure of Cassaine.

Frangulanine, a cyclopeptide alkaloid from *Hovenia dulcis* induces mitochondrial swelling in 0.5 M Kcl solution at the concentration at 6.5μM. The alkaloid showed ion selectivity on the induction of mitochondrial swelling. Ion specific pores are involved in sodium-potassium transport. Ion specific pores in the membrane are blocked by frangulanine, which acts as an ionophore for potassium. ***Vertine*** the principal alkaloid of *Heimia salicifolia* has anti-inflammatory activity.

Dicentra cucullaria contains isoquinoline alkaloids: including ***bicuculline, corlumine, protopine, cryptopine,*** and ***cularine.*** *Liriodendron tulipifera* contains isoquinoline alkaloids (0.1%), particularly of the aporphine type: including ***remerine, lysicamine, liriodenine, and lanugosine***. The alkaloid ***violine*** from *Viola odorata* has an emetine-like effect. *Quassia amara* contains indole alkaloids of:-beta-carboline type, including ***l-vinyl-4, 8-dimethoxy-betacarboline*** and canthinone type, including ***2-methoxy-6-one, 3-methylcanthine-5, 6-dione.***

Sambucus ebulus is reported to quinolizidine alkaloids: ***cytisine, methylcytisine, anagyrine, isosparteine, lupanine, and tinctorine***. *Lupinus luteus* contains quinolizidine alkaloids including ***sparteine, lupinine and p-cumaroyllupinine***. Lupinine and benzolylupinine are reported to be anthelmintic.

Areca alkaloids

The major stimulant alkaloid of *Areca catechu* is ***arecoline*** (up to 0.2%) (Fig. 5.59), the remainder of the alkaloid content (total about 0.45%) being composed of ***arecaidine, guvacine,*** and ***guvacoline.*** Arecoline is an agonist for muscarinic acetylcholine receptors. Arecoline has been employed in veterinary practice as a vermicide to eradicate worms.

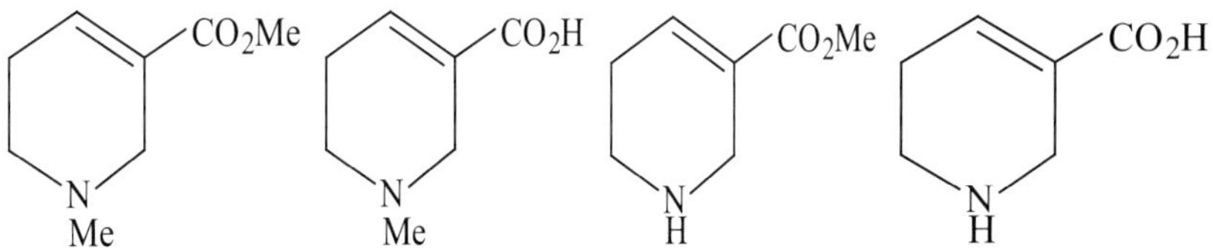

Arecoline Arecaidine Guvacoline Guvacine

Fig. 5.59 Structure of *Areca alkaloids.*

Ashwagandha alkaloids (see Solanaceae alkaloids also)

Withania somnifera contains alkaloids including ***anagrine, anaferine, cuscohygrine, hygrine, isopelletriene, pseudopelletriene, 3α-togloxytropine, 3-tropyltigloate, tropine and withasomnine and visamine*** (Fig. 5.60). Acute toxicity of *W. somnifera* alkaloids is modest. A 180-day study involving rats found unfavorable increases in catecholamine content of the heart and decreases in the adrenal glands.

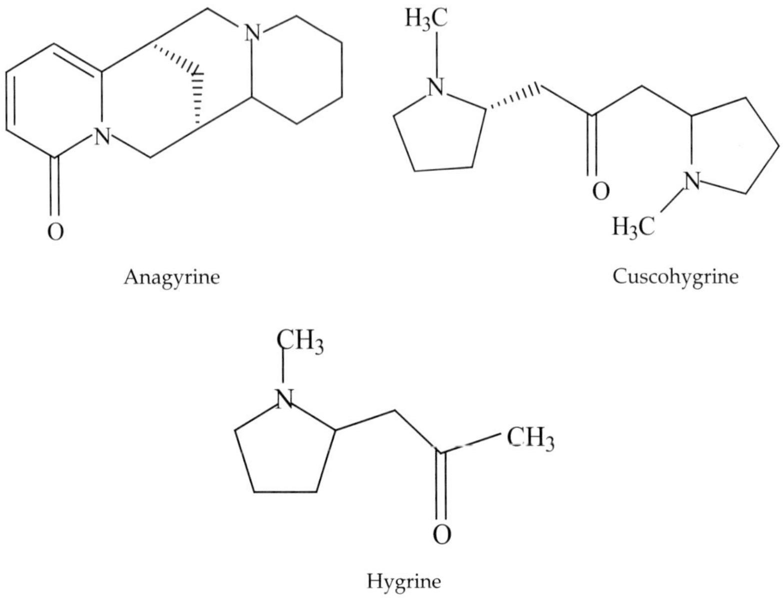

Fig. 5.60 Structure of Ashwagandha alkaloids.

Ephedra alkaloids

The plants typically contain 0.5–2.0% of alkaloids, according to species, and from 30–90% of the total alkaloids is (-)-ephedrine. Related structures, including thendiastereoisomeric (+)-pseudoephedrine and the demethyl analogues (-)-norephedrine and (+)-norpseudoephedrine are also present. In *Ephedra intermedia*, the proportion of pseudoephedrine exceeds that of ephedrine.

Ephedrine (Fig. 5.61) is an indirectly acting sympathomimetic amine with effects similar to noradrenaline. Lacking the phenolic groups of the catecholamines, it has only weak action on adrenoreceptors, but it is able to displace noradrenaline from storage vesicles in the nerve terminals, which can then act on receptors. It also has bronchodilator activity, giving relief in asthma. It has vasoconstrictor action on mucous membranes, making it an effective nasal decongestant.

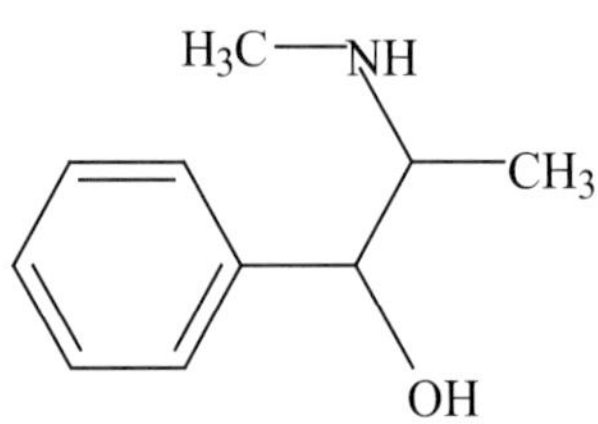

Fig. 5.61 Structure of Ephedrine.

Pseudoephedrine is also widely used in compound cough and cold preparations and as a decongestant.

Ergot alkaloids (Claviceps purpurea)

The ergot sclerotia contain from 0.15–0.5% alkaloids, and more than 50 have been characterized. The medicinally useful compounds are derivatives of (+) - lysergic acid. There are three classes of ergot alkaloids: the ergotamine group consisting of ***ergotamine, ergosine*** and the corresponding isomers the ergotoxine group consisting of ***ergocornine, ergocristine, ergokryptine*** and their isomers ***ergometrine*** and its isomer. ***Ergotamine*** (ergonovine) finds application in anti migraine prescriptions (Fig. 5.62). It is a partial agonist of α-adrenoceptors and 5-HT receptors. It is not suitable for obstetric use because it also produces a pronounced peripheral vasoconstrictor action. Ergometrine is used as an oxytocic, and is injected during the final stages of labor and immediately following childbirth, especially if hemorrhaging occurs. Bleeding is reduced because of its vasoconstrictor effects, and it is valuable after Caesarian operations.

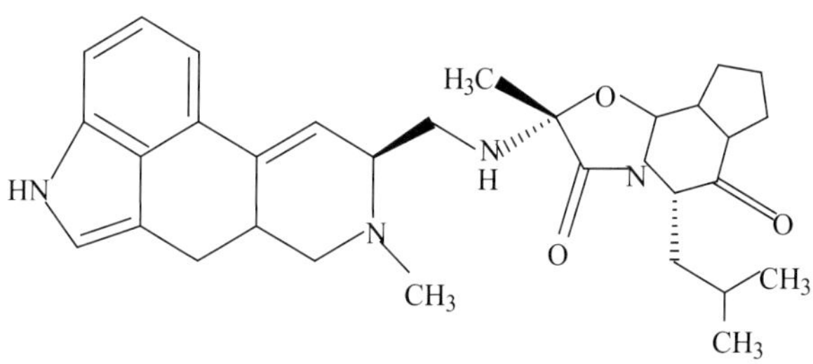

Fig. 5.62 Structure of Ergotamine.

Cevadilla alkaloids

They are obtained from *Schoenacaulon officinale.* The alkaloids include ***cevadine*** (veratrine), ***veratridine*** (veraine), ***sabadilline*** (cevadilline), ***sabadine and sabadinine***. Cevadine is toxic and used externally as an insecticidal.

Cinchona alkaloids

A considerable number of alkaloids have been characterized in the cinchona bark, four of which account for some 30–60% of the alkaloid content. These are quinine, quinidine, cinchonidine, and cinchonine (Fig. 5.63).

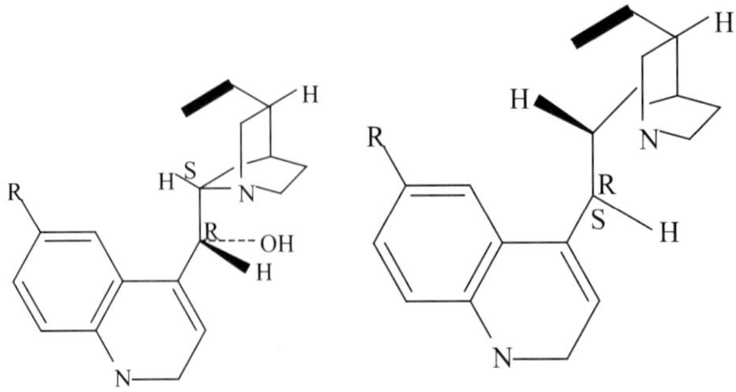

R= OMe, (-) quinine
R= OH. (-) cinchonidine

R= OMe, (+) quinidine
R= OH. (+) cinchonine

Fig. 5.63 Structure of Cinchona alkaloids.

Quinine is extremely bitter, and also possesses antipyretic, analgesic and anti-inflammatory properties. While quinine is still the drug of choice for the treatment of Falciparum malaria, it can be also used to treat nocturnal leg cramps and arthritis. It can cause various side-effects like nausea, vomiting, cinchonism, and rarely, pulmonary oedema.

Quinidine is used in the treatment of cardiac arrhythmia. Intravenous injection of quinidine is also used in the treatment of Plasmodium falciparum malaria. Thrombocytopenia is the major adverse effect.

Coca alkaloids

Coca leaves are obtained from species of *Erythroxylum coca* and *E. truxillense*. Coca leaf contains 0.7–2.5% of alkaloids, the chief component ***(-)-cocaine*** (Fig. 5.64). It is a potent central nervous system stimulant and appetite suppressant. For its euphoretic effect, cocaine is often used recreationally, and it is one of the most common drugs of abuse and addiction. Cocaine is used as a topical anaesthetic in eye, throat and nose surgery. In overdose, it leads to tachyarrhythmia and hypertension.

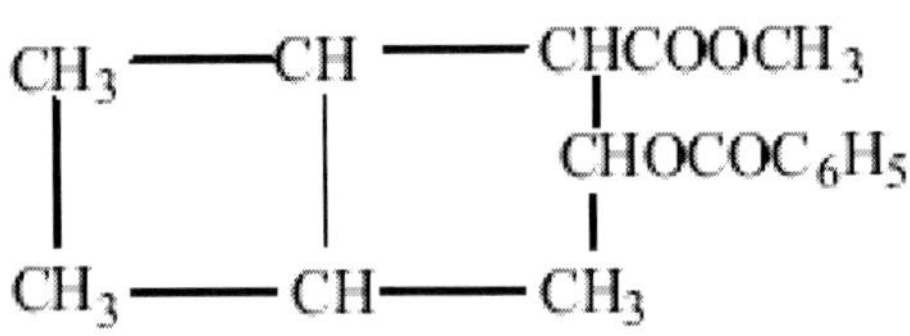

Fig. 5.64 Structure of Cocaine.

Conium alkaloids

The major alkaloid (about 90%) is the volatile liquid ***coniine***, with smaller amounts of structurally related piperidine alkaloids, including ***N-methylconiine*** and ***γ-coniceine.*** Coniine contributes to the foul smell of hemlock. It is a neurotoxin, causes respiratory paralysis and is toxic to all classes of livestock and humans. In 399 BC, Socrates was put to death by this poison (Fig. 5.65).

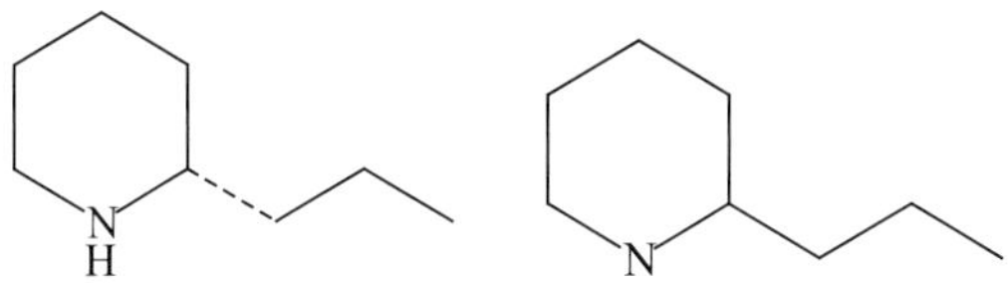

(+)-conine γ-coniceine

Fig. 5.65 Structure of Conium alkaloids.

Curare alkaloids

These are obtained from *Chondrodendron tomentosum.* ***(+) tubocurarine*** (Fig. 2.6) is the principal alkaloid. In addition, it contains ***chondrocurarine, isochondrodendrine, curine, curarine, cylleanine and tomentocurarine.*** (+) tubocurarine is a strong neuromuscular blocking agent.

Iboga alkaloids (Terbernanthe iboga)

The root bark contains up to 6% indole alkaloids, the principal component of which is ***ibogaine***. Ibogaine is a CNS stimulant, and is also psychoactive. In large doses, it can cause paralysis and respiratory arrest. Ibogaine is of interest as a potential drug for relieving heroin craving in drug addicts (Fig. 5.66).

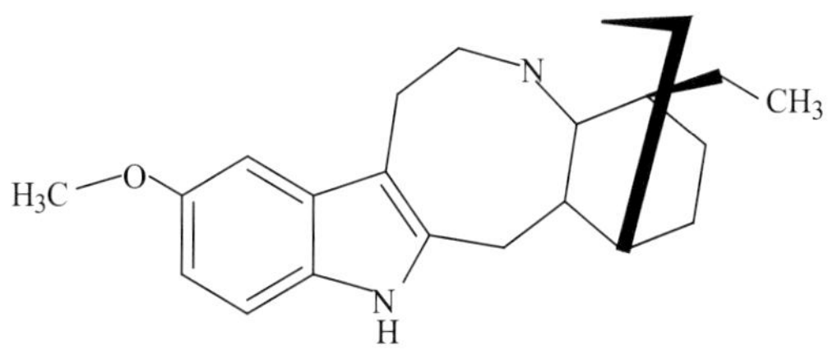

Fig. 5.66 Structure of Ibogaine.

Ipecac alkaloids

These are obtained from dried root of *Cephalis ipecacunaha*. The chief alkaloids are emetine, cephaline and psychotrine (Fig. 5.67). ***Cephaline*** is a reduction product of ***psychotrine***. The actions of emetine, cephaline are similar; psychotrine is inert. ***Emetine*** is specific for amoebic dysentery.

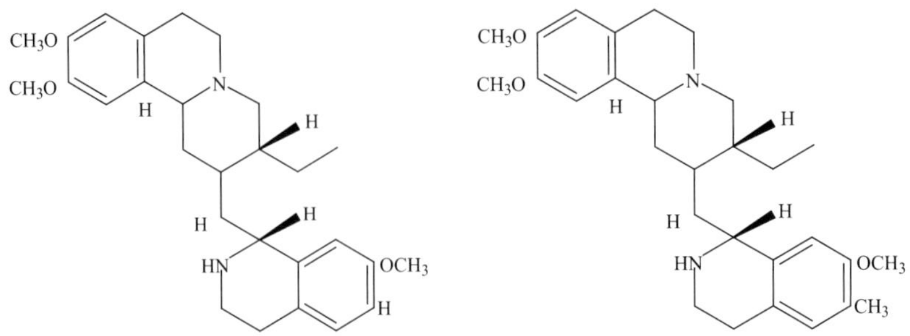

Fig. 5.67 Structure of Cephaline and Emetine.

Nux-vomica alkaloids

These seeds contain 1.5–5% of alkaloids, chiefly ***strychnine*** (about 1.2%) and ***brucine*** (about 1.6%). ***Strychnine*** (Fig. 2.6) is very toxic, affecting the CNS and causing convulsions. This is a result of binding to receptor sites in the spinal cord that normally accommodate glycine. Its only medicinal use is in very small doses as an appetite stimulant and general tonic, sometimes with iron salts if the patient is anaemic. Brucine (Fig. 5.14) is considerably less toxic. Distribution of strychnine and brucine in various species of Strychnos is tabulated below.

Table 5.1 It shows alkaloid % in various species of Strychnos.

Species	Strychnine	Brucine
Strychnos nux-blanda	x	x
Strychnos tieute	1.4%	traces
Strychnos nux-vomica	1.23%	1.55%
Strychnos lucida	0.3%	1.5–2.4%
Strychnos ignatii	1.23%	1.55%

Kurchi alkaloids

These are obtained from *Holarrhena antidysenterica*. They include ***connesine*** (Fig. 5.68), ***holarrhenine, kurchine, kurchicine conkurchine, conesimine, holarrhine, holarrhimine, conimine and isoconessimine***. Steroidal alkaloids, ***antidysentericine, reghloarrhenine-A, reghloarrhenine-B and reghloarrhenine- C*** have been isolated from *H. antidysenterica*. In castor oil-induced diarrhea, the alkaloids reduced the diarrhea.

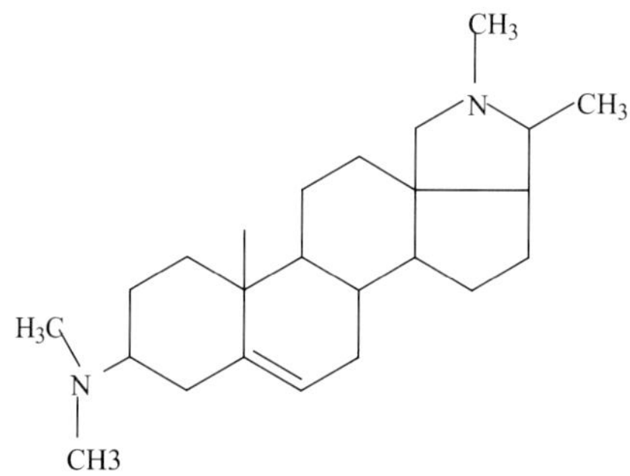

Fig. 5.68 Structure of Conessine.

Opium alkaloids

Although the ripe poppy (*Papaver somniferum*) capsule can contain up to 0.5% total alkaloids, opium represents a much more concentrated form and up to 25% of its mass is composed of alkaloids. Of the 40 alkaloids identified, about 6 represent almost all of the total alkaloid content. A typical commercial sample of opium would probably have a morphine content of about 12%. Powdered opium is standardized to contain 10% of anhydrous morphine.

Morphine (Fig. 2.6) is a powerful analgesic and narcotic, and remains one of the most valuable analgesics for relief of severe pain.

Codeine (Fig. 5.5) is a relatively safe non-addictive medium analgesic, but is too constipating for long-term use. Codeine also has valuable antitussive action, helping to relieve and prevent coughing. It effectively depresses the cough center, raising the threshold for sensory cough impulses.

Papaverine (Fig. 5.6) possesses spasmolytic and vasodilator activity.

Noscapine has good antitussive and cough suppressant activity comparable to that of codeine, but no analgesic or narcotic action.

Pilocarpus alkaloids

Pilocarpus microphyllus is currently the main source. The alkaloid content (0.5–1.0%) consists principally of the imidazole alkaloid ***pilocarpine***; together with small amounts of ***pilosine*** (Fig. 5.69). Pilocarpine salts are valuable in ophthalmic practice and are used in eyedrops as miotics and for the treatment of glaucoma. Pilocarpine is a cholinergic agent and stimulates the muscarinic receptors in the eye, causing constriction of the pupil and enhancement of outflow of aqueous humour.

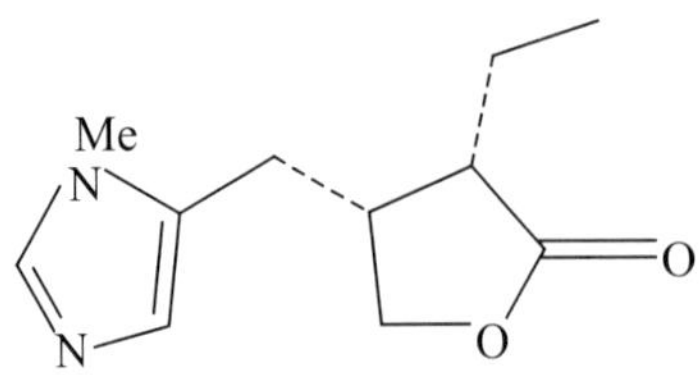

Pilocarpine

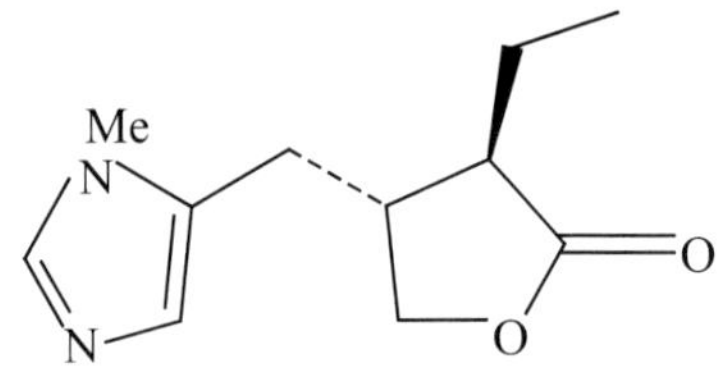

Isopilocarpine

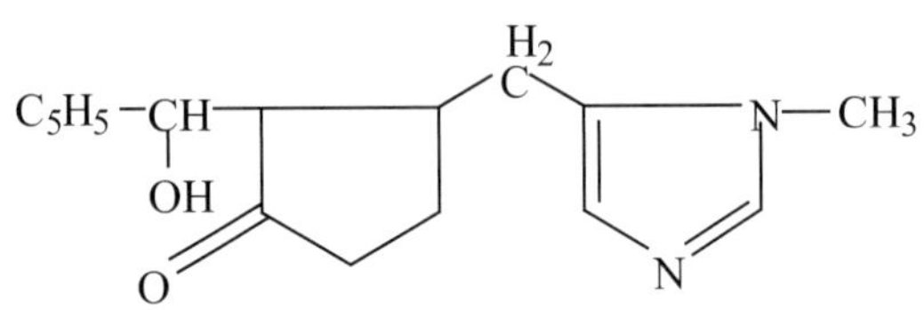

Pilosine

Fig. 5.69 Structure of *Pilocarpus* alkaloids.

Psilocybe alkaloids (Psilocybe maxicana)

The active hallucinogens, present at about 0.3%, are the tryptamine derivatives ***psilocybin*** and ***psilocin,*** which are structurally related to the neurotransmitter, 5-hydroxytryptamine, the factor contributing to their neurological effects (Fig. 5.70).

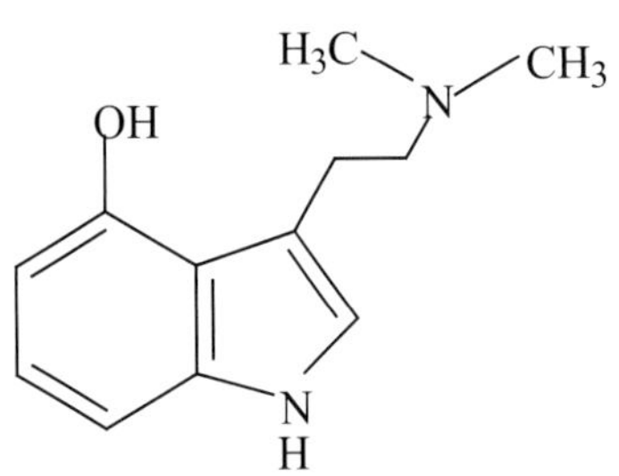

Psilocin

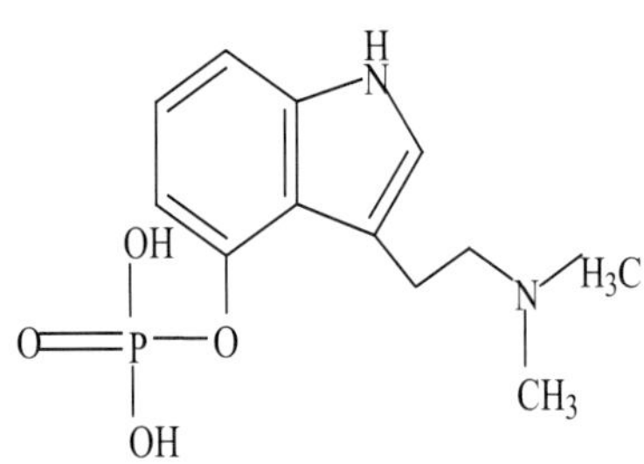

Psilocybin

Fig. 5.70 Structure of Psilocybe alkaloids.

Resperine alkaloids (Rauwolfia serpentina)

Reserpine and ***deserpidine*** (Fig. 5.71) have been widely used as antihypertensive and mild tranquillizers. They act by interfering with catecholamine storage, depleting levels of available neurotransmitters. Prolonged use of reserpine has been shown to lead to severe depression in some patients, a feature not so prevalent when the powdered root was employed. Both ***ajmalicine*** (Fig. 5.71) and ***ajmaline*** are used clinically in Europe. Ajmalicine is employed as an antihypertensive, whilst ajmaline is of value in the treatment of cardiac arrhythmias.

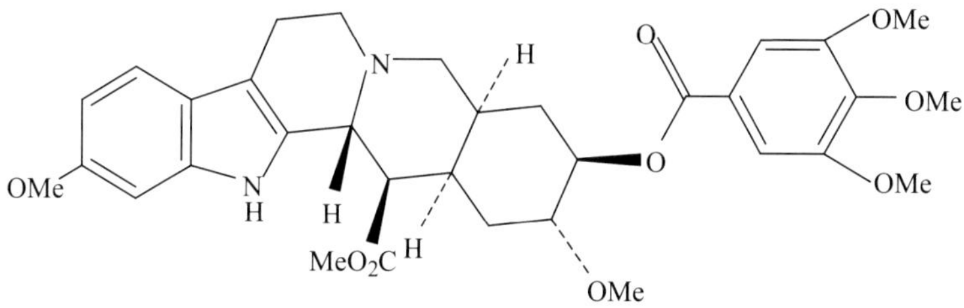

Reserpine

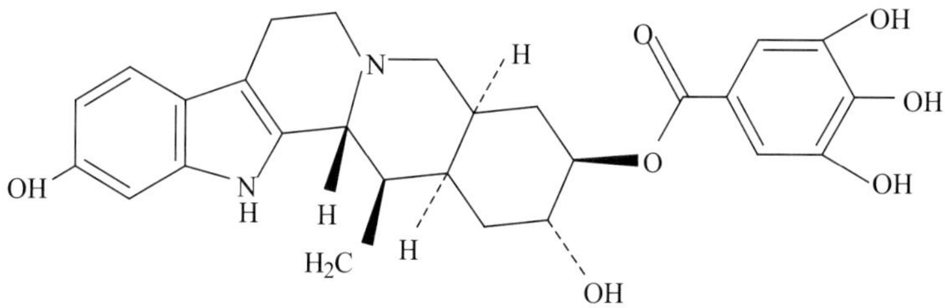

Desperidine

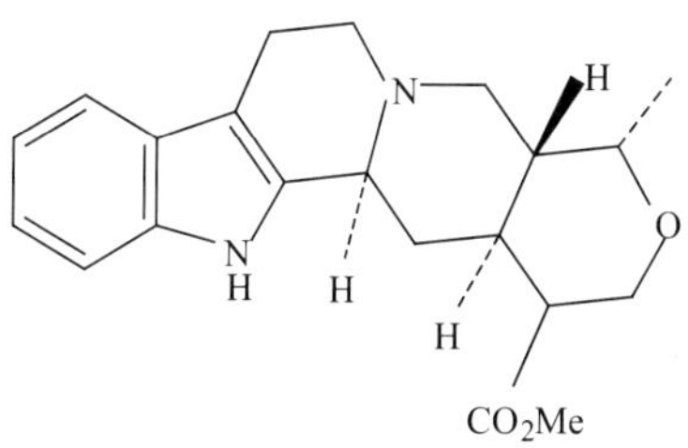

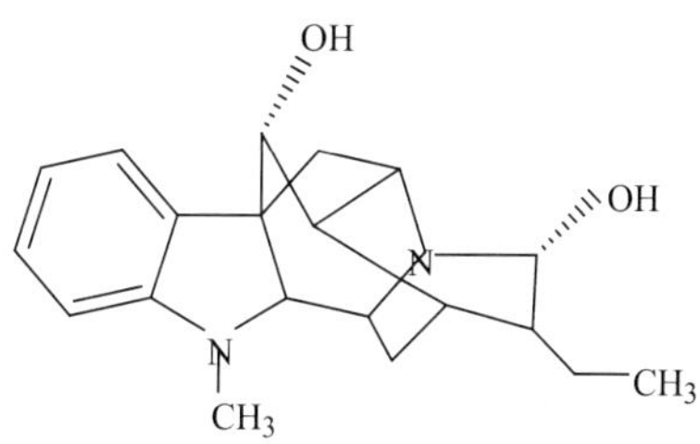

Ajamalicine Ajmaline

Fig. 5.71 Structure of Resperine alkaloids.

Tobacco alkaloids

Tobacco is the cured and dried leaves of *Nicotiana tabacum*. Tobacco leaves may contain from 0.6–9% of ***(-)-nicotine***, an oily, volatile liquid alkaloid, together with smaller amounts ***anabasine*** and ***nornicotine*** (Fig. 5.72). It also contains ***N-formyl nor-nicotine, cotinine, myosmine, nicotyrine, anabasine*** and ***nicotelline.***

In lower concentrations, nicotine is a stimulant, i.e. it increases activity, alertness and memory, and this is one of the main factors that contribute to the dependence-forming properties of tobacco smoking. Nicotine increases the heart-rate and blood pressure, and reduces appetite. In higher doses, nicotine acts as a depressant.

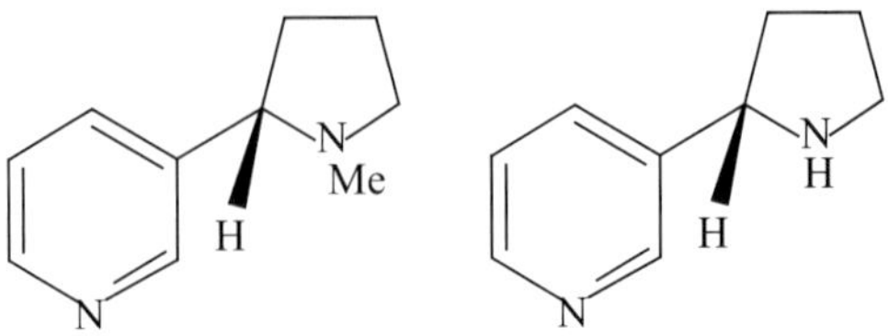

Nicotine Nornicotine

Fig. 5.72 Structure of *Tobacco alkaloids.*

Vasaka alkaloids

These are obtained from *Adhatoda vasica*. They include ***vasicine*** (Fig. 5.6), ***adhatodine,*** and ***vascinone*** (Fig. 5.73). Vasicine is a quinazoline alkaloid, bitter in taste and occurs in crystalline form. It is soluble in chloroform. Oxidation of vasicine leads to formation of vascinone. Other alkaloids include ***anisotine, 3-hydroxyanisotine, vanestine, desmethoxyaniflorine and desmethoxyvascinone***. Vasicine reduces ovalbumin and platelet activating factor induced allergic reactions. Vasicine has significant anti-inflammatory and abortifacient activity.

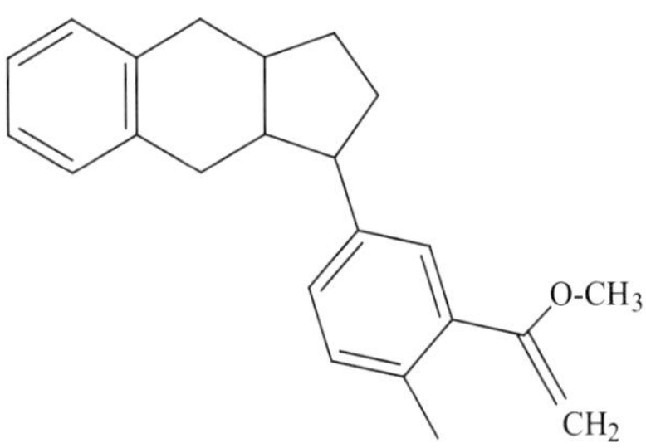

Adhatodine

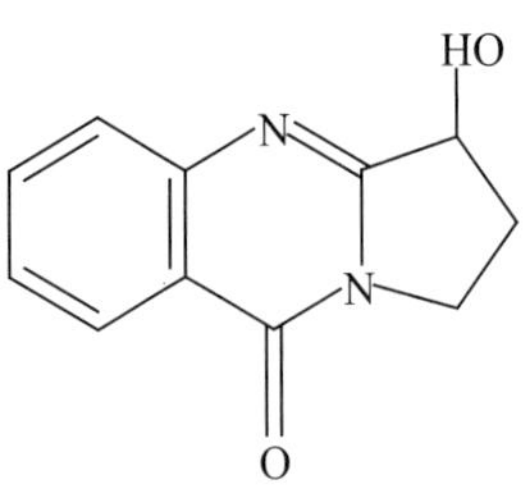

Vasicinone

Fig. 5.73 Structure of *Vasaka alkaloids.*

Vasicine yield from various samples in India ranged from 0.541 to 1.105% on a dry basis. Yeild as high as 2.18% was reported from a foreign sample of which more than half was the l-form and remainder dl-form of the alkaloid. The plant shows wide seasonal variation in vasicine content in its leaves. High concentration is attained twice in a year (3% in March and 1.4% in September), which concides with the flowering stage of the medicinal plant. During the vegetative stage, the plant contained very low concentration of the alkaloid.

Caltha alkaloids

Cathine and ***cathionine*** are obtained from *Caltha edulis* (Fig. 5.74). Cathine is analeptic and bronchodilator.

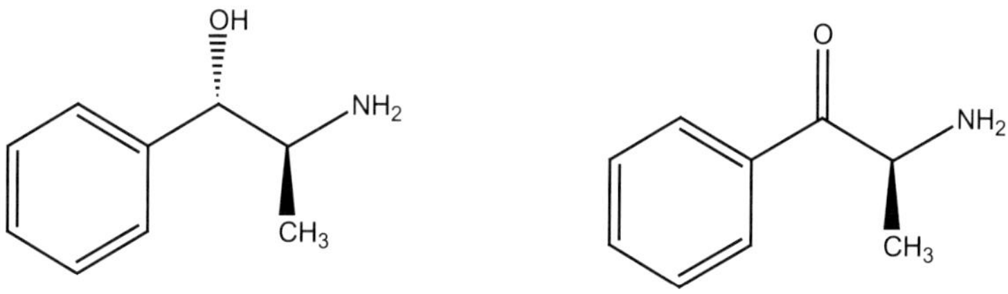

Cathine Cathionine

Fig. 5.74 Structure of Cathine and Cathionine.

Gelsemium alkaloids

Gelsemine and ***gelseminine*** are obtained from *Gelsemium semepervirens* (Fig. 5.75). They act like strychnine. Other alkaloids include 21-oxygelsemine, gelsemicine, gelsidine, gelsevirine, and sempervirine.

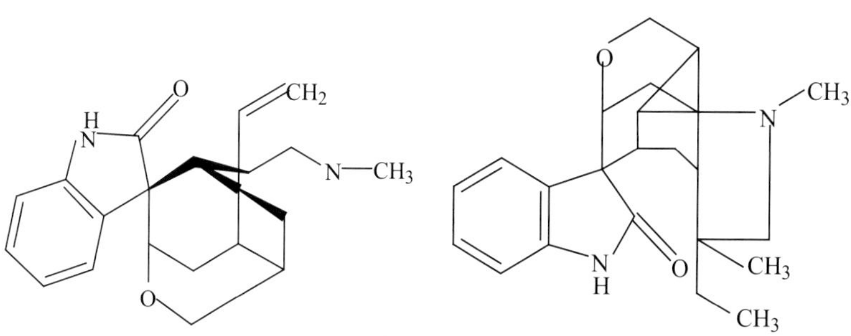

Fig. 5.75 Structure of Gelsemine and Gelseminine.

Harmal alkaloids

Harmaline (harmidine), ***harmine*** (banisterine) and ***harmalol*** are obtained from *Peganum harmala* (Fig. 5.76). They are oxytocic in action.

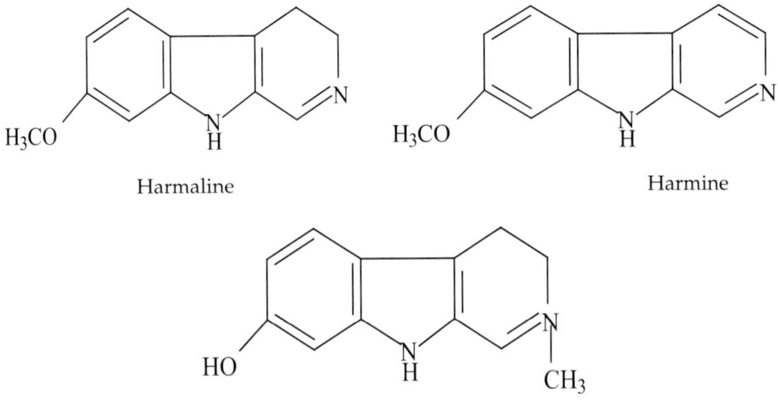

Fig. 5.76 Structure of Harmal alkaloids.

Jasminum alkaloids

Jasminine and ***jasminidine*** are obtained from *Jasminum grandiflorum* (Fig. 5.77).

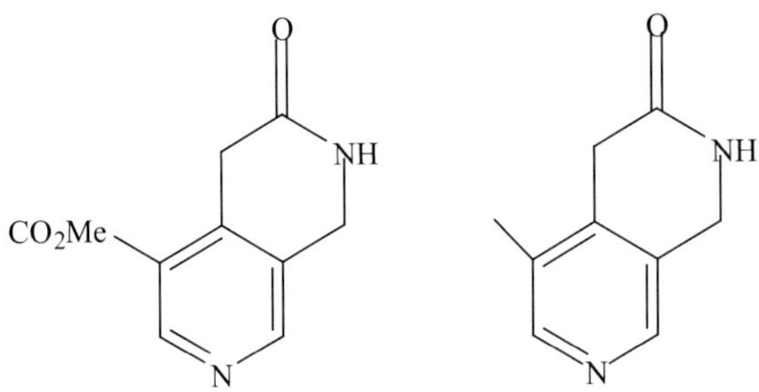

Fig. 5.77 Structure of Jasminine and Jasminidine.

Vinca alkaloids

Catharanthus roseus is known as the common or Madagascar periwinkle. Formerly, it was classified as *Vinca rosea.* It is a perennial evergreen herb in Apocynaceae and originally native to the island of Madagascar. In all 70 alkaloids have been identified in periwinkle. Some have anticancer and hypoglycemic properties. Some act as hemostatics (arrest bleeding).

Two groups began working independently on periwinkle in 1950s when they heard of a tea which Jamaicans drank to treat diabetes. Beer and Noble from the University of Western Ontario became interested in the plant as a possible 'oral insulin'—they isolated alkaloid—***vinblastine.***

Svoboda injected a crude extract of the whole periwinkle plant into mice that were infected with P-1534 leukemia. The mice (60–80%)

experienced a prolonged life. Lilly produced vinblastine as the drug Velban and also synthesized another alkaloid, ***vincristine*** (Fig. 5.78). In the first human test in 1960, a 49 yr old man dying of Hodgkin's disease, was walking within a week and 4 mon later the tumor disappeared.

Vinblastine (Fig. 5.78) has been especially effective for treating Hodgkin's disease. It is the first drug of choice in the treatment of many forms of leukemia and since the 1950's it has increased the survival rate of childhood leukemias by 80%. Vincristine has been especially effective for treating acute childhood leukemia, often with 99% remission rates.

Vinblastine, vincristine, and two semi-synthetic derivatives (vindesine and vinorelbine) all have the same mode of action. They inhibit mitosis in metaphase by binding tubulin. Vinblastine binds to tubulin dimmers in a 1:1 ratio and prevents microtubule formation. Other alkaloids which bind to tubulin include colchicine, maytansine and mescaline, but they have been studied much less than Vinblastine.Vinblastine also seems to fight cancer by interfering with glutamic acid metabolism. They are all administered intravenously once a week. The compounds can be fatal if they are administered any other way, and can cause a lot of tissue irritation if they leak out of the vein.

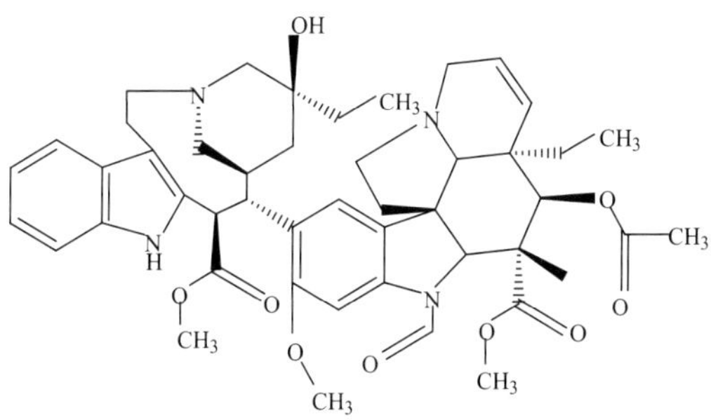

Vinblastine

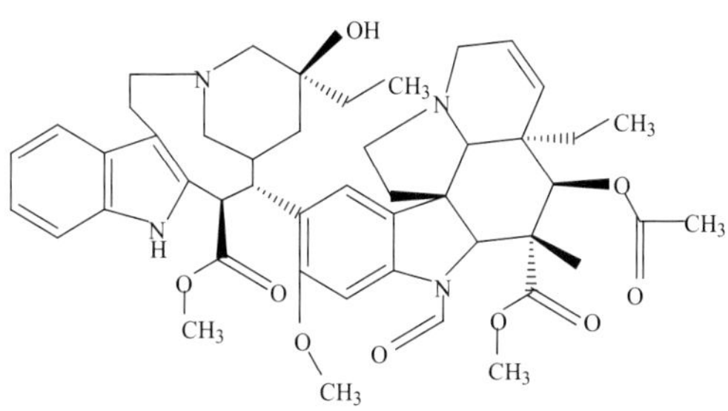

Vincristine

Fig. 5.78 Structure of Vinca alkaloids.

Cepharanthine makes memberane more stable by inhibiting peroxidation of their lipids.

Carbazole alkloids, ***glycoborinine, glycozoline*** and ***glycozolidine*** are reported from roots of *Glycomis arborea* (Fig. 5.79).

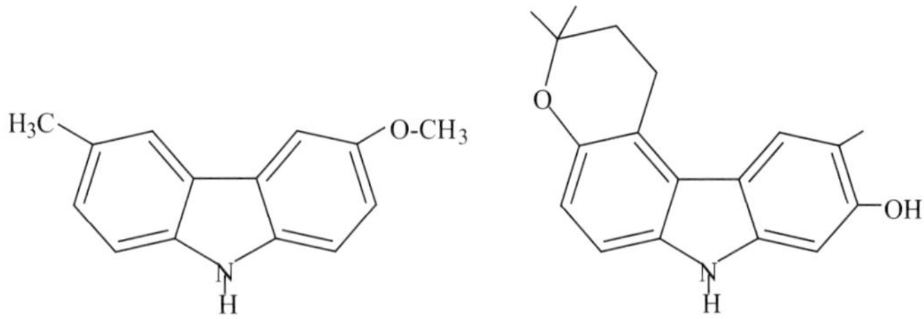

Fig. 5.79 Structure of Glycozoline and Glycozolidine.

Lunasia amara contains quinoline alkaloids, such as ***lunacridine***, which would be worth investigating for pharmacology (Fig. 5.80).

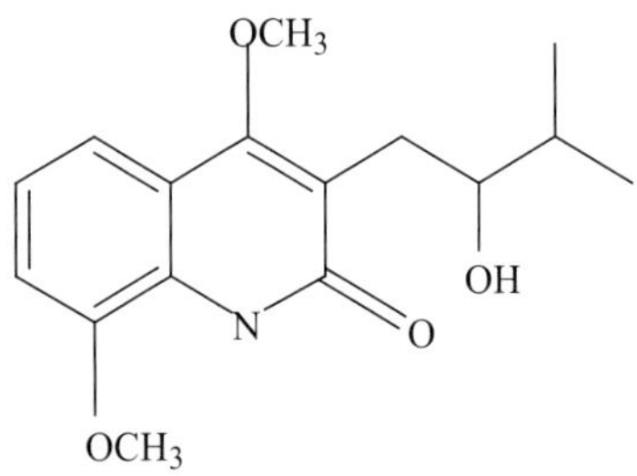

Fig. 5.80 Structure of Lunacridine.

Buchanine, a novel pyridine alkaloid has been reported from *Cryptolepis buchanani. Boldoa fragrans* contains isolqinoline alkaloid, ***boldine*** (Fig. 5.81). *Tribulus terrestris* contains β-carboline alskloid, ***tribulusterine*** (Fig. 5.82).

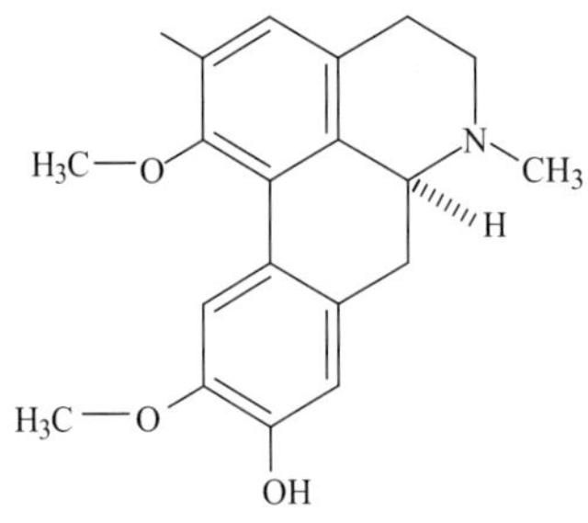

Fig. 5.81 Structure of Boldine.

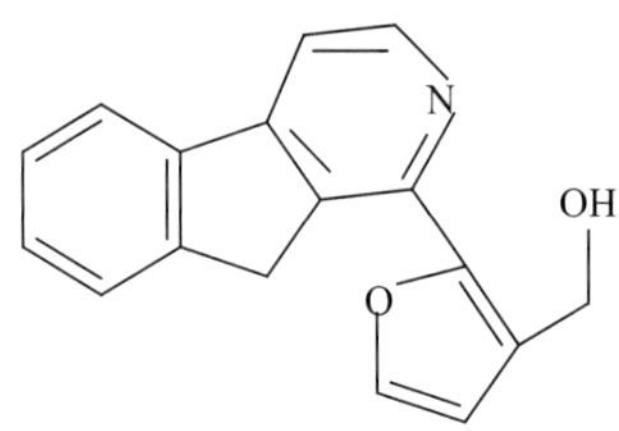

Fig. 5.82 Structure of Tribulusterine.

Crinasiatine, crisiaticidine-A, isocraugsodine, lycoriside and ***palmilycorine*** have been reported from *Crinum asiaticum* (Figs. 5.83, 5.84). The plant also contains *lycorine*.

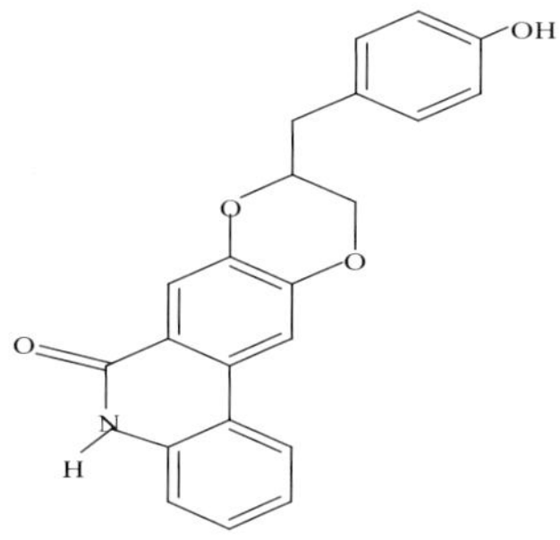

Fig. 5.83 Structure of Crinasiatine.

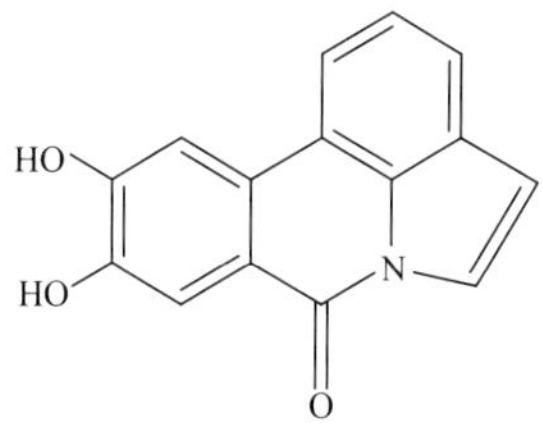

Fig. 5.84 Structure of Crisiaticidine-A.

5-hydroxymethyl-1-1 (1, 2, 3, 9-tetrahydro-pyrrolo [21-b] quinazolin-1-yl)-heptan-1-one is reported to be the bioactive alkaloid of *Sida cordifolia* (Fig. 5.85). It has analgesic and anti-inflammatory activites.

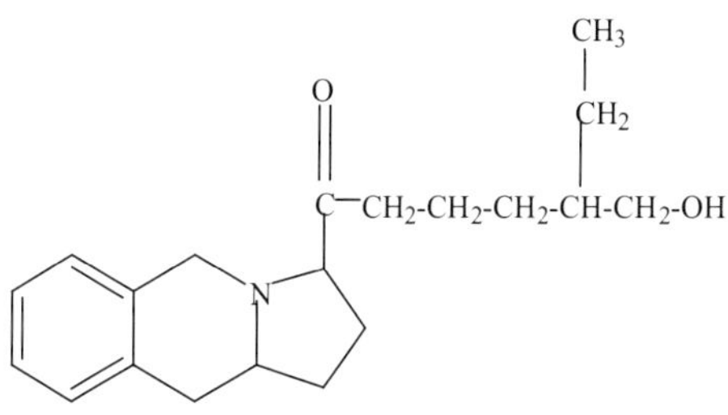

Fig. 5.85 Structure of bioactive alkaloid of *Sida cordifolia.*

Wilfordine, wilfortrine, wiforgine and wilforine are major alkaloids of immunomodulator Chinese drug, *Triptergyium wilfordi.* **Chakranine** is an alkaloid isolated from *Bragantia wallichii.* **Colycotomine** is alkaloid of *Acacia sinuata.* A great variety of isoquinoline alkaloids like ***protopine*** and ***stypoline*** have been isolated from *Corydalis govaniana* (Fig. 5.86). The plant also contains corydaline.

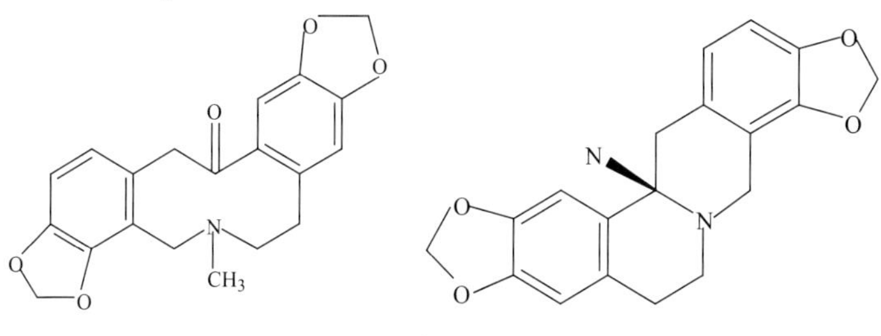

Fig. 5.86 Structure of Protopine and Stypoline.

Jateorrhiza columba (*Cocculus palmata*) contains several isoquioniline alkaloids including ***cocculine, columbamine, jatorrhizine, plamatine*** and ***umbellatine*** (Fig. 5.87). *Cocculus hirsutus* is reported to contain ***cohirstine, cohirsitinine, cohirsinine, jamatinine*** and ***haiderine***. ***Cosculine*** from the stem and leaves of *Cocculus pendulus,* is active against the cells derived from human epidermoid carcinoma of nasopharynx.

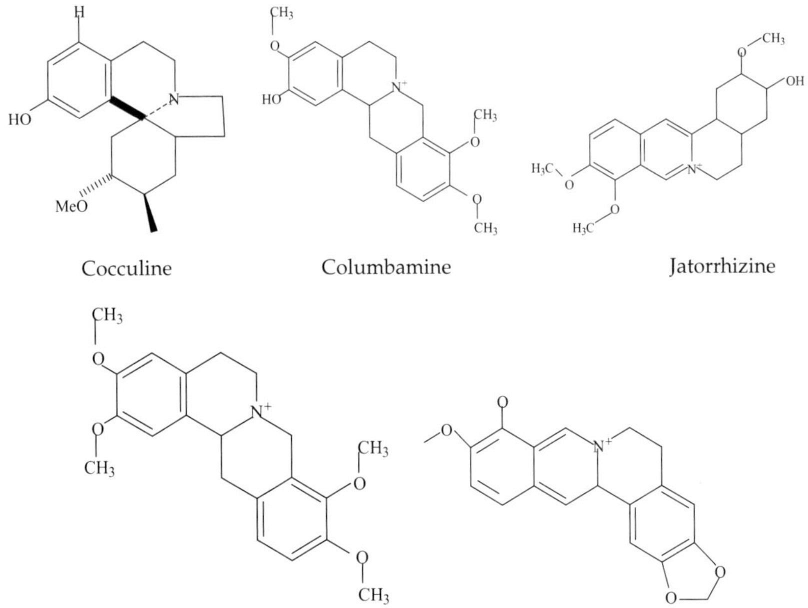

Cocculine Columbamine Jatorrhizine

Palmatine Umbellatine

Fig. 5.87 Structure of alkaloids of *Jateorrhiza columba.*

Ricinine (Fig.5.88) and ***N-demethyl-ricinine,*** isolated from leaf extract of *Ricinus communis* have significant hepatoprotective, choleretic and anticholestatic activity.

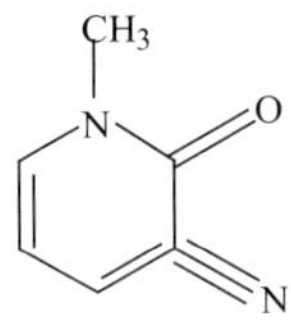

Fig. 5.88 Structure of Ricinine.

Major alkaloids of *Fumaria parviflora* include ***protopine, cryptopine, sanguinarine, fumaridine, fumaramine, parfumine and fumariline***. Sanguinarine is antimicrobial. ***Theobromine* and *theophylline*** from *Theobroma cocoa* are well-reputed bronchodilators (Fig. 5.89). ***Voacangine*** and ***mitragynine,*** isolated from *Voacanga africana* and *Mitragyna cilita,* respectively, are potential analgesic agents. ***Nitidine*** isolated from *Toddalia asiatica* has potential antiHIV activity (Fig. 5.89).

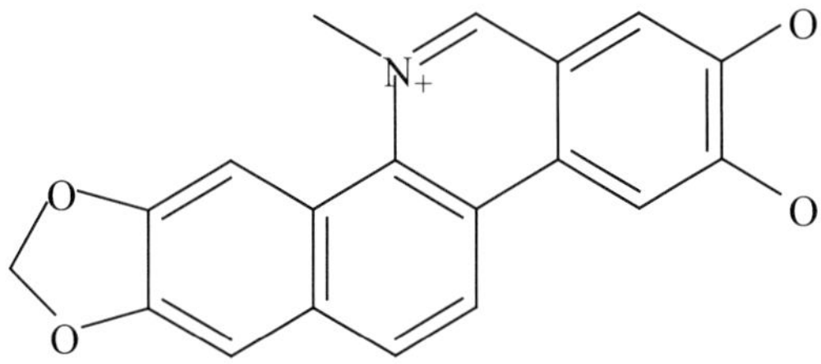

Fig. 5.89 Structure of Nitidine.

Huperzine-A from *Lycopodium (Huperzia) serrata* and ***galantamine*** obtained from *Galanthus nivalis,* respectively, have anticholinesterase activity and are used in Alzheimer's dementia (Figs. 5.90, 5.91).

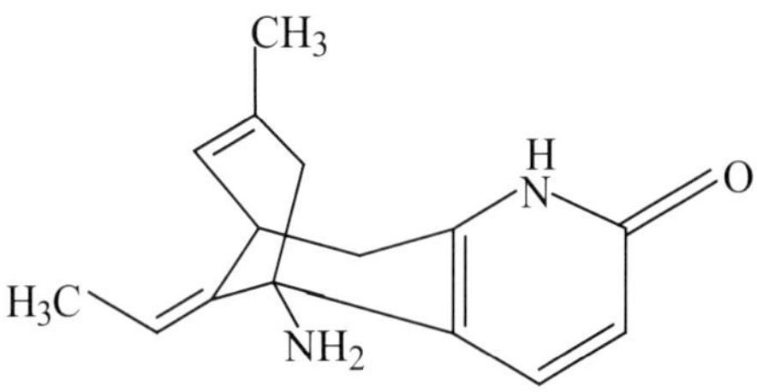

Fig. 5.90 Structure of Huperzine-A.

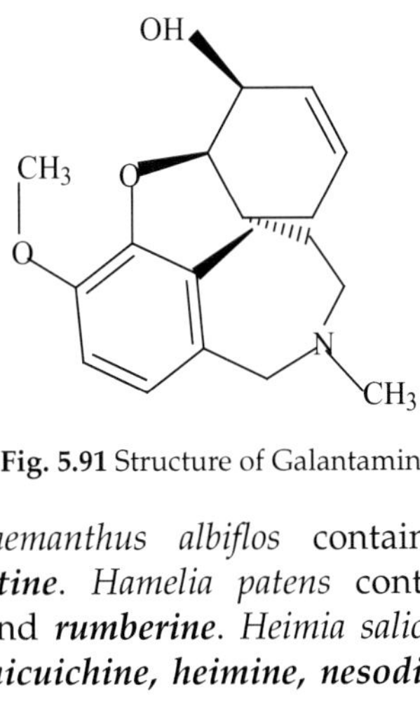

Fig. 5.91 Structure of Galantamine.

The bulb of *Haemanthus albiflos* contains alkaloids including ***lycorenine*** and ***tazettine***. *Hamelia patens* contain oxindole alkaloids including ***palmirine*** and ***rumberine***. *Heimia salicifolia* contains alkaloids including ***lythrine, sinicuichine, heimine, nesodine, sinine (lythrindine),***

vertine and ***cryogenine.*** Cryogenine decreases spontaneous motor activity, hypothermia, blepharoptosis and ataxia. Lythrine and vertine are diuretic. *Adina cordifolia* contains alkaloids including ***adifoline, cordifoline, 10-deoxy cordifoline*** and ***10-deoxy adifoline.*** Adifoline is a central nervous system depressant and hypotensive in experimental animals.

Solanaceous alkaloids

Solanaceae is one of the medicinally important families, many plants of these possess narcotic properties and constitute the drugs used in indigenous systems with great ethnobotanical diversity. The alkaloids isolated from the plants of Solanaceae have been successfully utilized in modern medicine particularly ophthalmic practice.

Atropa belladonna contains 0.3–0.6% of alkaloids, mainly (-)-hyoscyamine. Belladonna root has only slightly higher alkaloid content at 0.4% mainly (-)-hyoscyaminc. Minor alkaloids including (-)-hyoscine and cuscohygrine are also found in the root. The *Datura stramonium* leaf usually contains 0.2–0.45% of alkaloids, principally (-)-hyosycamine and (-)-hyoscine in a ratio of about 2:1. *Datura sanguinea,* yields leaf material with a high (0.8%) alkaloid content in which the principal component is (-)-hyoscine. *Datura meteloides* contain ***meteloidine*** (Fig. 5.92). The alkaloid content of hyoscyamus is relatively low at 0.045–0.14%, but this can be composed of similar proportions of (-)-hyoscine and (-)-hyosycamine. Egyptian henbane, *Hyosycamus muticus,* has much higher alkaloid content than *H. niger.*

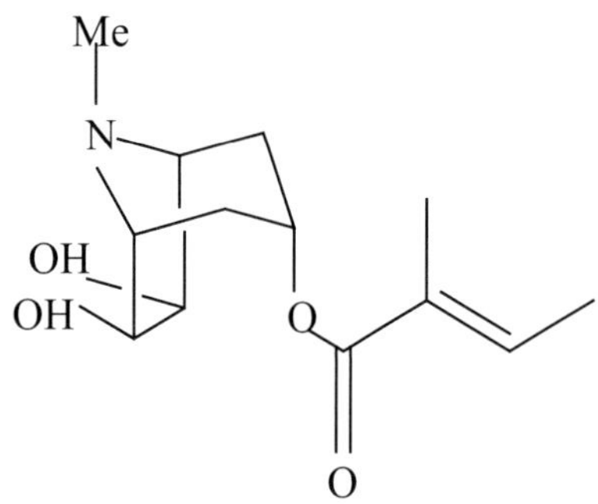

Fig. 5.92 Structure of Meteloidine.

The alkaloids distributed in Solanaceae are of tropine or pyridine type. Tropine alkaloids are formed by combination of Tropic acid and Tropine. Tropine alkaloids are tropane derivatives. Some alkaloids are pyridine (cuscohygrine) or piperdine (isopelliterine) derivatives. Depending upon the basic nucleus, the alkaloids found in Solaneacea are classified into the following types:

1. Soladulcidine type
2. Solasodine type
3. Tomatidenol type.

The majority of alkaloids have anticholinergic activity. They act by inhibiting acetylcholine. ***Atropine*** (Fig. 5.5) is used as a mydriatic (to dilate the pupil) in ophthalmic practice and as an antidote to morphine poisoning. ***Hyoscine*** (Fig. 5.5) is used as an antispasmodic and the salt available is hyoscine butyl bromide. Hyoscine also has a significant anti-emetic effect, and that is why it is used for the treatment of motion sickness.

Withania alkaloids deserve special mention. ***Isopelletrine*** has antihelmenthic activity whereas ***somniferine*** is a sedative. The total alkaloids of roots have a variety of pharmacological actions. In animal models, the alkaloids have shown hypotensive and analeptic activity. Solasodine has glucocorticoid-like effects, hypocholesterolaemic and antiatherosclerotic activity. Nicotine exists in liquid form that is resorbed through the skin.

The majority of alkaloids are toxic. 100 mg of atropine is sufficient for killing a person as it causes respiratory distress. ***Daturine*** alkaloid present in *Datura alba* and other species, accounts for the toxicology of *Datura*. Daturine is converted into atropine in the human body and it causes death from asphyxiation. ***Solasodine*** possesses embryotoxic activity and cumulative effects. The toxicity (LD50) in mice orally is 27.5 mg/kg. Solanine is a protoplasmic poison, which acts on amoeboid cells of the body and causes heamolysis. It has been used as an agricultural poison.

Scopolamine and ***hyoscyamine*** (Fig. 5.5) also possess toxic activity and poisoning causes delirium, which is due to central nervous system excitation. Nicotine is a well-known toxic substance. Common symptoms of poisoning with Solanaceous alkaloids include restlessness, delirium, mania, hallucinations, asphyxiation, sleep, urinary problems, constipation and death. *Cestrum diurnum* contains alkaloids including ***nicotine*** and ***nornicotine*** (Fig. 5.72) *Cestrum nocturnum* contains alkaloids including nornicotine, ***cotinine*** and ***myosmine.***

Table 5.2 indicates the lethal dose (LD50) of therapeutically important alkaloids of Solanaceae.

Table 5.2 LD50 values of therapeutically active alkaloids found in Solanaceae.

SNo.	Name of alkaloid	LD50
1.	Atropine	400mg/kg
2.	Hyoscyamine	375mg/kg mice (i.p)
3.	Hyoscine	700–1,300mg/kg mice (s.c).
4.	Homatropine	6783mg/kg
5.	Nicotine	0.3 mg/kg
6.	Solasonine (Solanine)	2.8mg/kg
7.	Solasodine	27.5mg/kg mice (orally)
8.	Tomantine (Lycopersicin)	80mg/kg mice (orally)

i.p =intraperotoneal s.c = subcutaneous.

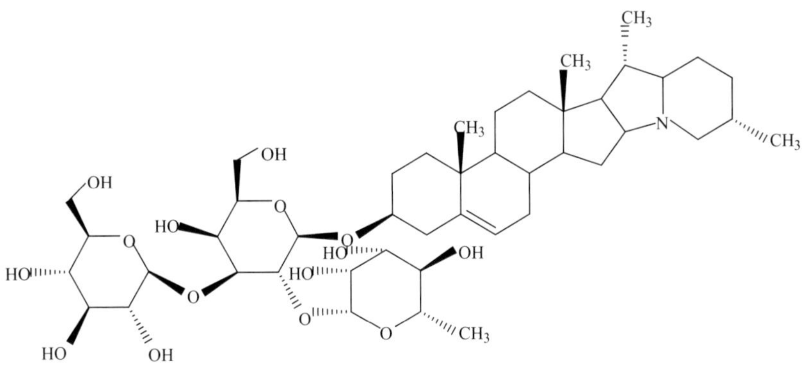

Solanine

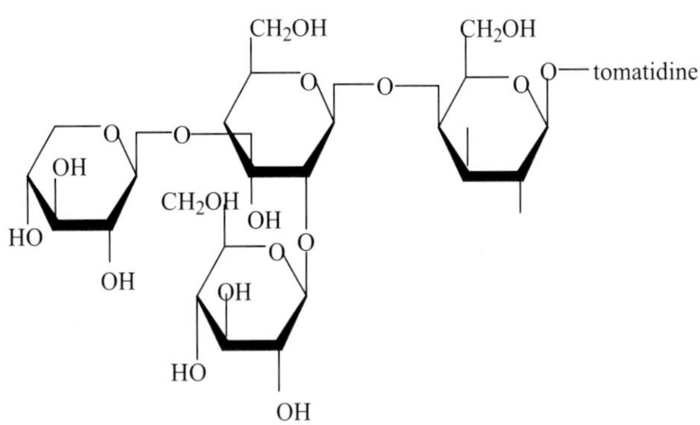

Tomantine

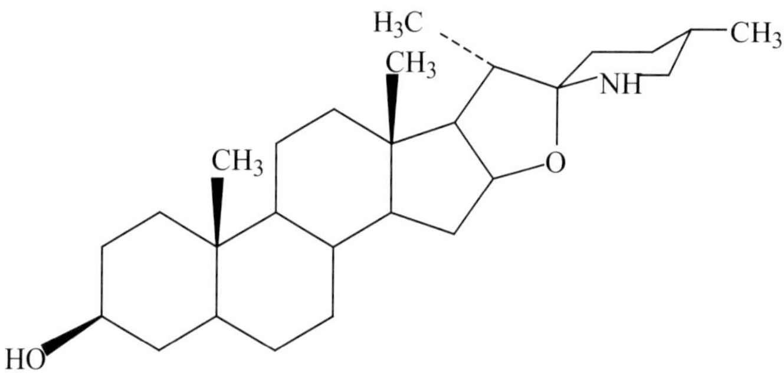

Tomatidine

Fig. 5.93 Alkaloids of Solanaceae (see Fig. 1.4 also).

Solanum dulcamara contains steroid alkaloid glycosides (0.07 to 0.4%). The alkaloid spectrum differs widely with the variety. Tomatidenol variety—α-solamarine, β-solamarine. Soladulcidine variety—soladulcidinetetraoside. Solasodine variety—solasonine and solamargine. The chief alkaloids of *Solanum nigrum* include solasonine, solamargine, and P-solamargine.

Lobelia alkaloids

Lobelia inflata contains about 0.2–0.4% of alkaloids, of which the piperidine derivative ***lobeline*** (Fig. 5.73) is the chief constituent. Minor alkaloids identified include ***lobelanine*** (Fig. 5.94).

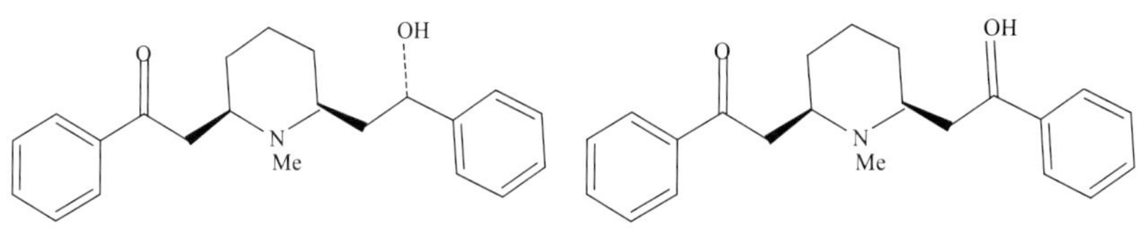

Lobeline Lobelanine

Fig. 5.94 Structure of Lobelia alkaloids.

Convolvulaceae alkaloids

Convolamine and ***convolvine*** are obtained from Convolvulus sp. They have anesthetic activity. ***Elymoclavine*** is obtained from *Ipomoea nil* (Fig. 5.95).

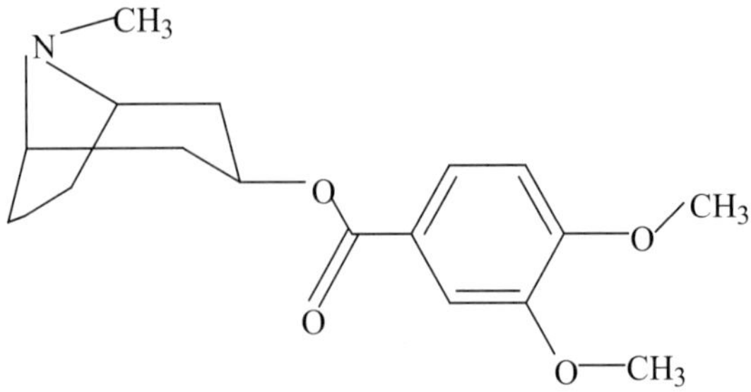

Convolamine

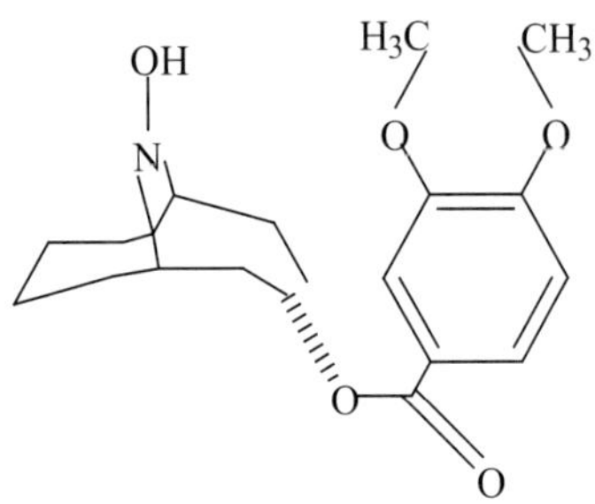

Convolvine

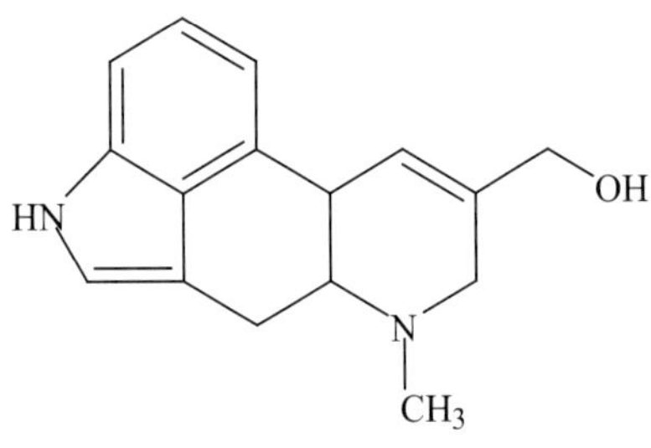

Elymoclavine

Fig. 5.95 Alkaloids of Convolvulaceae.

Pyrrolizidine alkaloids (Fig. 5.96)

Alkaloids are present in many plants ranging from non-poisonous forms to highly fatal toxic forms. Pyrrolizidine alkaloids are one such type, which are highly toxic principles present in Boraginaceae, Compositae, and Leguminosae families. The intoxication is caused by the oral intake of the plants containing the alkaloids either in food or medicinal form. Plants containing pyrrolizidine alkaloids are sometimes contaminated with agricultural crops also.

At least 100 types of pyrrolizidine alkaloids are present in plants. The majority of them are hepatotoxic. The varied manifestations of liver injury caused by pyrrolizidine alkaloids containing herbs are acute and chronic hepatitis, steatosis, hepatic fibrosis, zonal or diffuse hepatic necrosis, bile duct injury, veno-occlusive disease, acute liver failure, and carcinogenesis.

Teucrium chamaedrys (germander) causes severe, acute hepatocellular injury. Traditionally it has been used for reducing weight. An elevated level of aminotransferase enzyme has been observed after the intake of the drug.The toxic principle of the herb has not been detected but the presence of pyrrolizidimne alkaloids has been postulated.

Symphytum officinale (comfrey) contains a number of pyrolizidine alkaliods including ***echinatin, lycopsamine, 7-acetyllycopsamine, echimidine, lasiocarpine, symphytine and intremedine***. Due to the presence of pyrolizidine alkaloids, the use of the drug has been banned in number of countries.The comfrey alkaloids have hepatotoxic and carcinogenic potential. Some companies are introducing pyrrolizidine alkaloid free comfrey extracts.

Cynoglossum officinale (Hound's tongue) is also known to contain pyrrolizidine alkaloids including ***heliosupine, 7-angeloylheliotridine and acetylheloisupine***. The herb in experiments has shown that it induces paralysis in peripheral nerve endings in frogs. The exact cause of this is not known, but because of the presence of the pyrrolizidine alkaloids in the herb, the use should be abolished.

The following genera also contain pyrrolizidine alkaloids *Senecio longilobus, Lycopodium serratum, Heliotropium eichwaldii, Helitropium indicum, Larrea maxima* (chaparral), *Crotolaria assamica, Crotolaria juncea,Crotolaria retusa, Eupatorium* sp., *Alkanna tinctoria* (alkanna) *Mentha pulegium* and *Tussilaga farfara* (Colt's foot).

All human beings are believed to be susceptible to hepatotoxic pyrrolizidine alkaloids. Home remedies and consumption of herbal teas in large quantities can be a risk factor and are the most likely cause of alkaloid poisonings. Evidence of toxicity may not become apparent until sometime after the alkaloid is ingested. The acute illness has been compared to the Budd-Chiari syndrome (thrombosis of hepatic veins, leading to liver enlargement, portal hypertension, and ascites).

Early clinical signs include nausea and acute upper gastric pain, acute abdominal distension with prominent dilated veins on the abdominal wall, fever, and biochemical evidence of liver dysfunction. Fever and jaundice may be present. In some cases the lungs are affected; pulmonary edema and pleural effusions have been observed. Lung damage may be prominent and has been fatal. Chronic illness from ingestion of small amounts of the alkaloids over a long period proceeds through fibrosis of the liver to cirrhosis, which is indistinguishable from cirrhosis of other etiology.

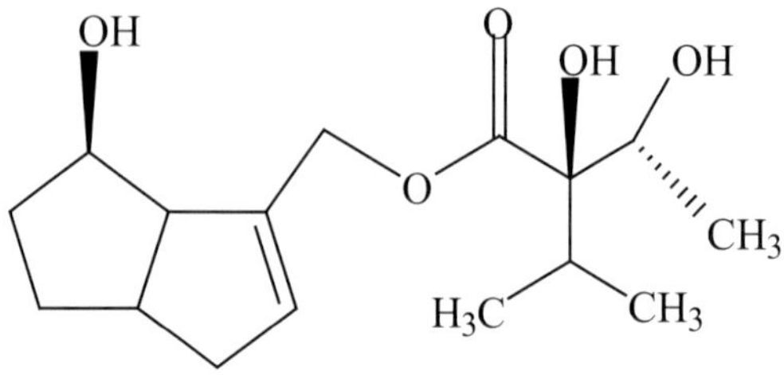

Indicine

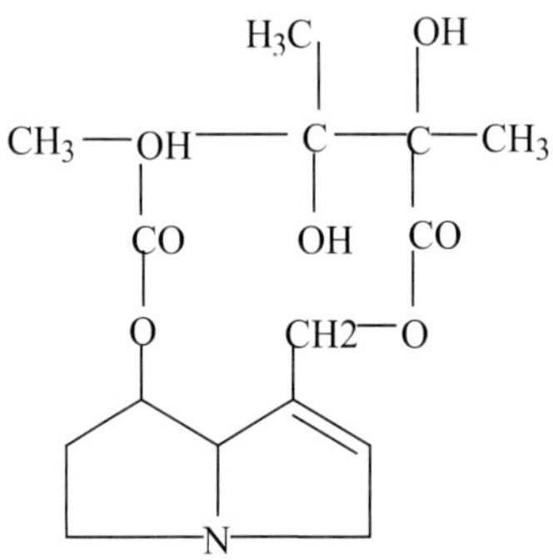

Monocrotaline

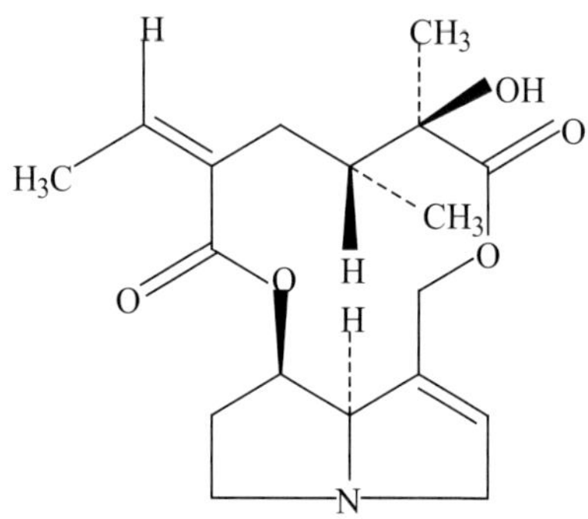

Senecionine

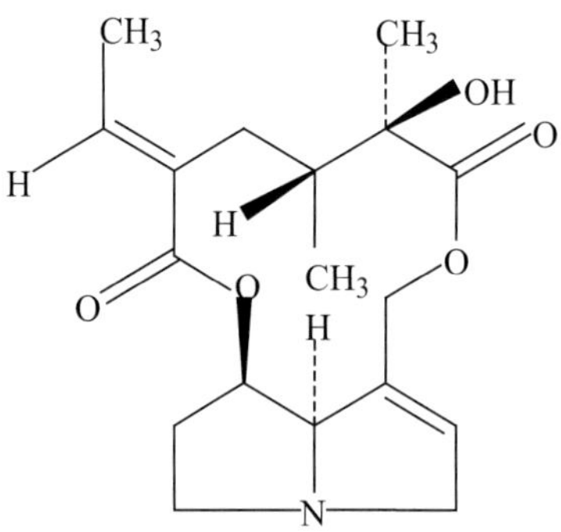

Seneciphylline

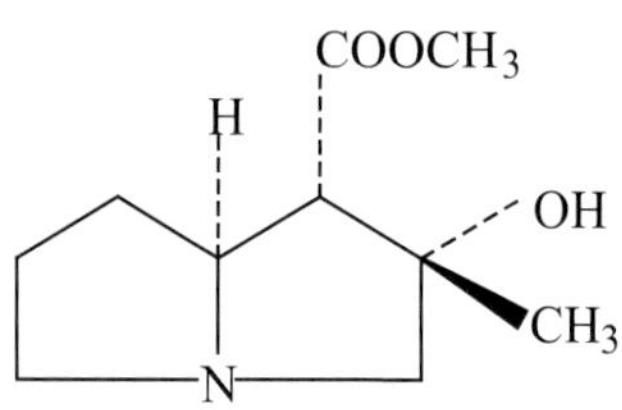

Tussilagine

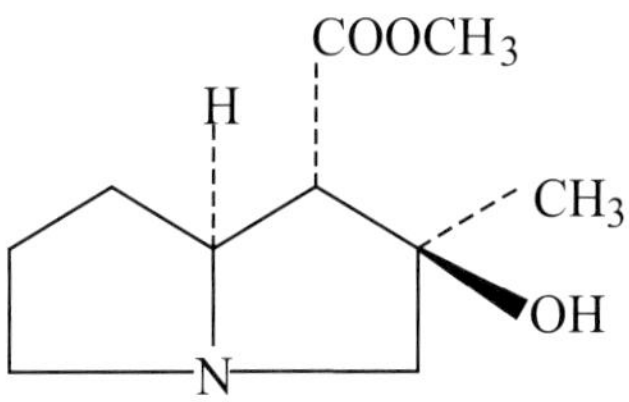

Isotussilagine

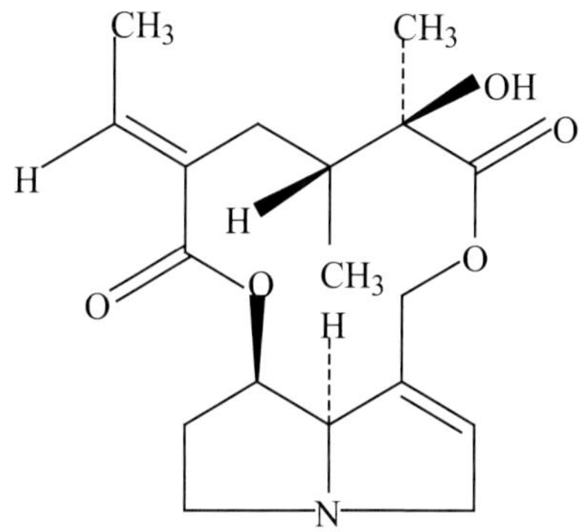

Intermedine

Fig. 5.96 Structure of common pyrrolizidine alkaloids.

Table 5.3 Summary of alkaloids studied from medicinal plants.

S.No	**Name of the alkaloid**	**Source**
1.	(+) – Tiliarine	*Tiliacora racemosa*
2.	Acalyphine	*Acalypha indica*
3.	Acetylmacroalstonine	*Alstonia macrophylla*
4.	Aconitine	*Aconitum napellus*
5.	Adhatodine	*Adhatoda vasica*
6.	Ajamlicine	*Rauwolfia serpentina*
7.	Ajmaline	*Rauwolfia serpentina*
8.	Alangine	*Alangium lamarckii*
9.	Anabasine	*Withania somnifera*
10.	Anagyrine	*Withania somnifera*
11.	Arbroflorine	*Kopsia arborea*
12.	Argemonine	*Argemone mexicana*
13.	Atisine	*Aconitum heterophyllum*
14.	Atropine	*Atropa belladonna*
15.	Berbamine	*Berberis aristata*
16.	Berberine	*Berberis aristata*
17.	Brucine	*Strychnos nux vomica*
18.	Bufotenine	*Mucuna pruriens*
19.	Carpaine	*Carica papaya*

Table 5.3 contd...

Table 5.3 contd...

S.No	Name of the alkaloid	Source
20.	Cerepegin	*Ceropegia juncea*
21.	Cimipronidine	*Cimicifuga racemosa*
22.	Cissampariene	*Cissampelos pareira*
23.	Codeine	*Papaver somniferum*
24.	Colchicine	*Colchicum leuteum*
25.	Conessine	*Halorrhena antidysenterica*
26.	Coptine	*Coptis teeta*
27.	Cosquisagenine	*Ruta graveolens*
28.	Cuscohygrine	*Withania somnifera*
29.	Cynaustraline	*Cynoglossum lanceolatum*
30.	Cynaustrine	*Cynoglossum lanceolatum*
31.	Daturine	*Datura alba, D. innoxia*
32.	Delphinine	*Peganum harmala*
33.	Desoxypeganine	*Alstonia scholaris*
34.	Ditamine	*Eclipta alba*
35.	Echitamine	*Alstonia scholaris*
36.	Ecliptine	*Alstonia scholaris*
37.	Elymoclavine	*Ipomoea nil*
38.	Ephedrine	*Ephedra sinica*
39.	Ergoclavine	*Claviceps purpurea*
40.	Ergotamine	*Claviceps purpurea*
41.	Erthyocentaurine	*Enicostemma hyssopifolium, Swertia lauri*
42.	Erytherine	*Erythrina indica*
43.	Fumarine	*Fumaria officinalis*
44.	Gagaminine	*Cynanchum wilfordii*
45.	Gentianadine	*Gentiana lutea*
46.	Gentianaine	*Gentiana lutea*
47.	Gentianamine	*Gentiana lutea*
48.	Gentianine	*Gentiana lutea, Gentiana kurroa, Swertia chiryata and Trigonella foneum-graceum*
49.	Gentiatibetine	*Gentiana lutea*

Table 5.3 contd...

Table 5.3 contd...

S.No	Name of the alkaloid	Source
50.	Gentitialutine	*Gentiana lutea*
51.	Graveoline	*Ruta graveolans*
52.	Gymnamine	*Gymnema sylvestre*
53.	Harmaline	*Peganum harmala*
54.	Harmine	*Peganum harmala*
55.	Herpestine	*Bacopa monneira*
56.	Hetisine	*Aconitum heterophyllum*
57.	Hetratisine	*Aconitum heterophyllum*
58.	Hygrine	*Withania somnifera*
59.	Hyoscine	*Hyoscyamus niger*
60.	Hyoscyamine	*Hyoscyamus niger*
61.	Icajine	*Strychnos nux vomica*
62.	Laurotetanine	*Litsoea sabifera*
63.	Macrocaprpamine	*Alstonia macrophylla*
64.	Mimosine	*Mimosa pudica*
65.	Momordicine	*Momordica charnatia*
66.	Monnierin	*Bacopa monneria*
67.	Moringine	*Moringa olifera*
68.	Morphine	*Papaver somniferum*
69.	Muscarine	*Agaricus muscaria*
70.	Napelline	*Aconitum ferox*
71.	Narceine	*Papaver somniferum*
72.	Narcotine	*Papaver somniferum*
73.	Nicotine	*Nicotina tabaccum*
74.	Nigellidine	*Nigella sativa*
75.	Noscapine	*Papaver somniferum*
76.	Nyctanthine	*Nyctanthes arbor-tristis*
77.	Oroxylin	*Oroxylum indicum*
78.	Oxycanthine	*Berberis aristata*
79.	Papaverine	*Papaver somniferum*
80.	Pareirubrine	*Cissampelos pareira*

Table 5.3 contd...

Table 5.3 contd...

S.No	Name of the alkaloid	Source
81.	Pelletrine	*Punica granatum*
82.	Pelosine	*Cissampelos pareira*
83.	Phyllanthine	*Phyllanthus niruri*
84.	Phyllurine	*Phyllanthus urinaria*
85.	Piperidine	*Piper nigrum*
86.	Piperine	*Piper nigrum*
87.	Precatorine	*Abrus precatorius*
88.	Premnazole	*Gmelina arborea, Premna integrifolia*
89.	Protopine	*Papaver somniferum*
90.	Pseudopelletrine	*Withania somnifera*
91.	Pseudostrychynine	*Strychnos nux vomica*
92.	Punarnavine	*Boehravia diffusa*
93.	Pyrethrin	*Anacyclus pyrethrum*
94.	Quinidine	*Cinchona officinalis*
95.	Quinine	*Cinchona officinalis*
96.	Rescinnamine	*Rauwolfia serpentina*
97.	Reserpine	*Rauwolfia serpentina*
98.	Ricinine	*Ricinus communis*
99.	Ruine	*Peganum harmala*
100.	Salvadorine	*Savadora presica*
101.	Saussurine	*Saussurea lappa*
102.	Serpentine	*Rauwolfia serpentina*
103.	Serpentinine	*Rauwolfia serpentina*
104.	Skimmianine	*Ruta graveolens*
105.	Solanine	*Solanum nigrum*
106.	Strychnine	*Strychnos nux vomica*
107.	Superbine	*Gloriosa superba*
108.	Theabine	*Papaver somniferum*
109.	Trigonelline	*Trigonella foneum- graceum*
110.	Tropine	*Withania somnifera*
111.	Tylophorine	*Tylophora asthamatica*

Table 5.3 contd...

Table 5.3 contd...

S.No	Name of the alkaloid	Source
112.	Vasicine	*Adhatoda vasica*
113.	Vasicinone	*Adhatoda vasica*
114.	Villalstonine	*Alstonia macrophylla*
115.	Vomicine	*Strychnos nux vomica*
116.	Withasomnine	*Newbouldia leavis, Withania somnifera*

5.4.3 Bitters (including quassonoids and limonoids)

Bitter principles are basically glycosides and are found commonly in plants of Genitiaceae. They are chemically unrelated but possess the common property of an intensely bitter taste. Although this group of drugs is not used today, but in earlier days were given to promote appetite and aid digestion. The bitters act on gustatory nerves, which results in increased flow of saliva and gastric juices (Fig. 5.97).

From the point of view of chemistry, the bitter principles contain lactone group. They may be diterpene lactones e.g., ***andrographolide*** or triterpenoids e.g., ***amarogentin***. The bitters have no action in general. The structure of ***tinosporinine***, the bitter constituent of *Tinospora malabarica* is shown in Fig. 5.97.

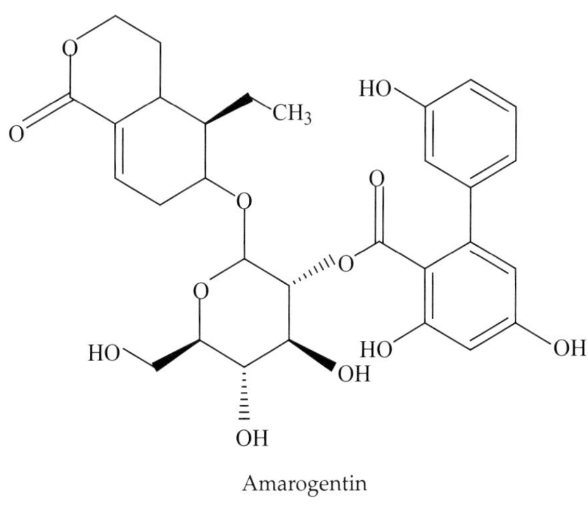

Amarogentin

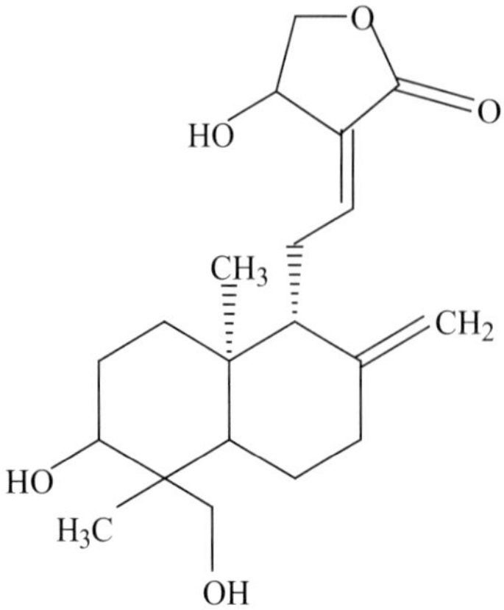

Andrographolide

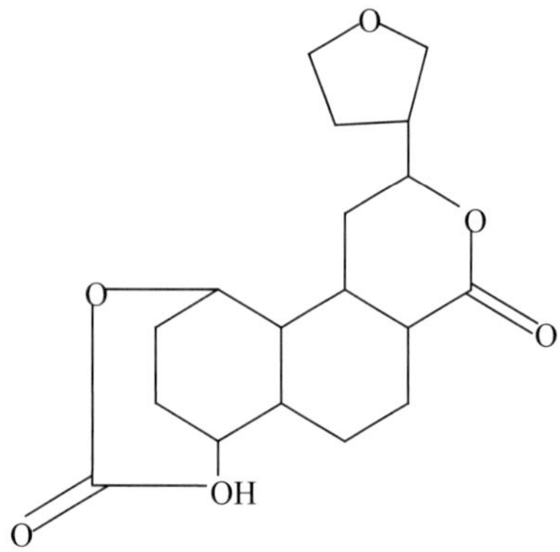

Calumbin

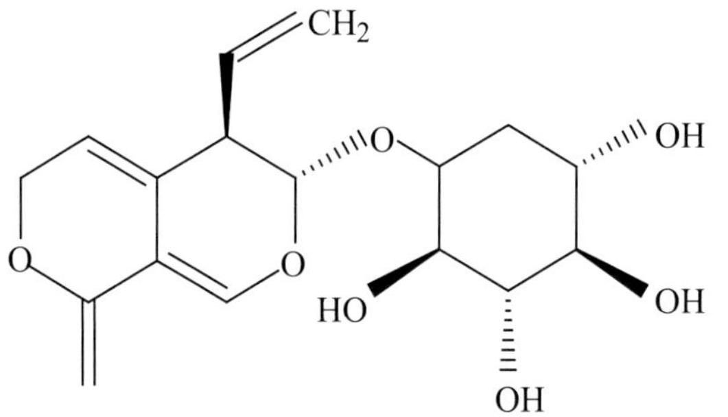

Gentiopicrin

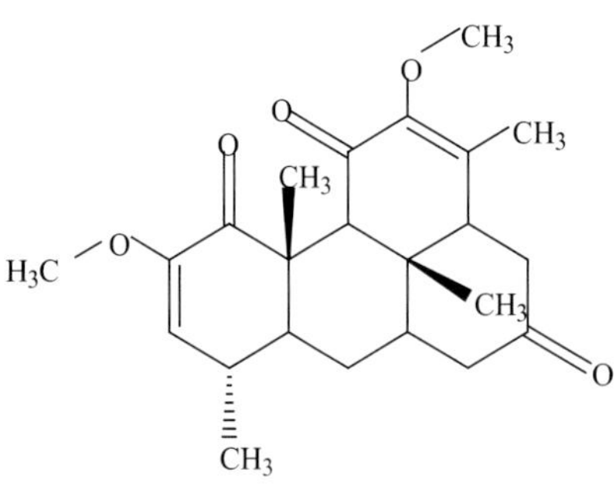

Quassin

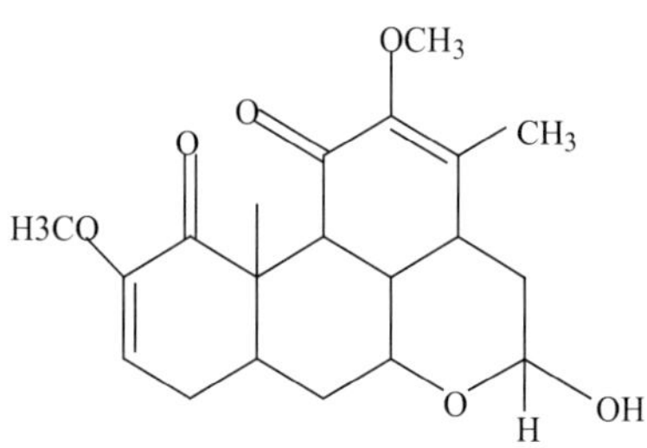

Neoquassin

Fig. 5.97 Structure of important bitter principles.

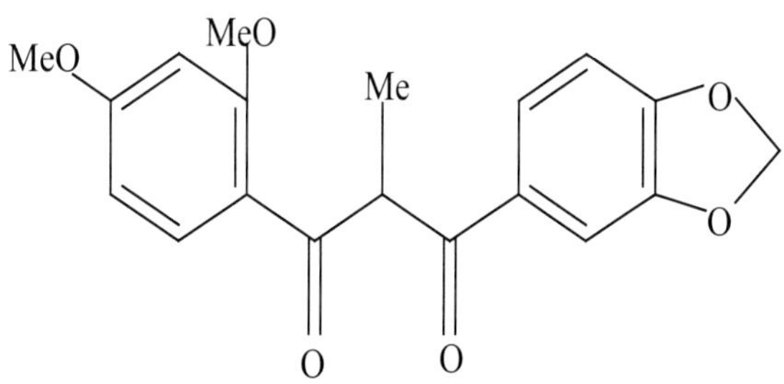

Fig. 5.98 Structure of Tinosporinone.

Some bitter principles are known to be astringent due to the presence of tannic acid. *Gentiana lutea* is the plant known to contain astringent bitter principles. They should not be prescribed with metals, as they are known to cause gastro-intestinal problems. Some such as, ***amarogentin*** has recently received importance because of antiprotozoal activity. ***Gentiopicrin*** is said to reduce thyroxine and metabolism (Fig. 5.99).

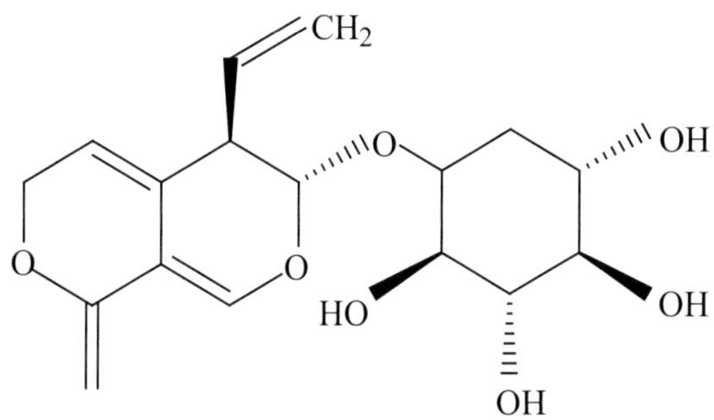

Fig. 5.99 Structure of Gentiopicrin.

Andrographolide has potent hepatoprotective activity. It is however missing in *A. echioides*. ***Ailanthone***, derived from *Ailanthus excelsa* is a typical example of quassonoid (Fig. 5.100). Isolation and structure of ***13,18-dehydroexcelsin*** (Fig. 5.101) is reported from *A. excelsa*. The bitter principle of *Polygala amara* is ***polygalin*** (polygamarin).

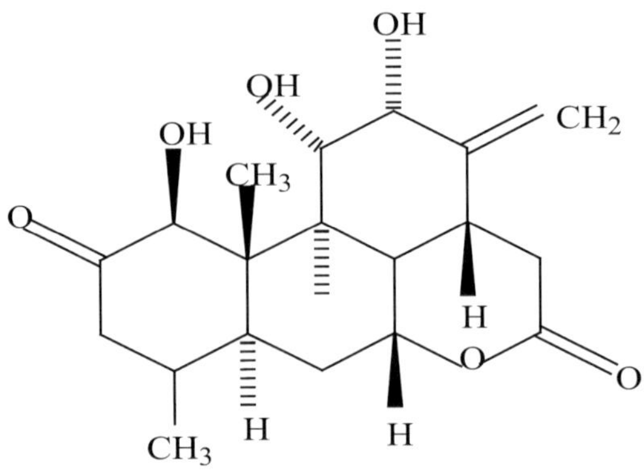

Fig. 5.100 Structure of Ailanthone.

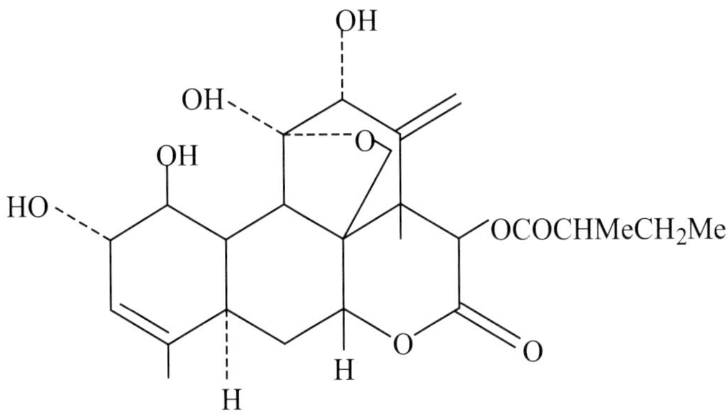

Fig. 5.101 Structure of 13,18-dehydroexcelsin.

Bakaynin and ***margosin*** are bitter principles of *Melia azadirachta*. Margosin is also found in *Azadirachta indica. Caesalpinia bonduc* contains bitter principles including ***α-, β- & γ- caesalpin. Kutkin*** is a combination of active herbal constituents-picrosides I, II and 1V and kutkoside- found in the herb *Picrorhiza kurroa,* which has been used in Ayurveda (Fig. 5.102). Kutkin's antioxidant activity has been shown to decrease levels of lipid peroxidases and hydroperoxidases, free radical producing agents, and help facilitate the recovery of SOD, a powerful antioxidant in the liver needed to prevent oxidative damage. Kutkin also possesses potent antibacterial and immunomodulating properties.

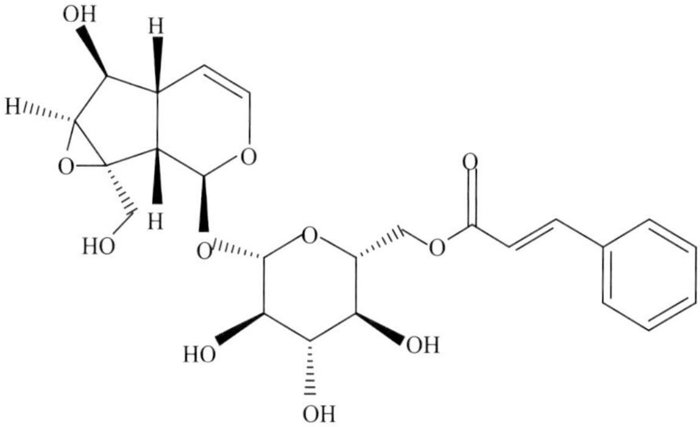

Picroside 1

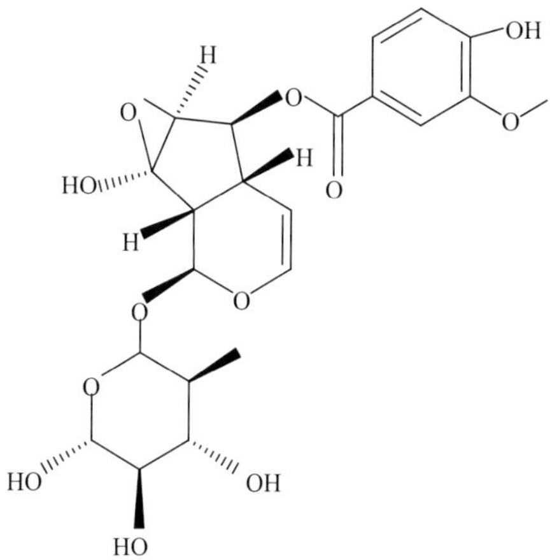

Picroside 11

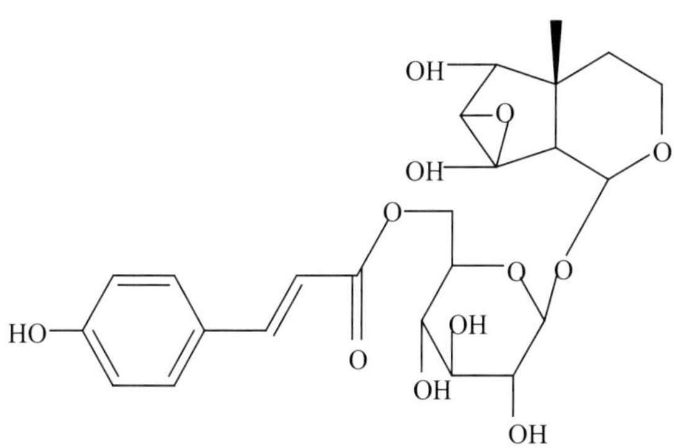

Picroside IV

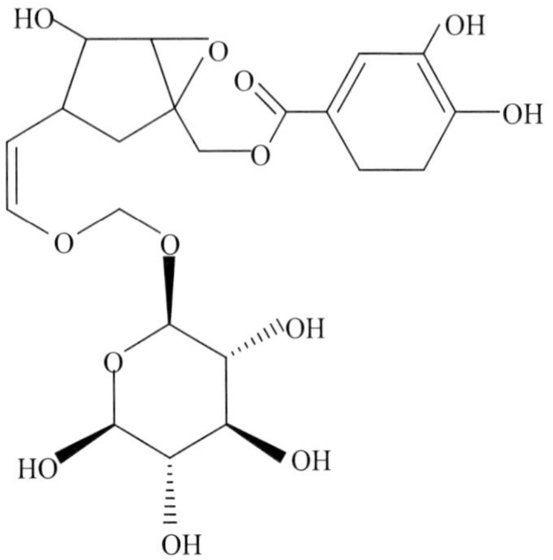

Kutkoside

Fig. 5.102 Iridoid glycosides of *Picrorhiza kurroa.*

Damianin is a brown bitter amorphous substance in *Turnera diffusa.* ***Aristolochic acid,*** the bitter principle of *Aristolochia indica* is reported to be nephrotoxic (Fig. 5.103). The sodium salt of aristolochic acid has been tried as an anti-inflammatory agent, but severe nephrotoxicity in humans and carcinogenicity in rodents aborted further developments.

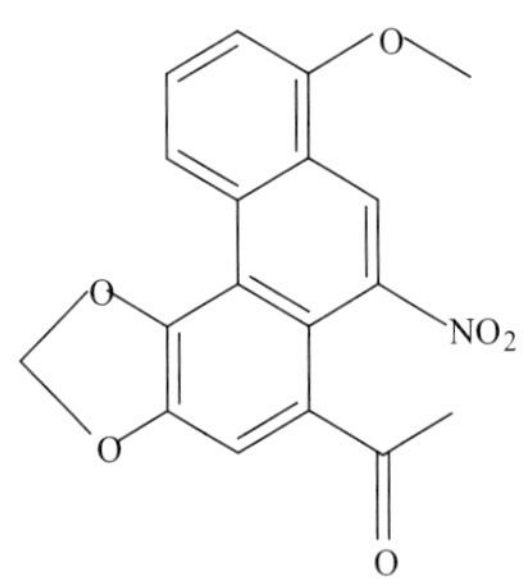

Fig. 5.103 Structure of Aristolochic acid.

Aphanamixis grandifolia, yields limonoids, including ***12-hydroxyamoorastatin,*** which has inhibited the growth of the murine P-388 lymphocytic leukemia cell lines (Fig. 5.104).

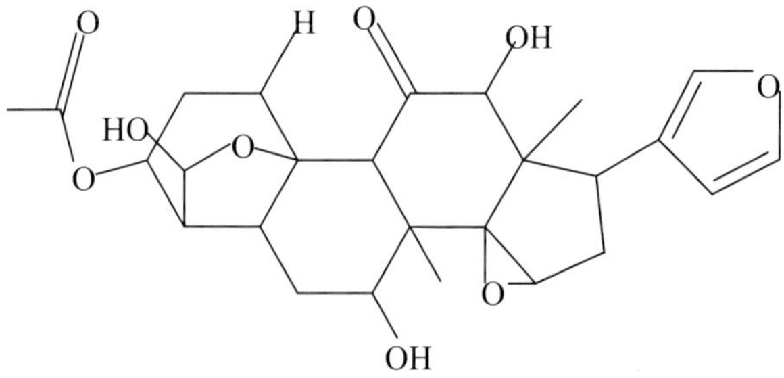

Fig. 5.104 Structure of 12-hydroxyamoorastatin.

Jangomolide, a novel limonoid and ***limonin***, have been reported from *Flacourtia jangomas* (Figs. 5.105, 5.106).

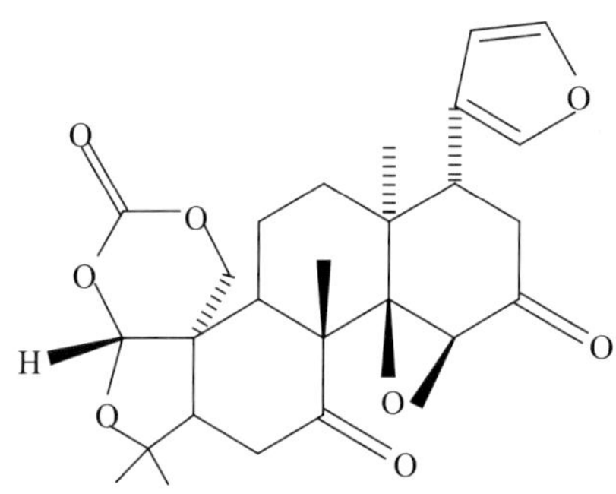

Fig. 5.105 Structure of Jangomolide.

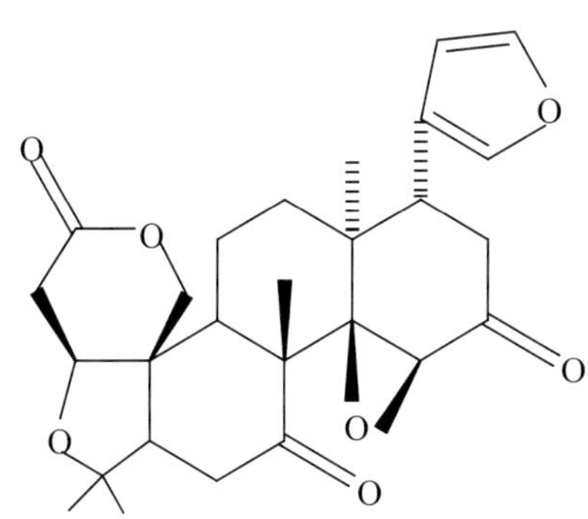

Fig. 5.106 Structure of Limonin.

Examples of cytotoxic quassinoids are bruceantin, longilactone, and 11-dehydroklaineanone. ***Bruceantin*** from *Brucea javanica* has attracted a great deal of interest on account of its ability to prevent the survival of a broad spectrum of cancer cells (Fig. 5.107). ***Longilactone*** from *Eurycoma longifolia* is cytotoxic and prevents the survival of the *Schistosoma* species at a dose of 200mg/mL (Fig. 5.108). ***11-Dehydroklaineanone*** from the same plant has inhibited the growth of the *Plasmodium* species cultured *in vitro* with an IC50 value as low as 2 μg/mL (Fig. 5.109).

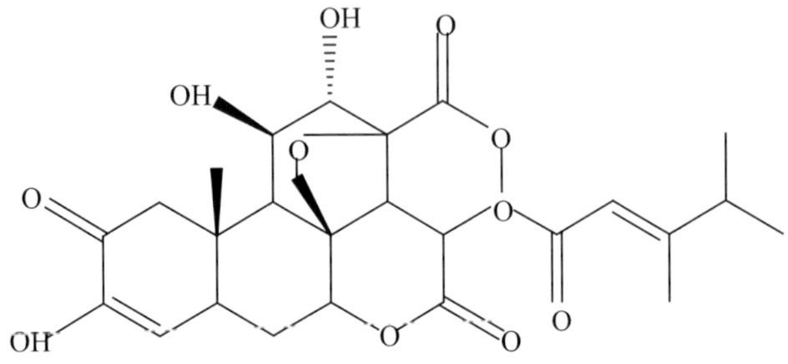

Fig. 5.107 Structure of Bruceantin.

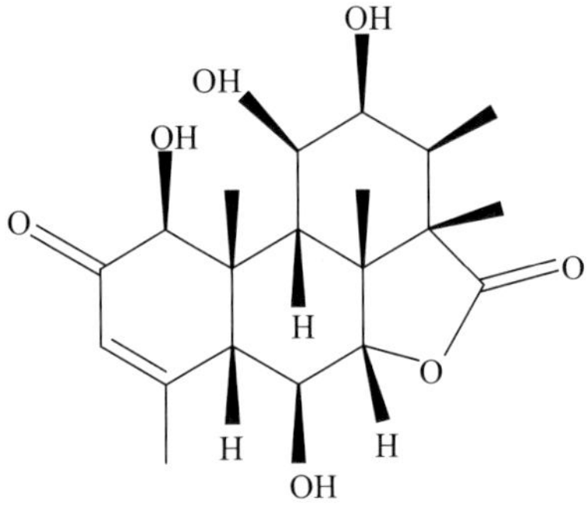

Fig. 5.108 Structure of Longilactone.

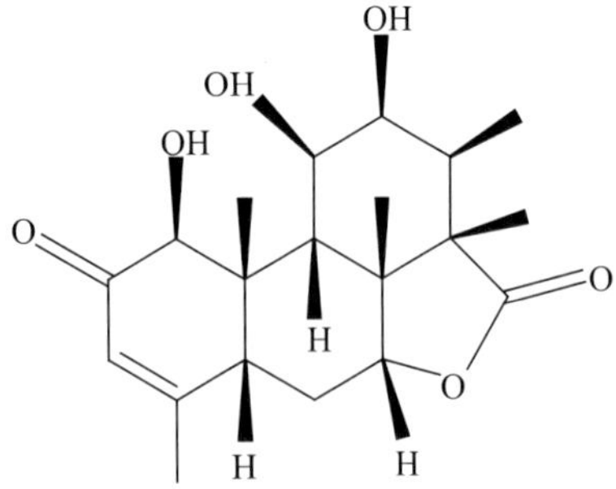

Fig. 5.109 Structure of 11-Dehydroklaineanone.

Quassia indica elaborates an interesting series of quassinoids, including ***samaderines, indaquassin, 2-O-glucosylsamaderine C,*** and ***simarinolide,*** which prevent the survival of the chloroquine-resistant K1 strain of *Plasmodium falciparum,* and exhibit cytotoxic properties, particularly the inhibition of endothelial cell–neutrophil leukocyte adhesion as well as anti inflammatory activity.

Bitters also possess aromatic property due to the presence of volatile oils, e.g.: *Citrus aurantium* (orange peel). They are sometimes used as flavoring agents. Major bitters, studied from medicinal plants are enumerated in Table 5.4.

Table 5.4 Bitters studied from medicinal plants.

S.No	Name of the bitter	Source
1.	Acorone	*Acorus calamus*
2.	Amarogentin	*Swertia chirata*
3.	Amphicoside	*Amphicome emodi*
4.	Andrographane	*Andrographis paniculata*
5.	Andrographolide	*Andrographis paniculata*
6.	Andrographone	*Andrographis paniculata*
7.	Bonducin	*Caesapinia bonducella*
8.	Celastrin	*Gunnera perpensa*
9.	Chaparrianone	*Hannoa klaineana*
10.	Gentiopicrin	*Swertia chirata*
11.	Homoandrographolide	*Andrographis paniculata*
12.	Klaineanone	*Hannoa klaineana*
13.	Kutkosides & Picrosides	*Picrorhiza kurroa*
14.	Neoandrographolide	*Andrographis paniculata*
15.	Tinosporinone	*Tinospora cordifolia*

5.4.4 Bioflavonoids

A flavonoid is identical to a bioflavonoid; furthermore, a flavonoid by definition is a polyphenol, since it contains more than one benzene ring in its structure. However, many regard polyphenols to be tannins. As seen above, green tea has an abundance of flavonoids and flavonals as well as tannin. These compounds have been well characterized in the herb and numerous papers support their use as antioxidants or free radical scavengers.

Flavonoids are an important group of polyphenols, widely distributed in plant flora. Four thousand flavonoids are known to exist and some of them are pigments in higher plants. ***Quercetin, kaempferol*** and ***quercitrin*** are common flavonoids present in nearly 70% of plants.

Flavonoids are derived from parent compounds known as flavans (Fig. 5.110).

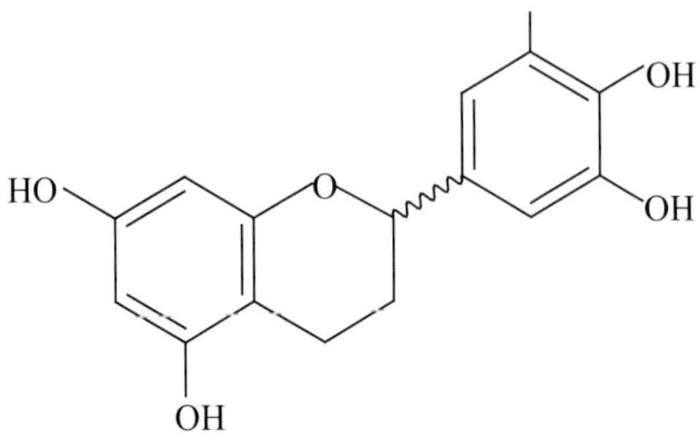

Fig. 5.110 Structure of Flavan.

The following diagram represents the flavonoid molecule.

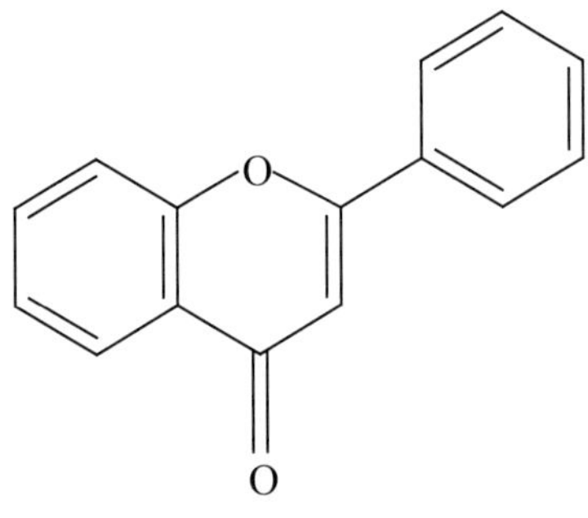

Fig. 5.111 Structure of Flavonoid Molecule.

Flavonoids are classified into five groups:

A. Flavones and flavonols (Fig. 5.112)

They are yellow and are usually found in a majority of plants. Leaves of *Calycopteris floribunda* contain flavanol ***calycopterin***, which is considered to be anthelmentic.

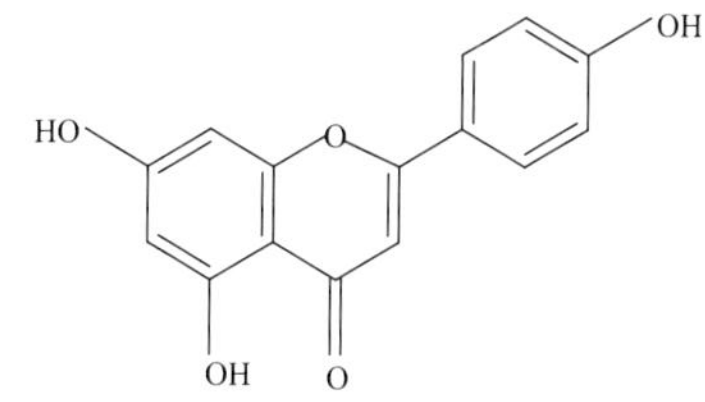

Apigenin

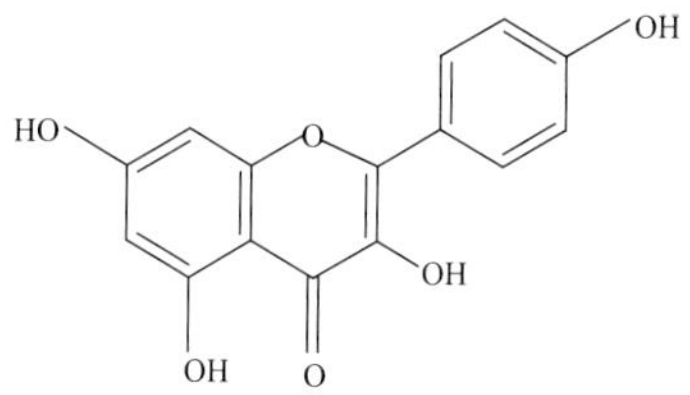

Kampferol

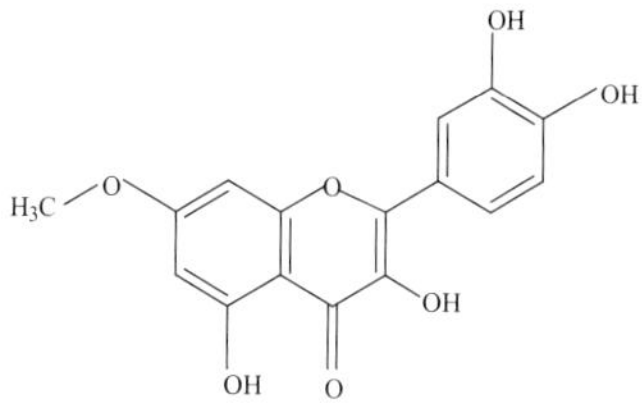

Rhamnetin

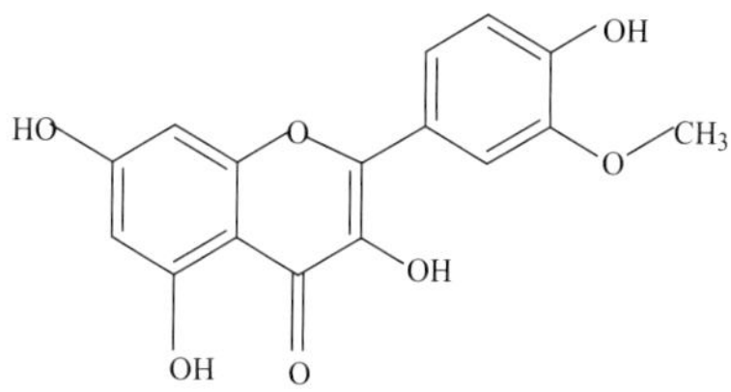

Isorhamnetin

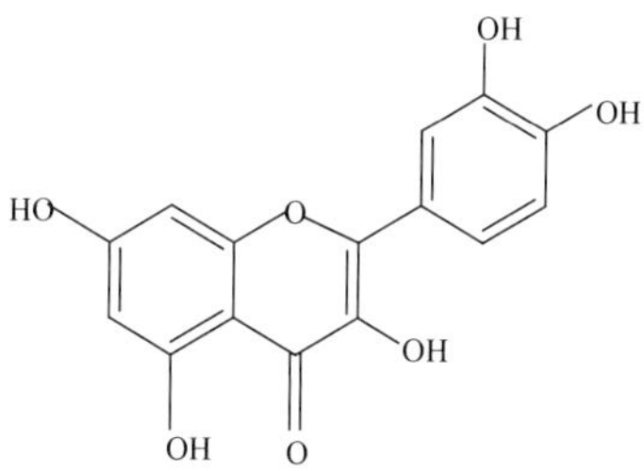

Quercetin

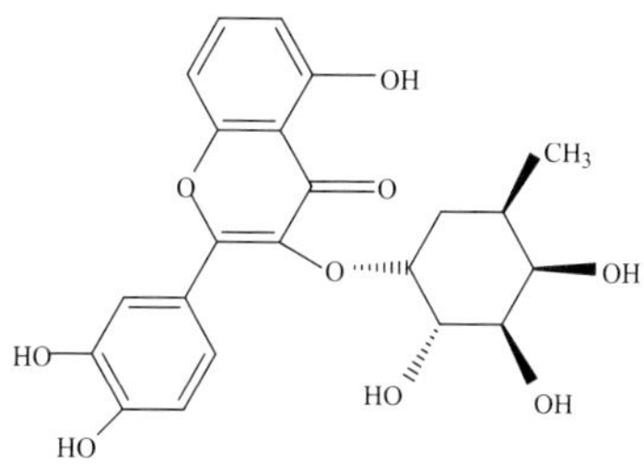

Querctirin

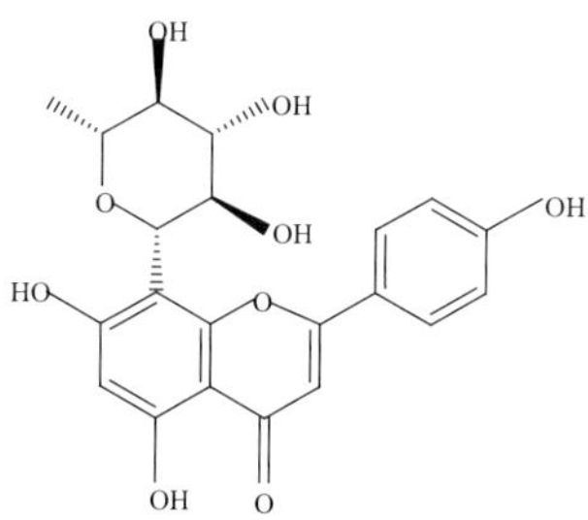

Vitexin

Fig. 5.112 Structure of Flavones and Flavonols.

Tetrahydroamentoflavone, from *Semecarpus anacardium*, inhibits the enzymatic activity of cyclooxygenase, with an IC50 value of 29.5 μM (COX-1). ***Lanaroflavone*** (Fig. 5.113), from *Campnosperma panamaense*, inhibited *Plasmodium falciparum* K1 chloroquine-resistant strain and *Leishmania donovani* cultured *in vitro* with IC values of 0.2g/mL and 3.9g/mL, respectively.

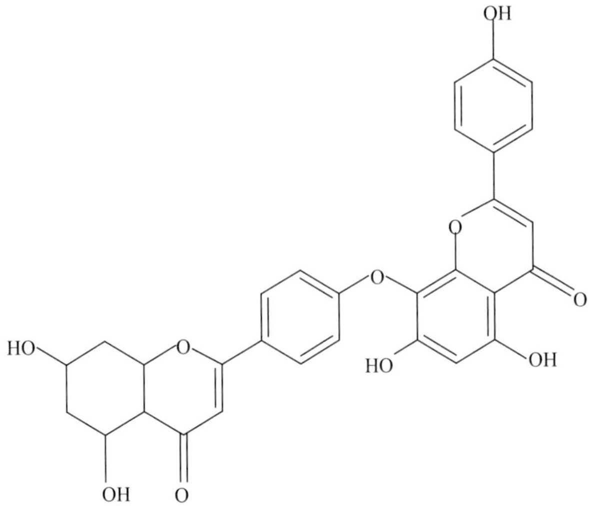

Fig. 5.113 Structure of Lanaroflavone.

5,3-dihydroxy-3,6,7,8,4-pentamethoxyflavone (Fig. 5.114) from *Polanisia dodecandra* inhibits a broad panel of cancer cells: central nervous system cancer (SF-268, SF-539, SNB-75, U-251), nonsmall- cell lung cancer (HOP-62, NCI-H266, NCI-H460, NCI-H522), small-cell lung cancer (DMS-114), ovarian cancer (OVCAR-3, SKOV- 3), colon cancer (HCT-116), renal cancer (UO-31), a melanoma cell line (SK-MEL-5), and leukemia cell lines (HL-60, SR), cultured *in vitro.*

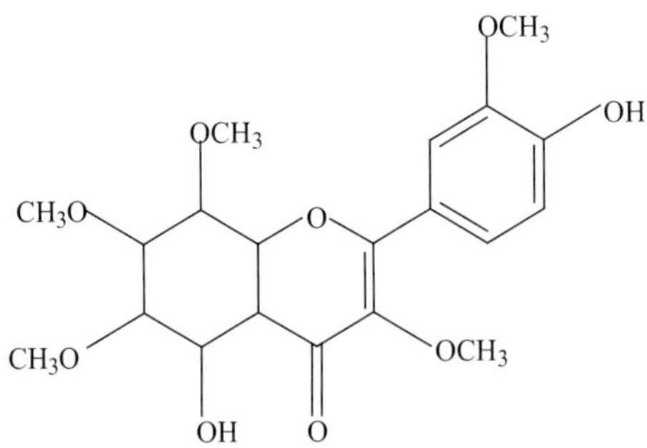

Fig. 5.114 Structure of 5,4′-Dihydroxy-3,6,7,8,4′-pentamethoxyflavone.

Acacetin (5,7-dihydroxy-4'methoxy-flavone), present in *Cirisium rhinoceros* inhibits the proliferation of the human liver and lung cancer cells, HepG2 and A549 cells, respectively by blocking apoptosis and cell cycle progression. (Fig. 5.115).

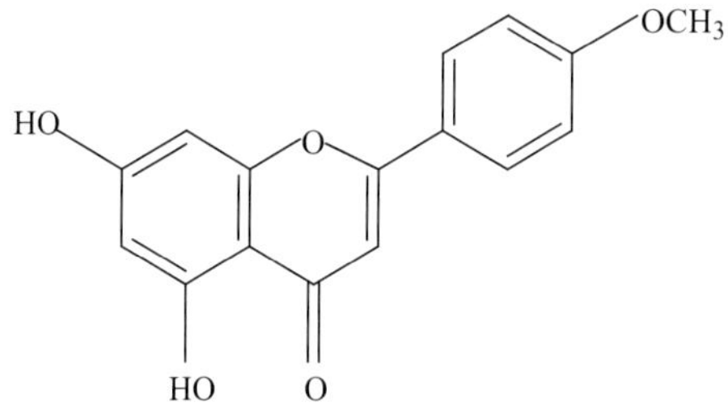

Fig. 5.115 Structure of 5,7-dihydroxy-4′methoxy-flavone.

The rhizomes of *Iris nepalensis* is reported to characteterize several isoflavones, including ***irisolone*** and ***irisolidone*** (Figs. 5.116, 5.117).

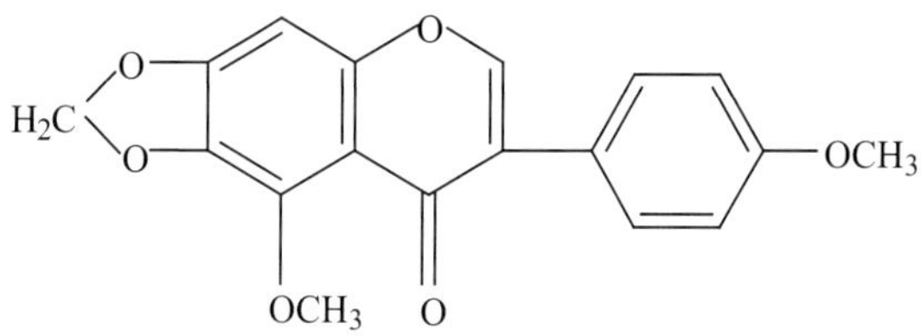

Fig. 5.116 Structure of Irisolone.

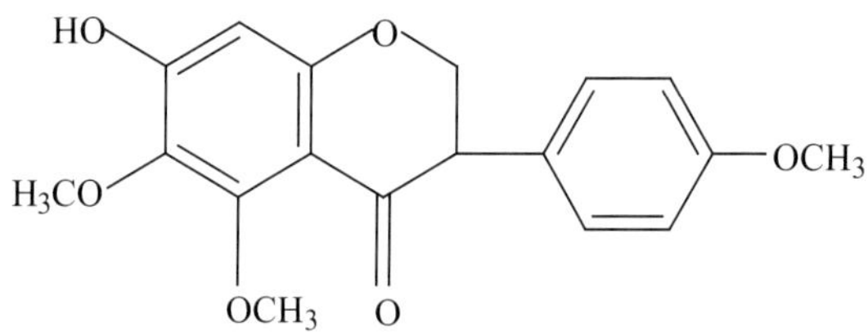

Fig. 5.117 Structure of Irisolidone.

B. Flavonols

These are found in plants of Rutaceae.

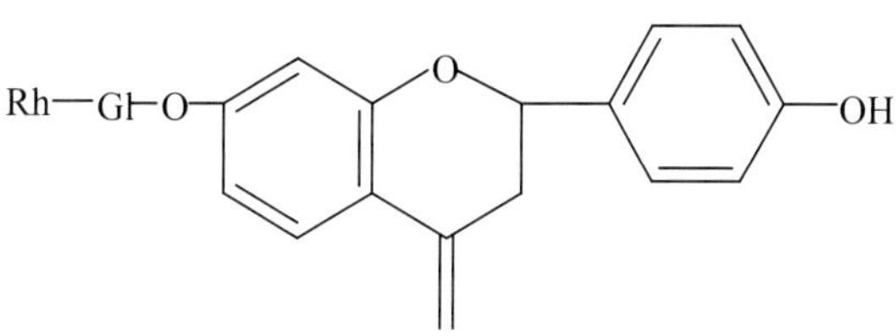

Fig. 5.118 Structure of Flavonol (Naringin).

C. Anthocyanidins

Anthocyanidins are derived from flavonols. In nature, they are found as glycosides and are called anthocyanins. They have characteristic colors ranging from red to blue and are responsible for the color of fruits. ***Cyanidin*** and ***delphinidin*** are similar examples (Fig. 5.119).

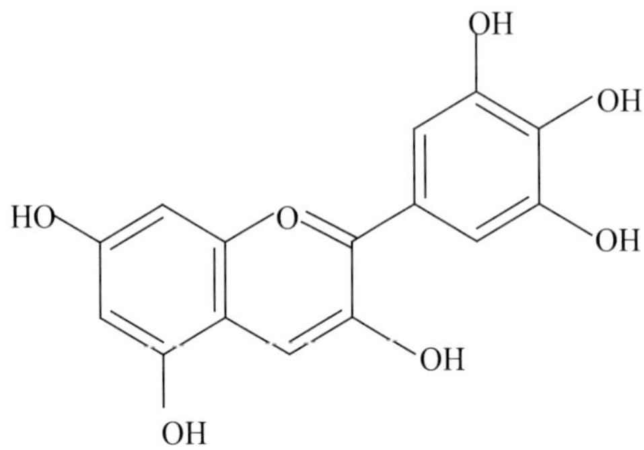

Cyanidin

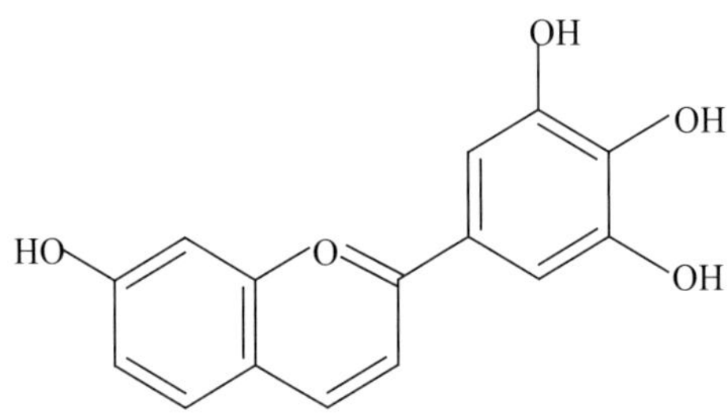

Delphinidin

Fig. 5.119 Structure of Anthocyanidins.

D. Proanthocyanidins

Proanthocyanidins on hydrolysis yield anthocyanidins. ***Procyanidin*** and ***prodelphinidin*** are common examples. Prodelphinidin isolated from *Rhynchosia minima* has demonstrated antibiotic activity.

E. Catechin and leucoanthocyanidins (Fig. 5.120)

Catechins are also derived from flavones. Leucoanthocyanidins have an additional hydroxy group.

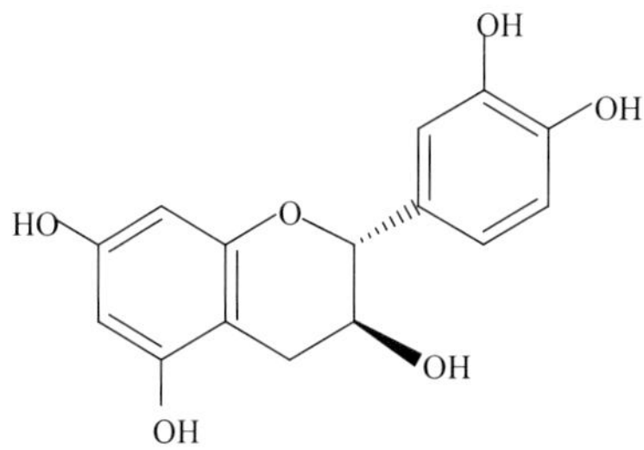

Fig. 5.120 Structure of Catechin.

Flavonolignans

Silymarin is a flavonol-lignan mixture obtained from seeds of *Silybum marianum* (Fig. 5.121). Silymarin is a mixture of silybin, isosilybin, silychristin and silydianin. Silybin A and B are collectively known as silibinin. Silymarin is a reputed hepatoprotective drug.

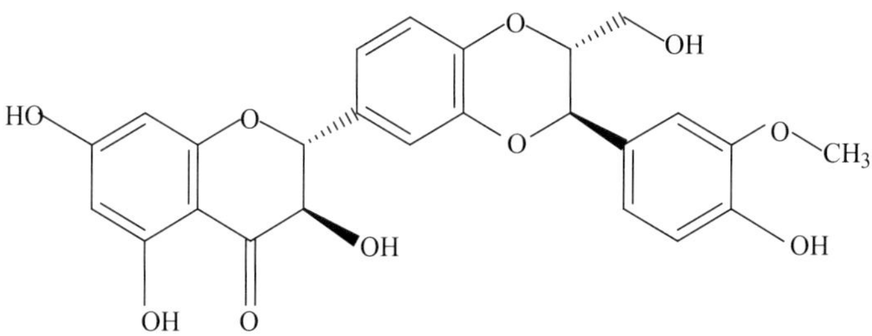

Fig. 5.121 . Structure of Silymarin.

Flavonolignans (***hydnowightin, hydnocarpin, and neohydnocarpin***) have been isolated from the seeds of *Hydnocarpus wightiana.* They are reported to be hypolipidemic, anti-inflammatory and antineoplastic (Figs. 5. 122,5.123).

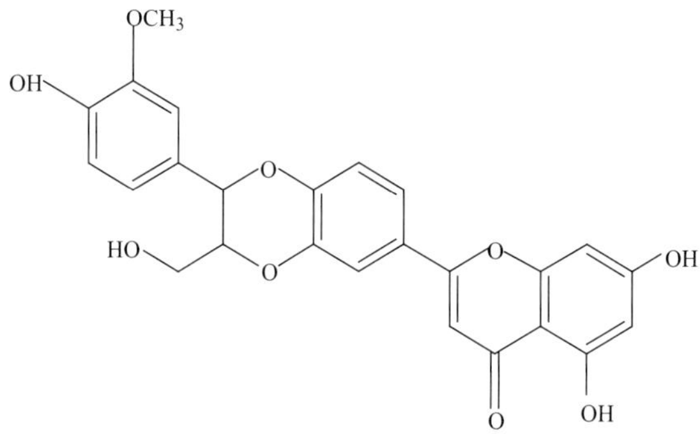

Fig. 5.122 Structure of Hydnocarpin.

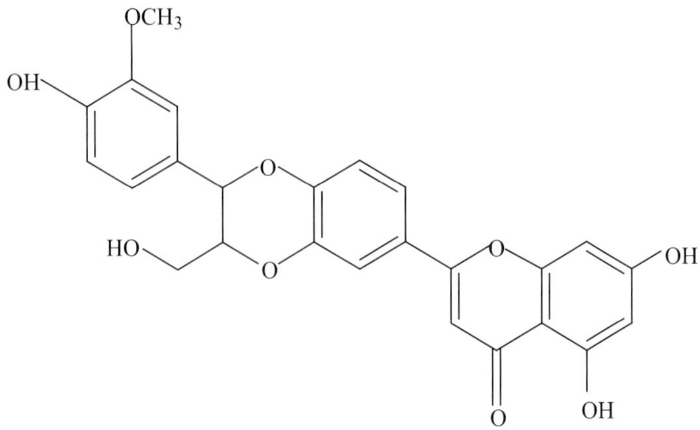

Fig. 5.123 Structure of Hydnowightin.

More about bioflavonoids

Catechin reported from *Artocarpus integra* has gastro protective activity. ***Cyanidanol,*** a stereoisomer of catechin from the seed coat of *Anacardium occidentale* has an inhibiting action on histidine decarboxylase. Flavones, ***oroxylin-a, baicalein*** **and** ***chrysin*** have been isolated from *Oroxylum indicum. Acacia auriculaeformis* is reported to contain flavone glycoside, ***auriculoside*** which is a CNS depressant.

Ononis spinosa contains flavonoids including ***formononetin*** (Fig. 5. 124), and ***biochanin A.*** (Fig. 5.125). Flavonoids, ***orientin*** (Fig. 5.126), ***and vicenin,*** isolated from the leaves of *Ocimum sanctum* have significant antioxidant activity. They increase the survival time in lethally irradiated mice.

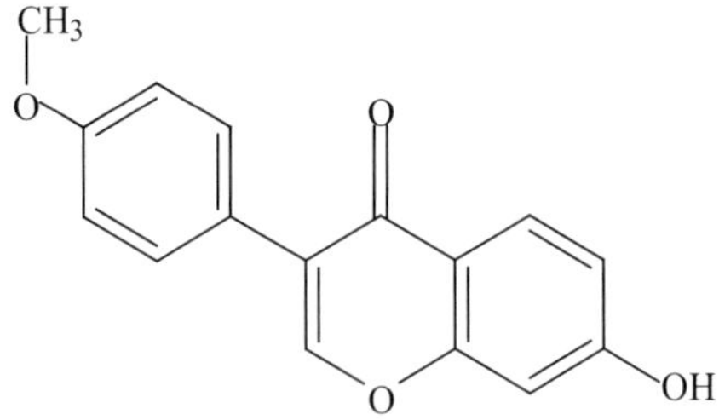

Fig. 5.124 Structure of Formononetin.

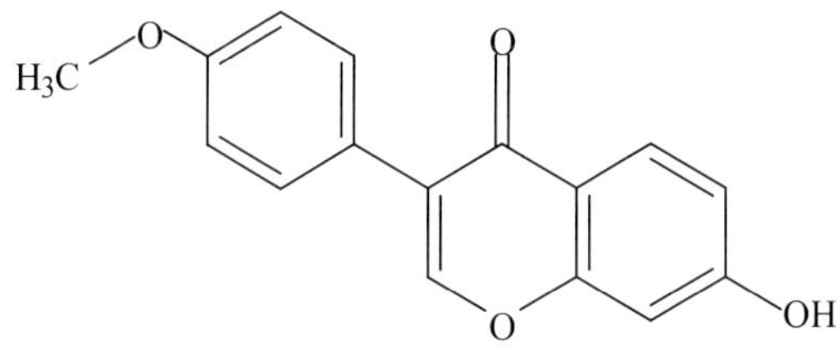

Fig. 5.125 Structure of Biochanin.

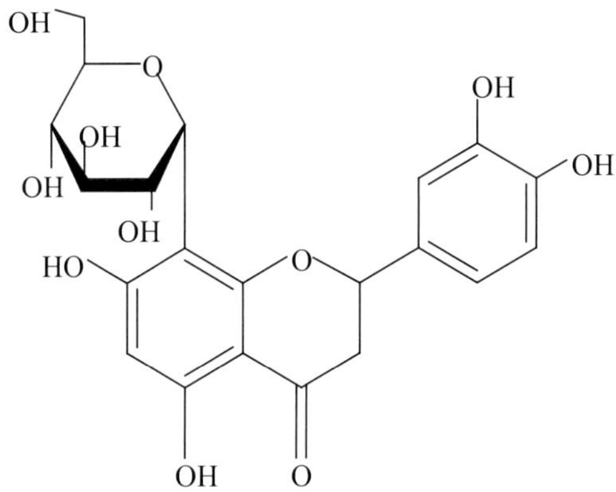

Fig. 5.126 Structure of Orientin.

Isoliquiritigenin, a flavonoid, present in licorice inhibits the proliferation of the human liver and lung cancer cells, HepG2 and A549 cels, respectively by blocking apoptosis and cell cycle progression (Fig. 5.127).

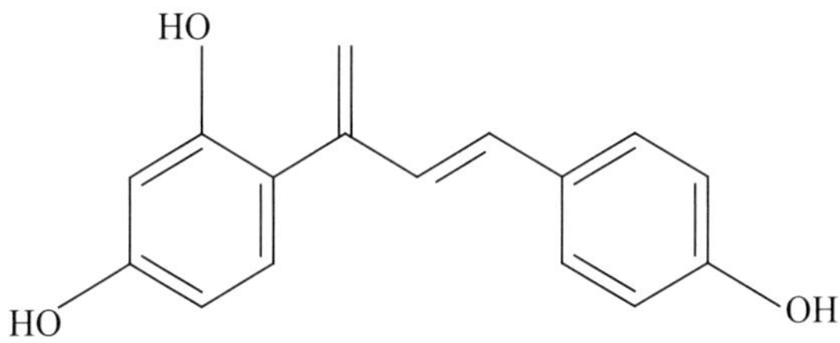

Fig. 5.127 Structure of Isoliquiritigenin.

Aspalathin is monomeric flavonoid found in *Aspalathus linearis* (Rooibos tea). It has significant antioxidant property. ***Chrysoeriol*** (3-methoxylutein) is a flavonoid found in snapdragons has anti-inflammatory and antioxidant properties (Fig. 5.128). ***Tricin*** is a flavone present in rice bran. It has an antioxidant property (Fig. 5.129).

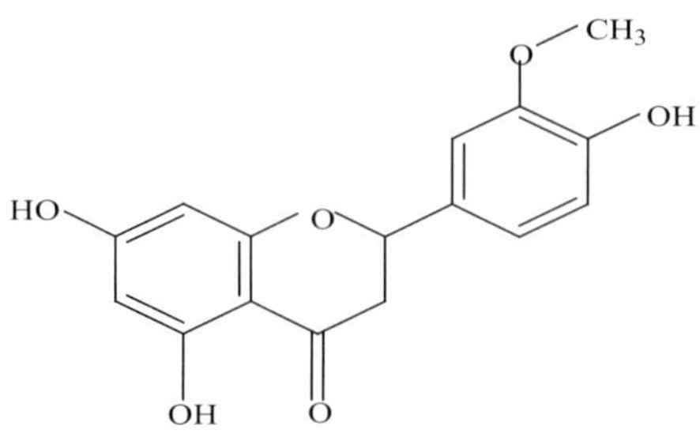

Fig. 5.128 Structure of Chrysoeriol.

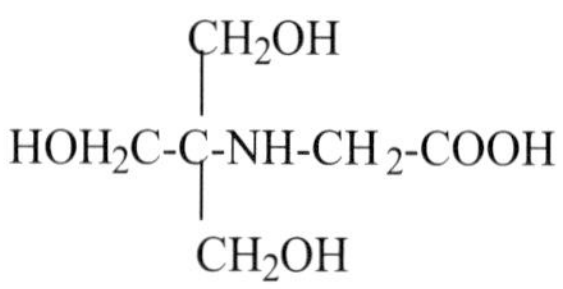

Fig. 5.129 Structure of Tricin.

Certain flavonoids of *Cananbis sativa* (***flavocannabiside and flavosativaside***) are weak inhibitors of lens aldose reductase, an enzyme implicated in the pathogenesis of cataracts in people suffering from diabetes and galactosemia (Figs. 5.130, 5.131).

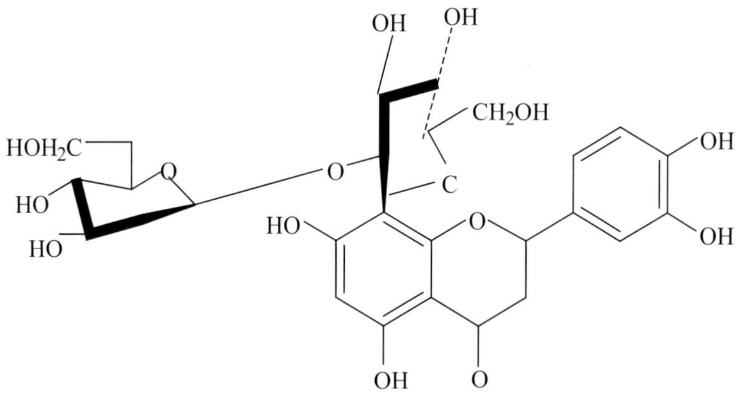

Fig. 5.130 Structure of Flavocannabiside.

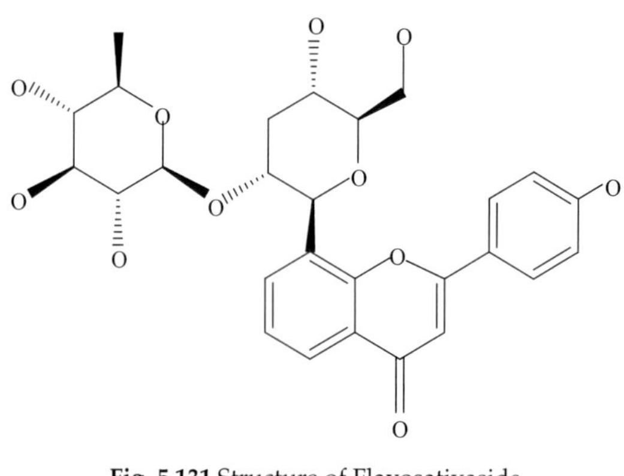

Fig. 5.131 Structure of Flavosativaside.

Flavonoids are widely distributed in the plant kingdom and have long been under investigation for antiparasitic activity without significant prospects for therapeutic value. With intensive studies on *Artemisia annua* the situation has changed dramatically. Flavonoids isolated from *A. annua* were not found active against *P. falciparum*, but demonstrated a marked and selective potentiating effect on the antiplasmodial activity of artemisinin. ***Artemetin*** (Fig. 5.132) and ***casticin*** (Fig. 5.133) act synergistically with artemisinin. Both are methoxylated flavones and are described as inhibitors of L-glutamine uptake by infected macrophages.

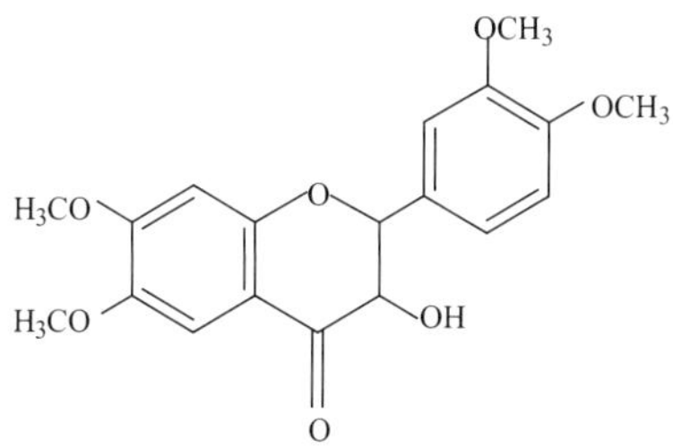

Fig. 5.132 Structure of Artemetin.

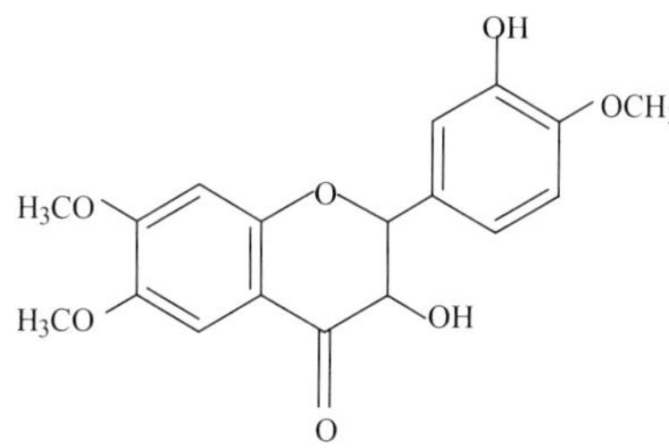

Fig. 5.133 Structure of Casticin.

From *Artemisia indica*, the flavonoids ***sakuranetin*** and ***7-methoxyaromadendrin*** (Fig. 5.134) also show high antiprotozoal activity. The mode of action of antiprotozoal flavonoids remains unclear. The effect on the generation of reactive oxygen species has been discussed, and so has sophisticated biochemical mechanisms like the inhibition of the P-glycoprotein-like transporter or modulation of protein phosphorylation on the SPK89 protein kinase in trypanosomes. Sakuranetin is also present in *Dodonaea viscosa*.

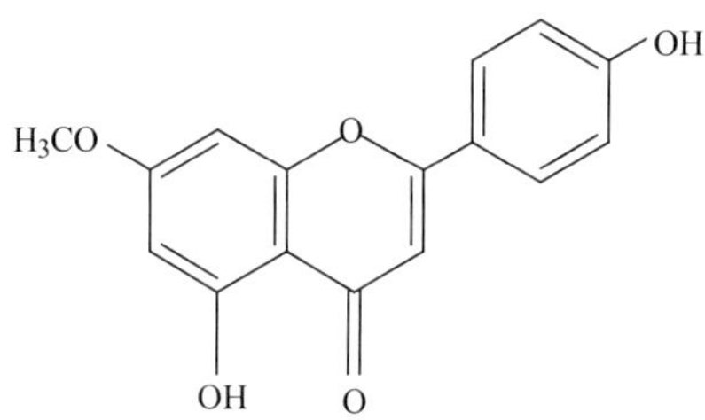

Sakuranetin

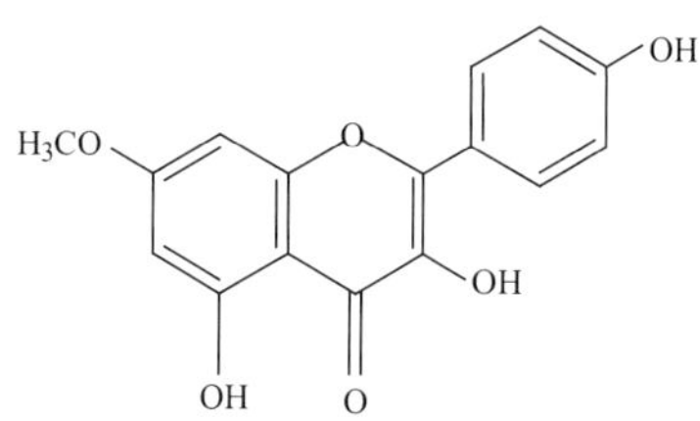

7-methoxyaromadendrin

Fig. 5.134 Structure of flavonoids of *Artemisia indica*.

Pinitol, present in *Bougenvillia spectabilis* has potent hypoglycemic activity (Fig. 5.135).

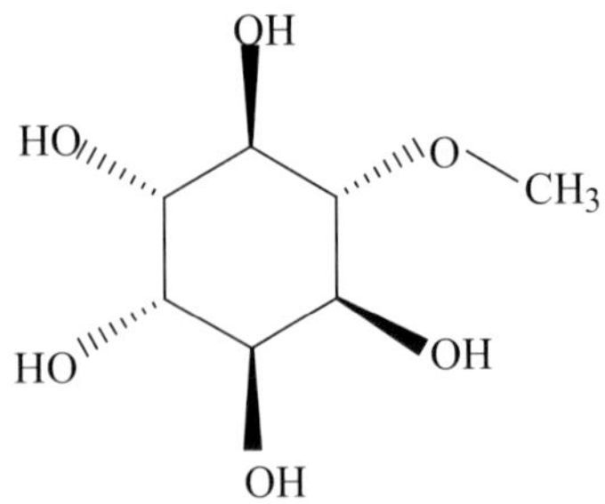

Fig. 5.135 Structure of Pinitol.

Flavonoids of *Ginkgo biloba* (Fig. 5.136)

G. biloba contains flavonoids (***bilobalide and ginkgolide A, B, and C***). Ginkgo has shown to increase production of adenosine triphosphate, resulting in increased cerebral glucose metabolism. Ginkgolide has been cited for its treatment of severe sepsis and for its effect of PAF receptor binding. Ginkgo inhibits platelet activating factor (PAF), thereby preventing PAF-induced clot formation and promoting clot breakdowns. Similarly, ginkgo inhibits PAF-induced bronchospasm. A standardized extract of *G. biloba* containing 24% ginkgo flavonglycosides of 40 mg is prescribed as peripheral vasodilator. ***Ginkgetin*** (Fig. 5.137) is reported to have anti-inflammatory, analgesic, anticancer and antiviral activites.

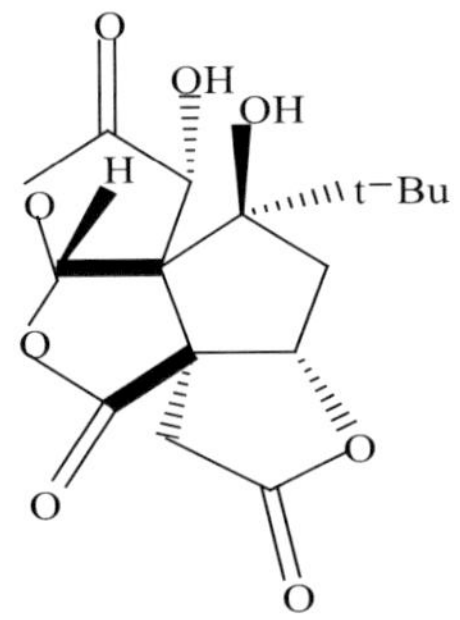

Structure of Bilobalide

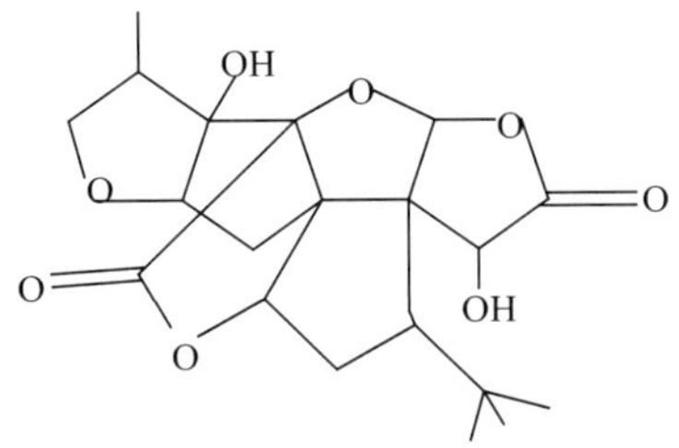

Structure of Ginkgolide

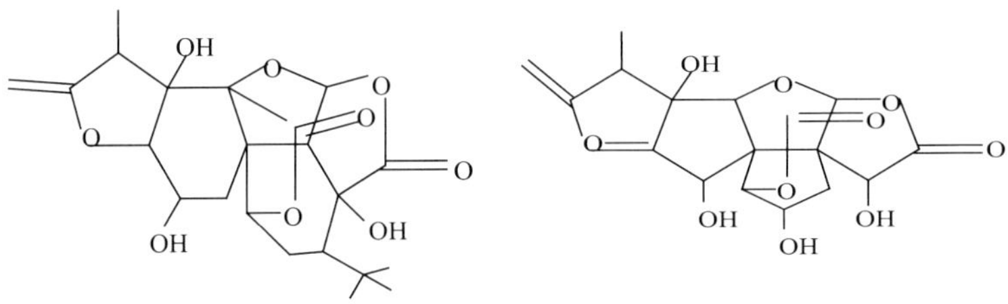

Structure of Ginkgolide B   Structure of Ginkgolide C

Fig. 5.136 Structure of Flavonoids of *Ginkgo biloba.*

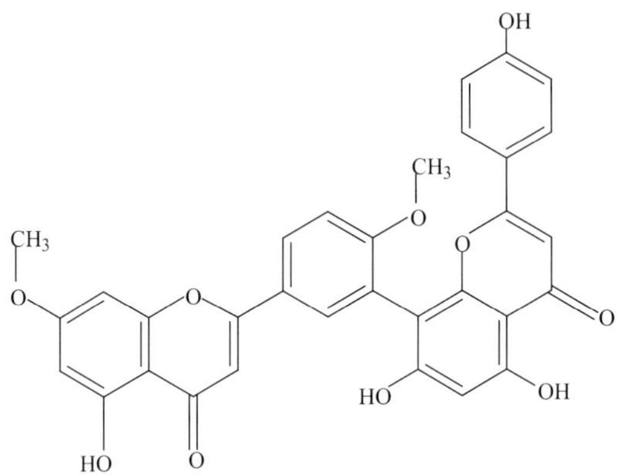

Fig. 5.137 Structure of Ginkgetin.

Eriodictyon californicum contains ***eriodictyonin, eriodictyol, chrysoeriodictyol, and xanthoeriodictyol.*** *Taxus baccata* contains ***sciadopytisin. Linarin, pectolinarin, and linariin*** are present in *Linaria vulgaris.* ***Cirsiliol*** and ***cirsimaritin*** are present in *Teucrium scorodonia. Baptisia tinctoria* contains isoflavonoids: ***formononetin baptigenin, pseudobaptigenin, (-)-maackiain, and their glycosides baptisin, pseudobaptisin and trifolirhizin.*** *Eupatorium perfoliatum* contains flavonoids including ***eupatorin, astragalin, rutin, and hyperoside.***

Thymus vulgaris contains flavonoids including ***luteolin, apigenin, naringenin, eriodictyol, cirsilineol, salvigenin, cirsimaritin, thymonine, and thymusine,*** partially present as glycosides. *Thuja occidentalis* contains including among others, ***quercitrin, mearusitrin, the biflavonoids hinoki flavone, and amentoflavone.*** *Vaccinium myrtillus* contains flavonoids including ***avicularin, hyperoside, isoquercitrin, quercitrin, meratine, and astragaline.***

Table 5.5 Flavonoids studied from medicinal plants.

S.No	Name of the flavonoid	Source
1.	Azharone	*Azadirachta indica*
2.	Echidonin	*Andrographis paniculata*
3.	Farreol	*Hildegardia barteri*
4.	Hildegardiol	*Hildegardia barteri*
5.	Hydroxymackiain	*Hildegardia barteri*
6.	Irosolidine	*Iris versicolor*
7.	Isochamaejasmin	*Stellera chamaejasme*
8.	Karanjin	*Pongamia glabra*
9.	Panicolin	*Andrographis paniculata*
10.	Planifolin	*Paepalanthus planifolius*
11.	Pterosupin	*Pterocarpus marsupium*
12.	Rutin	*Ruta graveolans*

5.4.5 Furanocoumarins

These are photosensitizing agents used in the treatment of pigment disorders. Ayurveda, the ancient science of India, has described the use of *Psoralia corylifolia* for the treatment of leucoderma. ***Psoralens*** isolated from the medicinal herb, are reputed drugs in the field of dermatology. Furanocoumarins are present in *Ficus carica, Apium graveolens, Ruta graveolens* and *Angelica gluca*. ***Marmelosin*** present in *Aegle marmelos* is the precursor compound for psoralen biosynthesis. ***Xanthotaxol*** from *Angelica archangelica* exhibited potent peripheral anti-OH-tryptamine acitivty in rat isolated fundal strip, also antihistaminic and antinicotinic activites on guinea-pig ileum.

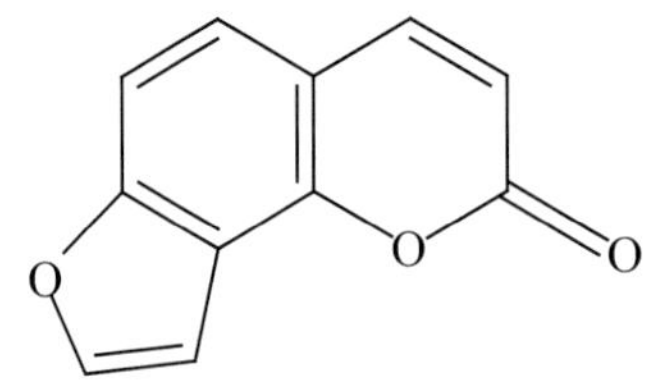

Angelicine

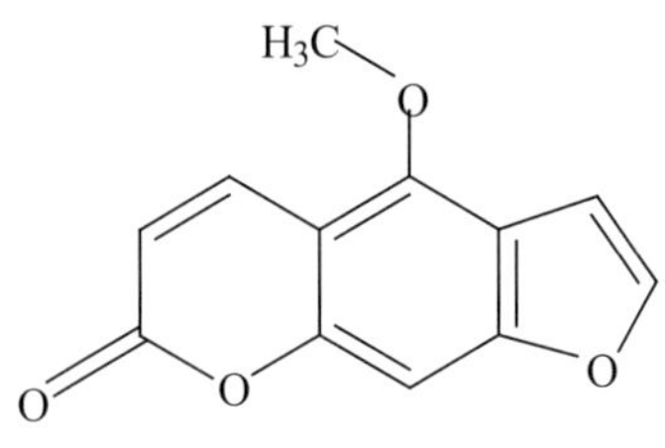

Bergepten

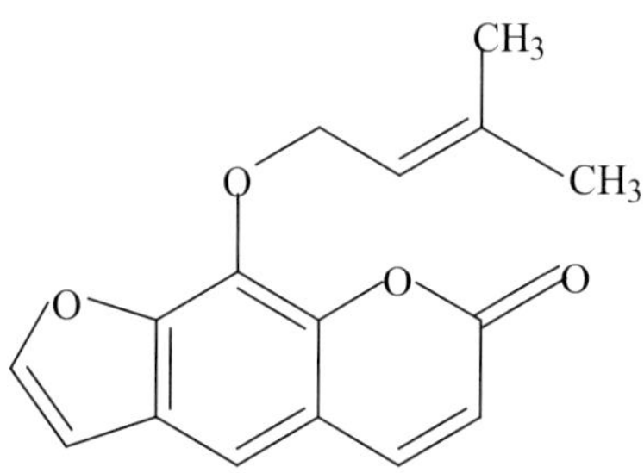

Imperatorin

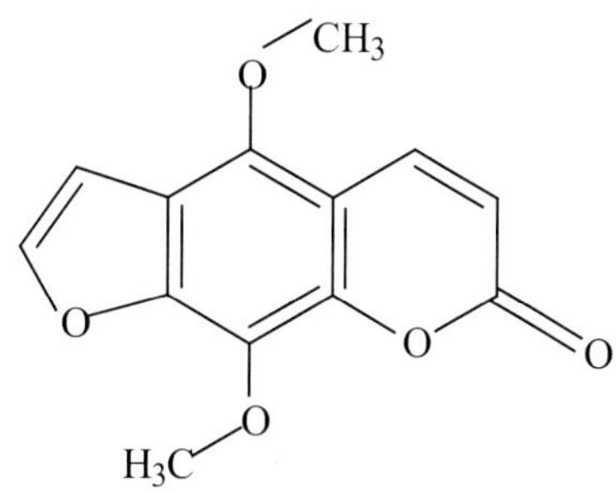

Isopimpinellin

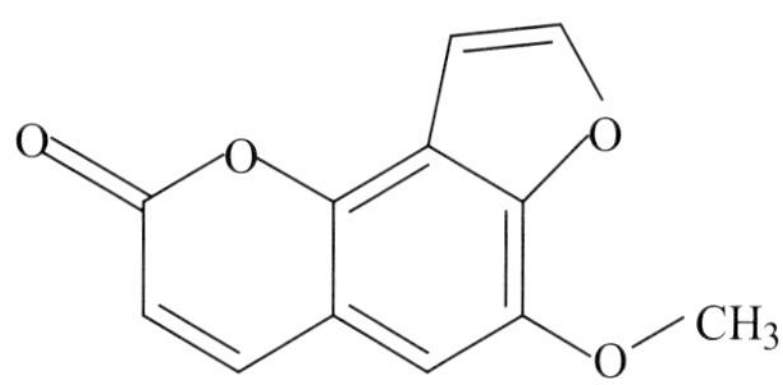

Sphondin

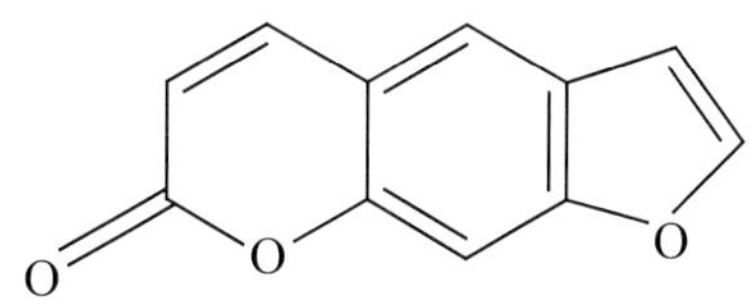

Psoralen

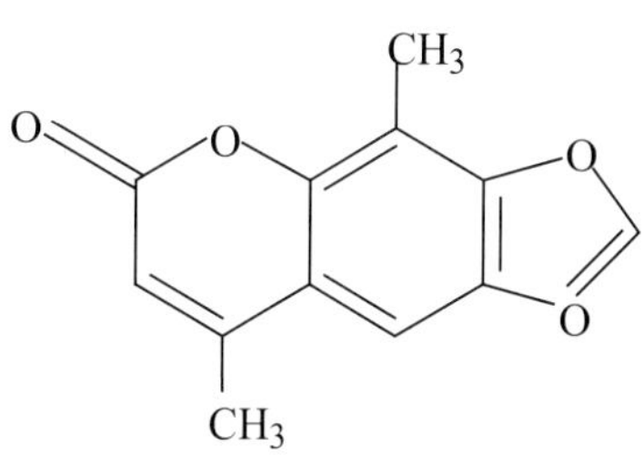

Trioxsalen

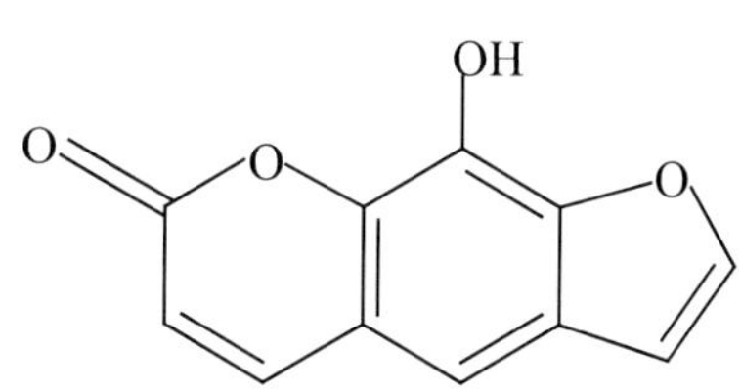

Xanthotaxol

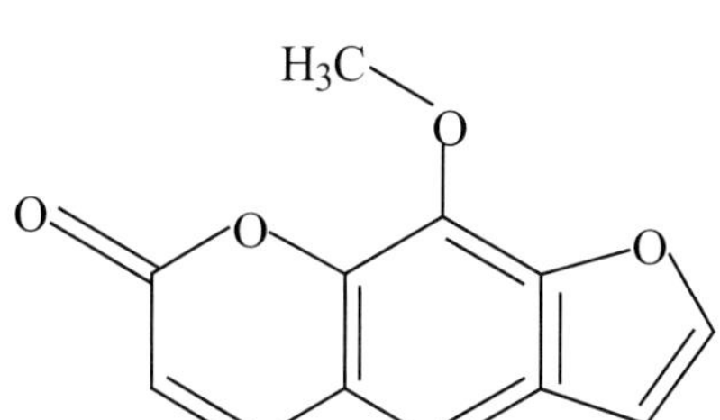

Xanthotoxin (ammonidin)

Fig. 5.138 Structure of common Furanocoumarins.

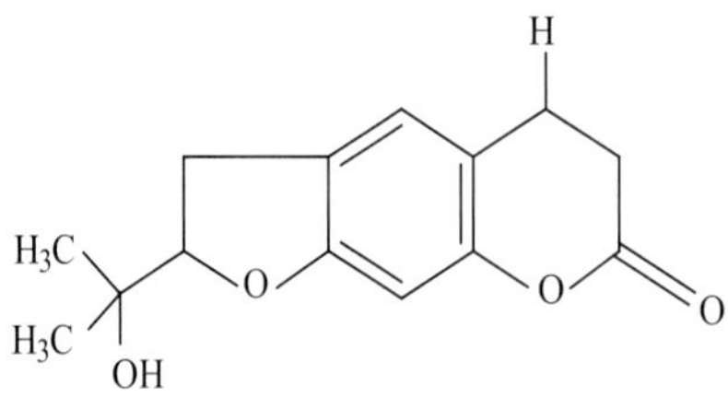

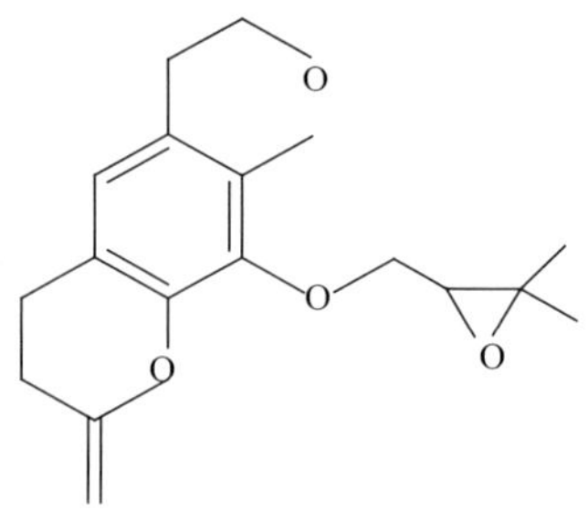

Marmelosin A Heraclenin

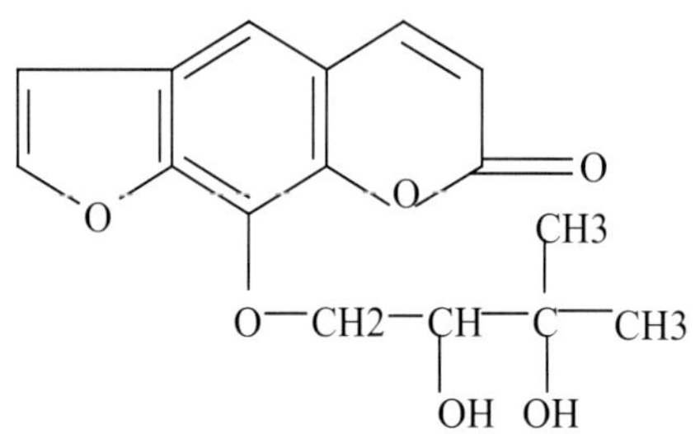

Heraclenol

Fig. 5.139 Structure of more Furanocoumarins.

Table 5.6 Furanocoumarins of Genus *Heracleum*.

Species	Coumarin
H. candicans	Heraclenin, heraclenol, bergapten, candicanin, candicopimaric acid
H. canescens	Xanthotoxin, imperatorin, heraclenin, bergapten, isopimpinellin, osthol and heraclenol
H.cachemiricum	Xanthotoxin, imperatorin, heraclenin, bergepten, and heraclenol
H.obtusifolium	Isoimperatorin
H. nepalense	Bergapten, isobergapten, sphondin and isopimpinellin
H.pinnatum	Xanthotoxin, imperatorin, heraclenin, bergapten, isopimpinellin, heraclenol and xanthotaxol
H. rigens	Heraclenin, isopimpinellin,osthol, imperatorin, bergapten and isobergapten
H.sublineare	Bergapten, isobergapten, sphondin and isopimpinellin
H. thomsoni	Psoralen, 5-hydroxyangelicin, heratomol, angelicin, lanatin, osthol, isobergapten, isopimpinellin, bergapten, sphondin and apterin
H.wallichii	Bergapten, isobergapten, sphondin, isopimpinellin, columbianetin, mrmesin nd vaginidiol

Source: RRL, Jammu.

Table 5.7 Furanocoumarins studied from medicinal plants.

S.No	Name of the furanocoumarins	Source
1.	Angelicine	*Angelica archangelica*
2.	Bergaptan	*Psoralia corylifolia*
3.	Heraclenin & Heraclenol	*Heracleum candicans, Opopanax chironium, Ruta montana,*
4.	Imperatorin	*Psoralia corylifolia*
5.	Isopimpinellin	*Psoralia corylifolia*
6.	Oxypeucedanin	*Ficus pumila*
7.	Psoralen	*Psoralia corylifolia*

5.4.6 Glycosides

These are water-soluble constituents, found in the cell sap. They are colorless, crystalline substances containing carbon, hydrogen and oxygen (Fig. 5.140). Some glycosides are peculiar in having nitrogen and sulfur. Glycosides are neutral in reaction. Chemically, glycosides contain a carbohydrate (glucose) and a non-carbohydrate part (aglycone or genin). Alcohol, glycerol or phenol represents aglycones. A glycoside can be readily hydrolyzed into its components with ferments or mineral acids.

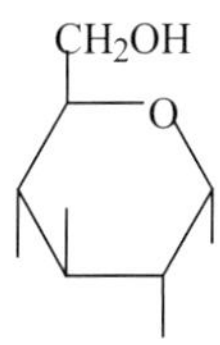

Fig. 5.140 Structure of glycoside.

Glycosides differ in their solubility in water. Some are soluble in ether and alcohol. ***Amygdalin*** found in almonds is a similar example to that of a glycoside (Fig. 5.141). ***Benzeldehyde*** is the decomposition product of amygdalin, responsible for the odor and taste of almonds. Glycosides are optically active and are levorotatory. *Adenanthera pavonina* contains HCN-glucoside known as lignoceric acid.

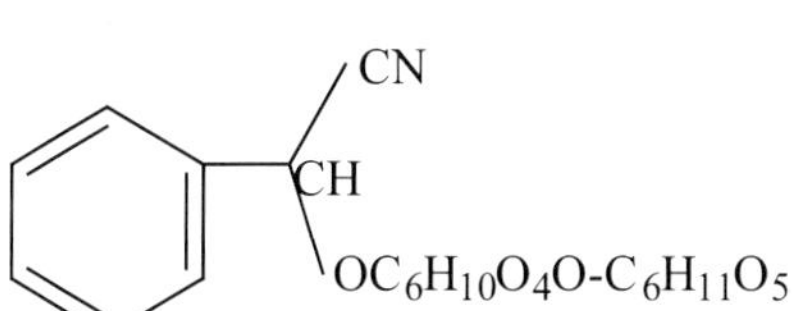

Fig. 5.141 Structure of Amygdalin.

Classification (Fig. 5.142)

Glycosides are classified on the basis of:

A. Type of sugar.
B. According to chemical nature of aglycone.
C. According to pharmacological action.

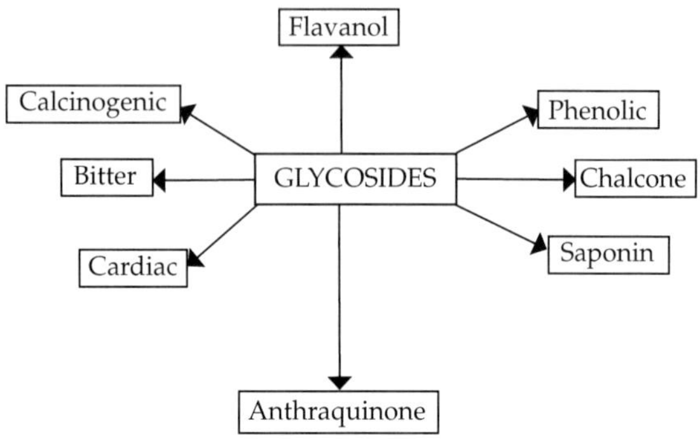

Fig. 5.142 Classification of Glycosides.

5.4.6.1 Cardiac glycosides

The group includes several medicinal herbs like *Digitalis purpurea, Thevetia nerifolia* and *Helleborus niger*. The active constituents of the plants are powerful glycosides and due to a specific action on the heart, they are known as cardiac glycosides or heart tonic.

Cardiac glycosides are divided in two groups:

A. Bufadienolides (Fig. 5.143): They are C 24 steroids. Hellbrien is a bufadienolide cardiac glycoside. On hydrolysis, it yields aglycone, hellbrigenin, which is more active than hellebrin.

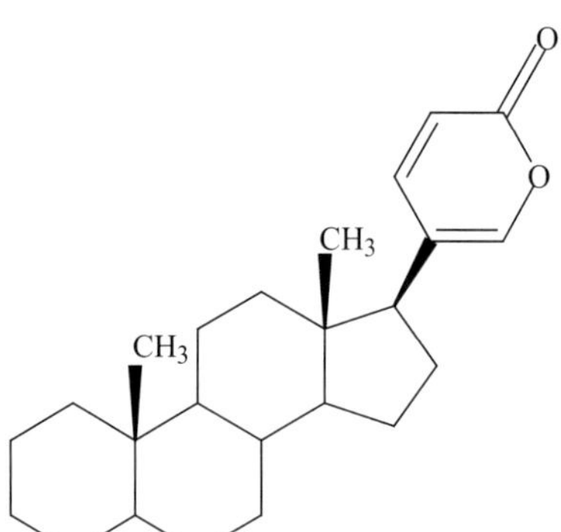

Fig. 5.143 Structure of bufadienolide.

B. Cardenolides (Fig. 5.144): They have C 17 position unsaturated lactone. They are cyclopentanophenantherne derivatives and have a hormonal nature. They act on the kidneys and heart. Digitalis glycosides belong to this group.

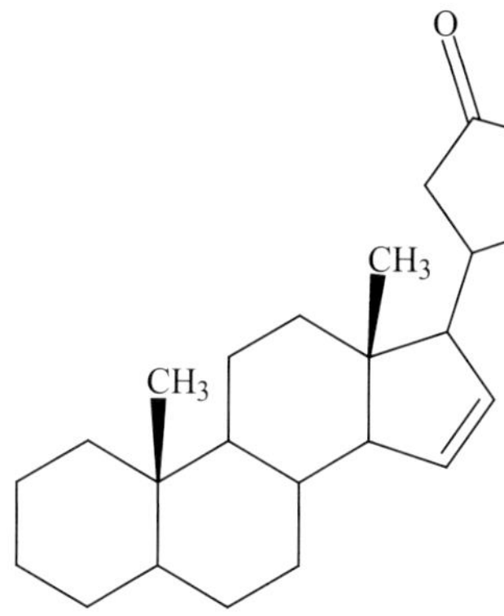

Fig. 5.144 Structure of cardenolides.

Digitalis glycosides deserve special mention. The glycosides are obtained from the leaves of *Digitalis purpurea,* when the medicinal plant is in the flowering stage. Then isolation of the glycosides in pure form is difficult and they are present in association with saponins. Stoll and Kries (1933) proved that the glycosides are of two types:

1. They are known as ***glycosides-A*** and ***digitoxin*** (Fig. 5.145) is the main glycoside. The aglycone for this group is digitoxigenin.

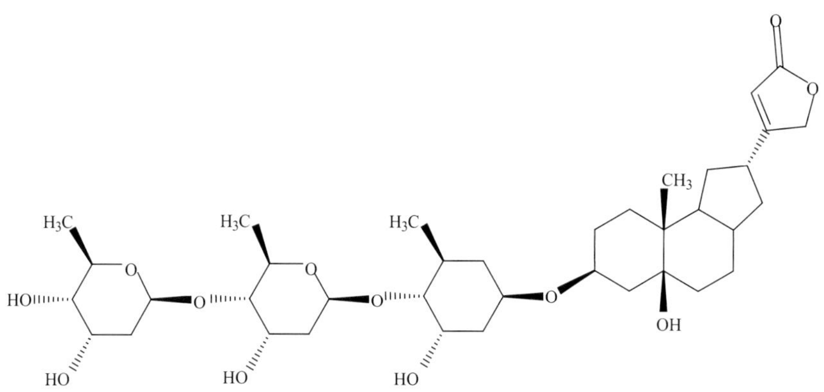

Fig. 5.145 Structure of Digitoxin.

2. The second type is known as ***glycoside-B*** and ***gitoxin*** is the main glycoside. The aglycone for this group is gitaligenin.

The glycosides obtained from *Digitalis lanata* are of three types:

1. Digilanid-A
2. Digilanid-B
3. Digilanid-C.

Some species of Digitalis are known to contain pregnane glycosides like ***digipupurin, diginin and digitalonin.*** Three glycosides, digitoxin, gitoxin and gitalin have been isolated in pure form for pharmacological application. ***Digoxin*** is utilized clinically as a heart tonic (Fig. 5.146).

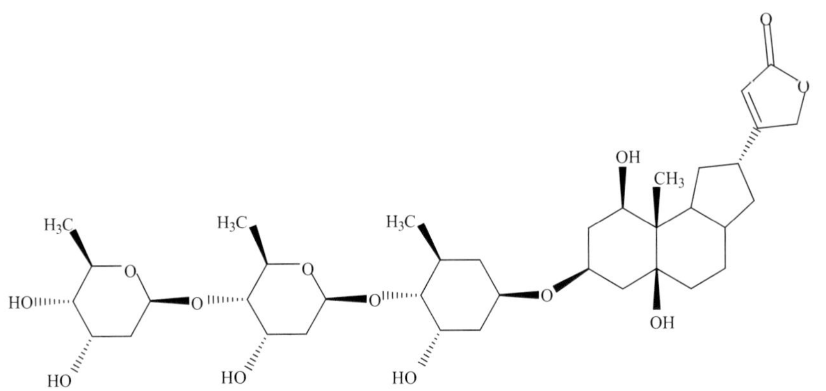

Fig. 5.146 Structure of Digoxin.

Cardiac glycoside of Asclepias currassavica

Asclepine

Glycosides of Terminalia arjuna

Arjunetin (Fig. 5.147) and ***fridelin*** (Fig. 5.148), new triterpene glycoside, ***arjunetosideandterminosideA,newcardenolide(16,17-dihydroneridienone 3-O-beta-D-glucopyranosyl-(1–>6)-O-beta-D-galactopyranoside,*** a novel naphthanol glycoside, ***arjunaphthanoloside***.

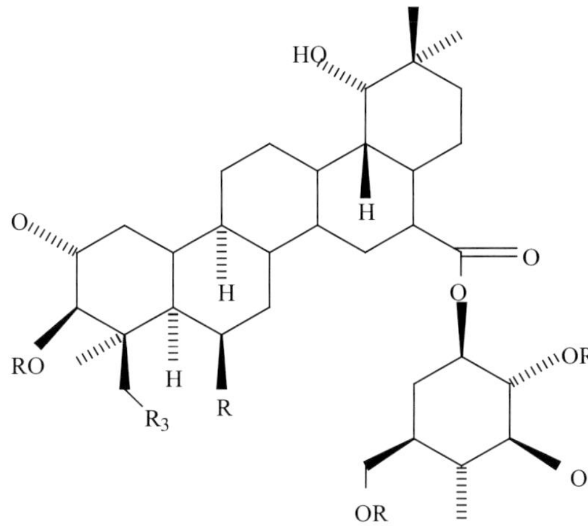

Fig. 5.147 Structure of Arjunetin.

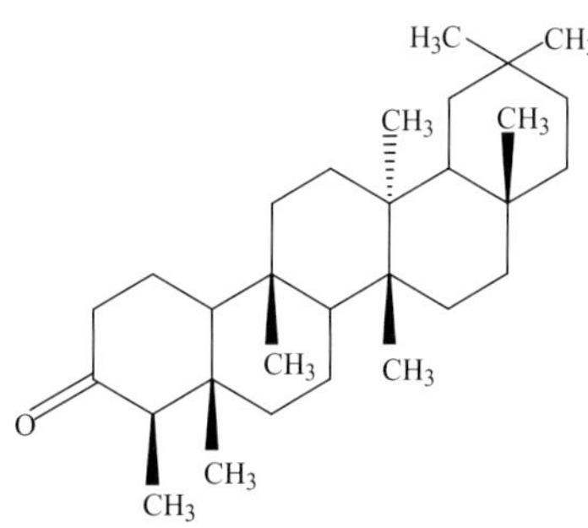

Fig. 5.148 Structure of Fridelin.

Cardiac glycosides of Nerium odorum

Neriodin, neriodorin, neriodorein and oleandrin (Fig. 5.149).

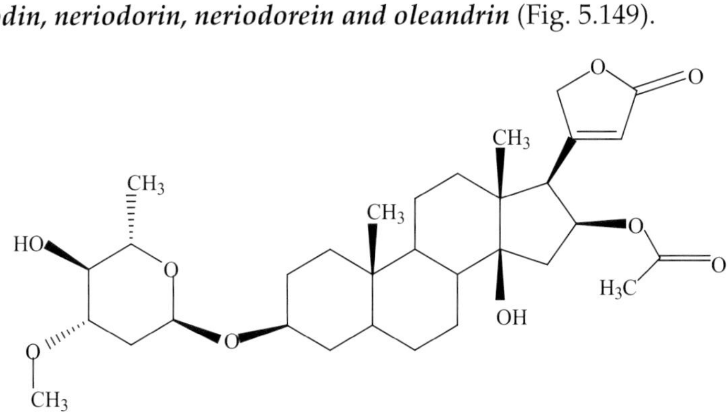

Fig. 5.149 Structure of Oleandrin.

Cardiac glycosides of Thevetia nerifolia

The cardioactive glycosides are triosides or monosides. Chief cardiac glycosides include ***thevetin A & B or cerebroside*** (triosides), ***cerebin and neriifolin*** (monosides), ***thevenerin*** (ruvoside) and ***perusitin*** (peruvosidic acid).

Cardiac glycosides of Stropanthus kombe (Fig. 5.150)

K-stropanthin, cymarin and cymarol

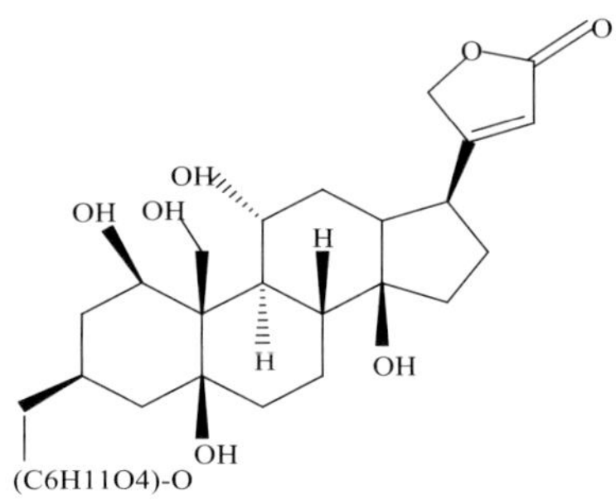

K-stropanthin

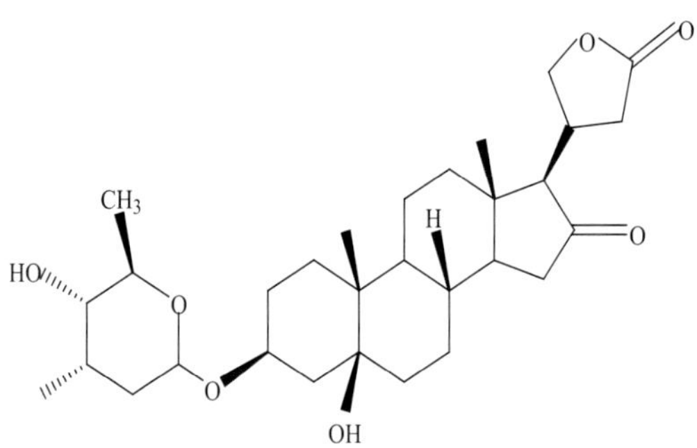

Cymarin

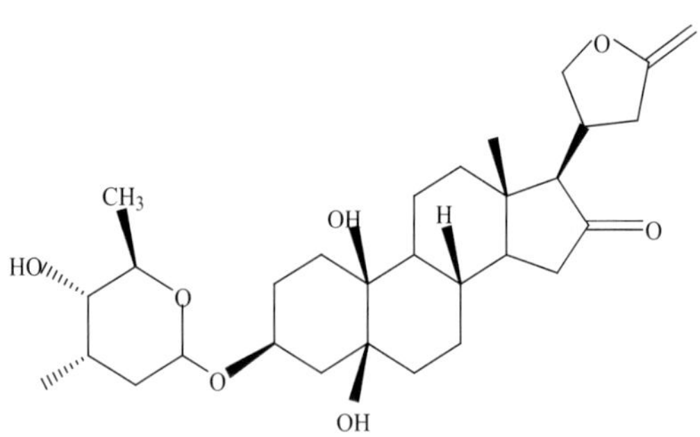

Cymarol

Fig. 5.150 Structure of cardiac glycosides of *Stropanthus kombe.*

Cardiac glycosides of Stropanthus gratus

Oubain and K-stropanthin

Cardiac glycosides of Stropanthus sarmentosus

Sermentogenin (Fig. 5.151)

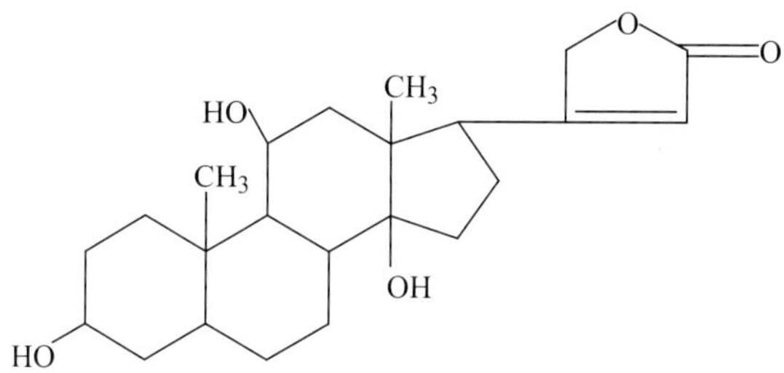

Fig. 5.151 Structure of Sermentogenin.

Cardiac glycosides of Urginea maritima

It contains crystalline glycosides: ***scillaren A*** (Fig. 5.152) and amorphous mixture of glycosides: ***scillaren B***. Scillarenin is reported to be antiviral.

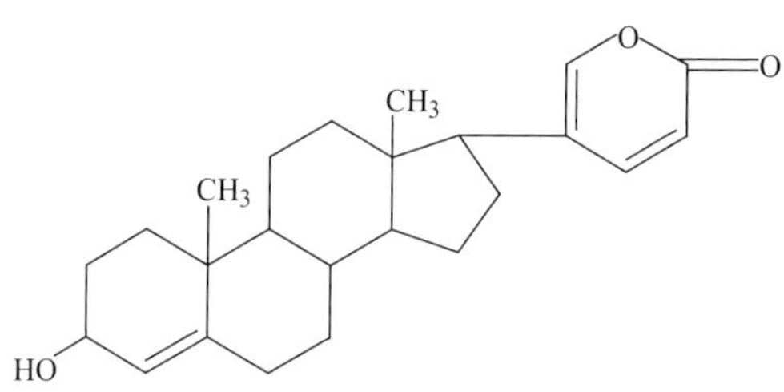

Fig. 5.152 Structure of scillaren A.

Cardiac glycosides of red variety of Urginea indica

Scilliroside and scillirubroside

Cardiac glycosides of Convallaria majalis (Fig. 5.153)

In the homeopathic system of medicine, C. *majalis* is a reputed remedy for various heart diseases. It also contains cardioactive steroid glycosides, which vary in the herb according to the geographical source, and the main active constituents are ***convalloside, convallatoxin* and *convallatoxol.*** In clinical studies, they have shown positive inotropic effect on the myocardium and lower the elevated left ventricular pressure as well pathologically raised venous pressure.

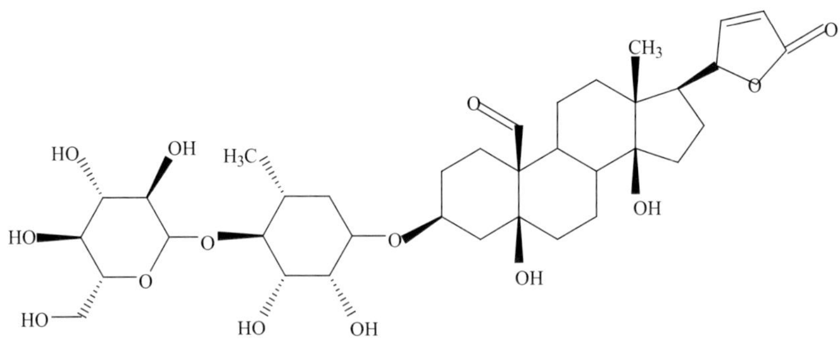

Convalloside

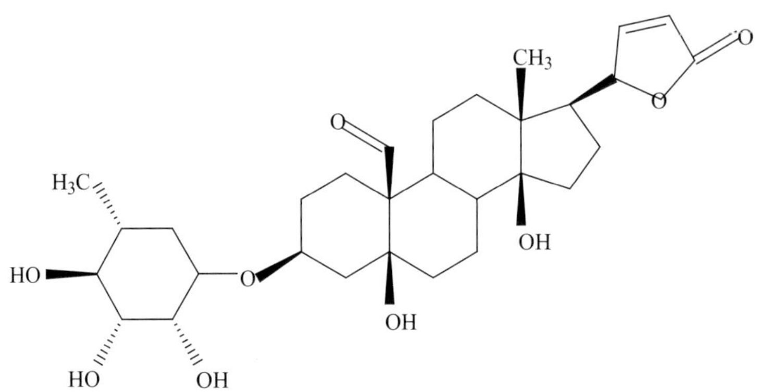

Convallatoxin

Fig. 5.153 Structure of *Cardiac glycosides of Convallaria majalis.*

Cardioactive glycosides of Adonis versalis

Adonidin or adonitoxin (Fig. 5.154), ***k-strophanthoside, k-strophanthoside β- and cymarin***

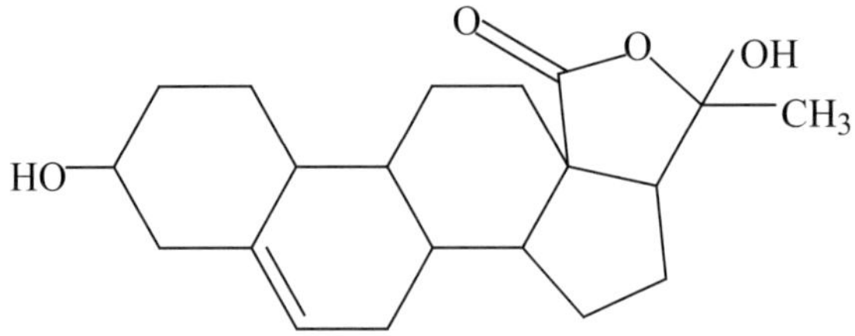

Fig. 5.154 Structure of Adonidin.

Cardioactive glycosides of Xysmalobium undulatum

Xysmalobium undulatum contains cardioactive steroid glycosides (cardenolides, mixture referred to as uzarone or xysmalobin): including among others ***uzarin*** (5.5%), ***xysmalorin*** (1.5%), ***allo-uzarine, allo-xysmalobin, urezin, uzaroside, ascleposide,*** **and** ***glucoascleposide.***

Cardioactive glycosides of Cheiranthus cheiri

Cheiranthus cheiri contains contains cardioactive steroid glycosides including cheirotoxin, erysimoside, glucoerysimoside and cheiroside A.

Cardioactive glycosides of Euonymus alata

Evonoside, evobioside, evomonoside, evolonoside, glucoevonoloside, and glucoevonogenin are present in seeds. Euatroside and euatromonoside are present in roots.

5.4.6.2 Anthracene glycosides

Anthracene glycosides are also known as anthracenosides. They are purgative in nature. On hydrolysis, they produce glycones like dianthrone, anthraquinone or anthrone. The sugars are arabinose, rhamnose or glucose. Anthraquinones (Fig. 5.155) are the active constituents and are responsible for the biological activity of the anthracene glycoside containing drugs. In addition to use in treating constipation, they are used for the treatment of skin disease like psoriasis and ringworm (Fig. 5.156).

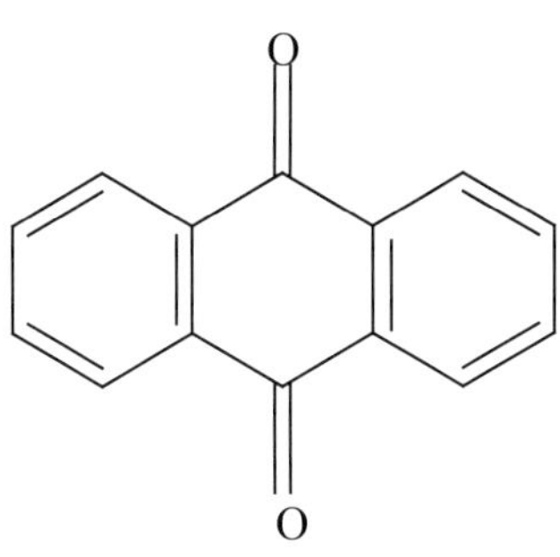

Fig. 5.155 Structure of Anthraquinone.

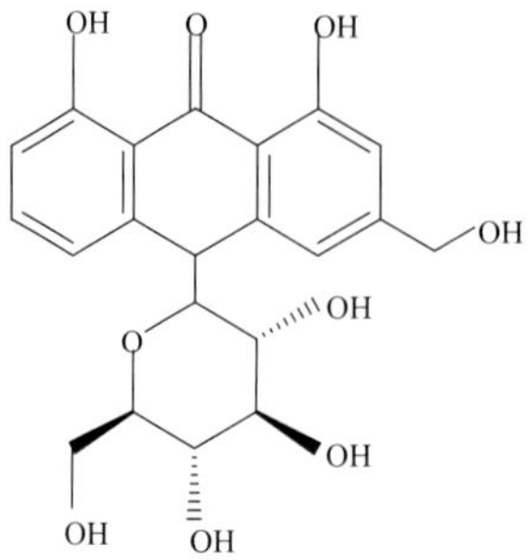

Aloin

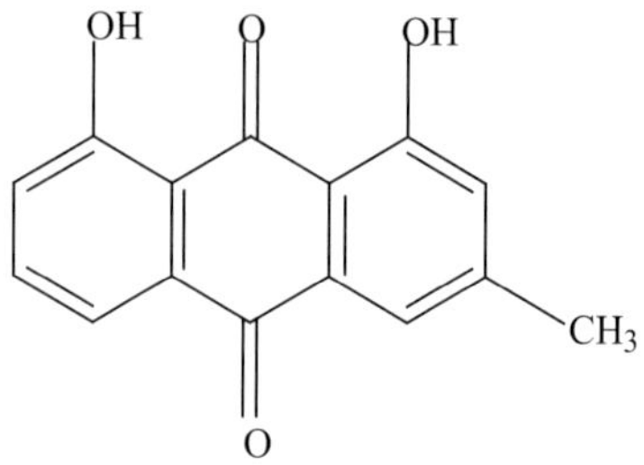

Chrysophanol

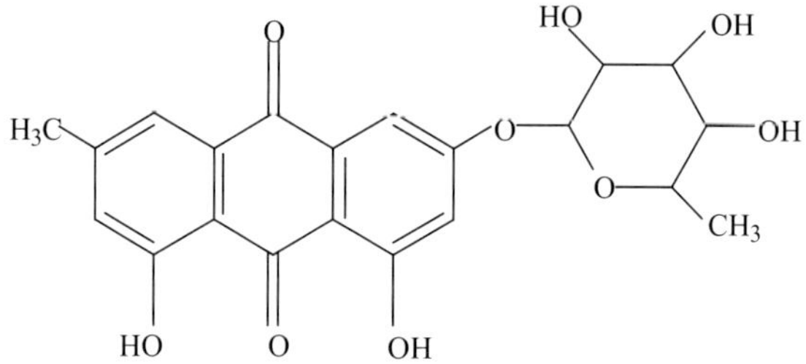

Frangulin A

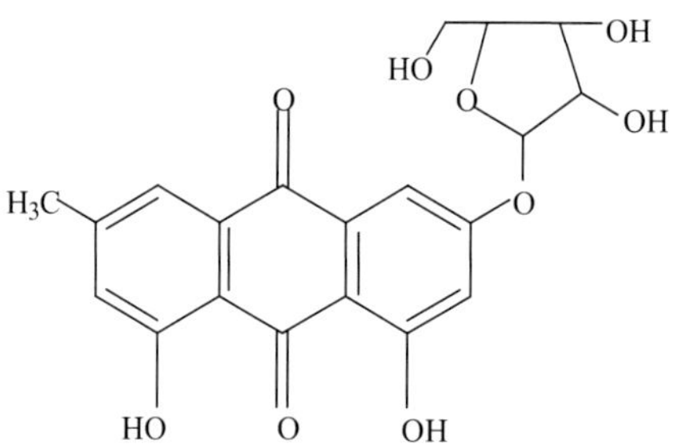

Frangulin B

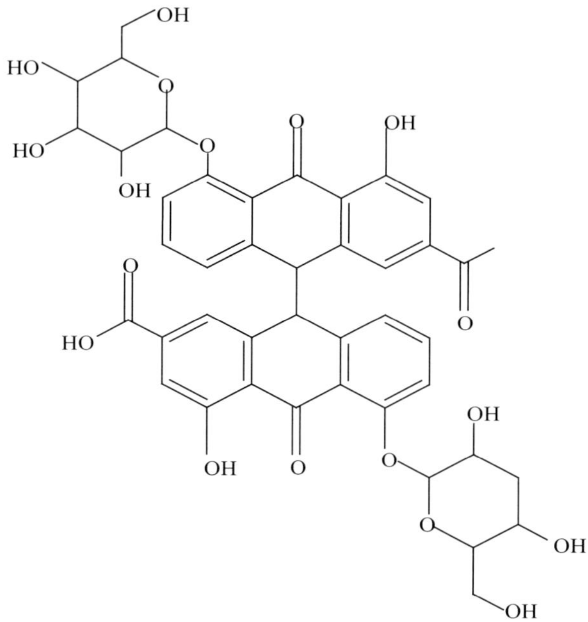

Sennoside (A and B)

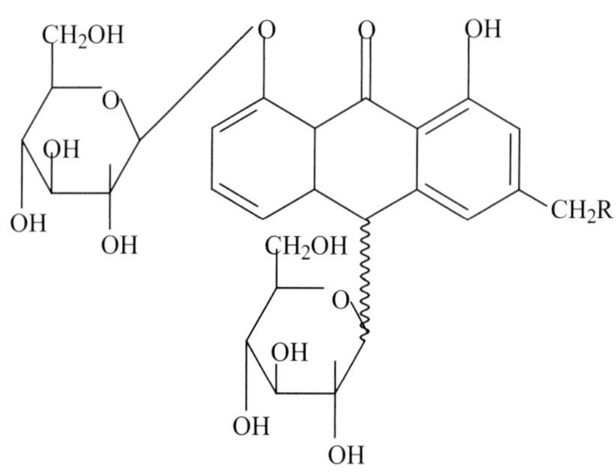

Cascarosides

Fig. 5.156 Structure of common anthracene glycosides.

Emodin, iso-emodin, aloe-emodin, and chrysophanol obtained from extracts of *Cascara sargada*, when administered separately, has little purgative effect. Yet the three given in admixture produces a good purgative action.

Aloe-emodin inhibits cell proliferation and induces apoptosis in two human liver cancer cell lines, Hep G2 and Hep 3B (Fig. 5.157).

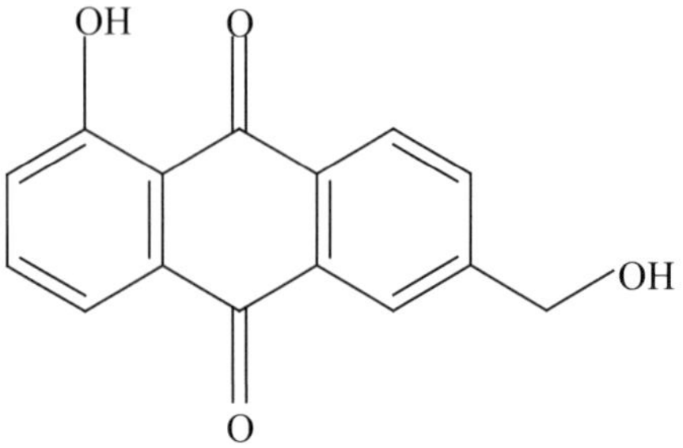

Fig. 5.157 Structure of Aloe-emodin.

Rhein, anthraquinione present in rhubarb, inhibited Hep G2 cell growth by inducing apoptosis and blocking cell cycle progression in G1 phase (Fig. 5.158).

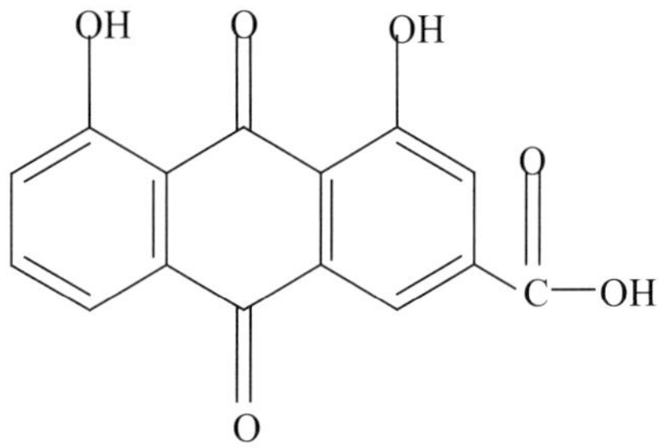

Fig. 5.158 Structure of Rhein.

Aloin from *Aloe vera* is a mixture of ***barbaloin, β-barbaloin and iso-barbaloin***. Barbaloin is present in all varieties of aloe and is a crystalline, water soluble glycoside. β-barbaloin is amorphous and iso-barbaloin is crystalline isomeric glycoside.

Anthraquinones including ***cordifoliol, cordifodiol and rubidian*** have been isolated from alcoholic extract of *Rubia cordifolia*. Rubidian has significant antioxidant activity. ***Morindin*** is anthraquinone glycoside present in *Morinda pubescens*. ***Peroxiscomicine***, anthracenone from *Karwinskia humboldtiana* is cytotoxic to mamillian tumors.

Damnacanthal obtained from *Damnacanthus indicus* and *Morinda citrifolia* is a potent and selective inhibiiter of P56 tyrosine kinase activity (Fig. 5.159) . It is also known to have antimalarial property.

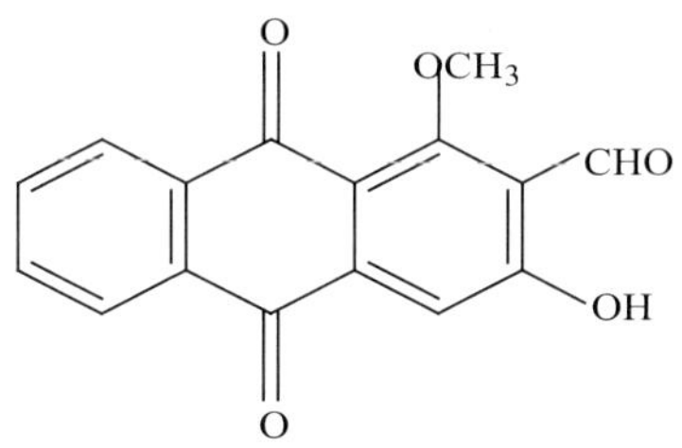

Fig. 5.159 Structure of Damnacanthal.

5.4.6.3 Chalcone glycosides

Chalcone (Fig. 5.160) is derived from a combination of one molecule of cinnamic acid and three molecules of acetate or malonate. Liquitrin (Fig. 5.161) and isoliquitrin (Fig. 5.162) are common examples. The core structure of the chalcones, chalcone, exhibits a significant chemopreventive action against two human breast cancer lines, MCF-7 and MDA-MB-231.

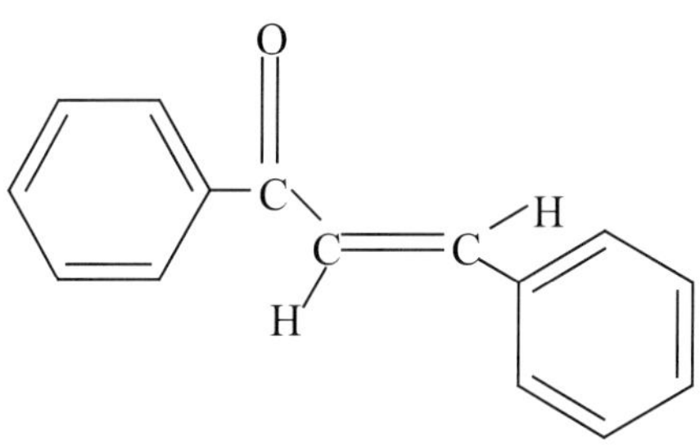

Fig. 5.160 Structure of Chalcone.

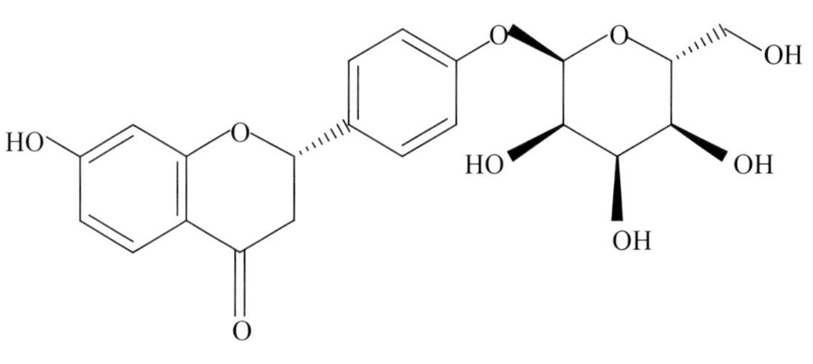

Fig. 5.161 Structure of Liquitrin.

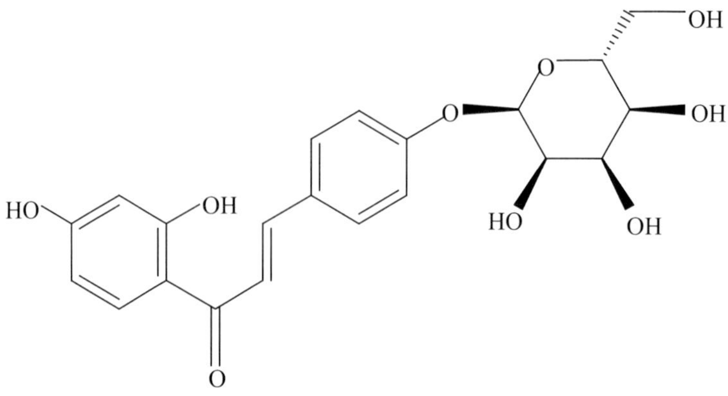

Fig. 5.162 Structure of Isoliquitrin.

5.4.6.4 Saponin glycosides

These have been discussed under saponins.

5.4.6.5 Apocartinoid glycosides

These are responsible for the color of the fruits and vegetables. They are found in *Crocus sativus* (saffron) in the form of ***crocin*** (Fig. 5.163).

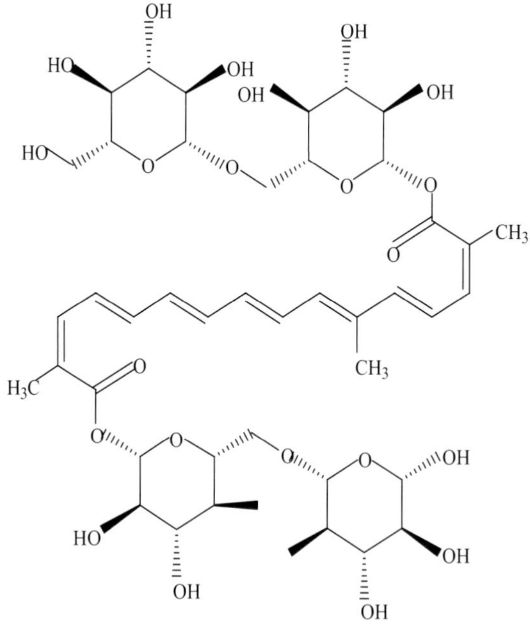

Fig. 5.163 Structure of Crocin.

5.4.6.6 Bitter glycosides

These have been discussed under bitters.

Table 5.8 Glycosides studied from medicinal plants.

S.No	Name of the glycosides	Source
1.	Acubin	*Plantago ovata*
2.	Aloin	*Aloe vera*
3.	Amarogentin	*Swertia chrayita*
4.	Amaroswerin	*Swertia chrayita*
5.	Androsin	*Picrorhiza kurroa*
6.	Apocynin	*Picrorhiza kurroa*
7.	Arjunetin	*Terminalia arjuna*
8.	Barringtonin	*Barringtonia acutangula*
9.	Calotropin	*Calotropis procera*
10.	Camellianoside	
11.	Chiratin	*Swertia chirayita*
12.	Chrysophanol	*Rheum emodi*
13.	Cichorin	*Cichoruim intybus*
14.	Denicunine	*Hemidesmus indicus*
15.	Ecliptasaponin	*Eclipta alba*
16.	Emodin	*Rheum emodi*
17.	Gentiopicroside or gentiopicrin	*Acrocomia mexicana, Anemarrhena asphodeloides, Anthocleista voglii, Centaurium eryhreae, Frasera carolinensis, Gentiana waltonii, Hintonia latifolia, Morinda lucida, Salvia lavandulifolia, Zygophyllum gaetulum*
18.	Glabrachalcone	*Pongamia glabra*
19.	Globularidin	*Globularia trichosantha*
20	Gymnemic acid	*Gymnema sylvestre*
21.	Liriodendrin	*Globularia trichosantha*
22.	Loganin	*Strychnos nux vomica*
	Pashnone	*Didymocarpus pedicellata*
23.	Physicon	*Aloe vera*
24.	Rhein	*Aloe vera*

Table 5.8 contd...

Table 5.8 contd...

S.No	Name of the glycosides	Source
25.	Rubidianin	*Rubia cordifolia*
26.	Scillaren-A	*Urginea indica*
27.	Sennosides	*Cassia angustifolia*
28.	Sewrtiamatrine	*Gentiana lutea*
29.	Sweroside	*Anthocleista djalonensis Gentiana lutea, Scabiosa columbiana*
	Swertianolin	*Swertia davidi*
30.	Thevetin	*Thevetia nerifolia*

5.4.6.7 More about glycosides

Spinonin a novel glycoside has been isolated from *Ononis spinosa subsp, leiosperma*. ***Jambulin*** is glucoside of *Syzygium cumini*. *Rhus cotinus* contains ***fustin***, which on hydrolysis yields fisetin (Fig. 5.164), a flavanol closely related to morin (Fig. 5.165), coloring matter found in *Chlorophora tinctoria*. *Phaselous lunatus* (Java bean) contains ***phaseolunatin*** (Fig. 5.166).

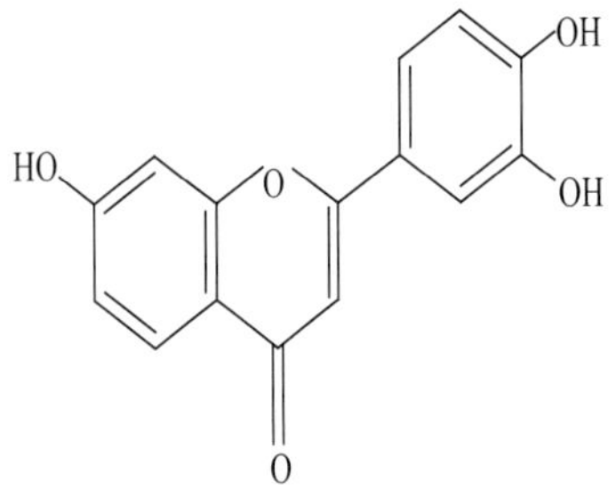

Fig. 5.164 Structure of Fisetin.

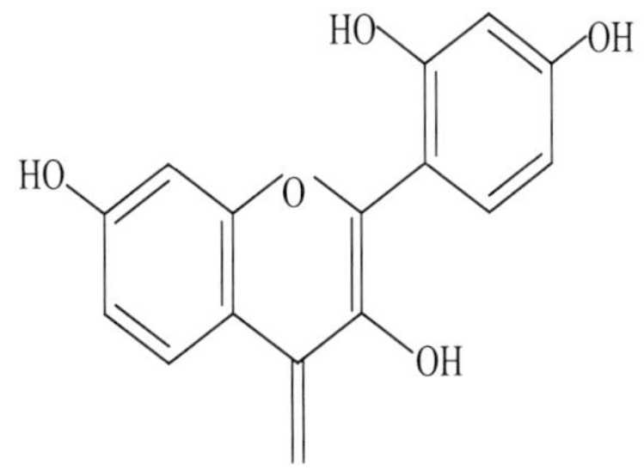

Fig. 5.165 Structure of Morin.

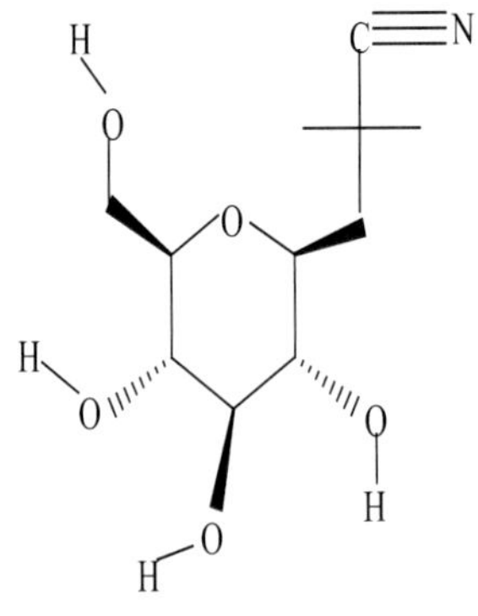

Fig. 5.166 Structure of Phaseolunatin.

Bitter iridoid glycoside, ***angustorine*** is present in *Galipea officinalis*. Iridoid glycosides including ***lamalbide, caryoptoside, alboside A and B*** are present in *Lamium album*. **Boldoglucin** is a glycoside present in *Boldo fragrans*. *Nasturtium officinale* contains a glucoside known as ***gluconastrutin***. ***Punarnavoside*** present in *Boerrhavia diffusa* has antifibriolytic activity. Bark of *Ficus benghalensis* contains hypoglycemic glucoside, ***bengalenoside***.

Barosma betulina contains a yellow, crystalline rhamno-glycoside, ***diosmin***. This is also found in *Conium maculatum*. ***Prulaurisin*** is obtained from *Prunus laurocerasus*. ***Adansonin*** is obtained from *Adansonia digitata*. ***Loganin*** (4–5%) is a glycoside present in *Strychnos nux vomica* (Fig. 5.167). It is identical to ***meliatoside*** isolated from *Menyanthes trifoliata*.

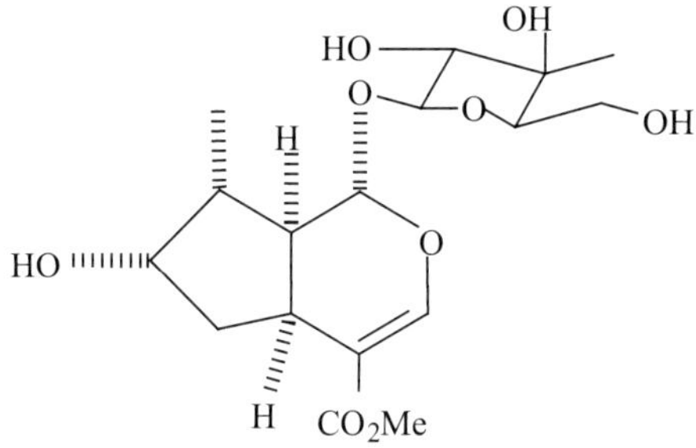

Fig. 5.167 Structure of Loganin.

Cornus officinalis is reported to contain iridoid glycosides including loganin, cornuside, sweroside, and morronoside. An unspecified glycoside, from *Caulophyllum thalictroides,* which has been localized from the drug and then injected into the ears of rabbits, causes a strong local irritation. Applying a solution into the rabbit's eyes, leads to inflammation. Glycoside is supposed to have an oxytoxic effect.

Bark of *Saraca ashoka* is reported to contain powerful oxytotic, a ***phenolic glycoside*** P_2. ***Barringtonin*** is a colorless glucoside, present in *Barringtonia acutangula*. ***Rhaponticin*** (Fig. 5.168), a crystalline glycoside present in *Rheum rhaponticum* has been reported to have estrogenic activity. ***Ziziphin***, a triterpene glycoside, present in *Ziziphus jujube* exhibits taste modifying property. ***Hodulcine*** (hoduloside), dammarene type glycoside is present in leaves of *Hovenia dulcis* (Japanese resin tree).

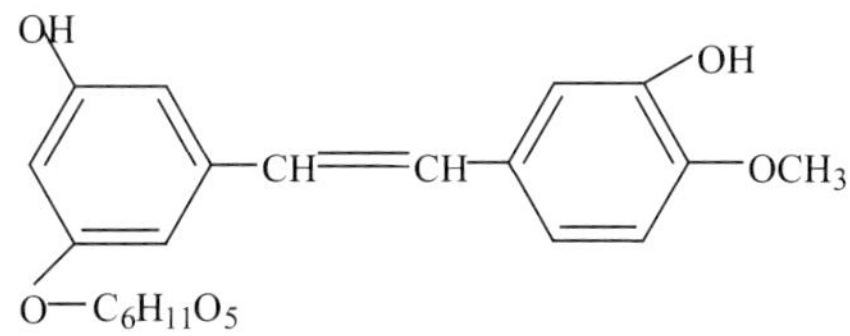

Fig. 5.168 Structure of Rhaponticin.

Gymnemic acid (Fig. 5.169) from *Gymnema sylvestre* is a mixture of at least nine closely related acidic glycosides. Ethyl acetate extract of gymnemic acid mixture has a paralyzing effect on taste glands. The plant also contains tritepene glycosides ***gymnestrogenin*** (Fig. 5.170) and ***gymnemagenin*** (Fig. 5.171).

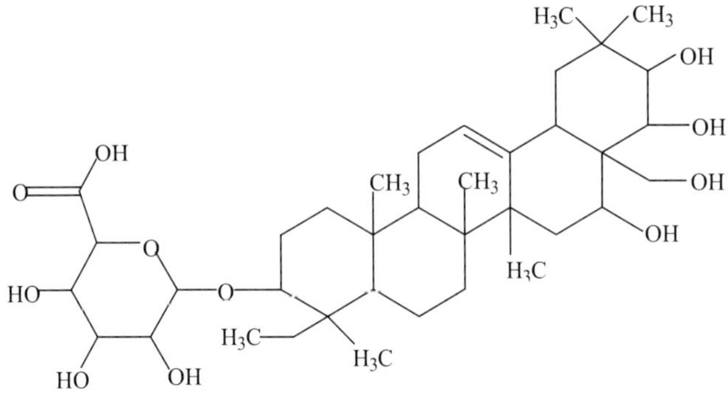

Fig. 5.169 Structure of Gymnemic acid.

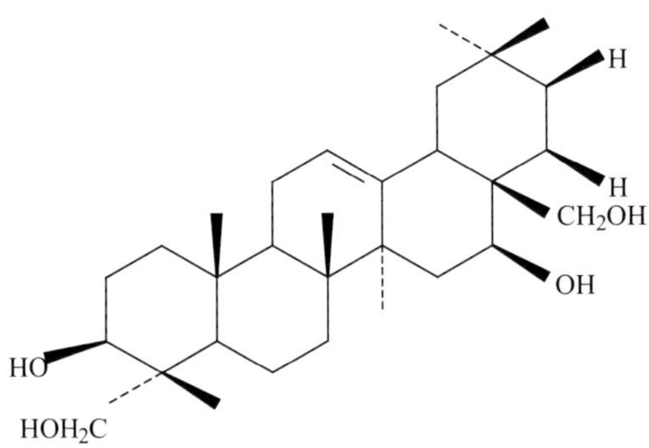

Fig. 5.170 Structure of Gymnestrogenin.

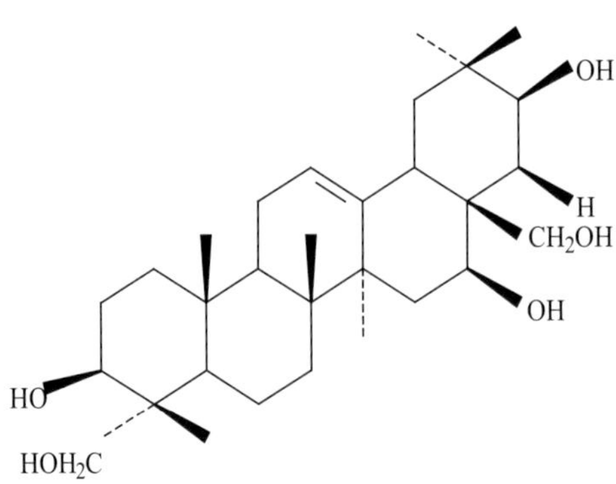

Fig. 5.171 Structure of Gymnemagenin.

Smilax ornata contains a crystalline glucoside, ***sarsasaponin***, yielding by hydrolysis, sarsasapogenin (Fig. 5.172) and dextrose. *Hemidesmus indicus* contains pregnane oligoglycosides ***medidesmine, hemisine, desmisine, denicunine and heminine***.

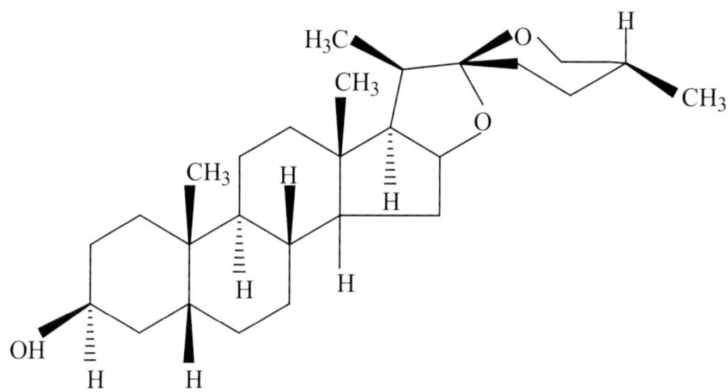

Fig. 5.172 Structure of Sarsasapogenin.

Vitex negundo is reported to contain glycosides, including ***agnuside, mussaenoside, negundoside*** (Fig. 5.173) ***and nishindoside***.

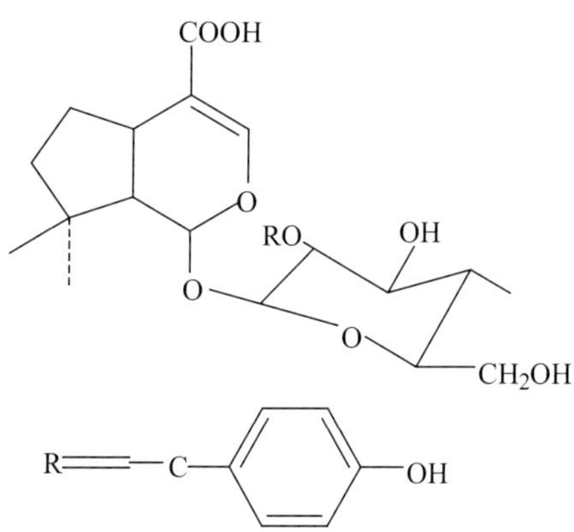

Fig. 5.173 Structure of Negundoside.

Calotropis procera is reported to contain glucosides (***calotropin, calotoxin and uscherin***). Calotropin (Fig. 5.174) has shown an antitumor effect *in vitro* on human epidermoid carcinoma cells of the rhinopharynx.

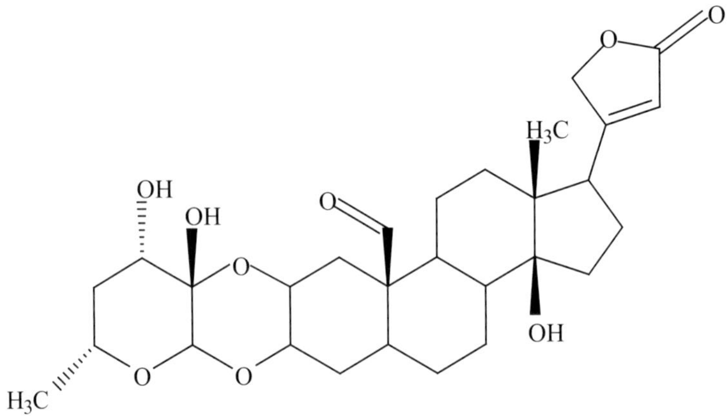

Fig. 5.174 Structure of Calotropin.

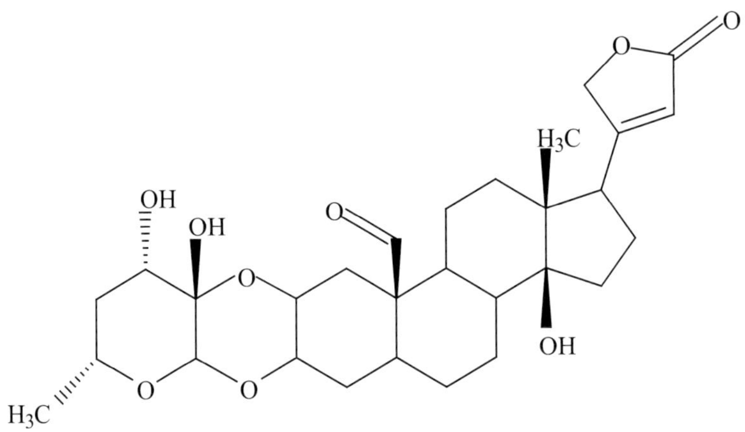

Fig. 5.175 Structure of Calotoxin.

Marsdenia cundurango contains pregnane glycosides A, A_{01}, A_1, B_0, C, C_1, E_0, and E_2. The mixture is known as ***conduragin***. It has potent anticancer property. Iridoid glucosides (***harpagide, harpagoside and procumbine***) from *Harpagophytum procumbens* (Fig. 5.176), native to Australia, have anti-inflamamtory and antirheumatic activity and are used in treatment of rheumatoid arthritis.

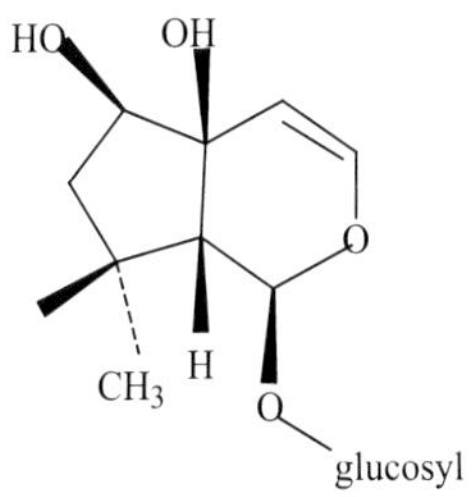

Harpagide

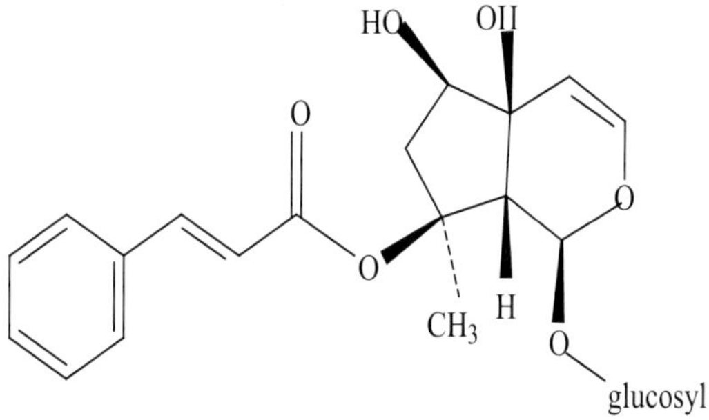

Harpagoside

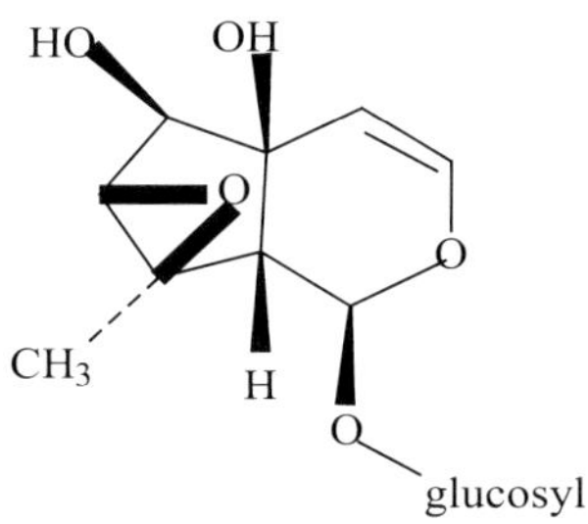

Procumbine

Fig. 5.176 Structure of *Harpagophytum procumbens* Iridoid glucosides.

Arbortristoside A & C obtained from alcoholic extract of *Nyctanthes arbortristis* demonstrated a significant anti-allergic effect comparable to disodium cromoglycate when tested in the experimental models of cutaneous anaphylaxis and mast cell degranulation in rats. ***Swertisioside*** from *Enicostemma littorale* has hypotensive activity. *Eclipta alba* is reported to contain triterpenoid glucosides (***ecliptasaponin C, ecliptasaponin D and eclabatin***).

Rubia cordifolia contains glycosides including ***lucidin, primeveroside, and ruberithric acid***. Glucosides, ***coreopsin, isocoreopsin, sulphurein, monospermoside, and isomonospermoside*** have been reported from *Butea monosperma. Sedum sarmentosum* contains a cyanogenetic glycoside, ***sarmentosin*** which is reported to lower SGPT levels in chronic viral hepatitis.

Glucosides-***sitoisdoside Vll and Vlll*** present in *Withania somnifera* exhibited marked anti-stress and anti-anxiety activity. Sitoindosides, when tested against subacute models of inflammation like granuloma formation and formalin-induced arthritis in rats, exhibited a significant anti-inflammatory activity, comparable to a 5mg/kg dose of hydrocortisone sodium succinate. *Solidago virgaurea* contains pehnolic glycoside ***leiocarposide,*** which is considered to be a diuretic and analgesic.

Oleuropein (Fig. 5.177), secoiridoid glycoside is present in, *Fraxinus excelsior, Olea europaea* and *Ligustrum obtusifoliu.* This compound has hypotensive, antioxidant, antiviral and antimicrobial properties. There is no known toxicity or contraindications for oleuropein.

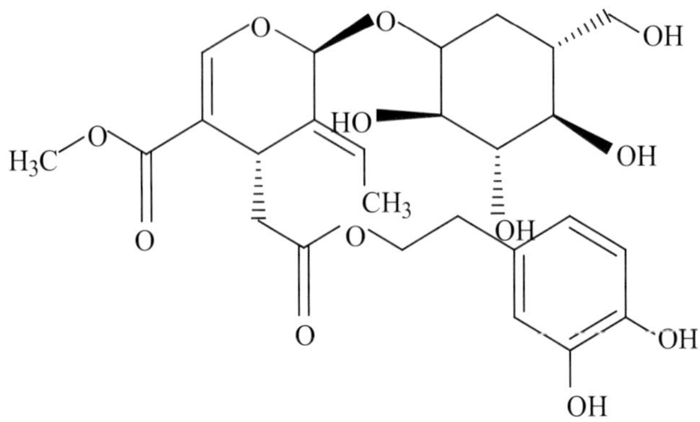

Fig. 5.177 Structure of Oleuropein.

Allamanda cathartica contains iridoid glucosides including ***allamandin, allamandicin*** **and** ***allamdin***. Allamandin (Fig. 5.178) is antileukemic.

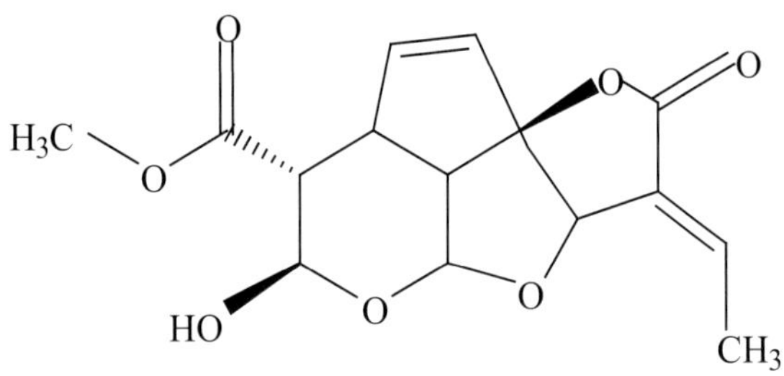

Fig. 5.178 Structure of Allamandin.

Tetrahydroswiertianolin (Fig. 5.179) from *Swertia japonica,* protects rodents against hepatic apoptosis induced by intraperitoneal injection of D-galactosamine (700mg/kg) and lipopolysaccharide (10mg/kg) via blockade of tumor necrosis factor (TNF)-α production at the transcriptional level.

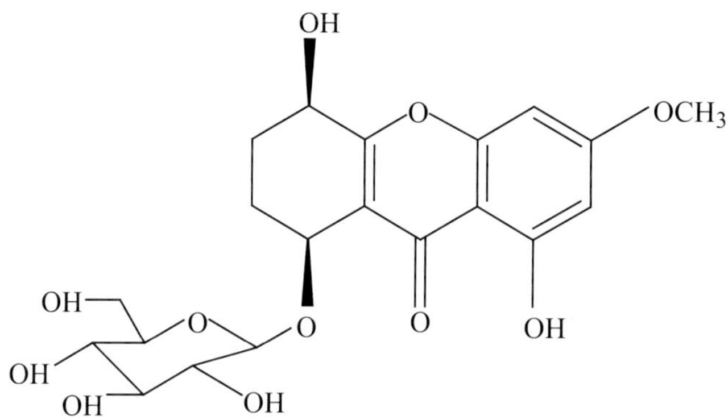

Fig. 5.179 Structure of Tetrahydroswiertianolin.

Ranunculosides (ranunculins)

They are volatile lactones, chemically similar to coumarins. However, they are less stable as compared to coumarins as they are not phenolic substances. On hydrolysis, they produce ranunculine and glucose. ***Ranunculin*** is then converted into unsaturated lactone: ***protoanemonine*** (very unstable compound). Upon drying, protoanemonine gets converted into ***anemonine***, which further converts into anemonic acid by itself. Ranunculin and protoanemonine have medicinal value (Fig. 5.180).

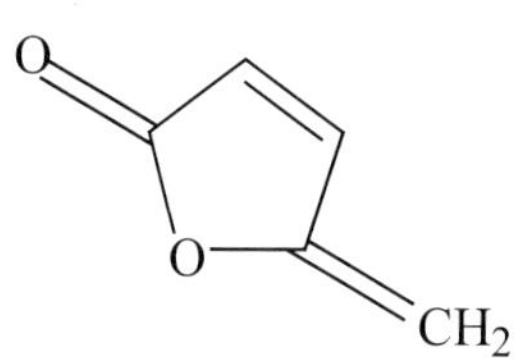

Protoanemonine

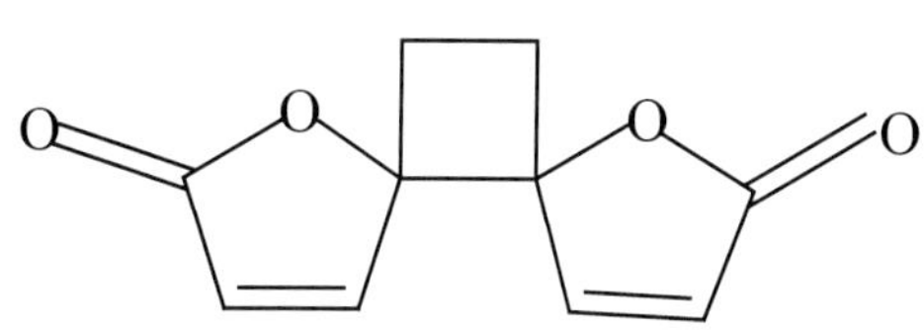

Anemonine

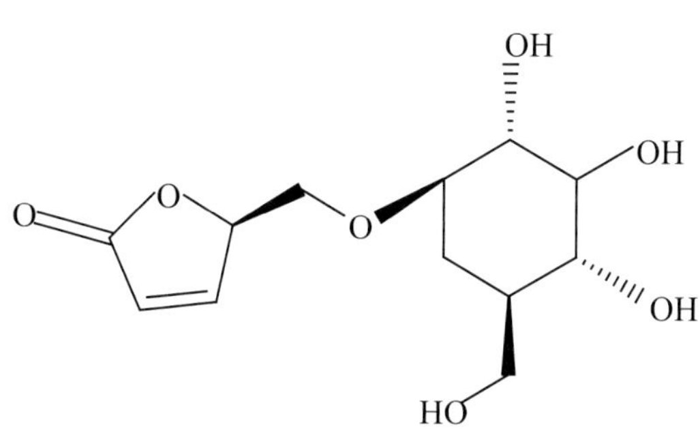

Ranunculin

Fig. 5.180 Structure of Ranunculosides.

Glycosides of Stevia rebaudiana

S. rebaudiana contains eight diterpene glycosides which are sweet principles. They are ***stevioside*** (Fig. 5.181), ***rebaudioside A to E, dulcoside A, and steviolbioside***. Stevioside and rebaudioside A, in pure state, are 300 and 450 times sweeter than sugar. Animal studies have demonstrated the hypoglycmeic and hypotensive activites of stevioside.

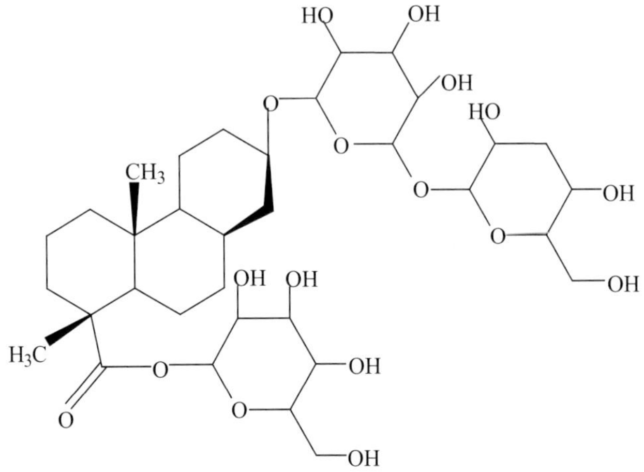

Fig. 5.181 Structure of Stevioside.

Pedicin from the *Fissitigma lanuginosum* inhibits tubulin assembly into microtubules with IC50 value of 300μ M. Other chalcones, ***fissistin,*** and ***isofissistin*** are cytotoxic against KB cells.

Genipin

It is an aglycone from iridoid glucoside, geniposide (Fig. 5.182). It is an excellent natural cross-linker. It blocks ativity of an uncoupling protein 2 (UCP2), which prevents insulin from being produced. LD_{50} i.v. 382 mg/kg in mice.

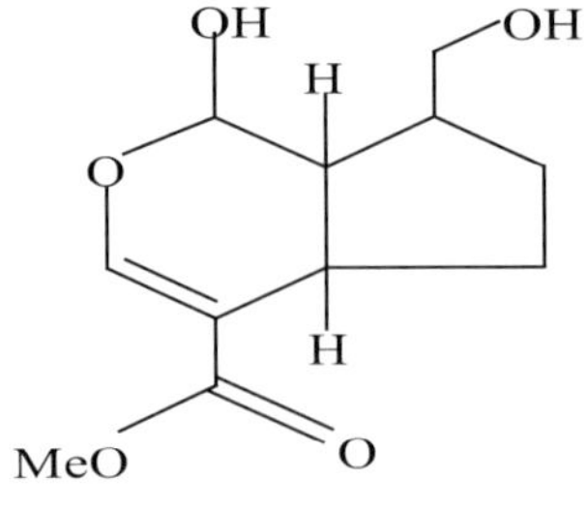

Fig. 5.182 Structure of Genipin.

Arbutin (Fig. 5.183)

It is a bitter neutral crystalline glucoside present in *Arctostaphylos Uva Ursi.* In addition *to arbutin,* ***methylarbutin,*** and ***hydroquinone*** are also present. The leaves typically contain 5–15% *arbutin* and up to 4% *methylarbutin.* Arbutin is metabolized by intestinal bacteria to aglycone hydroquinone, which is absorbed, conjugated to glucuronides and sulfate esters in the liver and excreted renally. Arbutin is rapidly absorbed after oral administration; renal excretion is also rapid, and 75% is excreted within the first 24 h.

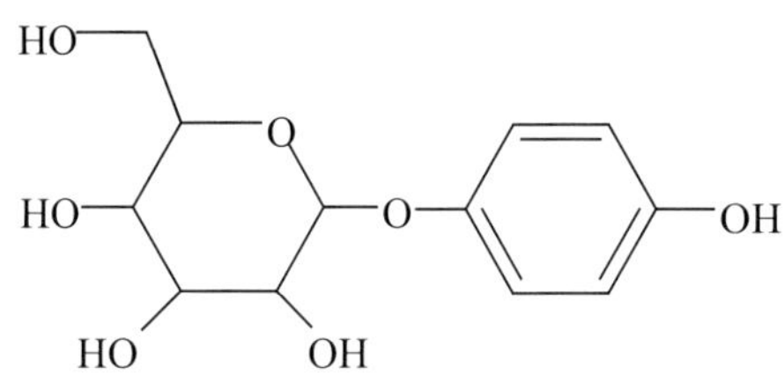

Fig. 5.183 Structure of Arbutin.

Arbutin has demonstrated *in vitro* bacteriostatic activity against *Bacillus subtilis, E. coli, Enterobacter, Helicobacter pylori, Klebsiella, Shigella sonnei,* and *Shigella flexneri.* Arbutin also repressed the expression of a key virulence factor for *Listeria monocytogenes.* Arbutin and its metabolites inhibited the growth of *Pseudomonas aeruginosa, Ureaplasma urealyticum* and *Mycoplasma hominis in vitro.*

Arbutin has been used to inhibit melanogenesis in treating hyperpigmented skin lesions such as chloasma. Hydroquinone is the active ingredient in a number of non-prescription and prescription strength products used to decrease hyperpigmentation.

Aescin (Fig. 5.184)

The principal extract and medicinal constituent of *Aesculus hippocastanum* (horse chestnut) seed is *aescin,* a mixture of triterpenoid saponin glycosides. It can be fractionated into beta-aescin, an easily crystallizable mixture, and alpha-aescin, which is water-soluble.

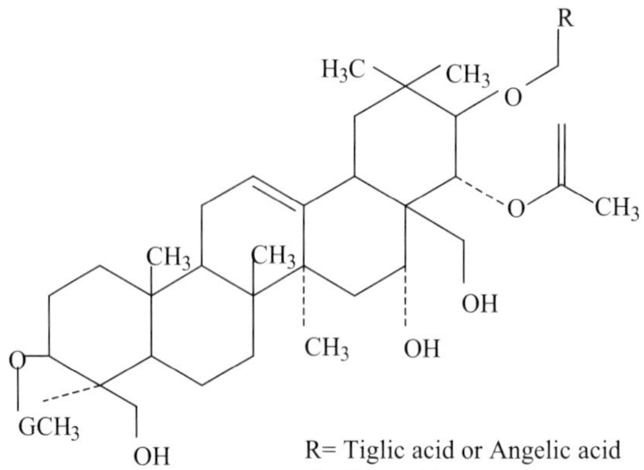

R= Tiglic acid or Angelic acid
G= Glycoside

Fig. 5.184 Structure of Aescin.

Aescin decreases the transcapillary filtration of water and proteins. It has been used to treat a wide variety of inflammatory and edematous conditions, to reduce swelling associated with bruises, fractures, brain trauma, post-operative and post-traumatic soft tissue swelling, and acute thrombophlebitis. Aescin reduces lysosomal enzyme activity by stabilizing lysosomal membranes and limiting enzyme release. Aescin also improves venous tone by enhancing the constricting effect of noradrenaline. Arterial vessels are not constricted and diastolic blood pressure is not affected.

Aescin stimulated the release of prostaglandins (PGF2-α) in the perfused isolated rat lung, and induced prostaglandin- mediated contractions in the isolated portal vein of rats and rabbits.

The use of horse chestnut seed in the treatment of chronic venous insufficiency has also been supported by placebo-controlled studies. A case series involving more than 5,000 patients with chronic venous insufficiency treated with standardized horse chestnut extract reported decreased subjective complaints. Another case series of 1183 patients with chronic venous insufficiency receiving an aescin preparation over 5 mon demonstrated a clear reduction in objective and subjective symptoms with few side effects.

5.4.7 Napthoquinones

These are phenolic compounds distributed in Rubiaceae and Verbenaceae. ***Plumbagin*** (5-hydroxy-2- methyl-1, 4-napthoquinone) from *Plumbago zeylanica* is a typical example of napthoquinone (Fig. 5.185). ***Shikonin*** obtained from *Lithospermum canascens* has hepatoprotective activity. Napthoquinones from Boraginaceae has shown hormone like activity.

Juglone obtained from *Junglans nigra* has sedative activity (Fig. 5.186). Napthoquinones from Boraginaceae has shown hormone like activity. ***Lapachol*** is a typical example of napthoquinone (Fig. 5.187).

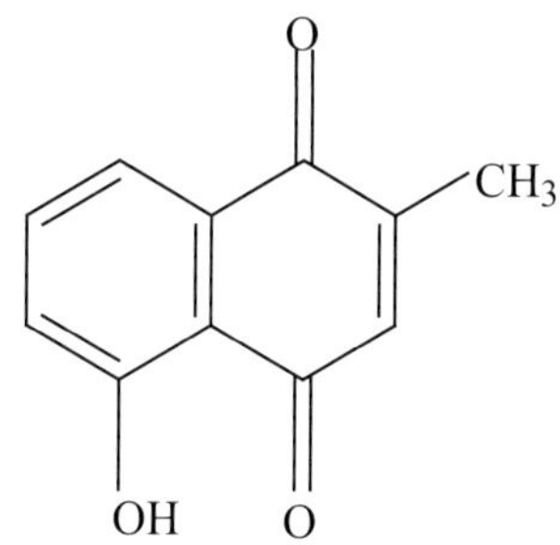

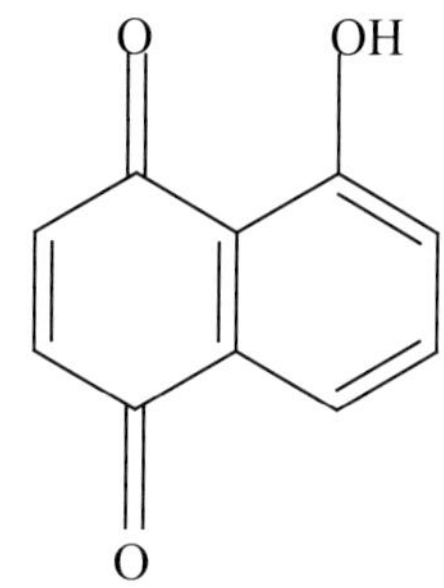

Fig. 5.185 Structure of Plumbagin. **Fig. 5.186** Structure of Juglone.

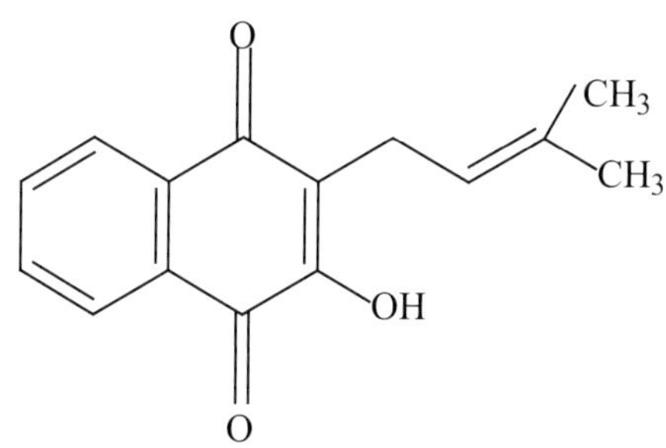

Fig. 5.187 Structure of Lapachol.

5.4.8 Neutral principles

Neutral principles are bodies of unidentified character. They are widely distributed in Asteraceae. Some neutral principles have been studied for structural determination and biological activity. ***Kalmeghin*** from *Andrographis paniculata* and ***picrotoxin*** (Fig. 5.188) from *Anamartia paniculata* are common examples. ***Charantin*** is a non-nitrogenous, neutral pure substance from *Momordica charnatia*. It has hypoglycemic, mild cholinergic blocking and non-specific antispasmodic activities. There is however a controversy as regards to the neutral principle status of charantin. According to modern literature, it is classified as steroid saponin. Charantin resolves into two compounds of well defined melting point.

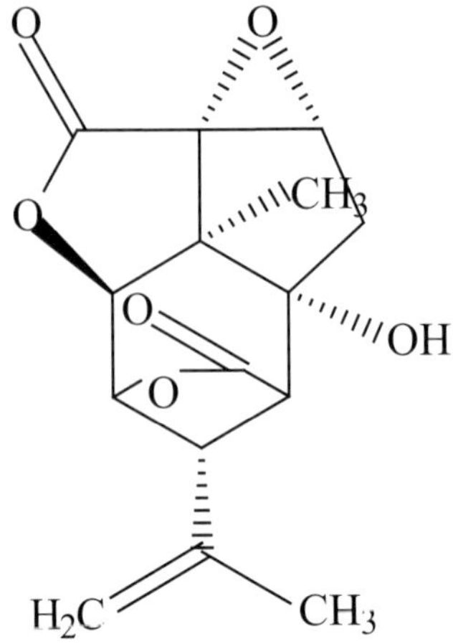

Fig. 5.188 Structure of Picrotoxin.

5.4.9 Phenolics

In the phenols group, cresols and the parent compound itself are the most important compounds as well as thymol. Natural compounds such as pyrocatechol, guaiacol and their derivatives are not toxic. A well-known pyrocatechol derivative is adrenaline. Monohydric phenols provide numerous natural scents (e.g. vanillin, thymol, carvacrol, zingiverone (in ginger), and salicylaldehyde).

Phenolic compounds are widely distributed in plant flora. They constitute an important part of glycosides (phenolic glycosides), flavonoids, and tannins. ***Curcumins*** (Fig. 5.189) are phenolics compounds from *Curcuma longa*. They have antioxidant, anti-inflammatory, anti cancer, and hepatoprotectiveactivities.Thepharmacologicalactivitiesofcucrcuminoids are due to unique molecular structure. In crude extracts of C. *longa* about 70–76% of curcumin is present alongwith 16% demethoxycurcumin and 8% bisdemethoxycurcumin. Tetracurcuminoids (Fig. 5.189) are derived from curcuminoids. This compound is, unlike the yellow curcuminoids.

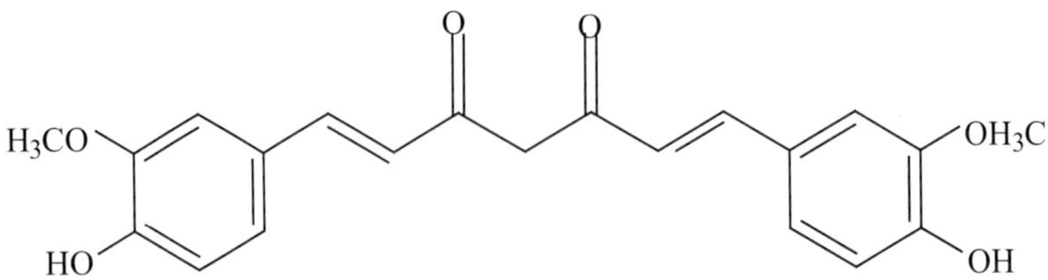

Fig. 5.189 Structure of Curcumin.

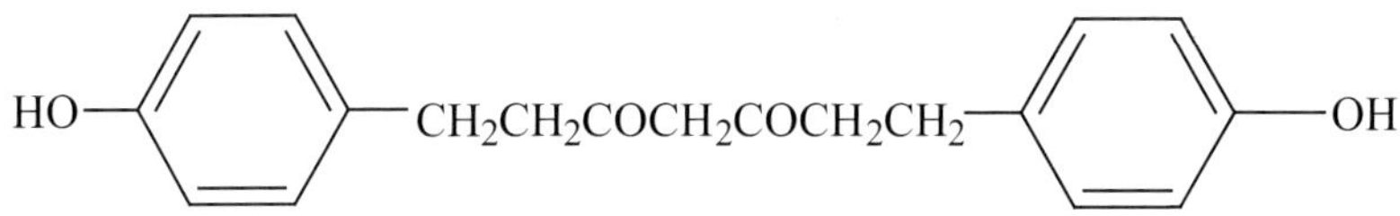

Tetrahydrobisdemethoxycurcumin

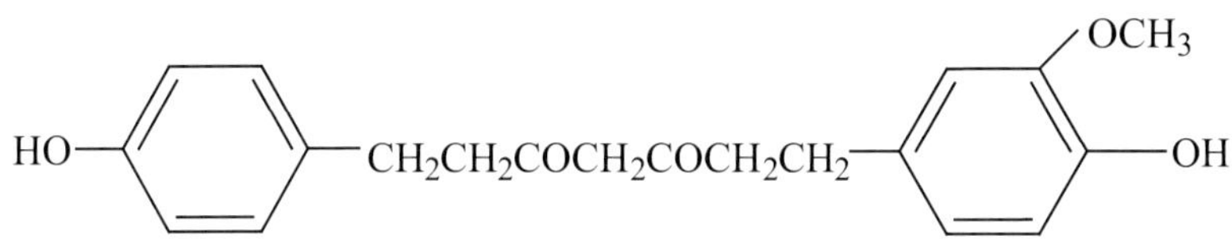

Tetrahydrocurcumin

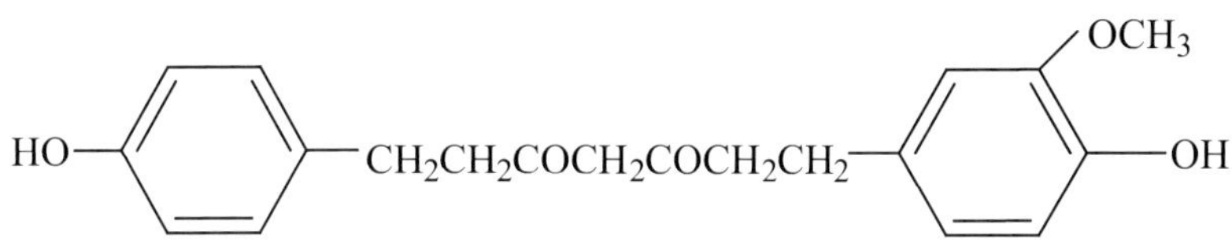

Tetrahydromethoxycurcumin

Fig. 5.190 Structure of Tetracurcuminoids.

Semimyrtucommulone from *Myrtus communis* has antioxidant activity. ***Kneracheline A and B***, from *Knema furfuracea* which inhibits the proliferation of bacteria cultured *in vitro*. ***3-undecylphenol and 3-(8Z-tridecenyl)-phenol*** from *Knema hookeriana*, inhibits the proliferation of *Bursaphelechus xylophilus* cultured *in vitro* with a maximum effective dose of 4.5mg/cotton ball and 20mg/cotton ball, respectively.

Meusone, a phenolic compound is found in *Mesua ferrea*. ***Habenariol*** from *Habenaria repens* has antioxidant activity. ***Bhilawanol and semecaprol*** are reported from *Semecarpus anacardium*. Oil of *Pongamia glabra* contains ***karanjin and pongamol***. ***Bakuchiol***, a phenolic is present in *Psoralia corylifolia* (bawachi) oil (Fig. 5.191).

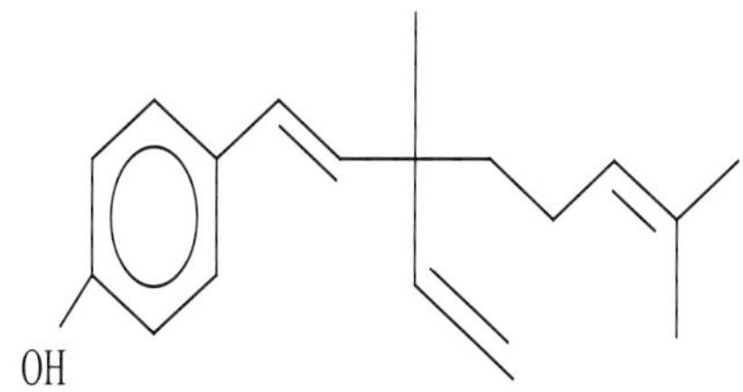

Fig. 5.191 Structure of Bakuchiol.

Flavonoids and tannins (variety of phenolics) have been discussed under respective headings.

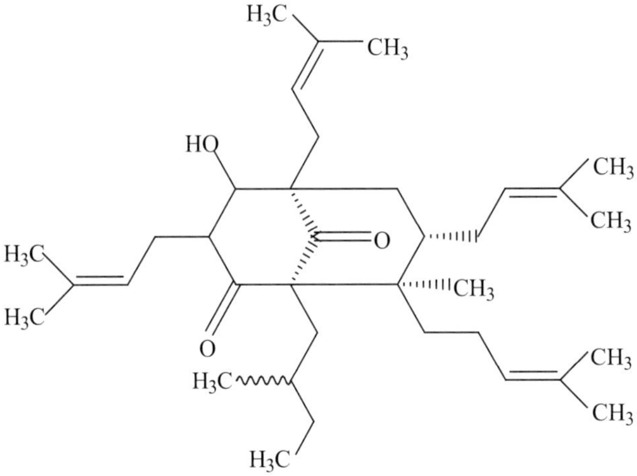

Adhyperforin

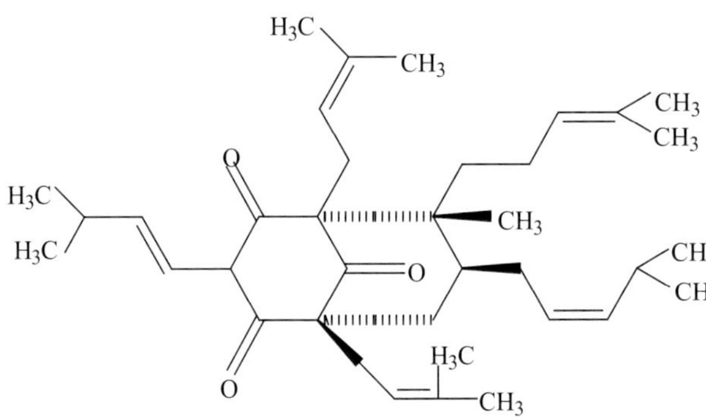

Hyperforin

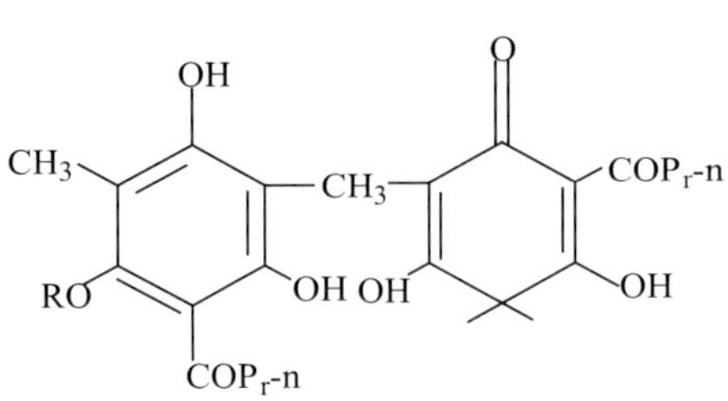

R=CH3

Filicic acid

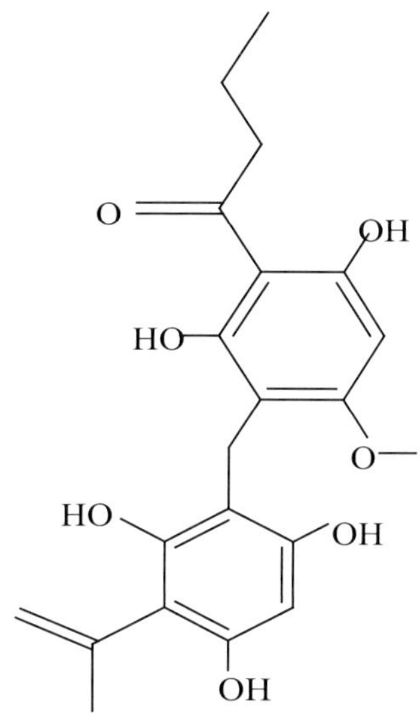

Margaspidin

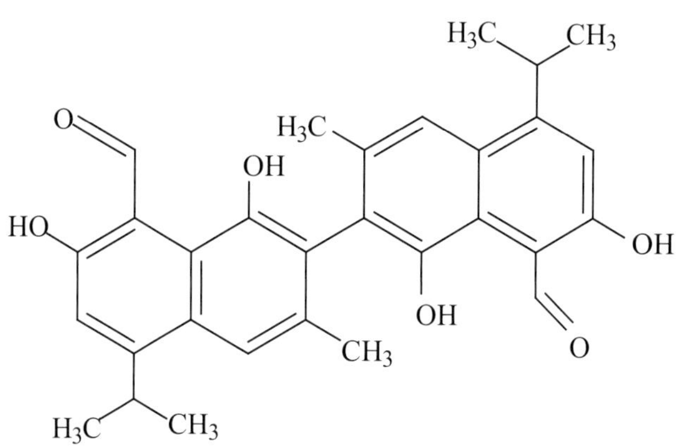

Gossypiol

Fig. 5.192 Structure of common phenolic compounds.

A variety known as polyphenols is found in fruits of some plants. They occur as natural color pigments and are responsible for the color of fruits. Phenols are classified in two groups:

A. Phenolic acids.
B. Flavonones, flavones, xanthones and catechins.

Caffeic acid (Fig. 5.193) is regarded as the most common of phenolic compounds distributed in plant flora. It is produced by hydroxylation of cinnamic acid. Caffeic acid is distributed in *Coffea arabica, Echinacea purpurea* and *Cichorium intybus*. ***Quinic acid*** is the degradation product of caffeic acid. ***Chlorogenic acid*** (Fig. 5.193) is a dark colored pigment and is the most abundant phenolic compound in plants next to caffeic acid. It is known to cause allergic dermatitis among humans. Other examples of caffeic acid derivatives are ***chicoric acid*** (Fig. 5.193), ***dicoffeayl tartaric acid and isocaffeic acid***.

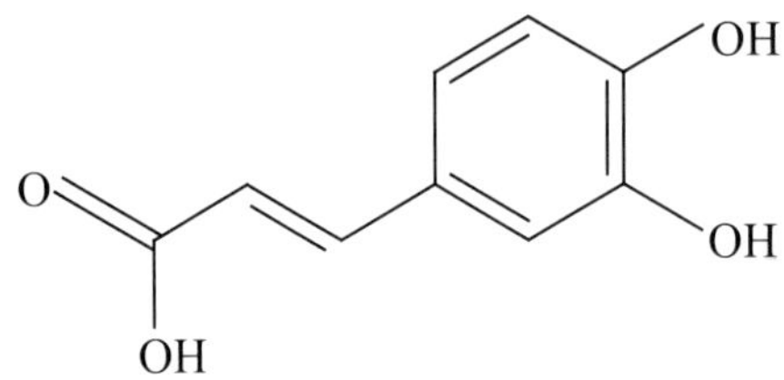

Caffeic acid

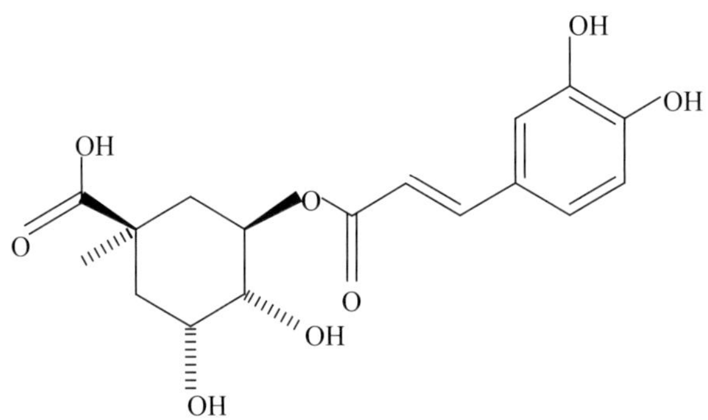

Chlorogenic acid

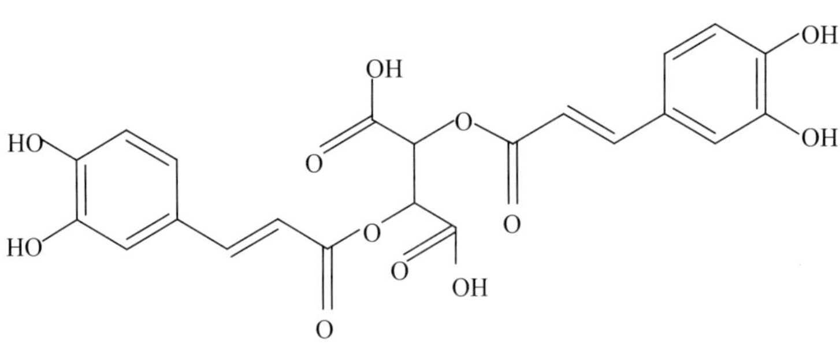

Cichoric acid

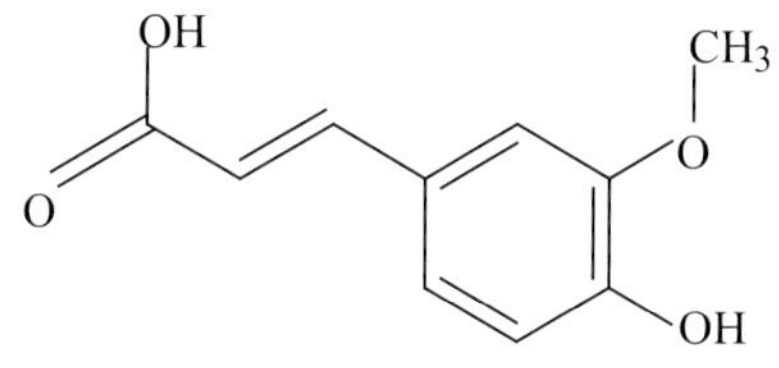

Ferulic acid

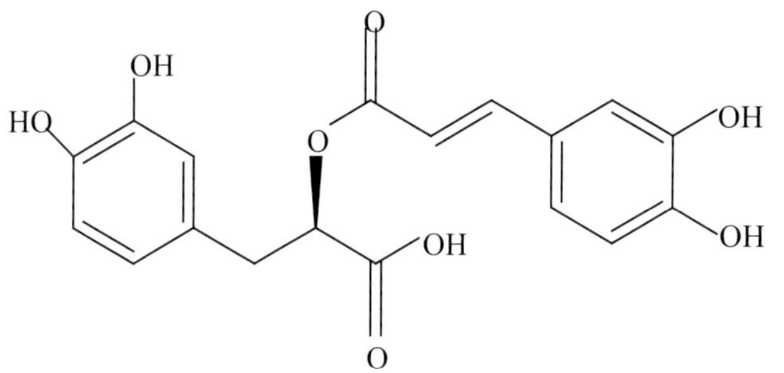

Rosmairinic acid

Fig. 5.193 Structure of phenolic acids.

The administration of phenolic acids to experimental animals produces a wide range of biological manifestations. Caffeic acid is observed as an inhibitor of dopa receptor. Caffeic and oxidized rosmairinic acids inhibit the effects produced by human lutenizing hormone of chorionic origin and by an equal part of lutenizing hormone and follicle stimulating hormone from human hypophysis. An injection of ferulic acid induces the liberation of follicle stimulating hormone from the pituitary and inhibits the liberation of prolactin. ***Ferulic acid*** (Fig. 5.193) also antagonizes the effect of androgens on the prostate of castrated rats.

Table 5.9 Phenolics studied from medicinal plants.

S.No	Name of the phenolic	Source
1.	Adhyperforin	*Hypericum perforatum*
2.	Curcumin	*Curcuma longa*
3.	Filicin or filicic acid	*Dryopteris filix-mas*
4.	Gossypiol	*Gossypium herbaceum*
5.	Hypeforin	*Hypericum perforatum*
6.	Margaspidin	*Dryopteris filix-mas*
7.	Semimyrtucommulone	*Myrtus communis*

The oleoresin obtained from rhizome of ginger contains pungent phenylalkylketones or vanillyl ketones, including ***[6]-gingerol, [6]-paradol, shogaol, and zingerone*** (Fig. 5.194). They have been reported to possess a strong anti-inflammatory activity. Shogaols are artifacts formed during storage, probably created by a dehydration reaction of the gingerols. The ratio of shogaols to gingerols sometimes is taken as an indication of product quality.

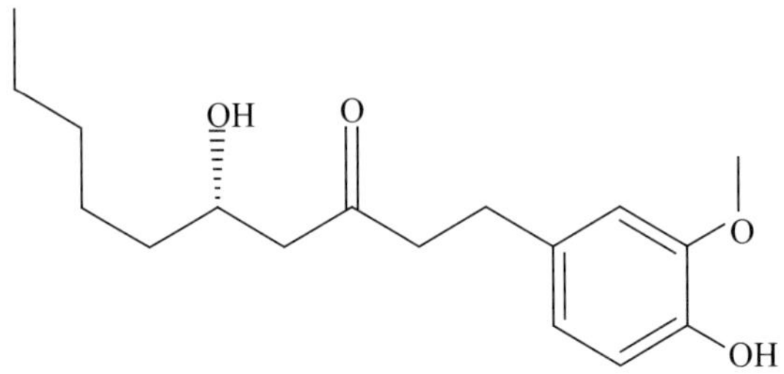

[6]-gingerol

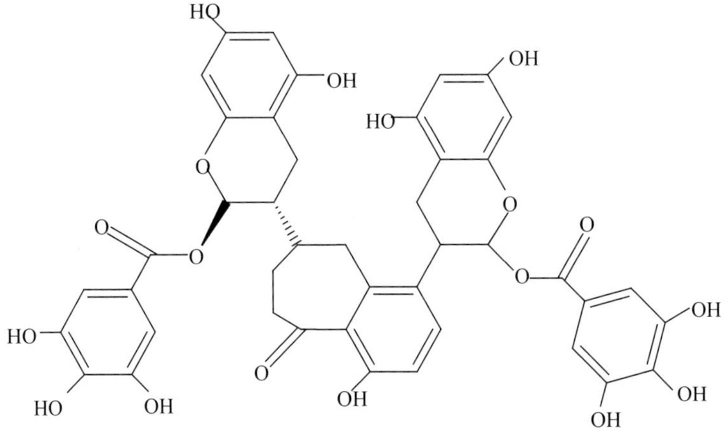

[8]-gingerol

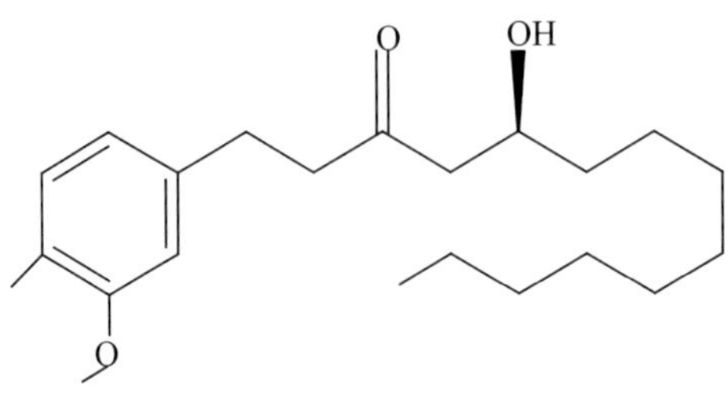

[10]-gingerol

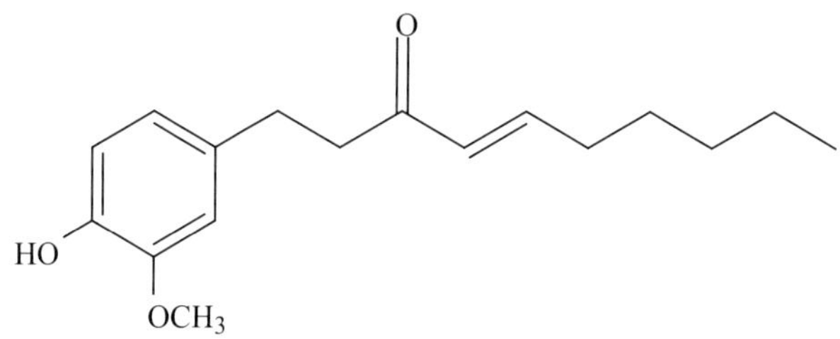

[6]-shogoal

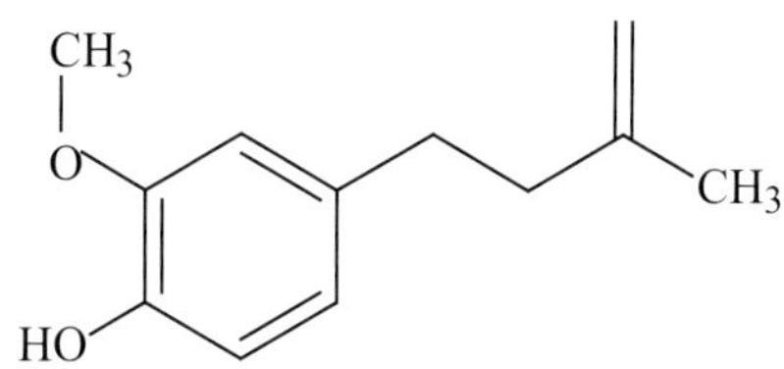

Zingerone

Fig. 5.194 Structure of major constituents of ginger oleoresin.

Gingerol and shogaol (oral; 70–140 mg/kg), are shown to cause vagal stimulation and hence a decrease in both blood pressure and heart rate. 6-gingerol and 6-shogaol suppressed gastric contraction but increased gastrointestinal motility and spontaneous peristaltic activity in laboratory animals.

Gingerol has been found to have a similar structure to acetyl salicylic acid, and these two compounds have similar effects on prostaglandin production. Some of the beneficial medicinal qualities claimed for ginger may stem from zingerone's effectiveness as an antioxidant. Zingerone reacts with free radicals that can cause tissue damage and inflammation.

Topical application of [6]-gingerol or [6]-paradol 30 min prior to 12-O-tetradecanoyl-phorbol-13-acetate (TPA) attenuated the skin papillomagenesis initiated by 7,12-dimethylbenz[a]anthracene in female ICR mice. ***Galanals A and B***, isolated from the flower buds of a Japanese ginger, *Zingiber mioga*, are potent apoptosis inducers in Human T lymphoma Jurkat cells.

Ginkgolic acid, alkylphenol present in *Ginkgo biloba*, has significant anxiolytic activity (Fig. 5.195).

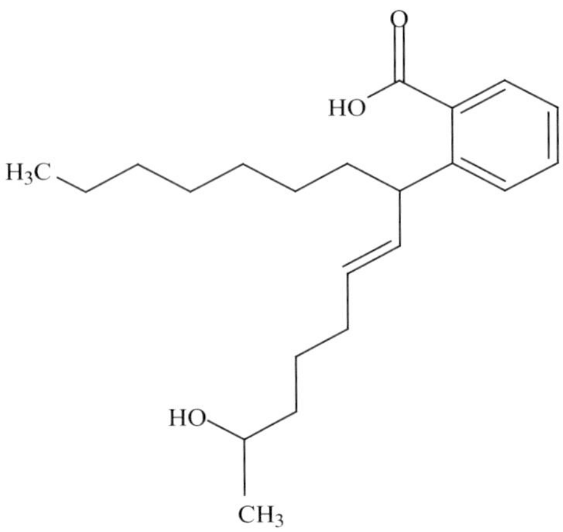

Fig. 5.195 Structure of ginkgolic acid.

Anacardic acid (Fig. 5.196) from the bark of *Ozoroa insignis* inhibits Hep-G2 (human hepatocellular carcinoma), MDA-MB-231 (human mammary adenocarcinoma), and 5637 (human primary bladder carcinoma).

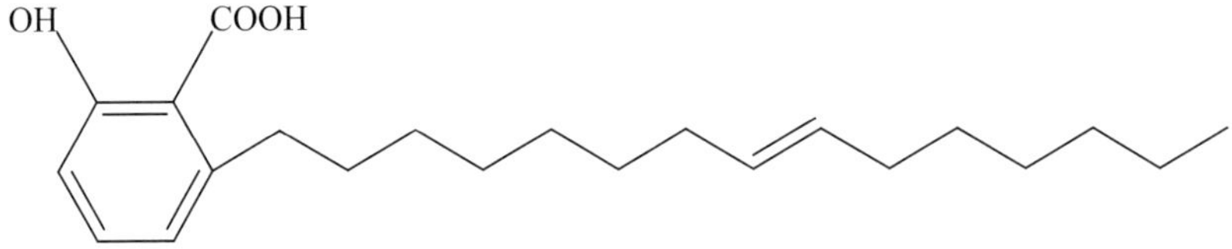

Fig. 5.196 Structure of Anacardic acid.

5.4.10 Resins, oleoresins and gum-resins

Resins are obtained by oxidization of volatile oils. Resins are brittle, non-volatile, solid substances. Sometimes resins are among the products of oxidization of terpenes. The chemical composition of resins is very complex and contains various compounds including acids. Resins are soluble in alkalies, alcohol and insoluble in water. They are obtained from plant exudates and are produced in special ducts.

Resin is obtained from *Ferula asfoetida* (Figs. 5.197, 5.199). Sometimes resins are produced as a result of injury of the plant part. ***Turpthein, convolvulin*** (jalapurgin), and ***jalapin*** are common resins found in Convolvulaceae (Fig. 5.200). Jalapin is mixture of acidic glycosides. Resin present in *Ipomoea hederaceae* (kaladana) is known as ***pharbiticin***.

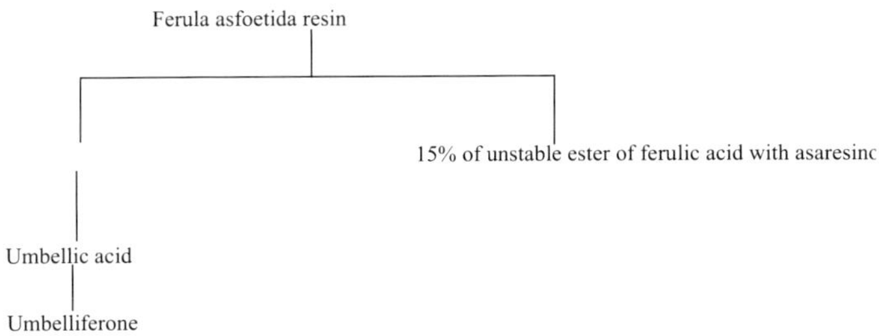

Fig. 5.197 Resin of *Ferula asfoetida.*

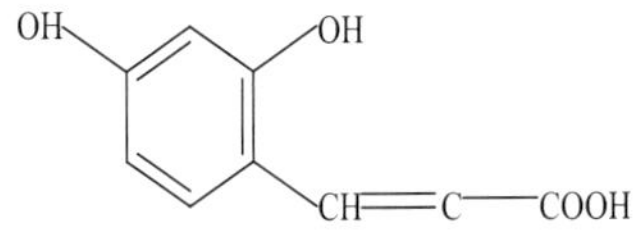

Fig. 5.198 Structure of Umbellic acid.

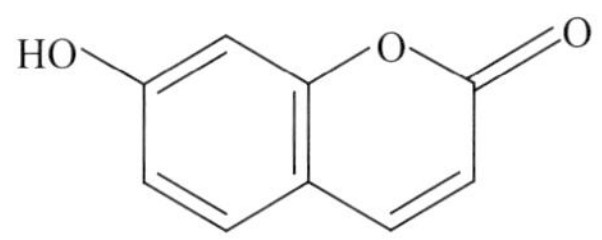

Fig. 5.199 Structure of Umbelliferone.

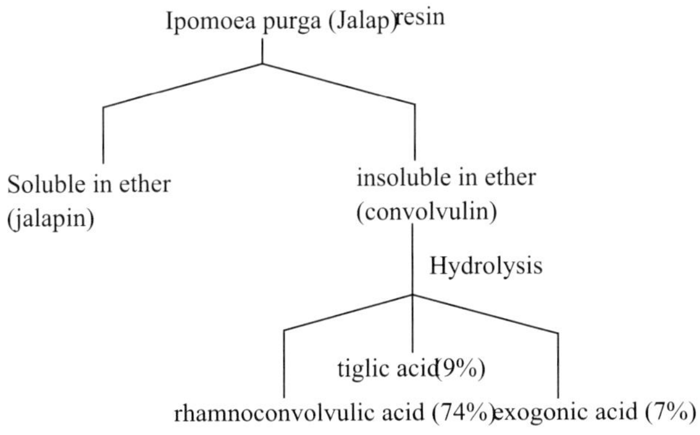

Fig. 5.200 Resin of *Ipomoea jalap.*

Guaiacum officinale contains 20–25% resin. Chief constituents are ***α- and β-guaiaconic acid*** (Fig. 5.201), ***guaiaretic acid, guaiacic acid, guaiacsapnoic acid,*** **and** ***guaiacsaponin.***

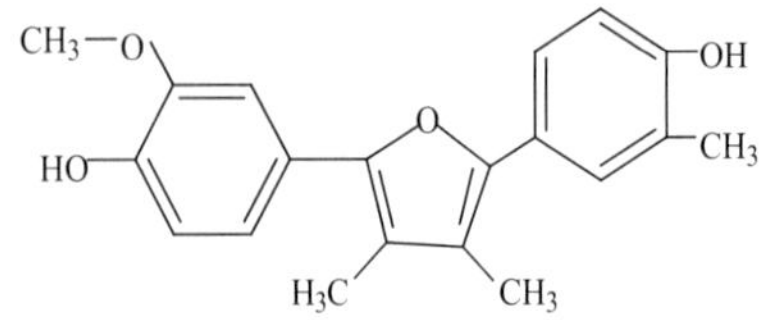

Fig. 5.201 Structure of α- guaiaconic acid.

Oleoresins are natural products of resin mixed with volatile oils. Oleoresin are obtained from traditional Ayurvedic drugs, *Commiphora mukul* and *Commiphora molmol* are typical examples (Figs. 5.202, 5,203). ***Myrobalanin,*** oleoresin obtained from *Terminalia chebula* is soluble in alcohol, ether, petroleum spirit and oil of turpentine.

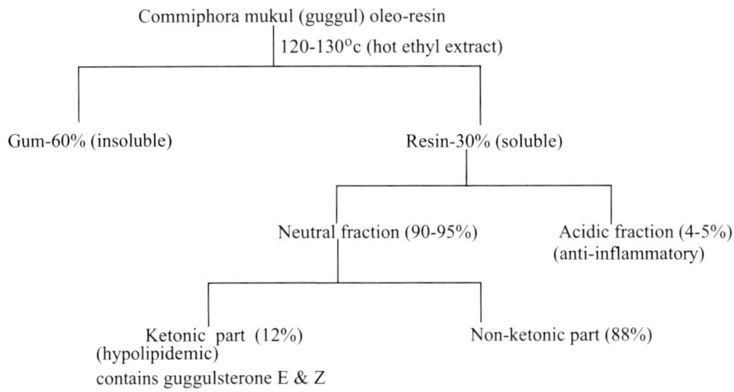

Fig. 5.202 Oleoresin of *Commiphora mukul.*

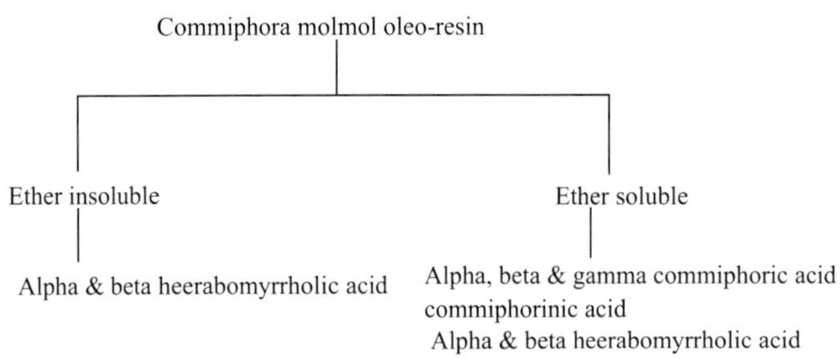

Fig. 5.203 Oleoresin of *Commiphora molmol.*

Toxicodendron (Rhus sp) contains ***urshiols,*** phenolic compounds, found in oleo-resin fraction and are derivatives of pentadecyl catechol or heptadecyl catechol (Fig. 5.204).

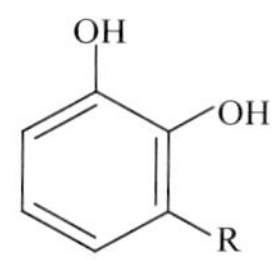

Fig. 5.204 Structure of Urshiol.

Gum-resins are plant exudates and are mixtures of gum and resin and often volatile oils. When gum–resins are dissolved in water, the gum becomes soluble and the resin is kept in suspension. Asafetida is a one such example. Gum resin from *Holarrhena antidysenterica* contains ***lettoresinol-A and lettoresinol-B***.

Balsams are combinations of resins or oleoresins with aromatic acids like benzoic acid or cinnamic acid or both. They are viscous and obtained from the trunk of certain plants.

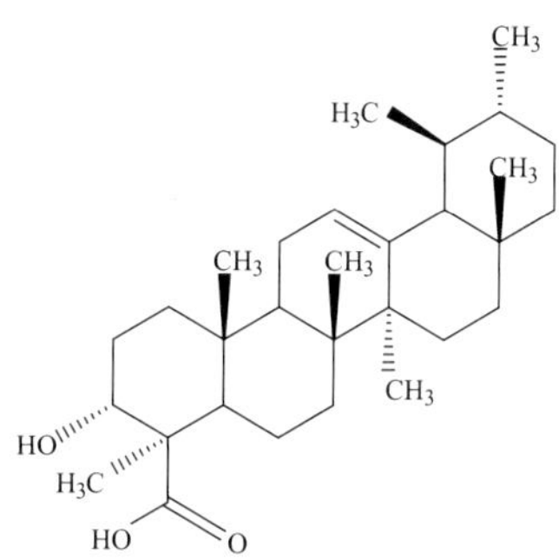

Boswellic acid

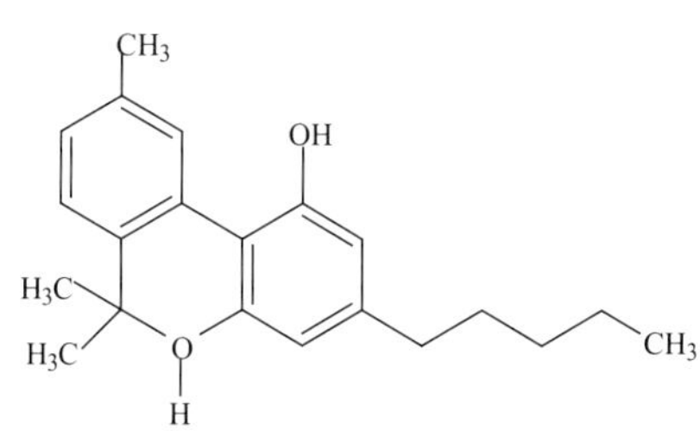

Cannabinol

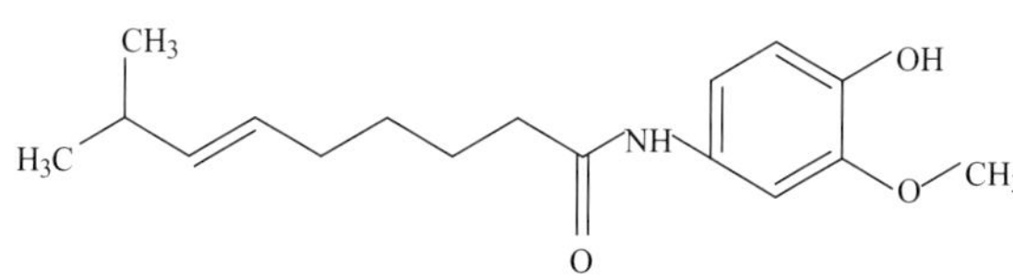

Capsiacin

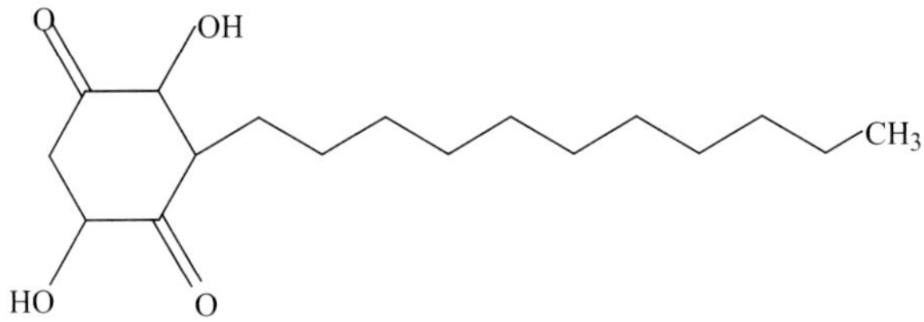

Embelin

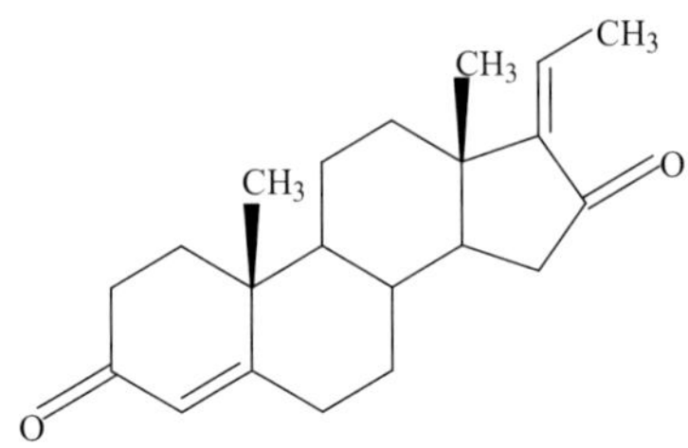

Guggulsterone

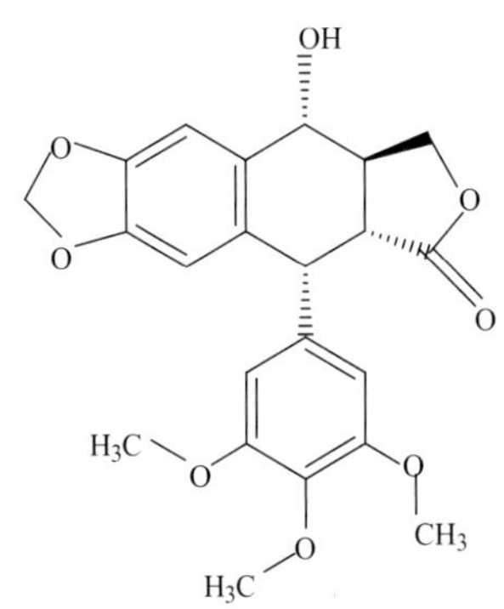

Podophyllin

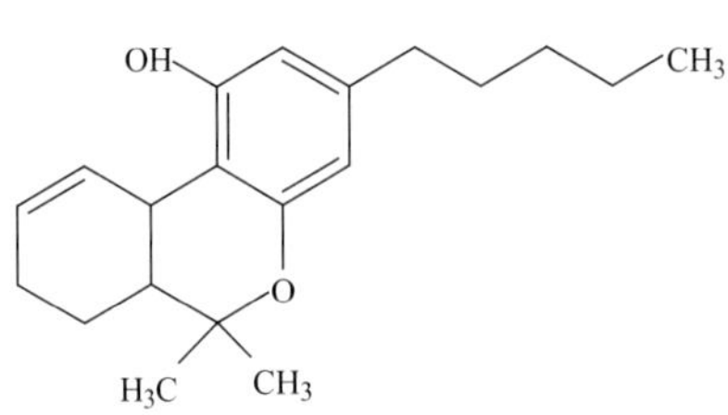

THC

Fig. 5.205 Structure of significant resins, oleoresins and gum-resins.

Table 5.10 Resins, oleoresins & gum-resins from medicinal plants.

S.No	Name of the Resins, oleoresins & gum-resins	Source
1.	Boswellic acid	*Boswellia serrata*
2.	Cannabinol	*Cannabis indica*
3.	Capsiacin	*Capsicum annum*
4.	Chavicin	*Piper nigrum*
5.	Embelin	*Embelia ribes*
6.	Galangin	*Alpinia officinarum*
7.	Gingerol	*Zingiber officinale*
8.	Guggulsterones	*Commiphora mukul*
9.	Podophyllin	*Podophyllum hexandrum*
10.	Turpethin	*Ipomoea turpethum*

5.4.11 Saponins

Saponins are glycosides found in a number of plants. Saponins are regarded as high molecular weight compounds in which, a sugar molecule is combined with triterpene or steroid aglycone. The saponins are classified in two groups:

A. Steroid saponins (Fig. 5.206).
B. Triterpene saponins (Fig. 5.207).

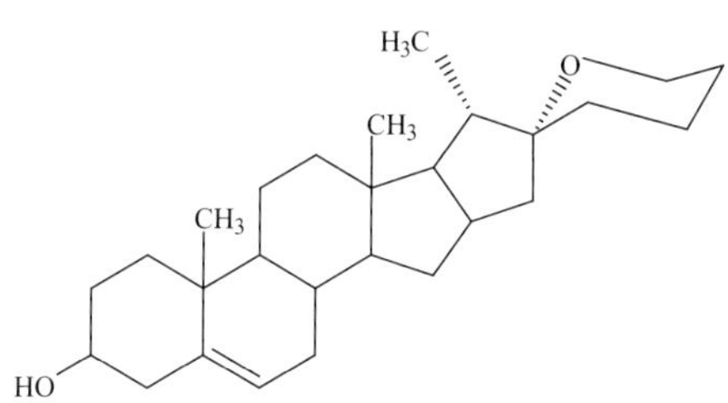

Disogenin

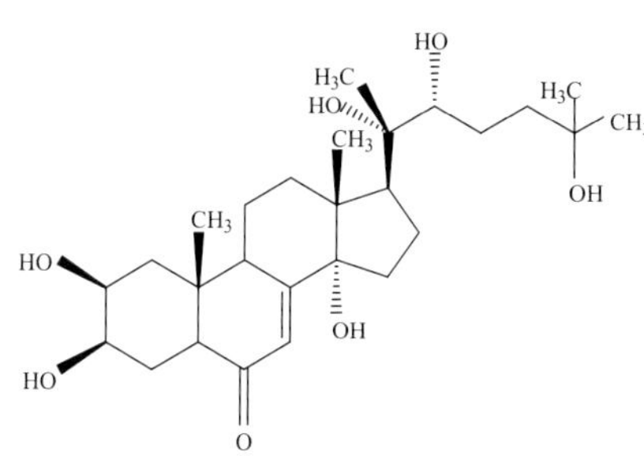

Ecdysterone

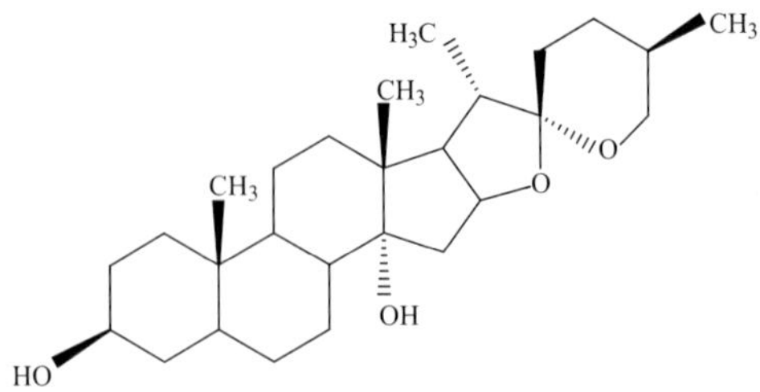

Tigogenin

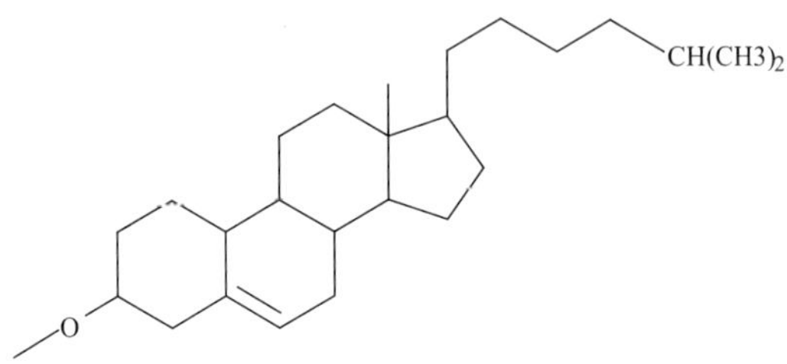

Charantin

Fig. 5.206 Structure of significant steroid saponins.

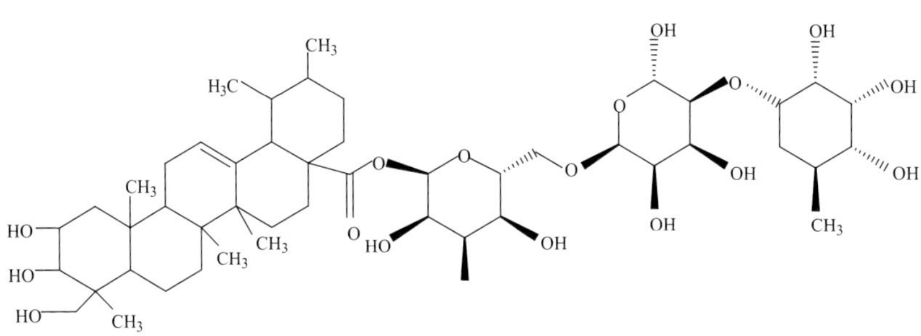

Asiaticoside

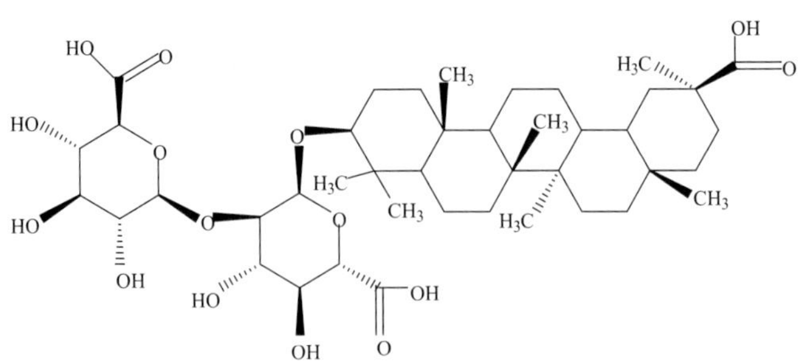

Glycyrrhizin

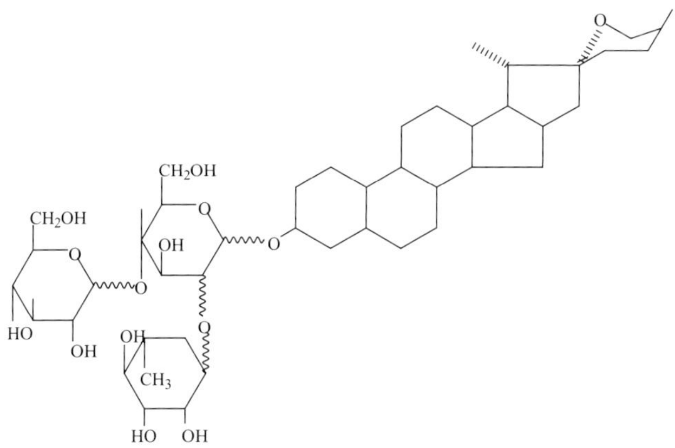

Shatavarin IV

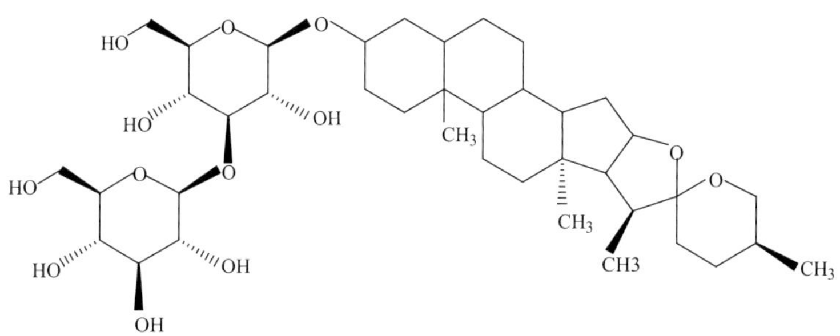

Asparagoside D

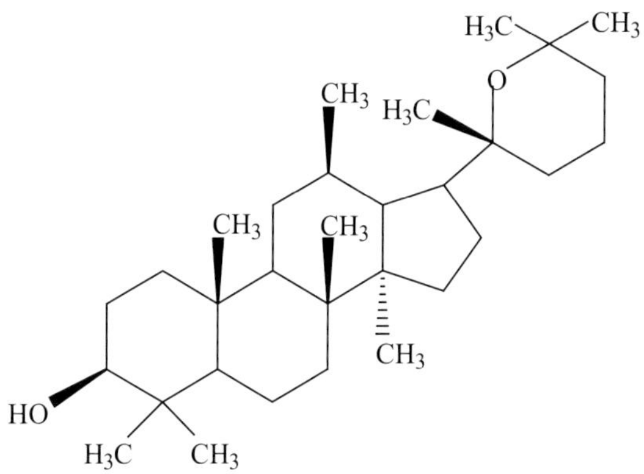

Panaxadiol (Ginseng saponin)

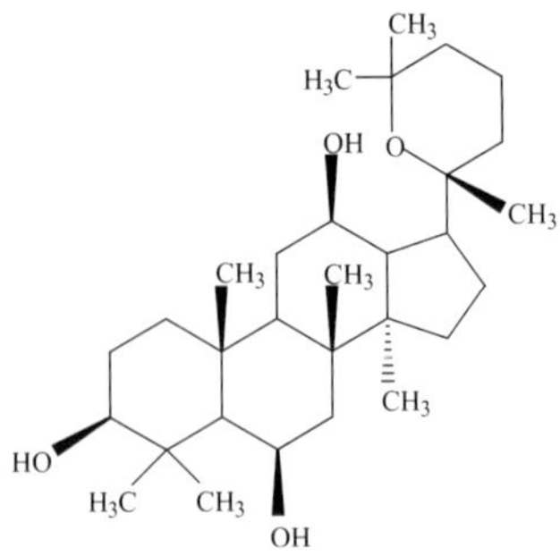

Panaxatriol (Ginseng saponin)

Fig. 5.207 Structure of significant triterpene saponins.

Triterpenoid saponins (***bacoside A, A_1-$_3$, and B***) are present in *Bacopa monneira* (Fig. 5.208). Bacoside A and B are optical isomers. The bacoside-A content is about 2.5–3.0%. Other bacosides have been reported which are dammarane-type saponins known as ***bacopasaponins*** (Fig. 5.209).

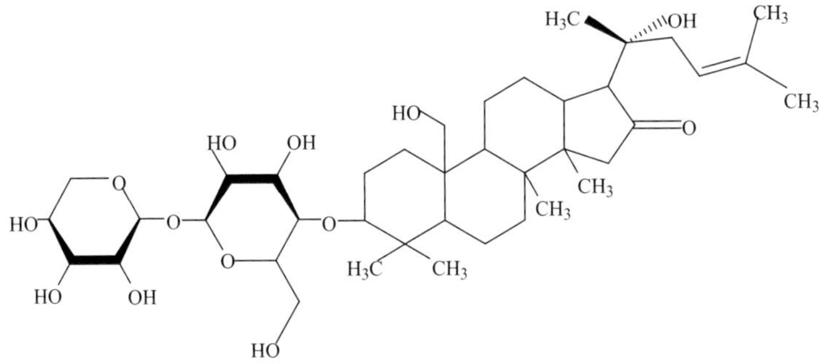

Bacoside A

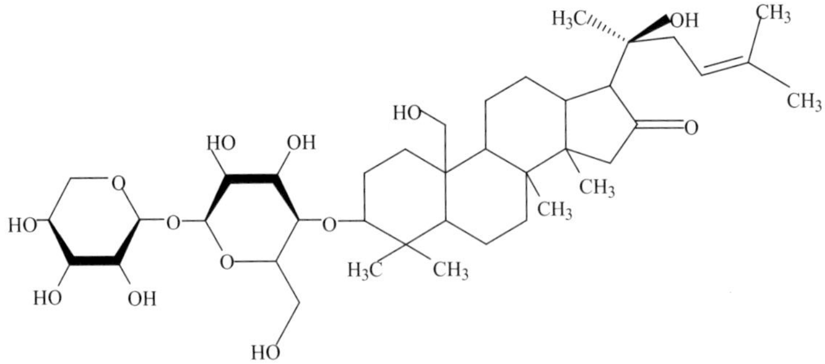

Bacoside B

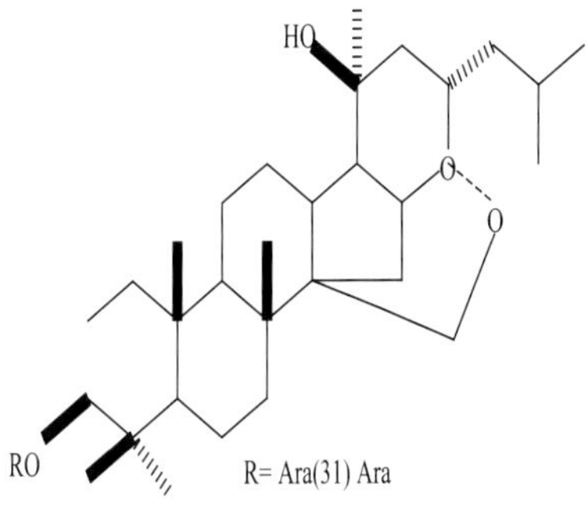

Bacoside A_1

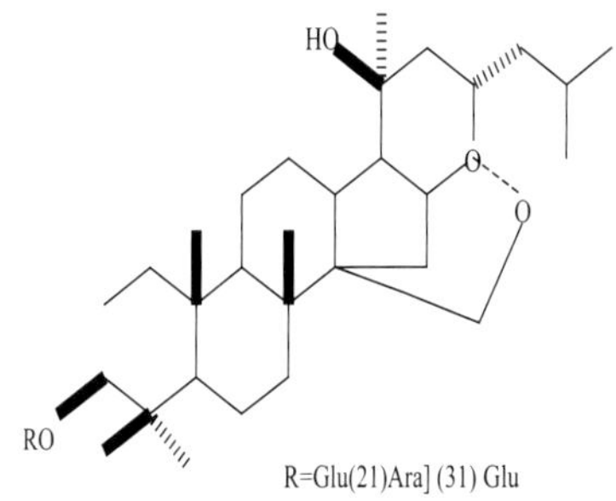

Bacoside A_3

Fig. 5.208 Major triterpenoid saponins of *Bacopa monneira.*

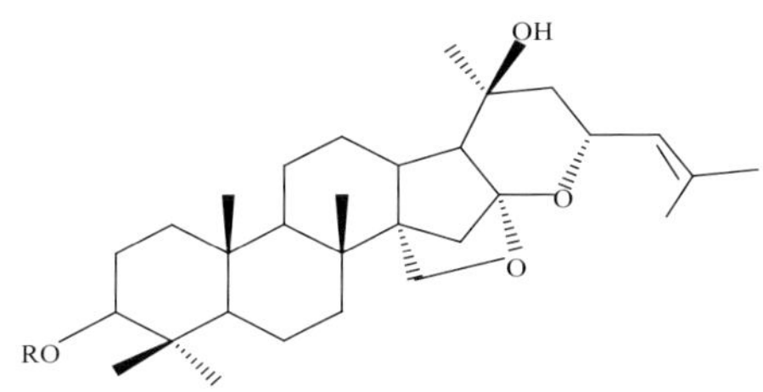

Alpha-L-arabinofuranosyl (1→2)-[6-O-sulphonyl-Beta-D-glucopyranosyl-(1→3)]-Alpha-L-arabinopyranosyl

Bacopaside 1

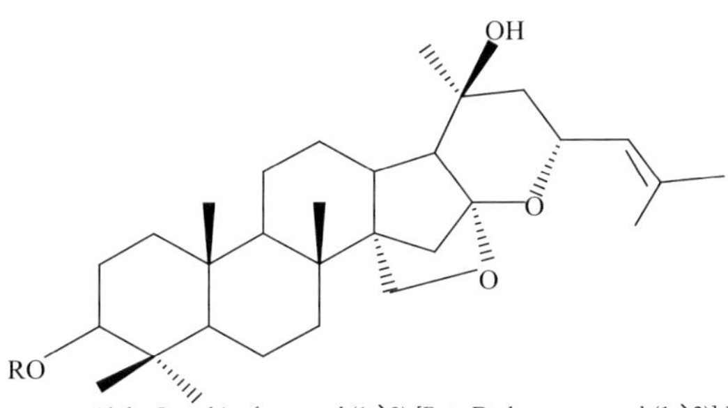

Alpha-L-arabinofuranosyl (1→2)-[Beta-D-glucopyranosyl-(1→3)]-Beta-D-glucopyranosyl Bacopaside 11

Fig. 5.209 Major bacopasaponins of *Bacopa monneira.*

Saponins have a characteristic feature of frothing. The term saponin is derived from *Saponaria vaccaria* (*Quillaja saponaria*), a plant, which abounds in saponins and was once used as soap. *Quillaja saponaria* is known to contain toxic glycosides ***quillajic acid*** (Fig. 5.210) ***and senegin***. Quillajic acid is strenutatory and senegin is toxic. Senegin is also present in *Polygala senega*.

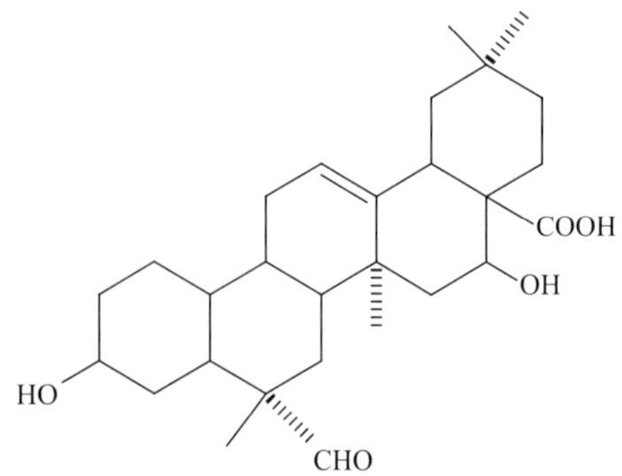

Fig. 5.210 Structure of Quillajic acid.

Saponins are soluble in water and insoluble in ether. Saponins, like glycosides on hydrolysis, give aglycones. *Cholorophytum borivilianum* (safed musli) is reported to contain furostane saponins and spirostane saponins.

Ruscogenins present in *Yucca liliaceae* are used in treating pain and inflammation of arthritis and rheumatism. *Yucca filamentosa* is the other known species, which contains *Ruscogenin* as an active principle. It also contains ***protoyuccoside C, yuccoside B, yuccoside E, yuccoside C,*** aglycones including sarsapogenin, tigogenin. Ruscogenins are also present in *Ruscus aculeatus* and in animal tests; there was an increase in venous tone and an electrolyte-like reaction on the cell wall of the capillaries.

Medicago sativa contains triterpene saponins ***sojasapogenols A-E***. *Anemarrhena asphodeloides* is reported to contain steroid saponins ***sarsapogenin, markogenin and neogitonin.*** A variety of experiments have been able to demonstrate antipyretic and cortisone-like effects for the drug with its steroid saponin content. In addition, inhibitions of platelet aggregation, of Na, K-ATP-ase and of DNA-polymerase were observed. The ***timosaponin A-III*** isolated from the drug reduced the serum levels of a 1-fetoprotein in animal experiments.

Platycodon grandiflorum is reported to contain triterpene saponins including ***platycodin, platycodoside C, aglycone platycodigenin,*** including ***glycosides of polygalic acid, platycogenic acids A to C. Entanin*** saponin from seeds of *Entada pursaetha* is antitumor. It inhibits Walker 256 tumors in rats without death.

Saponins are extremely poisonous, as they cause heamolysis of blood and are known to cause cattle poisoning. Some saponins are important from the therapeutic point of view, as in *Digitalis pupurea,* the presence of saponins is necessary for activity of cardiac glycosides. Recent studies have shown that saponins have hypolipidemic and anticancer activity.

Phenoside, a dammarane type saponin, isolated from extract of *Gynostemma pentaohyllum* (gypenoside) is a novel insulin-releasing substance (Fig. 5.211). *Bupleureum falcatum*, Chinese medicine, contains oleanene saponins called ***saikasaponins a, b$_{1\text{-}4}$, c, d, and f***. Saikasaponins a and d are more active than b_1, b_2 and c. The former pair has stronger haemolytic and anti-inflammatory properties.

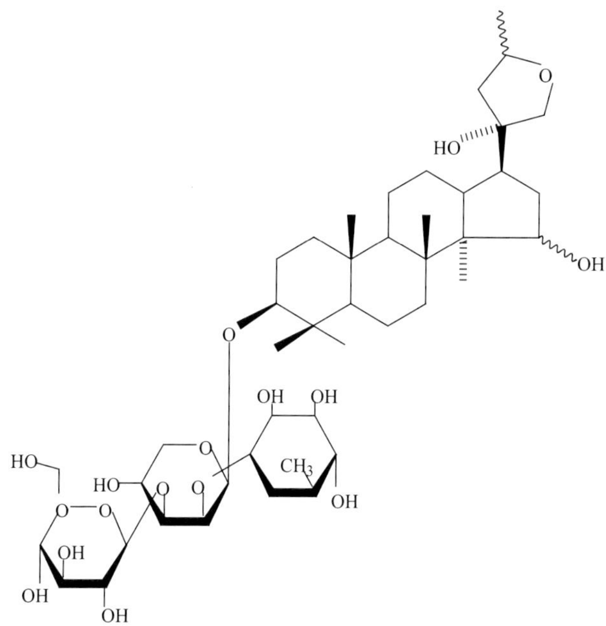

Fig. 5.211 Structure of Phenoside.

Table 5.11 Saponins from medicinal plants.

S.No	Name of the Saponins	Source
1.	Asiaticoside	*Centella asiatica*
2.	Asparagosides	*Asparagus officinalis*
3.	Bacoside	*Bacopa monneira*
4.	Disogenin	*Dioscorea floribunda*
5.	Ecdysterone	*Achyranthes aspera*
6.	Glycyrrhizin	*Glychyrhiza glabra*
7.	Tigogenin	*Costus speciosus*
8.	Uttronin-A	*Solanum nigrum*

5.4.12 Tannins

These are widely distributed in plant flora. They are phenolic compounds of high molecular weight. Tannins are soluble in water and alcohol and are found in the root, bark, stem and outer layers of plant tissue. Tannins have a characteristic feature to tan, i.e. to convert things into leather. The tannins are acidic in reaction and is attributed to the presence of phenolic or carboxylic group. Tannins form complex with proteins, carbohydrates, gelatin and alkaloids.

Tannins are classified in two classes:

1. Hydrolysable tannins.
2. Condensed tannins.

Hydrolysable tannins, upon hydrolysis, produce ***gallic acid*** (Fig. 5.212) and ***ellagic acid*** (Fig. 5.213). Depending on the type of acid produced, the hydrolysable tannins arecalled gallotannins or egallitannins. On heating, they form pyrogallic acid.

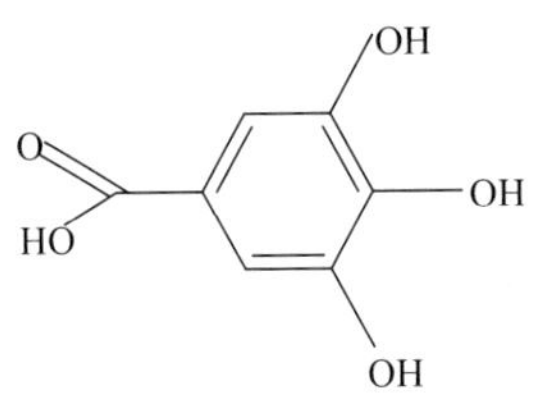

Fig. 5.212 Structure of Gallic Acid.

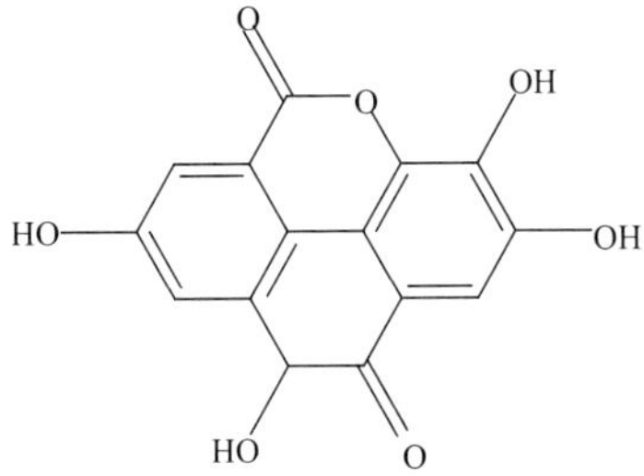

Fig. 5.213 Structure of Ellagic Acid.

Aceritannin (Fig. 5.214) and ***hamemellitannin*** (Fig. 5.215), obtained from *Acer* sp. and *Hamamelis virginiana*, respectively, are examples of gallotannins.

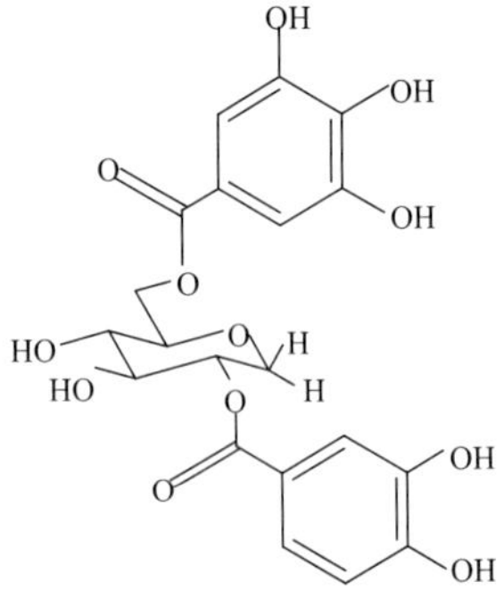

Fig. 5.214 Structure of Aceritannin.

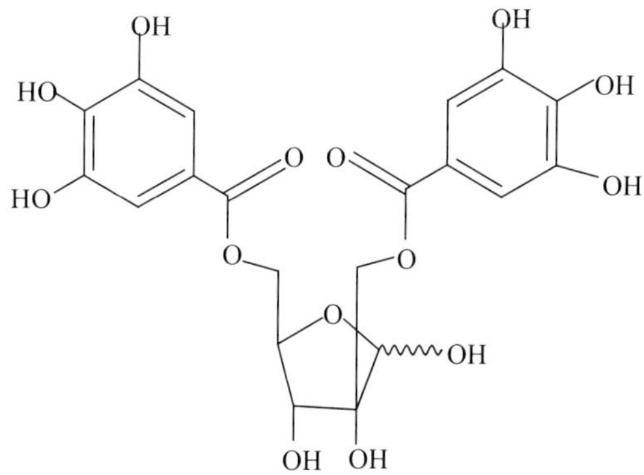

Fig. 5.215 Structure of Hamemellitanin.

Amariin, hydrolysable tannin is found in leaves and seeds of *Colutea nepalensis*. ***Oenthein-B*** is hydrolysable tannin is found in *Oenothera erythrosepala*. They are also found in *Rheum emodi, Aegle marmelos* and *Quercus infectoria*. Ellagic acid isolated from *Terminalia arjuna* has antimutagenic activity. A dichloromethane-methanol extract of *Anisophyllea apetala* provided 3′-methyl-3,4-O,O-methylidene-4′-O-β-D-glucopyranosyl ellagic acid, which showed some DNA damaging effect *in vitro*, and potently inhibited the survival of yeast.

Casuarina stricta has been reported to contain elagitannins including ***casuarinin, casuriin, stachyurin, casuarictin and strictinin. Casuarinin*** (Fig. 5.216) isolated from *Terminalia arjuna* has been studied as antiviral against antiherpes simplex virus type 2 and anticancer activity in non-small cell lung cancerA 549 cell line.

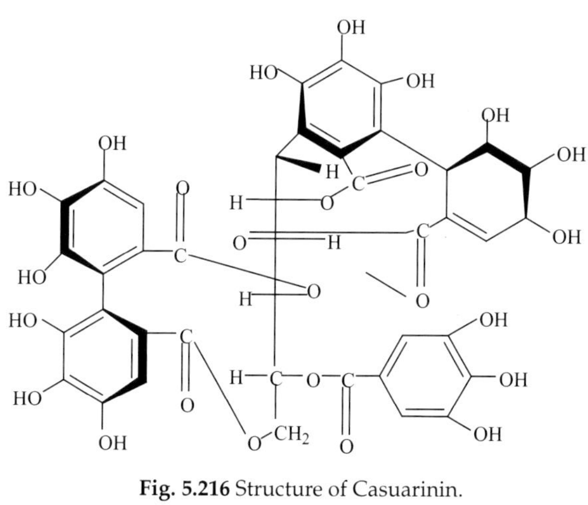

Fig. 5.216 Structure of Casuarinin.

Condensed tannins are formed by condensation of catechin units. These types of tannins are also known as phalbatannins. They are powerful antioxidants and are present in *Vitis vinifera, Acacia catechu, Lawsonia inermis* and *Pinus sylvestris*. Some tannins are potential carcinogenic agents.

Acacia catechu (black catechu) contains 10% ***acacetechin***, which on oxidation yields, catechu tannic acid (Fig. 5.217). *Uncaria gambier* (pale catechu) is good source of tannin. *Pterocarpus marsupium* contains glucosidal tannin, known as ***kinotannic acid*** (70–80%).

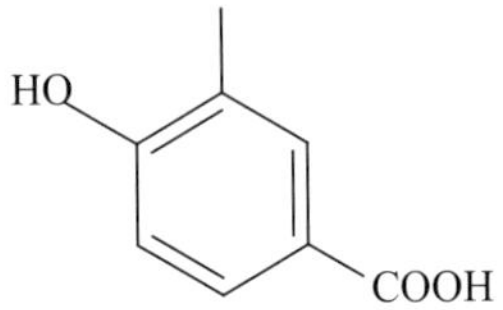

Fig. 5.217 Structure of catechu tannic acid.

Tannins are used as antiseptic and this activity is due to presence of the phenolic group. Tannin-rich medicinal plants are used as healing agents in a number of diseases. In Ayurveda, formulations based on tannin-rich plants have been used for the treatment of diseases like leucorrhoea, rhinnorhoea and diarrhea.

Triphala (combination of fruits of *Terminalia chebula, Terminalia belerica and Emblica officinalis*) is an example similar to that of tannin containing formulation. *Terminalia chebula* contains 30% tannins with ***chebulinic acid*** (Fig. 5.218) and ***chebulagic acid*** (Fig. 5.219), as the principal constituents. *Terminalia belerica* contains 17% tannins. Recently, elagitannin (***peduncugalin***) was isolated from the fruits of *Emblica officinalis* (Fig. 5.220). ***Neochebulic acid*** is minor isomer of chebulic acid. Chebulic acid is closely allied to ellagic acid and reported to be hepatoprotective, chebulinic acid is an antioxidant while chebulagic acid is immunosuppressant, respectively.

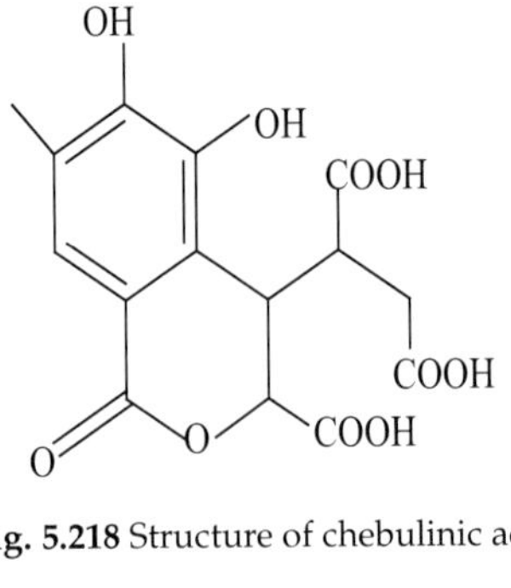

Fig. 5.218 Structure of chebulinic acid.

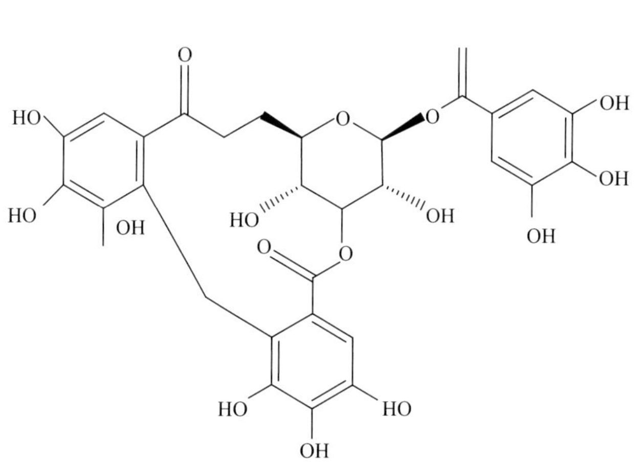

Fig. 5.219 Structure of chebulagic acid.

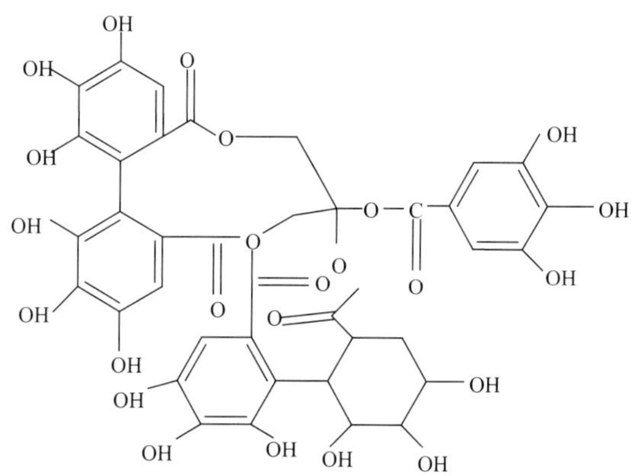

Fig. 5.220 Structure of Peduncugalin.

A group which contains useful detoxifying chemicals is the polyphenols, which can also include many of the tannins. There are numerous references to support the use of these materials. The best known, polyphenols probably come from green tea or *Thea viridis*, a fast growing additive in today's skin care repertoire. Tea contains high levels of tannins or phenolic materials (approx.10–25%) which consist of catechin (flavanol) and gallic acid units and corialgin (Fig. 5.221).

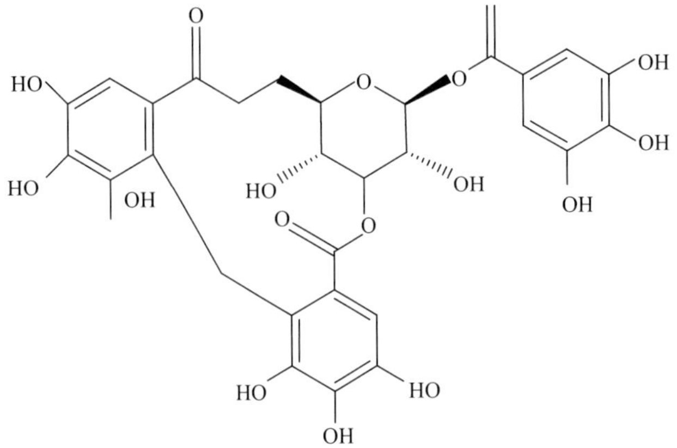

Fig. 5.221 Structure of Corialgin.

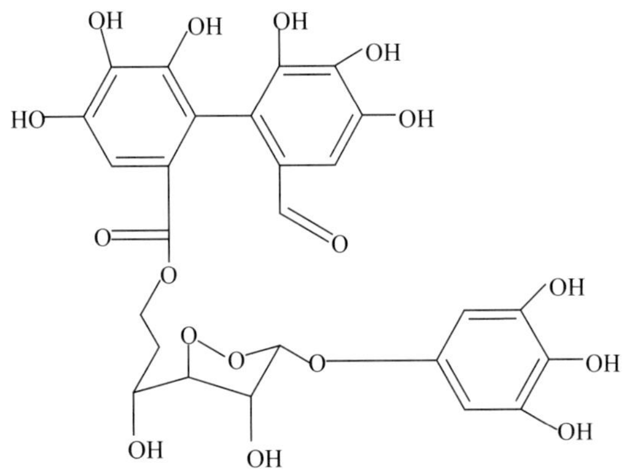

Fig. 5.222 Structure of Isocorialgin.

Family Melastomataceae elaborate an unusual series of hydrolyzable tannin oligomers such as ***nobotannin B*** (Fig. 5.223), which has exhibited anti-Human Immunodeficiency Virus (HIV) property *in vitro*. ***Procyanidin B-2*** (Fig. 5.224) and ***castalagin*** (Fig. 5.225) are ubiquitous in the family and are known to lower blood pressure in spontaneously hypertensive rats dose-dependently through decrease of sympathetic tone.

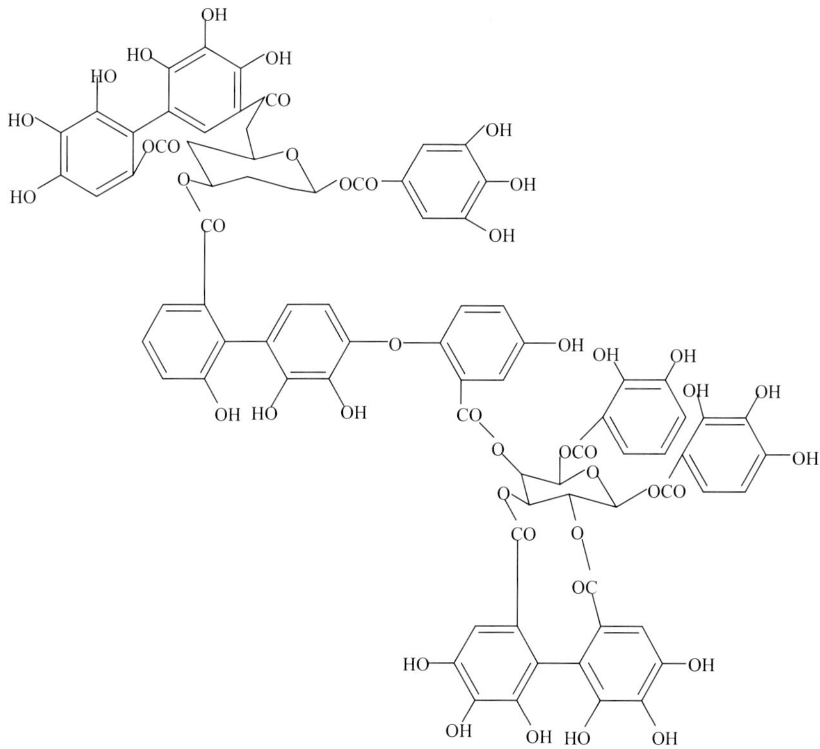

Fig. 5.223 Structure of Nobotannin B.

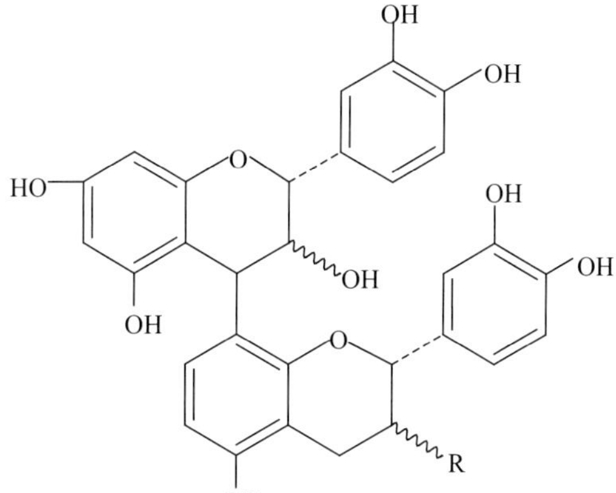

Fig. 5.224 Structure of procyanidin B-2.

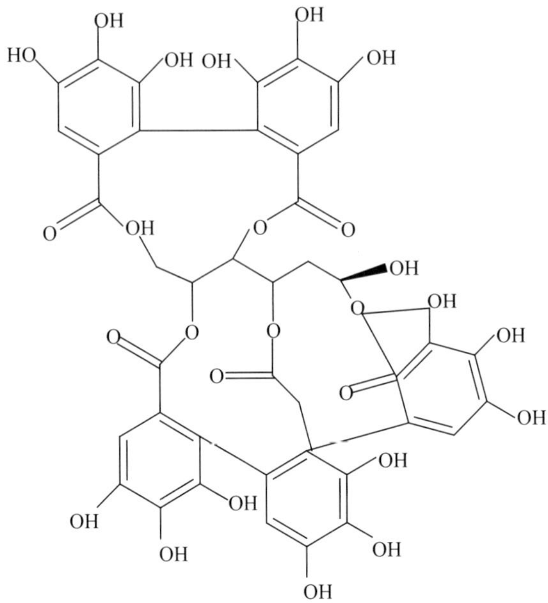

Fig. 5.225 Structure of Castalagin.

Green tea infusion contains intact catechin polyphenols, which give rise to its bitterness and astringency. Six catechin polyphenols have been isolated from green tea; (-)-***epigallocatechin, (-)-epicatechin, (-)-epigallocatechin-3-O-gallate (EGCG), gallocatechin-3-O-gallate (GCG), methyl-epigallocatechin-3-O-gallate, and (-)-epicatechin-3-O-gallate (ECG).*** These substances were tested for their antioxidant activity, and the gallic acid esters EGCG and EGC were found to be the strongest antioxidants, with EGCG being over 200 times more active than vitamin E in an *in vitro* model. In another test, EGCG was more active against fat rancidity (lipid peroxidation) than vitamin C or vitamin E, and also exhibited synergistic action with those vitamins.

Tea contains the polyphenol ECGC, or (-)-epigallocatechin gallate (Fig. 5.225), as major components and it is these in which the researchers are interested. The polyphenols are powerful antioxidants capable of scavenging H_2O_2 and superoxide anions and thus preventing free radical damage to the body. This is a mechanism that has been associated with cancer and other disorders.

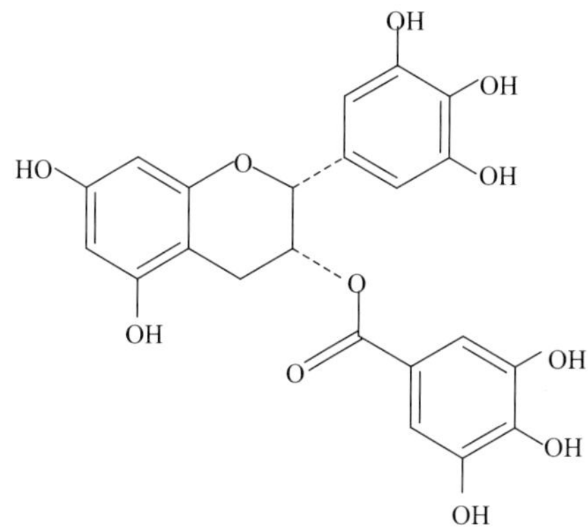

Fig. 5.226 Structure of (-)-epigallocatechin gallate.

ECGC also has an astringent effect and may inhibit cell membrane phosphorylation. The researchers do not know whether the polyphenols inhibit the initiation or the promotion of tumors. Following the oral feeding of a polyphenolic fraction, isolated from green tea, (GTP) in drinking water, an increase in the activities of antioxidant and phase II enzymes in skin, small bowel, liver, and lung of female SKH-1 hairless mice was observed. ***Prodelphinidin B-2 3,O-gallate*** (Fig. 5.227), a proanthocyanidin gallate has been demonstrated for anti-proliferative activity in human non-small cell lung cancer A549 cells by G0/G1 arrest and apoptosis induction.

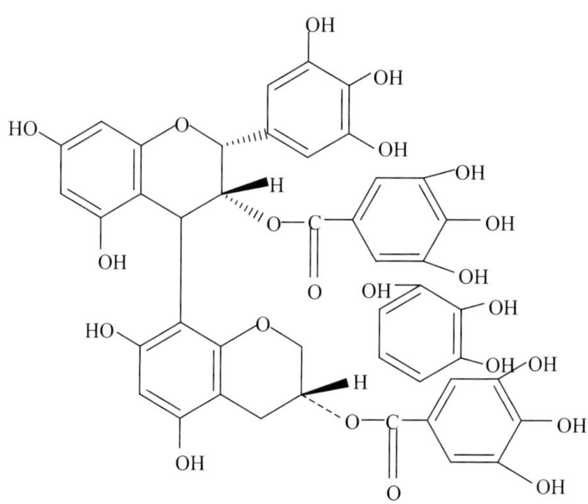

Fig. 5.227 Structure of Prodelphinidin B-2 3,*O*-gallate.

Prodelphinidin B-2 3,3′-di-O-gallate (Fig. 5.228), a proanthocyanidin gallate from *Myrica rubra* has been demonstrated to arrest cell cycle at

G0/G1 phase and induce apoptosis in MCF-7 and A549 cells by p21/WAF1 and Fas/APO-1/Fas ligand pathway, respectively.

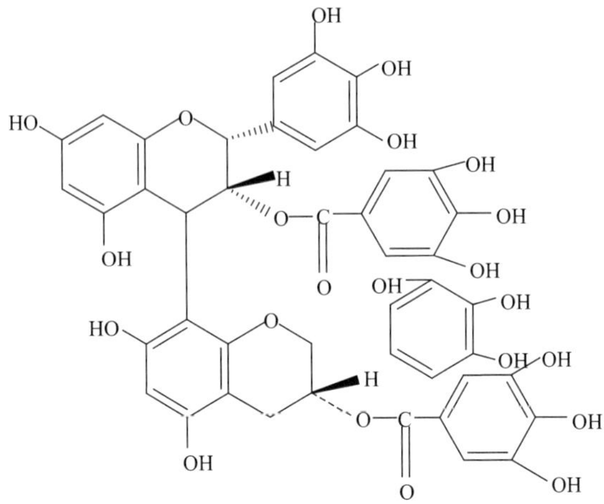

Fig. 5.228 Structure of Prodelphinidin B-2 3,3′-di-*O*-gallate.

Putranjivain A (Fig. 5.229), isolated from whole plant of *Euphorbia jolkini,* has been demonstrated for anti-proliferative activity in human breast adenocarcinoma by blocking cells by G0/G1 arrest and apoptosis induction.

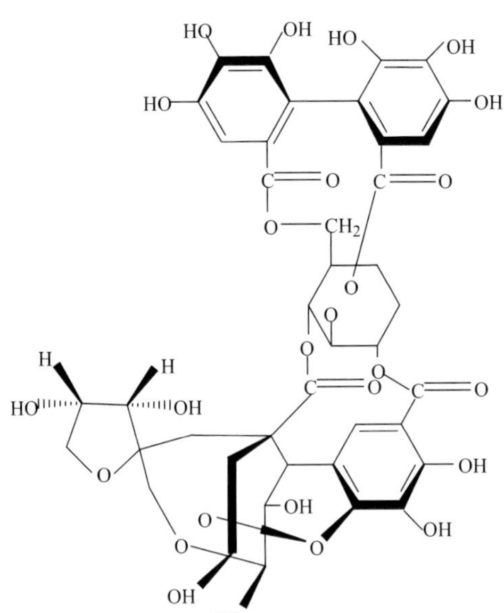

Fig. 5.229 Structure of Putranjivain A.

Punicalagin are tannins, large polyphenol compounds found in *Punica granatum*. They are the largest molecules found intact in rat plasma after oral ingestion. They were found to show no toxic effects in rats that were given a 6 diet of punicalagin for 37 d. The compound has *in vitro* antiproliferative, antioxidant and apoptotic activities.

Methyl gallate (Fig. 5.230) from *Toona sinensis* protects DNA in canine Cocker Spaniel kidney cell line (MDCK) cells against hydrogen peroxide-induced oxidative stress.

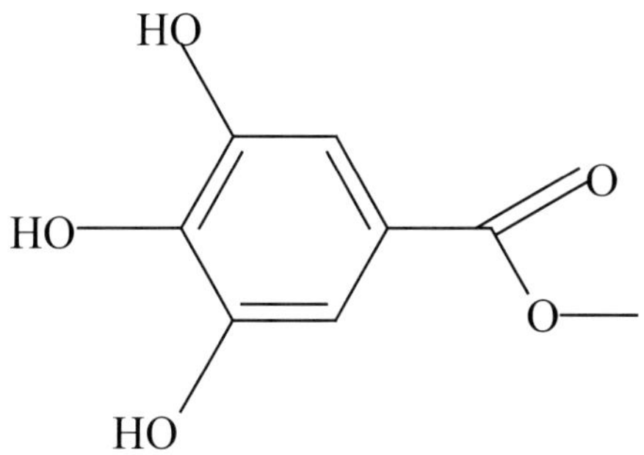

Fig. 5.230 Structure of methyl gallate.

Ellagitannins ***oenothein A*** and ***eonothein B*** from *Epilobium* species has inhibitory activity on α-reductase and aromatase.

Table 5.12 Tannins from medicinal plants.

S.No	Name of the tannins	Source
1.	Chebulic acid	*Terminalia chebula*
2.	Chebulinic acid	*Terminalia chebula*
3.	Chebulagic acid	*Terminalia chebula*
4.	Corialgin	*Terminalia chebula*
5.	Ellagic acid	*Widely distributed*
6.	Gallic acid	*Widely distributed*
7.	Neochebulinic acid	*Terminalia chebula*
8.	Oenothein-B	*Oenothera erythrosepala*
9.	Punicalagin	*Terminalia chebula*
10.	Tannic acid	*Widely distributed*
11.	Terchebin	*Emblica officinalis, Terminalia chebula*
12.	Terflavin-A	*Terminalia cattapa T.chebula, T.macroptera*
13.	Termigrandin	*Terminalia chebula*

5.4.13 Terpenes (see resins, oleoresins and gum-resins)

Terpenes are among the most widespread and chemically diverse groups of natural products. They are flammable unsaturated hydrocarbons, existing in liquid form. They are found in essential oils, resins or oleoresins. They are classified as mono, di, tri and sesquiterpenoids. The function of terpene is generally considered to be both ecological and physiological.

Terpenoids includes hydrocarbons of plant origin of general formula $(C_5H_8)n$. Terpenoids of *Boswellia serrata* are shown in Fig. 5.231. Methylated cyclohexane contains a large number of terpenoids including cyclogeraniol, ionones, and irones. Ionones are reported from *Boronia megastigma* oil. Irons are present in orris root. Borneols and isoborneols occur in *Dryobalanops camphore*.

Terpenoids of Boswellia serrata

Tetracyclic tritepene acids	Pentacyclic triterpene acids:
3-alpha-acetotirucall 8-24-dione-21-oic acid	beta-boswellic acid
3-ketotirucall 8-24-dione-21-oic acid	acetyl-beta-boswellic acid
3-alpha-hydroxytirucall 8-24-dione-21-oic acid	acetyl-11-keto-beta-boswellic acid
3-beta-acetotirucall 8-24-dione-21-oic acid	11-keto-beta-boswellic acid

Fig. 5.231 Terpenoids of *Boswellia serrata.*

Terpenes are classified by the number of 5-carbon units they contain.

5.4.13.1 Hemiterpene: C_5

A familiar example is ***angelic acid.***

5.4.13.2 Monoterpene: C_{10}

These are major component of many essential oils (Fig. 5.232).

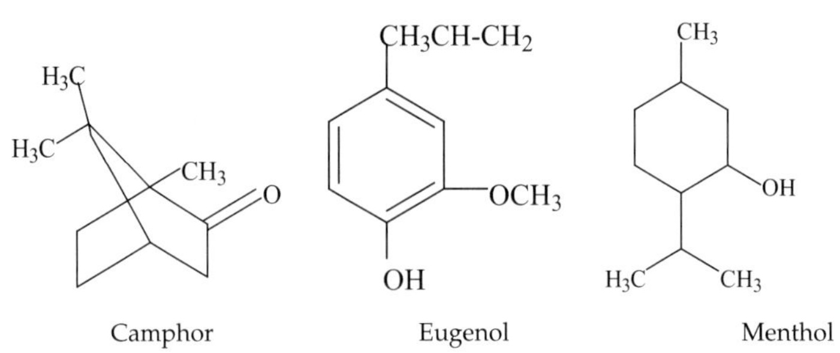

Fig. 5.232 Structure of common monoterpenes.

Terpinen-4-ol (Fig.5.233) is the primary antibacterial component of tea tree oil from *Melaleuca alternifolia*. ***Thujone*** has achieved notoriety as the neurotoxic agent in wormwood oil from *Artemisia absinthium* used in the preparation of the drink absinthe, now banned in most countries (Fig. 5.234).

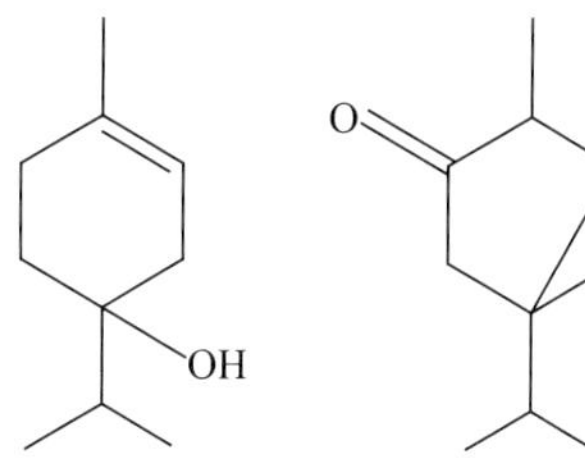

Terpinen-4-ol Thujone

Fig. 5.234 Medicinally important monoterpenes.

Pulegone is a major constituent of oil of pennyroyal from *Mentha pulegium*, which has a folklore history as an abortifacient (Fig. 5.235).

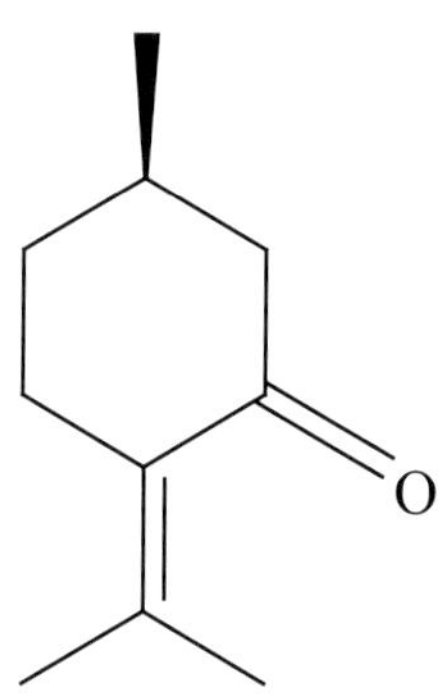

Fig. 5.235 Structure of (+)-Pulegone.

5.4.13.3 Sesquiterpene: C_{15}

Like monoterpenes, these are major components of many essential oils. The sesquiterpene acts as irritants when applied externally and when consumed internally their action resembles that of gastrointestinal tract irritant. A number of sesquiterpene lactones have been isolated and broadly they have antimicrobial (particularly antiprotozoal) and neutotoxic action.

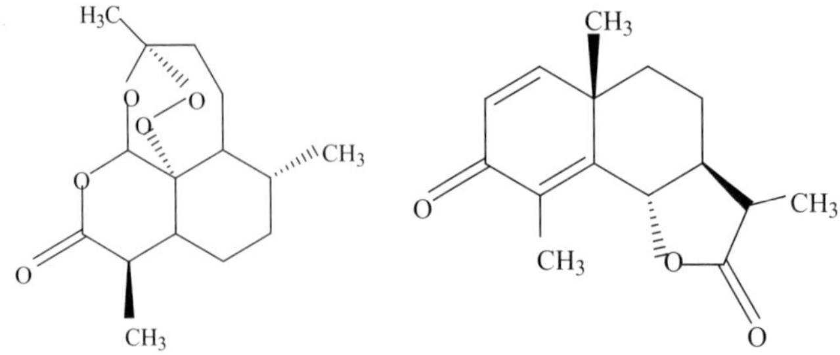

Artemisnin Santonin

Fig. 5.236 Structure of common sesquiterpenes.

Sesquiterpene lactone, ***palasonin***, isolated from *Butea monosperma has* anthelmintic activity (Fig. 5.237). It inhibits glucose uptake and depletes the glycogen content in Ascaridia galli.

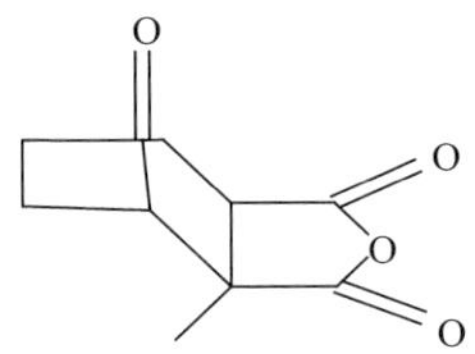

Fig. 5.237 Structure of Palasonin.

Parthenolide (Fig. 5.238), sesquiterpene lactone, found in *Tanacetum parthenium* has anti-inflammatory, antisecretory and spasmolytic activities (Fig. 5.165). It also inhibits the release of serotonin through various mediators and inhibits activation of MAP kinase. It has been used in the treatment of migraine.

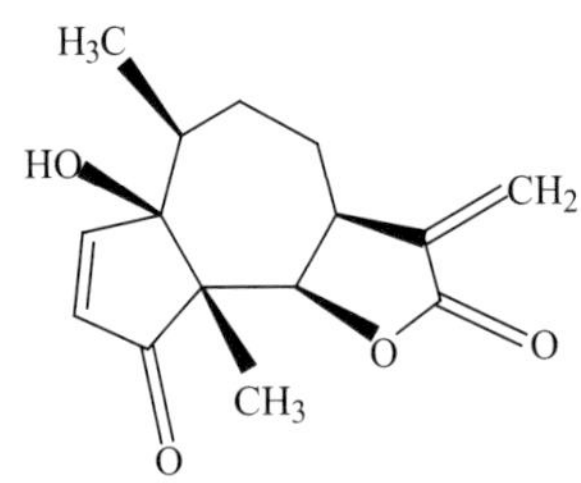

Fig. 5.238 Structure of Parthenolide.

Arglabin isolated from *Artemisia myriantha* has immunomodulator activity. ***Inuviscolide*** isolated from *Inula viscosa* has anti-inflammatory activity. ***Lychnophorolide A***, sesquiterpene lactone from *Eremanthus eriopus* is reported to be analgesic. ***Khusinol*** and ***khusol*** are sesquiterpene alcohol from *Vetiveria zizanoides* oil (Fig. 5.239).

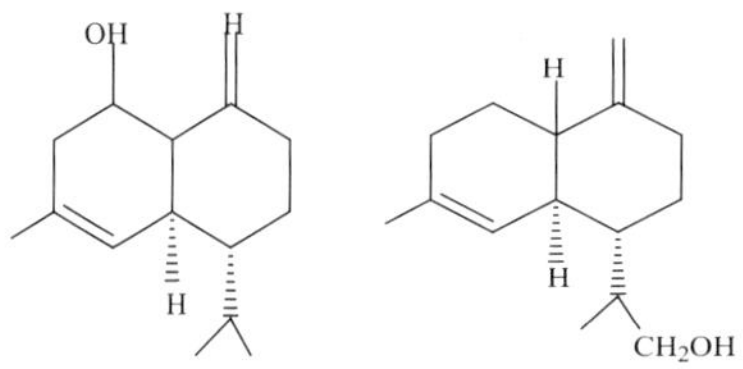

Fig. 5.239 Structure of Khusinol and Khusol.

Himachalol (Fig. 5.240), a sesquiterpene alcohol obtained from hexane soluble extract of the wood of *Cedrus deodara* was found to have significant anti-allergic activity. The effect was comparable to disodium cromoglycate in experimental models. ***Jatamansone*** **(valeranone)** and ***nardostachone*** are the principal sesquiterpenes of *Nardostachys jatamansi* (Figs. 5.241, 5.242). Jatamansone has antihypertensive, sedative and tranquilizing activities.

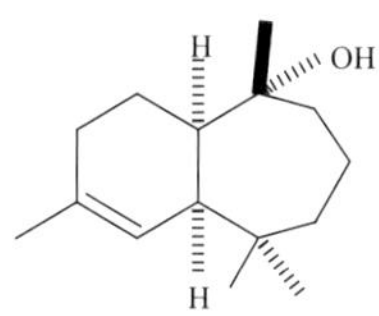

Fig. 5.239 Structure of Himachalol.

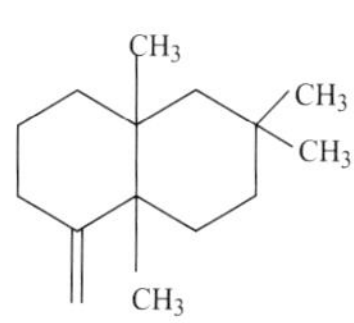

Fig. 5.240 Structure of Jatamansone.

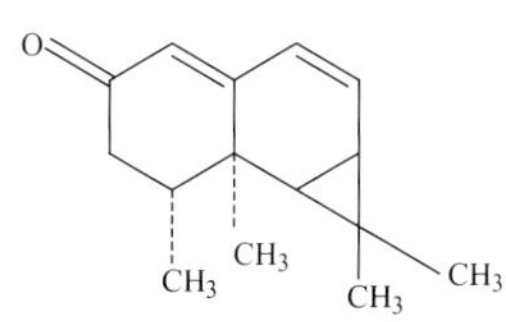

Fig. 5.241 Structure of Nardostachone.

5.4.13.4 Diterpene: C_{20}

These are classically considered to be resins. ***Taxol***, the anticancer agent, is familiar example (Fig. 5.242).

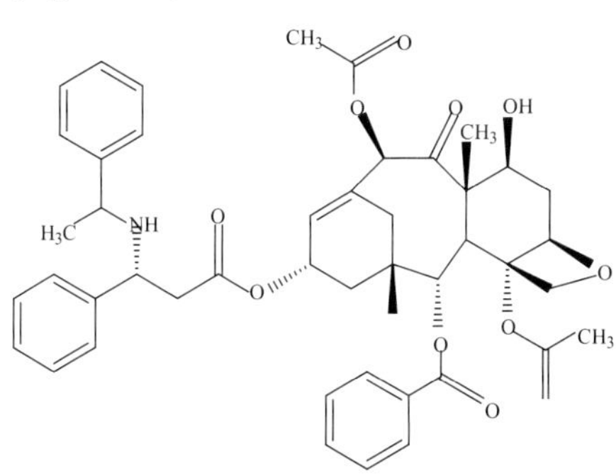

Fig. 5.242 Structure of Taxol.

Taxol is found predominantly in the bark of *T. brevifolia*, but in relatively small amounts (about 0.01–0.02%). Up to 0.033% of taxol has been recorded in some samples of leaves and twigs, but generally the taxol content is much lower than in the bark. The content of some other taxane derivatives in the bark is considerably higher, e.g. up to 0.2% ***baccatin III.*** Other taxane derivatives characterized include ***10-deacetyltaxol, 10-deacetylbaccatin III, cephalomannine*** and ***10-deacetylcephalomannine.***

Paclitaxel (taxol) is being used clinically in the treatment of ovarian and breast cancers, non-small-cell lung cancer, small-cell lung cancer, and cancers of the head and neck. Taxol has also been shown to bind to a second target, a protein which normally blocks the process of apoptosis (cell death). Inhibition of this protein allows apoptosis to proceed.

Pondicin and ***oridonin*** isolated from *Rabdosia rubscens* are anticancer. ***Dehydrocrotonin*** isolated from *Croton cajucara* is antileukemic and anti-ulcerogenic. ***Scopadulin,*** tetracyclic diterpene from *Scopolia parviflora* has antiviral activity. *Ballota nigra* contains diterpenes including ***marrubiin, 7-acetoxymarrubiin, ballotinon, ballotenol*** and ***ballonigrin.***

Salvinorin A, ditepene found in *Salvia divinorum* is reported to be useful against alcoholic addiction (Fig. 5.243).

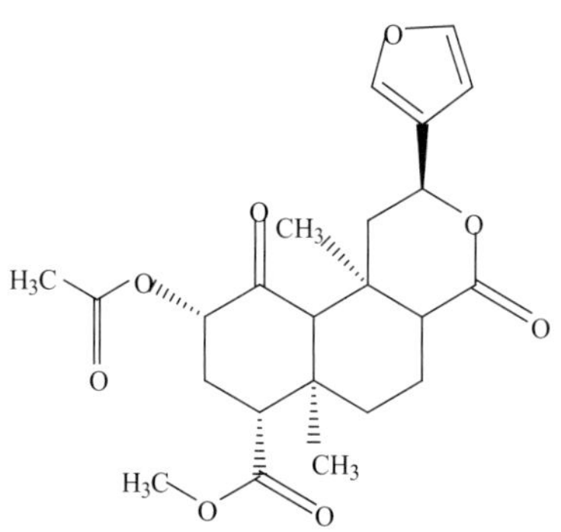

Fig. 5.243 Structure of Salvinorin A.

Forskolin is a labdane diterpene found in *Coleus forskolii*. It is used to raise the levels of cyclic AMP (Fig. 5.244). It resensitizes cell receptors by activating the enzyme adenyl cyclase and increasing intracellular levels of cAMP, which is an important signal carrier that is necessary for proper biological response of cells to hormones and other extracellular signals.

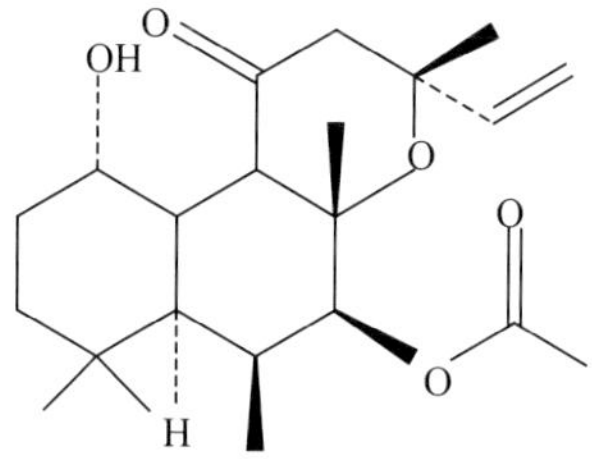

Fig. 5.244 Structure of Forskolin.

5.4.13.5 Triterpene: C_{30}

This group includes common triterpenes, steroids, sterols, and cardiac glycosides.

Common triterpenes: ***Amyrins, ursolic acid and oleanic acid*** (Fig. 5.245).

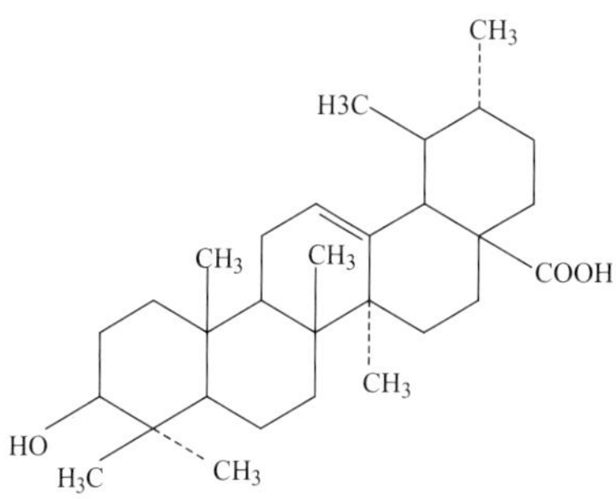

Ursolic acid

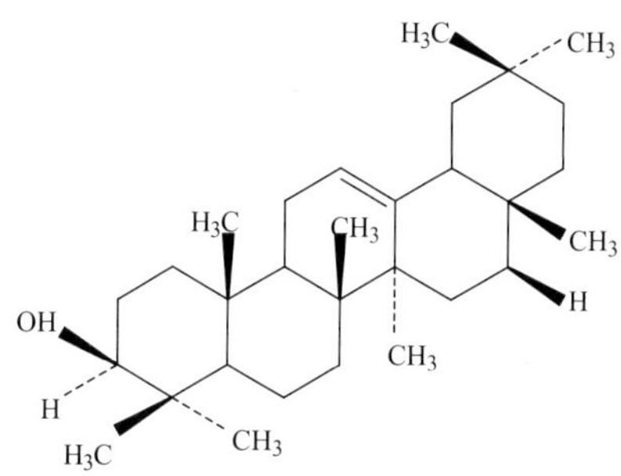

α- amyrin

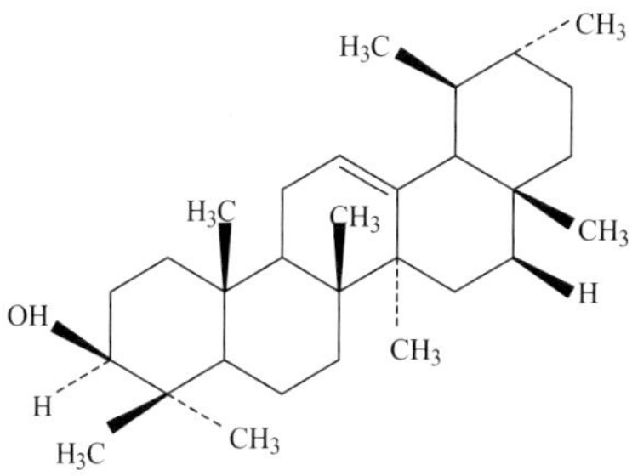

β-amyrin

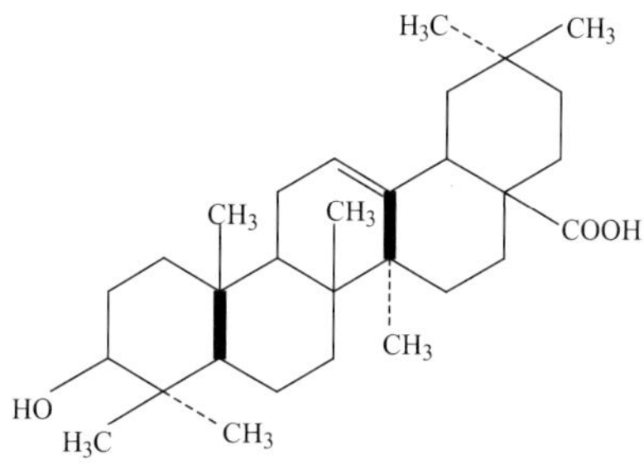

Oleanoic acid

Fig. 5.245 Structure of common triterpenes.

Healianol, triterpene alcohol from *Calendula officinalis* has anti-inflammatory activity. Triterpenoids ***lupeol*** and ***varunol*** have been isolated from the root and stem bark of *Cratavea nurvala*. A triterpenoid,

3 α-hyrdoxy-D-friedoolean-1-ene has been reported from *Adhatoda vasica*. ***Galphimine-B*** from *Galphinia glauca* has sedative activity. ***Abrisapogenol E*** and ***abrisapogenaldacetal*** from *Rabinia pseudoacacia* are cytotoxic. Triterpenoids from *Pourouma guainensis* are antileshmanial. ***Azadirachtin*** (Fig. 5.246) isolated from *Azadirachta indica* is potent insecticidal. ***Nimbin, nimbinin*** and ***nimbidin*** are other triterpenes derived from neem.

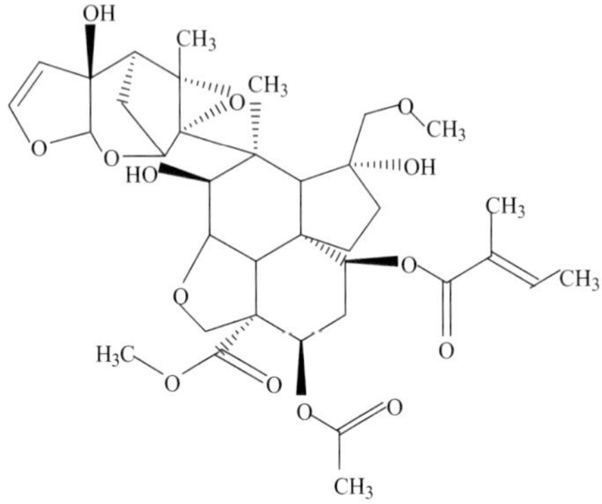

Fig. 5.246 Strucuture of Azadirachtin.

Saponins: See Section 2.3.10.

Sterols: ***β-sitosterol, stigmasterol, taraxasterol and campesterol*** (Fig. 5.247).

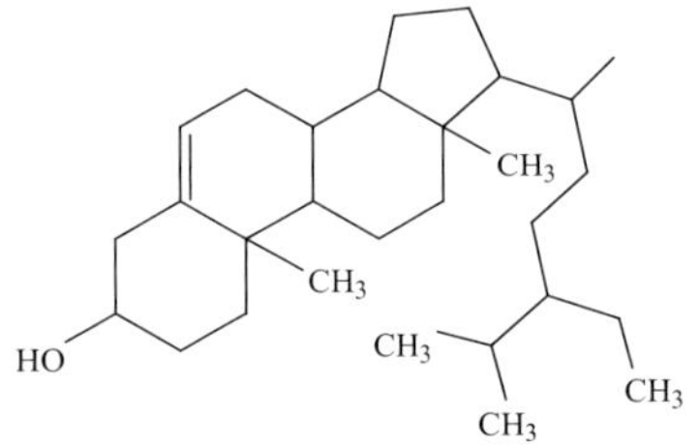

β-sitosterol

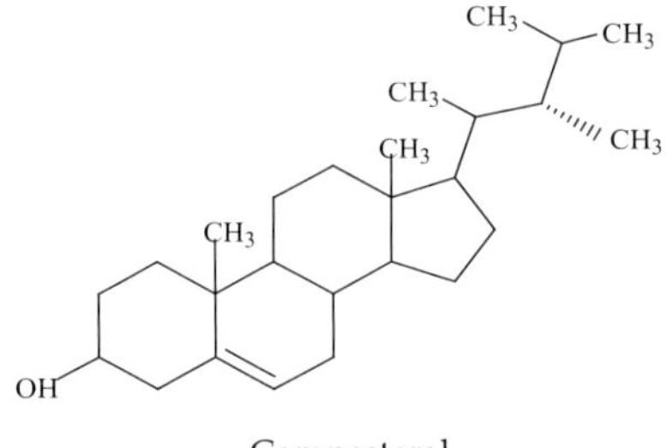

Campesterol

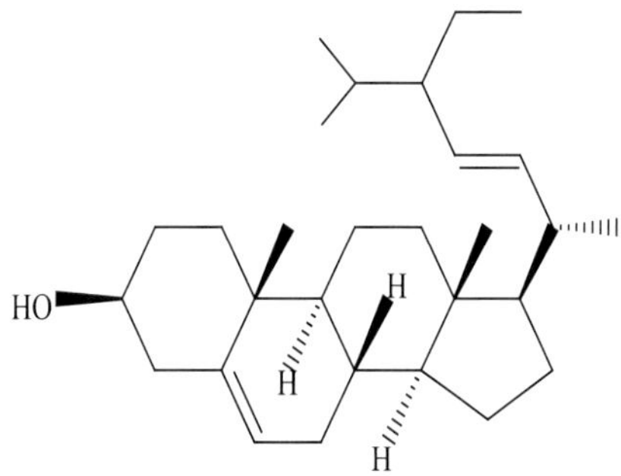

Stigmasterol

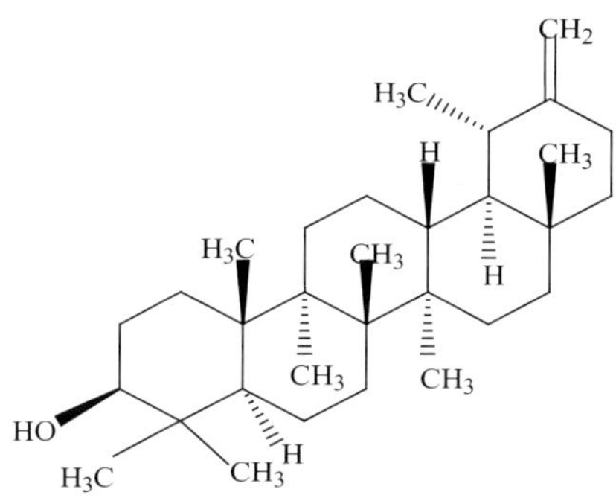

Taraxasterol

Fig. 5.247 Structure of common sterols.

Phytosterols compete with dietary cholesterol for uptake in the intestines. They have demonstrated the ability to block the uptake of cholesterol and facitilate its excretion from the body. Contradicting this, some studies have reported inefficency of morning sterols in competing with dietary cholesterol.

Phytostans (stanols)

These are saturated phytosterols. Thay are widely distributed in plants. They have cholesterol lowering activity.

Table 5.13 Terpenes from medicinal plants.

S.No	Name of the terpene	Source
1.	2-α-hydroxymicromeric acid	*Terminalia chebula*
2.	Absinthin	*Artemisia absinthum*
3.	Ajoene	*Allium sativum*
4.	Allicin	*Allium sativum*
5.	Amberttolide	*Hibiscus abelmoschus*
6.	Angelic acid	*Angelica glauca*
7.	Asarone	*Acorus calamus*
8.	Azadirachtin	*Azadirachta indica*
9.	Betulinic acid	*Nelumbo nucifra, Betula utilis*
10.	Calliterpenone	*Callicapra macrophylla*
11.	Camphor	*Cinnamomum camphora.*
12.	Cineol	*Eucalyptus globulus*
13.	Elephantopin	*Elephantpous scaber*
14.	Grandiflorenic acid	*Aspilia mossambicensis*
15.	Helenin	*Inula racemosa*
16.	Himachalol	*Cedrus deodara*
17.	Jatamansin	*Nardostachys jatamansi*
18.	Kaurenoic acid	*Aspilia mossambicensis*
19.	Menthol	*Mentha spicata*
20.	Myricadiol	*Myrica nagi*
21.	Palasonin	*Butea frondosa*
22.	Safranal	*Crocus sativus*
23.	Salacinol	*Salacia reticulata*
24.	Santonin	*Artemisia santonica*
25.	Stemmin C	*Solenostemma arghel*
26.	Stemnoside A	*Solenostemma arghel*
27.	Stemnoside B	*Solenostemma arghel*
28.	Thymol	*Thymus vulgaris*
29.	Valerianic acid	*Valeriana officinalis*

5.4.14 Withanolides

These are a group of naturally occurring oxygenated ergostane type steroids having lactone in side chain and 2-en-1-one system in ring A (Fig. 5.248).

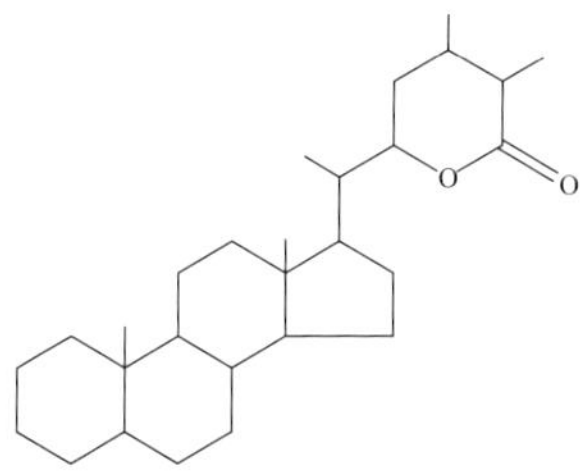

Fig. 5.248 Basic withanolide skeleton.

Withanolides isolated from the leaves of *Withania somnifera* have shown significant antitumor and immunomodulator activity. ***Coagulin F and coagulin G*** isolated from *Withania coagulens* have anti fungal properties. ***3 beta–hydroxyl 2, -3-dihydrowithanolide F,*** present in fruits, have significant hepatoprotective and anti-inflammatory activity. ***Withaferin A*** (Fig. 5.176) and withanolide E have immunosupressant activity.

Withanolides of W. somnifera

Withanolides were first isolated from *W. somnifera* hence the name withanolide. Chemically, they are a group of naturally occurring oxygenated ergostane type steroids having lactone in side chain and 2-en-1-one system in ring A. Velde and Lavie reported ***1α, 3β, 20-trihydroxy witha-5, 24-dienolide*** from *W. somnifera* chemotype ***3. 7α, 27-hydroxy-1-oxo- witha-2, 5-24-trienolide*** and ***7α, 27-dihydroxy-1-oxo- witha-2, 5-24-trienolide*** have been reported from *W. somnifera* chemotype 3.

Withanolide S has been reported from from the leaves of *W. somnifera*. Kirson, Glotter and Withanolides Q and R have been reported from the offspring of Indian chemotype 1 and 3 of *W. somnifera*. Withasomnilide, withasomniferabolide, somniferanolide and somnwithanolide have been reported from the stem bark of *W. somnifera*. A new compound, chlorohyrdin 11 has been reported. 27-O-glucosides (sitoinoside IX and X) were reported from from roots of *W. somnifera*. Withasomidienone have been reported. The withanolides reported from *W. somnifera* are tabulated below:

Table 5.14 Major withanolides of *W.somnifera*.

SNo	Name	Pharmacological activity
1.	20β-hydroxy-1-oxo-(22*R*)-witha-2,5,24-trienolide, witha-coagulin and a known withanolide, 17β-hydroxy-14α, 20α-epoxy-1-oxo-(22*R*)-witha-3,5,24-trienolide	
2.	27-deoxywithaferin A	
3.	27-hydroxy withanolide A	

Table 5.14 contd...

Table 5.14 contd...

4.	3ß-hydroxy-2,3-dihydro withanolide F	Hepatoprotective
5.	3-β-hydroxy-2,3-dihydro-withanolide	Antibacterial, antitumor, immunomodulator and anti-inflammatory
6.	4β-hydroxy-withanolide	
7.	Ashwagandholine	
8.	chlorohyrdin II	
9.	Coagulin	
10.	Compound WS-1	Hypno-sedative
11.	Compound WS-2	Antibacterial
12.	Jaborosalactone	
13.	Sitoindside-IX,X	Immunomodulator and C.N.S effects
	sominolide, soinone, withasomnilide, withasomniferabolide, somniferanolide and somnwithanolide	
14.	Somniferanolide	
15.	Somnwithanolide	
16.	Withacoagin	
17.	Withaferinil	Antitumor
18.	Withaniol	
19.	Withanoferin	
20.	Withanolide D	Antitumor
21.	Withanolide E	Antifeedent
22.	Withanolide Q and R	
23.	Withanolide S and T	
24.	Withanolide sulfoxide	Antitumor
25.	Withanolide Y	
26.	Withanolide Z	Not confirmed
27.	Withanolide-5 ,20α,(R)-dihydroxy-6α,7α- epoxy-1-oxo-5 α -with a-2,24-dienolide	Immunomodulator
28.	Withanone	Antitumor
29.	Withanoside	
30.	Withaphysanolide A	
31.	Withasomidienone	
32.	Withasomniferabolide	
33.	Withasomnilide	
34.	Withastramonolide	

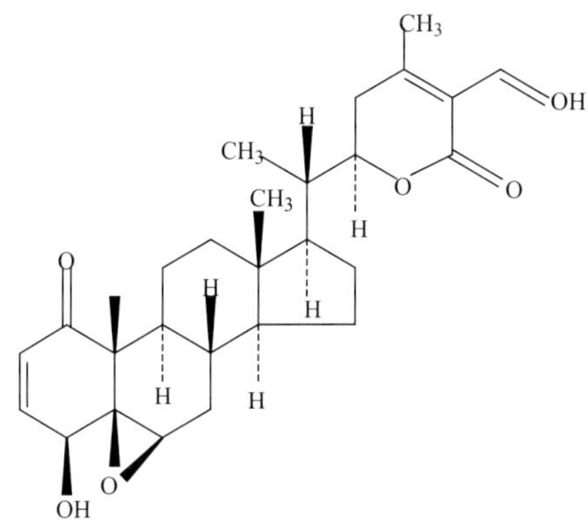

Withaferin A

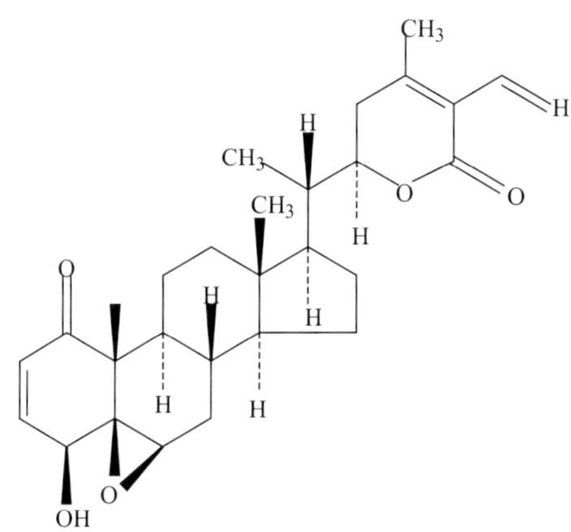

27-deoxywithaferin A

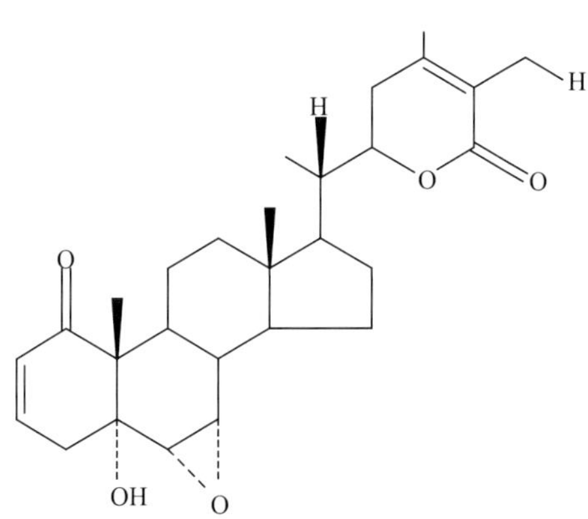

Withanone

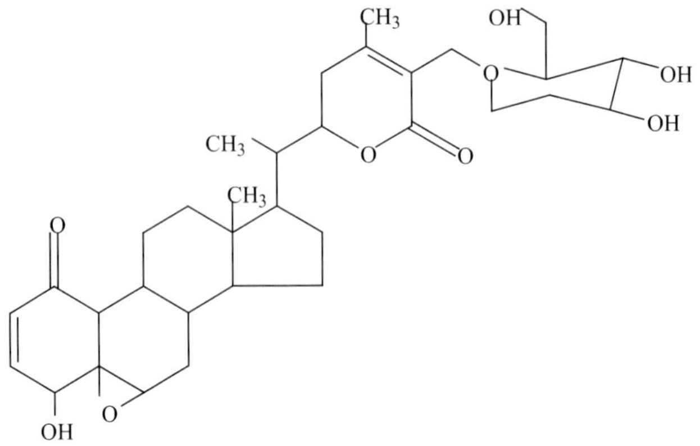

Sitoindoside IX

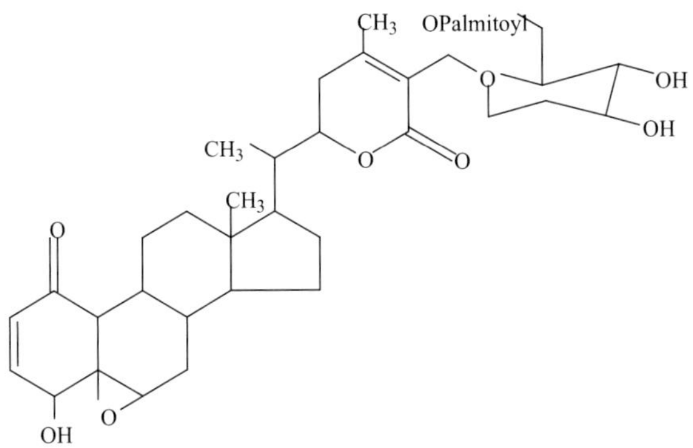

Sitoindoside X

Fig. 5.249 Structure of major withanolides of *W.somnifera.*

Table 5.15 Withanolides from other medicinal plants.

S. No	Name of the withanolide	Source
1.	20beta-hydroxy-1-oxo-(22R)-witha-2,5,24-trienolide; Withacoagulin; 17beta-hydroxy-14-alpha, 20alpha-epoxy-1-oxo-(22R)-witha-3-,5,24-trienolide	*Withania coagulans*
2.	6alpha,7alpha-epoxy-5alpha,17beta-dihydroxy-1-oxowitha-2,24-dienolide and 17-Epiacnistin-A	*Discopodium penninervium*
3.	Bracteosin A, bracteosin B and bracteosin C	*Ajuga bracteosa*
4.	Chantriolides A and B	*Tacca chantrieri*
5.	Cilistadiol, cilistepoxide and cilistol-A	*Solanum sisymbiifolium*

Table 5.15 contd...

Table 5.15 contd...

S. No	Name of the withanolide	Source
6.	Cinerolide, 24,25-dihydrowithanolide S, visconolide, physalactone, withanolide S and 4beta-hydroxywithanolide E	*Physalis cinerascens*
7.	Daturalactone-4; Nic-3 (hyoscyamilactol); 16alpha-acetoxyhyoscyamilactol	*Hyoscyamus niger*
8.	Daturametelins C, D, E, F and G	*Datura metel*
9.	Ixocarpalactone A, 2,3-dihydro-3β- methoxyixocarpalactone A,2,3-dihydro-3β-methoxyixocarpalactone B, 2,3-dihydroixocarpalactone B, and 4β,7β,20*R*-trihydroxy-1-oxowitha-2,5-dien-22,26-olide	*Physalis philadelphica*
10.	Physachenolides A - E.	*Physalis chenopodifolia*
11.	Physagulins L, M and N, Physagulin D and quercetin 3-O-rhamnosyl-(1-->6)-galactoside	*Physalis angulata*
12.	Pplantagiolides A-E and chantriolide A	*Tacca plantaginea*
13.	Subtrifloralactones A-E,F-L, new C-18 oxygenated withanolide, 13β-hydroxymethylsubtrifloralactone E and philadelphicalactone A	*Deprea subtriflora*
14.	Two new withanolide glycosides and 18-acetoxywithanolide D and 18-acetoxy-5,6-deoxy-5-withenolide D	*Dunalia brachyacantha*
15.	Withaferin A and witharistatin	*Withania aristata*
16.	Withangulatin A and withangulatin I	*Physalis angulata*
17.	Withanolide 1 and 2	*Jaborosa caulescens (var. caulescens and* var. *bipinnatifida)*
	Withanolide-7	*Physalis minima*
18.	Withanone	*Discopodium penninervium, Exodeconus maritimus, Withania somnifera*
19.	Witharifeen and daturalicin	*Datura innoxia*

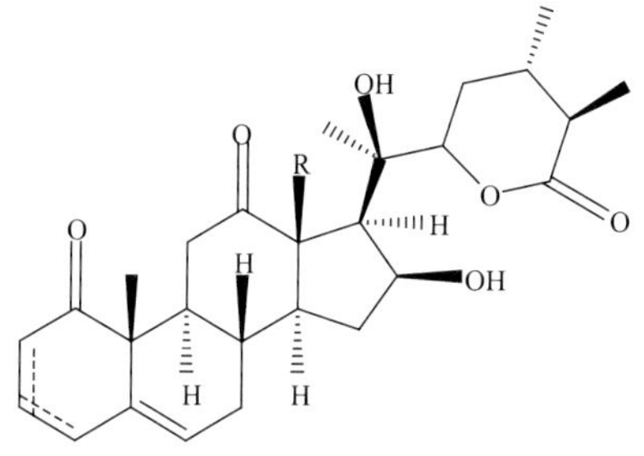

13 β-hydroxymethylsubtrifloralactone E

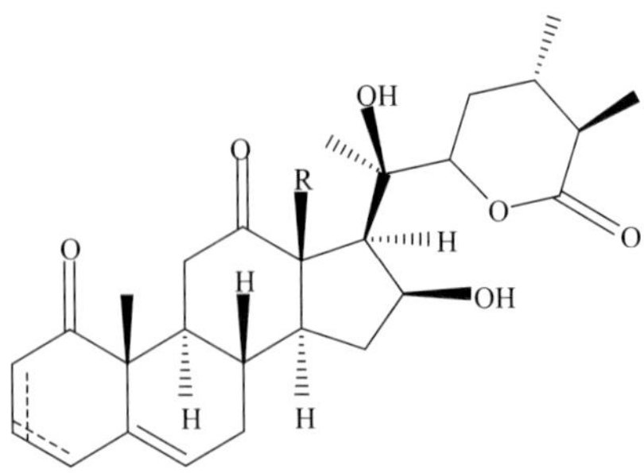

ixocarpalactone A

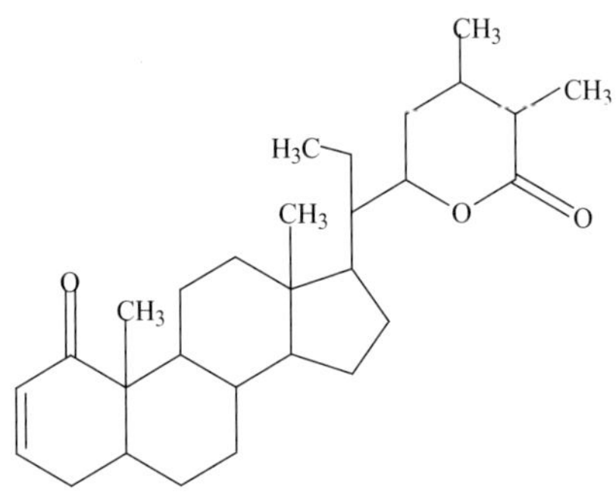

Daturametelin

Fig. 5.250 Structure of bioactive withanolides.

5.4.15 Xanthones

These are phenolic product of a triketide precursor and benzoid unit. ***Swerchinin*** from *Swertia chirata* lowered blood glucose in normal and streptozocin induced mild and severe diabetes in rats (Fig. 5.251). The effect of swerchirin was comparable to tolbutamide. Swerchirin has demonstrated a protective effect on hematopoiesis in 6-Co-irridiated mice. *S.chirata* also contains a novel dimeric xanthone, ***chiratanin***.

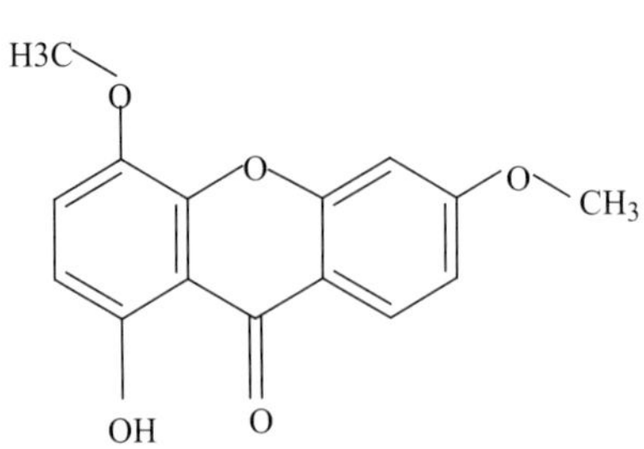

Fig. 5.251 Structure of Swerchirin.

Gentiacaulin isolated from *Swertia kochiana* has vasodilator activity. ***Gentisin*** (Fig. 5.252) and ***isogentisin*** (Fig. 5.253) from *Gentiana lutea* has significant mutagenic activity. ***Gentisein*** (Fig. 5.254) is also isolated from *G. lutea.*

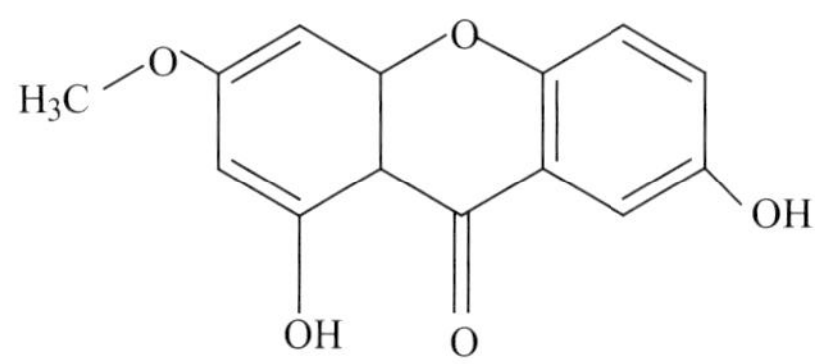

Fig. 5.252 Structure of Gentisin.

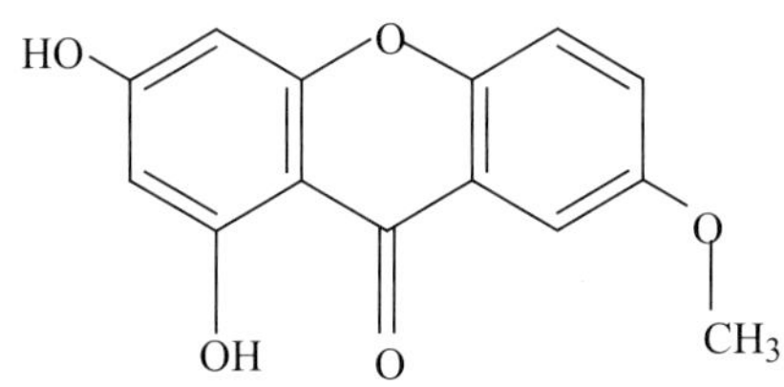

Fig. 5.253 Structure of Isogentisin.

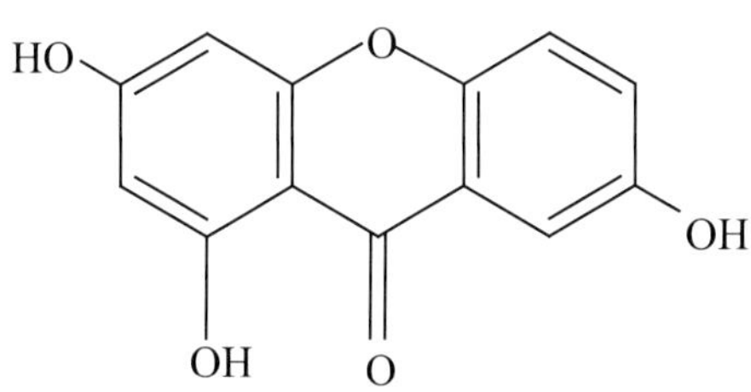

Fig. 5.254 Structure of Gentisein.

The ethanol extract of the bark of *Garcinia dulcis* furnished five xanthones; ***garciniaxanthone*** showed inhibitory effects on the growth of *P. falciparum* with IC50 value of 0.96 mg/ml85.

Table 5.16 Xanthones from medicinal plants.

S.No	Name of the xanthone	Source
1.	Decussatin	*Swertia decussata*
2.	Dixylopyranoside	*Swertia thomsonii, S. franchetiana*
3.	Gentiacaulein	*Gentiana kochiana*
4.	Gentiakochianin	*Gentiana kochiana*
5.	Gentisin	*Gentiana lutea*
6.	Swerchirin	*Swertia chirayita, S.japonica and S. mussotii*
7.	Swertiapuniside	*Swertia punicea*

5.4.16 Miscllaneous compounds

5.4.16.1 Diarylheptanoids

Diarylheptanoids are rare compounds found in the family Zingiberaceae. They are found in rhizomes of *Zingiber officinale, Alpinia galangal* and *Hedychium spicatum*. Animal studies have shown diarylheptanoids to be potent anti-inflammatory agents as they selectively inhibit cycloxygenase enzyme responsible for inflammation. Gingerenone A and B and galangol are common examples. The pungent odor of *Z. officinale* and *A. galangal* is due to the presence of these compounds. This group of compounds has not been thoroughly investigated for pharmacological activity.

5.4.16.2 Isoflavones

Isoflavones are found in *Glycine max* (soybean). Clinical research has demonstrated soy isoflavones to be effective in menstrual diseases. They also have an antioxidant activity . Isoflavones belong to a group of compounds known as phyto-estrogens. The isoflavone content of soybean is 10 to 20 times more than other plants. ***Genistein*** is a potent tyrosine inhibitor, where as ***daidezein*** also has an antioxidant activity (Fig. 5.255). Recently isoflavones have been investigated for hypolipidemic activity.

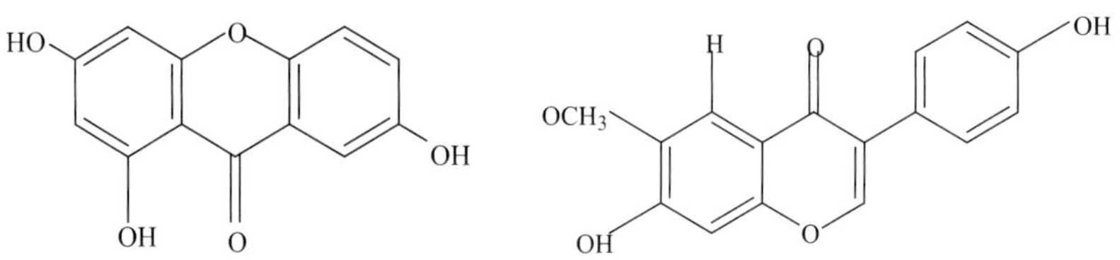

Genistein

Daidezein

Fig. 5.255 Structure of soy isoflavones.

5.4.16.3 Furochromones

Furochromones are a group of coumarins, derived from γ-benzopyrone. They are related to furanocoumarins and are present in plants of the family Apiaceae and Rutaceae. *Ammi visnaga* is a potential source of furochromones like ***khellin*** and ***visnagin***. Khellin was once used as coronary vasodilator drug (Fig. 5.256).

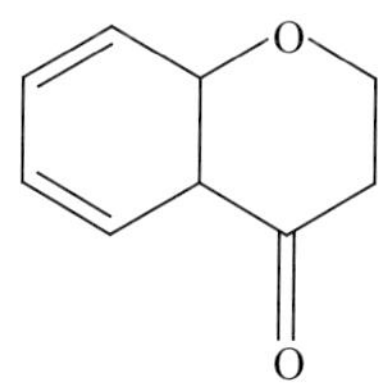

Benzopyrone

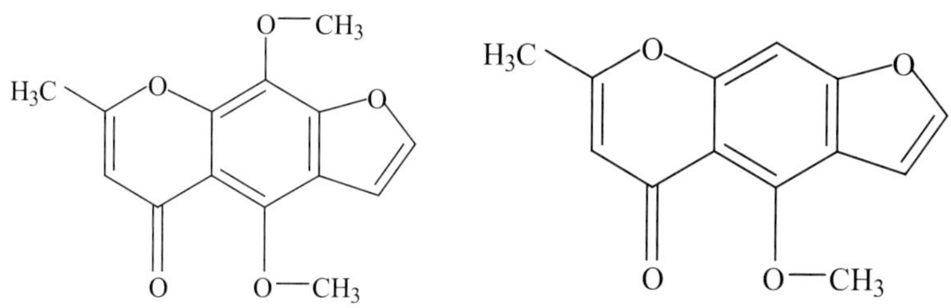

Khellin

Visnagin

Fig. 5.256 Structure of common furochromones.

5.4.16.4 Phenylpropanoids

These contain a three-carbon side chain attached to a phenol. Common examples are hydroxycoumarins, lignans and phenyl-propenes. Volatile oil of *Myristica fragrans* contains phenylpropanoids, including ***myristicin, saffrole*** and ***elemicin*** (Fig. 5.257).

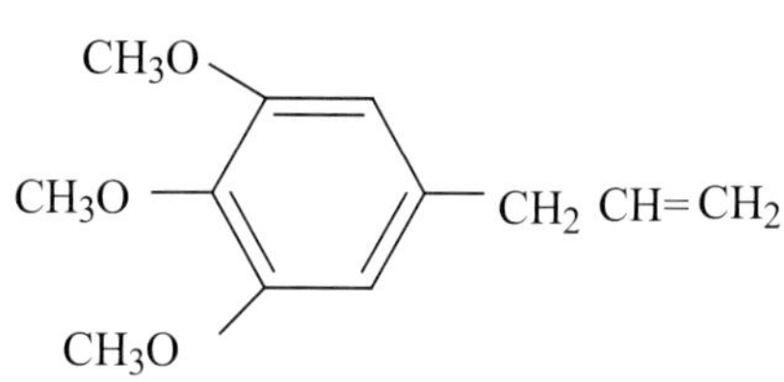

Elemicin

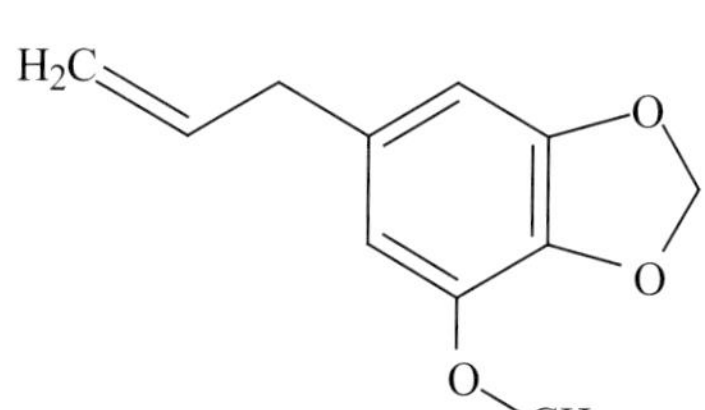

Myristicin

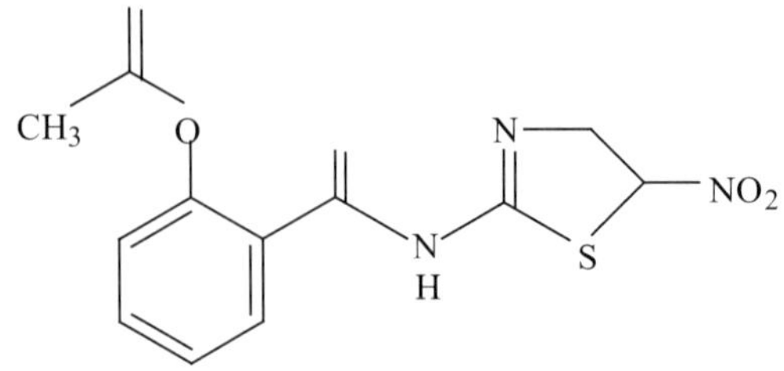

Saffrole

Fig. 5.257 Phenylpropanoids of *Myristica fragrans.*

Hydroxycoumarins

Hydroxycoumarins represents another group of coumarins (Figs. 5.258, 5.259), which are widely distributed in Apiaceae and Gramineae. They are found in *Aesculus hippocastanum*, *Angelica pubescens*, *Artemisia lactiflora* and *Crategeus oxycantha*. ***Callophylloide*** and ***jacareubin*** coumarins derived from *Calophyllum inophyllum* have antiarrhythmic, coronary vasodilator, anticoagulant, anti-inflammatory and antiarthritic and gastro protective activites, respectively. *Calophyllum apetalum* contains coumarin named ***aperalolide.***

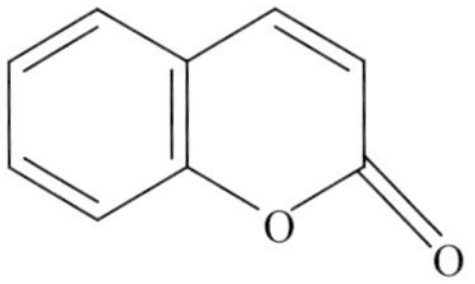

Fig. 5.258 Structure of coumarin.

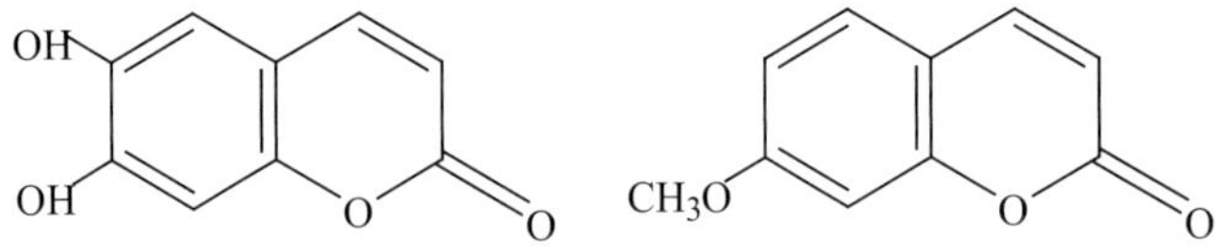

Aesculetin Herniarin

Fig. 5.259 Structure of common hydroxycoumarins.

Lignans

Sargentodoxaceae, lignan from *Sargenodoxa cuneata* has anthelmintic activity. ***Termilignan***, ***thannilignan*** and ***anolignan B*** isolated from *Terminalia belerica* demonstrated significant anti HIV, antimalarial and antifungal effects *in vitro*. ***Sesamin*** and ***sesamolin*** are lignan derivatives found in seasame oil. The flavonol lignans ***hydnocarpin, hydnowightin***

and neohydnocarpin isolated from the seeds of *Hydnocarpus* sp are lipid lowering, anti-inflammatory and antitumoral in animal experiments.

Schisandra chinensis is used in chronic persisent hepatitis. The active principles are various biphenylcotenoic lignans called ***waweizisu C, wuweizichun B, schisantherin A,B,C,D*** which all lower SGPT levels in chronic viral hepatitis. Lignans of *Phyllanthus niruri* deserve special mention. They have antiviral and hepatoprotective properties. ***Phyllanthin, hypophyllanthin*** and ***isontetralin*** are major lignans (Fig. 5.252).

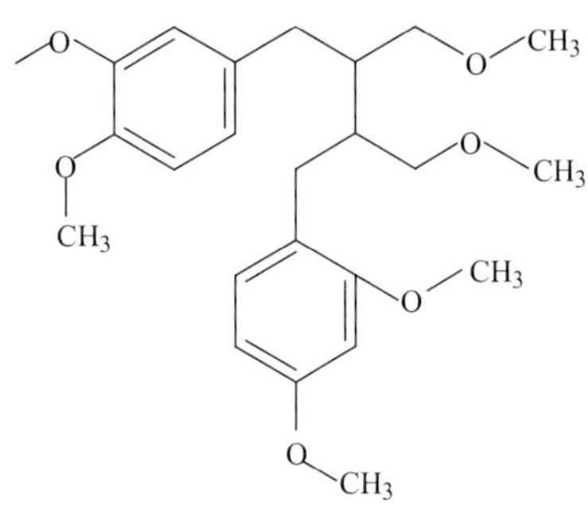

Fig 5.260 Structure of Phyllanthin.

7-hydroxymatairesinol, lignan found in *Picea abies* (Norway spruce) is precursor of enterolactone (Fig. 5.261). It is antioxidant and anticancer. ***Lariciresinol*** obtained from flaxseed is reported to be antioxidant, anticarcinogenic, and antidiabetic (Fig. 5.262).

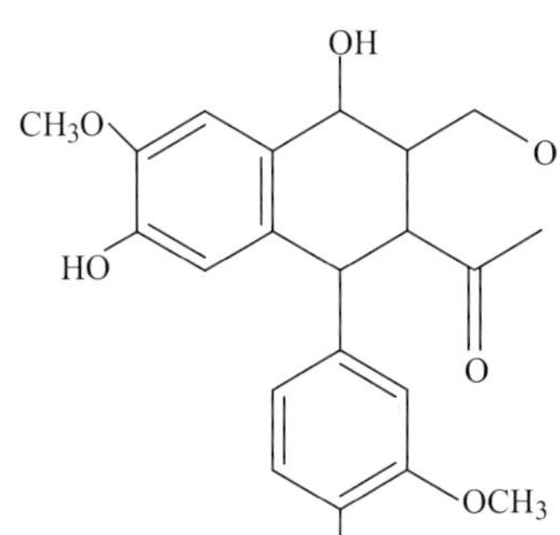

Fig. 5.261 Structure of 7-hydroxymatairesinol.

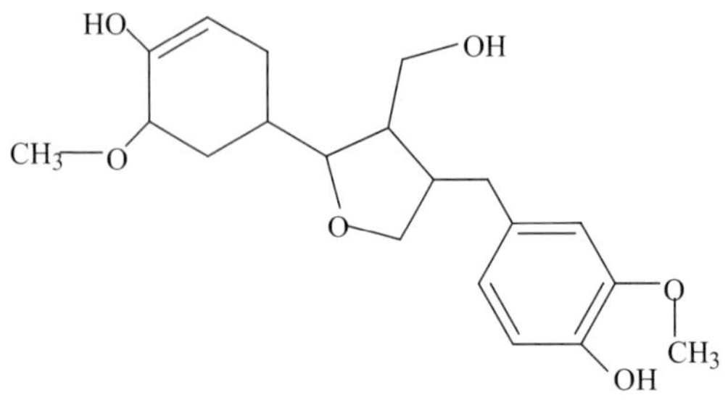

Fig. 5.262 Structure of Lariciresinol.

Lignans of *Myristica argentea* including ***erythro-austrobailignan-6 and meso-dihydroguaiaretic acid, myristargenol A, and myristargenol B*** from the aril of the seeds, show some levels of activity against *Streptococcus mutans.* Erythro-austrobailignan-6, meso-dihydroguaiaretic acid, and ***nectandrin-B*** (Fig. 5.185) exert an antiproliferative effect on MCF-7 cells as well as antioxidant activity on the 1,1-diphenyl-2-picrylhydrazyl radical. In addition, Nectandrin-B inhibits the enzymatic activity of 17β hydroxysteroid dehydrogenase and antiaromatase activities (Fig. 5.263).

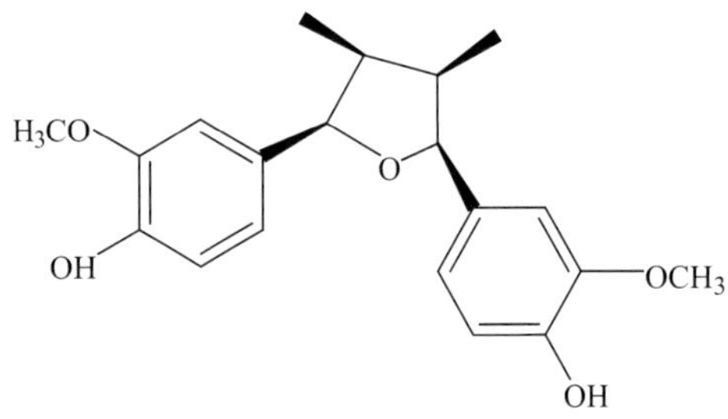

Fig. 5.263 Structure of Nectandrin-B.

Tetrahydroyfuran lignans like ***grandisin*** (Fig. 5.264) and ***veraguensin*** (Fig. 5.265) have been demonstrated to prevent the transmission of Trypanosoma cruzi by blood transfusion.

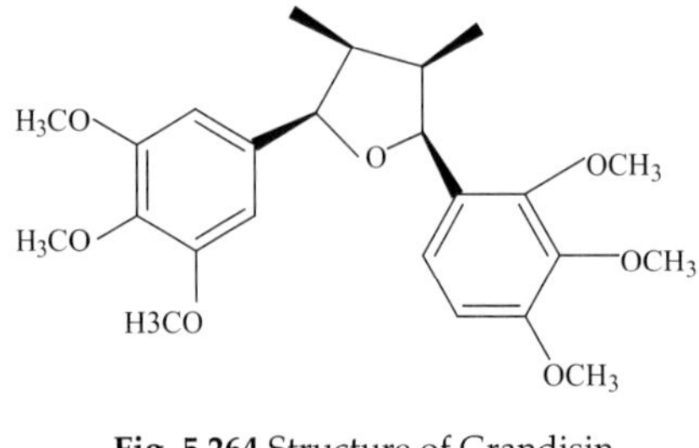

Fig. 5.264 Structure of Grandisin.

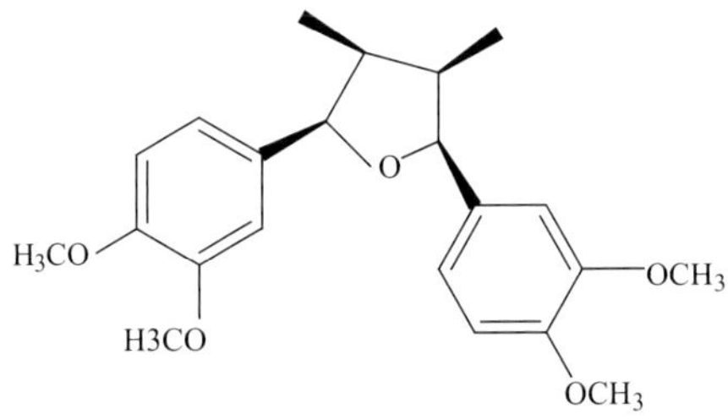

Fig. 5.265 Structure of Veraguensin.

Urtica dioica contains lignans ***including secoisolariciresinol-9-O-glucoside, neo-olivil,*** and ***neo-olivil-4-O-glucoside.***

Phenyl-propenes

Eugenol is common example (Fig. 5.266). It is used as a dental analgesic.

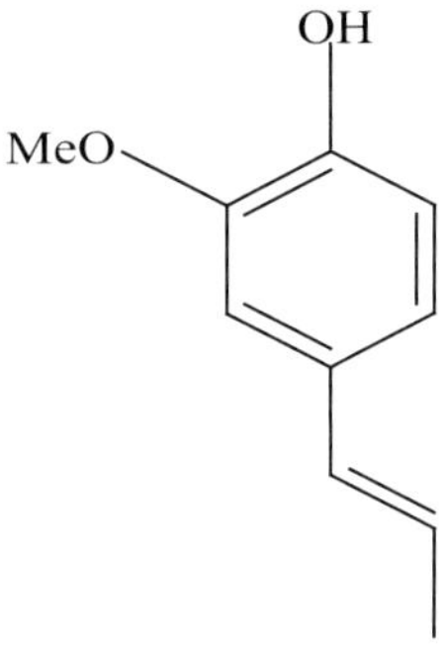

Fig. 5.266 Structure of Eugenol.

5.4.16.5 Napthodianthrones

Napthodianthrones are derivatives of anthracene. Anthracenes have been discussed under glycosides. The main reason for discussing napthodianthrones separately is that they have gained importance because of the pharmacological significance of hypericin present in *Hypericum perforatum*. Napthodianthrones are known to cause photosensitivity. Napthodianthrone, fagopyrin is present in *Fagopyrum esculentum* (Fig. 5.267).

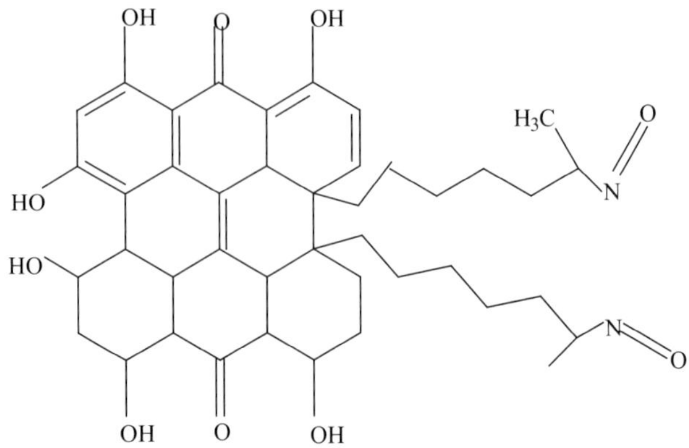

Fig. 5.267 Structure of Fagopyrin.

5.4.16.6 Pyranocoumarins

Pyranocoumarins are a class of coumarin compounds. They are rare compounds and only a few of them have been investigated. ***Costataolide-A*** (Fig. 5.268) isolated from *Caulophyllum acuminata* has anti -HIV activity. Pyranocoumarin, ***luvangetin*** isolated from *Aegle* marmelos has gastroprotective activity. ***Ceylatin*** has been reported from ***Atalantia ceylanica*** (Fig. 5.269).

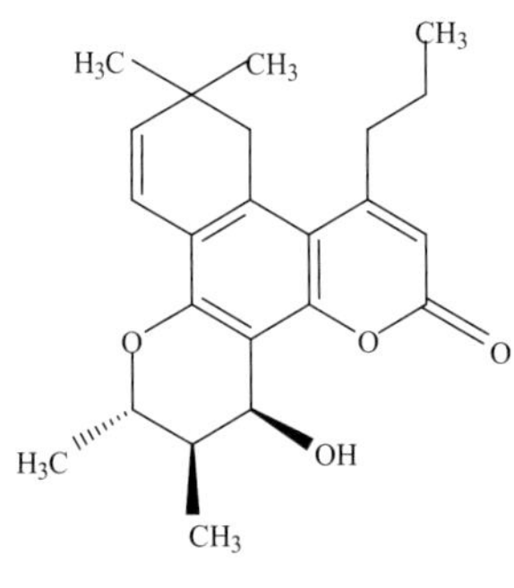

Fig. 5.268 Structure of Costatolide-A.

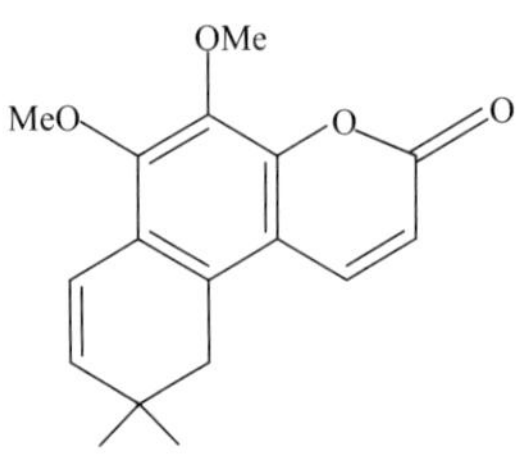

Fig. 5.269 Structure of Ceylatin.

5.4.16.7 Organosulfur compounds

Garlic contains a number of bioactive compounds (Fig. 5.270). The most distinguishing are the organosulfur compounds ***alliin*** (S-allyl-L-cysteine sulfoxide, an odorless sulfur-containing amino acid derivative), and allicin (diallyldisulfide-oxide, responsible for the pungent odor of garlic) and ajoene.

When garlic is crushed or ground, the enzyme alliinase converts alliin to allicin. Allicin is thought to be responsible for many of the therapeutic benefits of garlic. Several investigations suggest that these sulfur compounds may inhibit the induction and growth of cancer. These benefits do not appear to be limited to a specific carcinogen, tissue or species. ***Ajoene*** has antiplatelet activity.

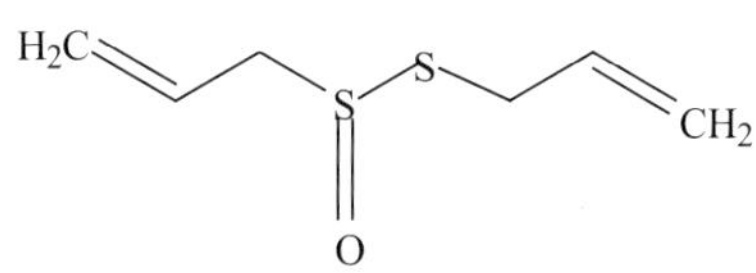

Allicin

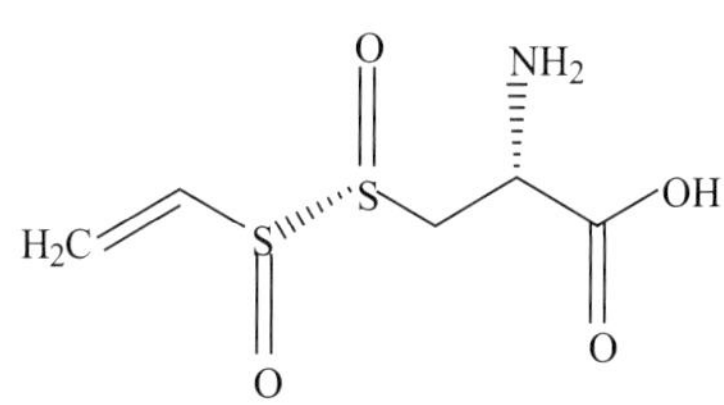

Alliin

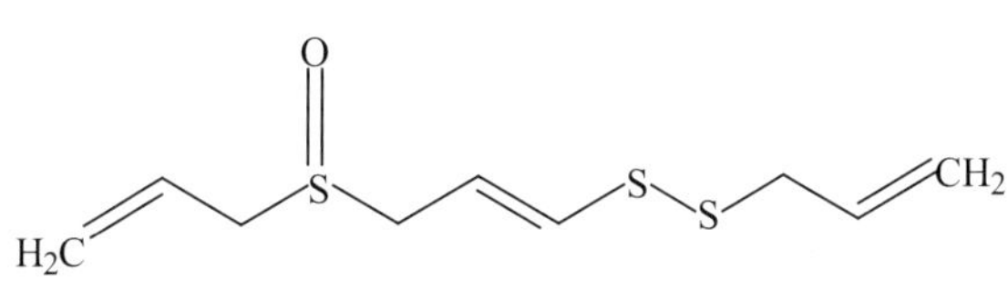

Ajoene

Fig. 5.270 Structures of organosulfur compounds of garlic.

References and further reading

Agarwal A, and Venkataraman BV. 2003. Trypsin inhibitory effect of wedelolactone and demethylwedelolactone. Phytother. Res. 17, 420–1.

Agarwal RR, and Dutt S. 1935. Chemical investigation of *Boerhaavia diffusa* Linn. Part 11. The isolation of an alkaloid 'Punarnavine'. Proc. UP Acad. Sci. 5, 240–242.

Agarwal RR, and Dutt SS. 1936. Chemical examination of punarnava or *B. diffusa* Linn. 2. Isolation of an alkaloid punarnavine. Chem. Abst. 30: 3585.

Agarwal SK, Singh SS, Verma S, and Kumar S. 2000. Antifungal activity of anthraquinone derivatives from *Rheum emodi*. J. Ethnopharmacol. 72, 43–6.

Ahmad WS. 1940. A note on the chemical examination of *Celastrus paniculatus*. Curr. Sci. 134–135.

Ahmed I, Chandranath I, Sharma AK, Adeghate E, Pallot DJ, and Singh J. 1999. Mechanism of hypoglycemic action of *Momordica charantia* fruit juice in normal and diabetic rats. The. J. Physiol. 520, 25.

Alen Y, Nakajima S, Nitoda T, Baba N, Kanzaki H, and Kawazu K. 2000a. Antinematodal activity of some tropical rain forest plants against the pinewood nematode, *Bursaphelenchus xylophilus*. Z. Naturforsch. 55, 295.

Alen Y, Nakajima S, Nitoda T, Baba N, Kanzaki H, and Kawazu K. 2000b. Two antinematodal phenolics from *Knema hookeriana*,a Sumatran rain forest plant. Z. Naturforsch. 55, 300.

Ali A, Kaur G, Hamid H, Abdullah T, Ali M, Niwa M, and Alam MS. 2003. Terminoside A, a new triterpene glycoside from the bark of *Terminalia arjuna* inhibits nitric oxide production in murine macrophages. J. Asian Nat. Prod. Res. 5, 137–42.

Amatya G, and Sthapit-VM. 1994. A note on *Nardostachys jatamansi*. J. Herbs, Spi. Med. Plants. 2, 39–47.

Ammon HP, Safayhi H, Mack T, and Sabieraj J. 1993. Mechanism of anti-inflammatory actions of curcumin and boswellic acids. J. Ethnopharmacol. 38, 113–119.

Anderson MM, O'Neill MJ, Phillipson JD, and Warhurst DC. 1991. *In vitro* cytotoxicity of a series of quassinoids from *Brucea javanica* fruits against KB cells. Planta Med. 57, 62.

Annoni F, Mauri A, Marincola F, and Resele LF. 1979. Venotonic activity of escin on the human saphenous vein. Arzneimittel-Forschung. 29, 672–5.

Annuk H, Hirmo S, Turi E, Mikelsaar M, Arak E, and Wadstrom T. 1999. Effect on cell surface hydrophobicity and susceptibility of Helicobacter pylori to medicinal plant extracts. Microbiol. Lett. 172, 41–5.

Anwar F. Popli S, Srivastava RM, and Khare MP. 1968. Studies in medicinal plants. Part 3. Protoberberine alkaloids from the roots of *Cissampelos pareira* Linn. Cell. Mol. Life Sci. 24: 999.

Arora RB, Singh M, and Kanta C. 1962. Tranquillizing activity of Jatamansone, a sesquiterpene from *Nardostachys jatamansi*. Life Sci. 6, 225.

Arora RB, Sharma JN, and Bhatia ML. 1967. Pharmacological evaluation of Peruvoside, a new cardiac glycoside from Thevetia neriifolia with a note on its clinical trials in patients with congestive cardiac failure. Ind. J. Exp. Biol. 5, 31–36.

Arora RB, Basu N, and Kakpoor V. 1971. Anti-inflammatory studies on *Curcuma longa* L. Ind. J. Med. Res. 59: 1289.

Arora S, Kaur K, and Kaur S. 2003. Indian medicinal plants as a reservoir of protective phytochemicals. Tera. Carcino. Muta. 23, 295–300.

Atal CK. 1980. Chemistry and Pharmacology of Vasicine-A new Oxytocic and Abortifacient. Regional Research Laboratory, Jammu, India.

Atta-Ur-Rahmann, Ali RA, Gilani A, Choudhary MI, Aftab K, Sener B, and Turkz S. 1993. Isolation of antihypertensive alkaloids from rhizomes of *Veratrum album*. Planta. Med. 59, 569.

Atul VS, and Kumaraguru Dhananjayan R. 1999. Effect of aegeline and lupeol-two cardioactive principles form the leaves of *Aegle marmelos* Corr. J. Pharmacol. 51, 252.

Babbar OP, Chhatwal VK and Ray IB. 1982. Effect of berberine chloride eye drops on clinically positive trachoma patients. Ind. J. Med. Res.76,83–88.

Badami S, Manorhara Reddy SA, Kumar EP, Vijayan P, and Suresh B. 2003. Antitumor activity of total alkaloid fraction of *Solanum pseudocapsicum* leaves. Phytother. Res. 17, 1001–4.

Bae EA, Han MJ, Kim NJ, and Kim DH. 1998. Anti-helicobacter pylori activity of herbal medicines. Biol. Pharmaceut. Bull. 21, 990–992.

Bajad S, Bedi KL, Singla AK, and Johri RK. 2001. Antidiarrheal activity of piperine in mice. Planta. Med. 67, 284–287.

Balanehru S, and Nagarajan B. 1991. Protective effect of oleanolic acid and ursolic acid against lipid peroxidation. Biochem. Int. 24, 981–990.

Banerjee R, Misra G, Nigam SN, Singh S, and Saxena RC. 1979. Steroid and terpenoid saponins as spermicidal agents. Ind. Drugs. 17, 6–8.

Banerjee S, Bandyopadhyay SK, Mukherjee PK, Mukherjee A, and Sikdar S. 1991. Further Studies on the anti-inflammatory activites of *Ricinus communis* in albino rats. Ind. J. Pharmacol. 23, 149–152.

Barik BR, Bhaumik T, Patra A, Roy S, Susan T, Alam M, Kundu AB, and Chatterjee A. 1976. Premnazole, an isoxazole alkaloid of *Premna integrifolia* L. and *Gmelina arborea* L. with anti-inflammatory activity. Fitoterapia. 639, 395.

Basu NK. 1957. The chemical constitution of achyranthine. J. Proc. Inorg. Chem. 29, 73–76.

Bell EA. 1980. The possible significance of secondary compounds in plants. In: E.A. Bell and B.V. Carlwood (eds) Secondary Plant Products Springer-Veriag, New York. pp. 11–21.

Berti F, Omini C, and Longiave D. 1977. The mode of action of aescin and the release of prostaglandins. Prostaglandins. 14, 241–9.

Bhakuni DS, Singh AN, and Sudha J. 1978. Biosynthesis of Oxyacanthine. J. Chem. Soc. Perkin Trans. 1, 1318.

Bhakuni OS, Dhar ML, Dhar MM, Dhawan BN, and Mehrotra BN. 1969. Screening of Indian plants for biological activity. Part II. Ind. J. Exper. Biol. 7, 250–262.

Bhan P, Dev S, Bass LS, Tagle B, and Clardy J. 1982. The stereochemistry of himachalol. J. Chem. Res. 2, 344–345.

Bhatnagar AK, Popli SP. 1967. Chemical examination of the roots of *Cissampelos pareira* Linn. V. Structure and stereochemistry. Experientia. 15.

Bhattacharya SK, Ghosal S, Chaudhuri RK, Singh AK, and Sharma PV. 1974. Chemical constituents of gentianaceae. XI. Antipsychotic activity of gentianine. J. Pharma. Sci. 63, 1341–2.

Bhattacharya SK, Reddy PK, Ghosal S, Singh AK, and Sharma PV. 1976. Chemical constituents of Gentianaceae XIX: CNS-depressant effects of swertiamarin. J. Pharma. Sci. 65, 1547–9.

Bhide MB, and Naik PY. 1980. Antiasthmatic potentiality of vasicinone-an alkaloid from *A. vasica* Nees. Abstracted in Proc. of Asian Sym. in Medicinal Plants and spices, Bangkok.

Bishayee A, and Chatterjee M. 1994. Hypolipidemic and antiatherosclerotic effects of oral *Gymnema sylvestre* R. Br. leaf extract in albino rats fed on a high fat diet. Phytother. Res. 8, 118–20.

Blake O, Booth R, and Corrigan D. 1994. The tannin content of herbal teas. Brit. J. Phytother. 3, 124–27.

Bos R, Woerdenborg HJ, van Putten FM, Hendriks H, and Scheffer JJ. 1993. Seasonal variation of the essential oil, valerenic acid derivatives, and valepotriates in *Valeriana officinalis* roots. Planta. Med. 59, 98.

Bosch J, Bonjoch J, and Amat M. 1996. The *Strychnos* alkaloids. In: GA Cordell (ed) The Alkaloids, Chemistry and Pharmacology Vol. 48. Academic Press, San Diego. pp. 75–189.

Bowman RM, Gray GA, and Grundon MF. 1973. Quinoline alkaloids. XV. Reactions of a quinoline isoprenyl epoxide with hydride reagents. Asymmetric synthesis and stereochemistry of lunacridine and related Lunasia alkaloids. J. Chem. Soc. 10, 1051.

Bruggeman IM and van der Hoeven JC.1985. Induction of SCEs by some pyrrolizidine alkaloids in V79 Chinese Hamster cells co-cultured with chick embryo hepatocytes. Mut. Res. 142, 209–212.

Carson CF, and Riley TV. 1995. Antimicrobial activity of the major components of the essential oil of *Melaleuca alternifolia*. J. App. Bacteriol. 78, 264–9.

Chakraborti SA, Mukherji G, and Datta PC. 1998. Comparative pharmacognosy of *Strychnos nux vomica* and *Strychnos potatorum* stem barks. Int. J. Crude. Drug. Res. 26, 121–126.

Chakraborty A, and Brantner AH. 2001. Study of alkaloids from *Adhatoda vasica* Nees on their anti-inflammatory activity. Phytother. Res. 15, 532–4.

Chakarborty A, Chowdhury BK, and Bhattacharya P. 1995. Clausenol and Clausenine-two carbazole alkaloids from *Clausena anisata*. Phytochemistry. 40, 295–8.

Chakarvarty AK, Sarkar T, Masuda K, and Shiojima K. 1963. Carbazole alkaloids from *Glycosmis arborea* (Roxb.) DC. Ann. Biochem. Exper. Med. 23, 123–126.

Chakravarty AK, Sarkar T, Masuda K, and Shiojima K. 1999. Carbazole alkaloids from roots of *Glycosmis arborea*. Phytochemistry. 50, 1263–1266.

Cham BE, and Daunter B. 1990. Solasodine glycosides. Selective toxicity for cancer cells and inhibition of cytotoxicity by rhamnose in mice with sarcoma 180. Can. Lett. 55, 221–225.

Chan MY. 1977. The effect of berberine on bilirubin excretion in the rat. Comp. Med. East West. 5, 161–168.

Chandan BK, Sharma AK, and Anand KK. 1991. *Boerhaavia diffusa*: a study of its hepatoprotective activity. J. Ethnopharmacol. 31, 299–307.

Chandra JS, and Sabir M. 1978. Modified method for isolation of palasonin—the anthelmintic principle of *Butea frondosa* seeds. Ind. J. Pharm. Sci. 40, 97–98.

Chatterjee A, and Dutta CP. 1967. Alkaloids of *Piper longum* L. Structure and synthesis of piperlongumine and piperlonguminine. Tetrahedron. 23, 1769–81.

Chatterjee M, and Dey CD. 1962. Pharmacological studies on chaksine chloride. IV. Action on some isolated tissues. Bull. Calcutta School Trop. Med. 10, 113–4.

Chauhan NS. 1999. Medicinal and Aromatic Plants of Himachal Pradesh. Indus Publishing Company, New Delhi. pp. 59–62.

Cheema MA, and Priddle OD. 1965. Pharmacological investigations of isochaksine, an alkaloid isolated from the seeds of *Cassia absus* Linn. (Chaksu). Arch. Int. Pharmacodyn. Ther. 158, 307–13.

Chen CX, Jin RM, Li YK, Zhong J, Yue L, Chen SC, and Zhou JX. 1992. Inhibitory effect of rhynchophylline on platelet aggregation and thrombosis. Chung. Kuo. Yao. Li. Hsueh. Pao. 13, 126–30.

Cheng DH, Ren H and Tang XC. 1996. Huperzine A, a novel promising acetylcholinesterase inhibitor. Neuroreport. 8, 97–101.

Chopra IC. 1958. Indigenous Drugs of India. Academic Publishers. Kolkata, West Bengal, India.

Cicslak J, Kuduk J, and Rulko F. 1964. *Gentiana lutea* alkaloids. Acta. Pol. Pharm. 21: 265–273.

Cozantis DA, and Toivakka E. 1978. Galanthamine. J. Ame. Med. Assoc. 240, 108.

Crozier A, Lean MEJ, Mcdonald MS, and Black C. 2001. Quantitative analysis of the flavonoids content of commercial tomatoes, onions, lettuce and celery. J. Agri. Food Chem. 45, 590–595.

Cuendet M, and Pezzuto JM. 2004. Antitumor activity of bruceantin: an old drug with new promise. J. Nat. Prod. 67, 269.

Damtoft S.1992. Iridoid glucosides from *Lamium album*. Phytochemistry. 31, 175.

Dasgupta B, and Basu K. 2005. Chemical investigation of *Abroma augusta* Linn. identity of abromine with betaine. Cell. Mol. Life Sci. 26, 477–78.

Dasgupta B, Basu K, and Dasgupta S. 1968. Chemical investigation of *Pluchea lanceolata*. II. Identity of pluchine with betaine hydrochloride. Experientia. 24, 882.

Datta SK, Sharma BN, and Sharma PV. 1978. Buchanine, a novel pyridine alkaloid from *Cryptolepis buchanani*. Phytochemistry. 17, 2047–48.

Deepak D, Srivastava S, and Khare A. 1997. Pregnane glycosides from *Hemidesmus indicus*. Phytochemistry. 44, 145–51.

Denyer CV, Jackson P, Loakes DM, Ellis MR, and Young DA. 1994. Isolation of antirhinoviral sesquiterpenes from ginger (Zingiber officinale). J. Nat. Prod. 57, 658–62.

Deodhar SD, Sethi R and Srimal RC. 1980. Preliminary studies on anti-rheumatic activity of curcumin. Ind.J. Med. Res. 71, 632–34.

Desai HK, Gawad DH, Govindachari TR, Joshi BS, Kamat VN, Modi JD, Parthasarathy PC, Radhakrishnan J, Shanbag MN, and Sighaye AR. 1973. Chemical invesigation of Indian plants, Part VII. Ind. J. Chem. 11, 840–842.

Dev S. 1989. Guggulsetrols: a new class of naturally occurring lipids. Pur. Appl. Chem. 61, 353–356.

Devi PU, Bisht KS, and Vinitha M. 1998. A comparative study of radioprotection by *Ocimum* flavonoids and synthetic aminothiol protectors in the mouse. Brit. J. Radiol. 71, 782–784.

Devi U, Solmon FE, and Sharda AC. 1999. Plumbagin, A Plant Napthoquinone with Antitumour and Radimodifying Properties. Pharm. Biol. 37, 231–236.

Dewick Paul M. 2001. Medicinal Natural Products—a biosynthetic approach. John Wiley & Sons, UK. 0-471-97477-3 (hardback).

Dey AC, Singh B, and Singh MP. 1994. Indian Medicinal Plants & Ayurvedic preparations. Dehra Dun, India.

De-YiX, Hong C, and Yuan-Yuan L. 1990. The antihypertensive effects by Stevioside in the conscious normal and hypertensive rats. Euro. J. Pharmacol. 183, 1822.

Dikshith TS, Raizada RB, and Mulchandani NB. 1990. Toxicity of pure alkaloid of *Tylophora asthamatica* in male rat. Ind. J. Exper. Biol. 28, 208–212.

Dixit VP. 1986. Antifertility effects of solasodine (C27 H 43 0 2 N) obtained from *Solanum xanthocarpum* berries in male rats and dogs. J. Ster. Biochem. 25, 27.

Dongre SH, Badami S, Ashok G, Ravi S, and Kottani R. 2007. *In vitro* cytotoxic properties of O-methylsolanocapsine isolated from *Solanum pseudocapsicum* leaves. Ind. J. Pharmacol. 39, 208–9.

Ducrey B. 1997. Inhibition of S alpha-Reductase and aromatase by ellagitannins oenothein A and eonothein B from Epilobium species. Planta. Med. 63, 111–114.

Duke J A. 1992. Handbook of phytochemical constituents of GRAS herbs and other economic plants. CRC Press, Boca Raton, FL.

Dunaouou C, Belle K, Oulad-Ali A, Anton R, and David B. 1997. Triterpenes and steroles from *Ruscus aculeatus*. Planta. Med. 62, 189–190.

Duwiejua M, Zeitlin IJ, Waterman PG, and et al. 1993. Anti-inflammatory Activity of Resins from Some Species of the Plant Family Burseraceae. Planta. Med. 59, 12–16.

Elford BC, Roberts MF, Phillipson JD, and Wilson RJ. 1987. Potentiation of the antimalarial activity of qinghaosu by methoxylated flavones. Trans. Soc. Trop. Med. Hyg. 81, 434–436.

Enghofer E, Seibel K, and Hammersen F. 1984. Antiexudative effect of *Aesculus hippocastanum* (horse chestnut) extract. Therapiewoche. 34, 4130–4144.

Erdelmeier CAJ. 1998. Hyperforin, possibly the major non-nitrogenous secondary metabolite of *Hypericum perforatum* L. Pharmacopsych. 31, 2–6.

Evans WC. 1990. *Datura*, a commercial source of hyoscine. Pharmaceut. J. 244, 651–653.

Ferreira FS, and Janick J. 1996. Distribution of artemisinin in *Artemisia annua*. In: J. Janick (ed) Progress in new crops. ASHS Press, Arlington, VA. pp. 579–584.

Filleur F, Le Bail JC, Duroux JL, Simon A, and Chulia AJ. 2001. Antiproliferative, antiaromatase, anti-17beta-HSD and antioxidant activities of lignans isolated from *Myristica argentea*. Planta Med. 67, 700.

Filleur, F, Pouget C, Allais DP, Kaouadji M, and Chulia AJ. 2002. Lignans and neolignans from *Myristica argentea* Warb. Nat. Prod. Lett. 16, 1.

Gale Jr. M, Carter V, and Parsons M. 1994. Cell cycle-specific induction of an 89 kDa serine/threonine protein kinase activity in Trypanosoma brucei. J. Cell. Sci. 107, 1825–1832.

Gear RW, Gordon NC, Heller PH, and Levine JD. 1995. Enhancement of morphine analgesia by the alpha 2-adrenergic antagonist yohimbine. Neuroscience. 66, 5–8.

Gentry EJ, Jampani HB, Keshavarz-Shokri A, Morton MD, Velde DV, Telkepalli H, Mitscher LA, Shawar R, Humble D, and Baker W. 1998. Antitubercular natural products: berberine from the roots of commercial Hydrastis canadensis powder. J. Nat. Prod. 61, 1187–1193.

Ghosal S, Saini KS, Razdan S, and Kumar Y. 1985. Chemical constituents of Amaryllidaceae. Part 12. Crinasiatine, a novel alkaloid from *Crinum asiaticum*. J. Chem. Res. 3, 100–101.

Ghosal S, Amrithalingam S, Mukhopadhyay M, Sarkar MK, and Das PK. 2004. Effect of lycoriside, an acylglucosyloxy alkaloid, on mast cells. Pharmaceut. Res. 3, 240–243.

Ghosh MN, and Singh H. 1974. Inhibitory effects of a pyrrolizidine alkaloid, crotalaburnine, on rat paw edema and cotton-pellet granuloma. Brit. J. Pharmacol. 51, 503.

Gopalakrishnan C, Shankaranarayan D, Kameswaran L and Natarajan S. 1979. Pharmacological investigations of tylophorine, the major alkaloid of *Tylophora indica*. Ind. J. Med.Res. 69, 513–520.

Gopalakrishnana G, Banumathi B, and Suresh G. 1997. Evaluation of the antifungal activity of natural xanthones from *Garcinia mangostana* and their synthetic derivatives. J. Nat. Prod. 60, 519–24.

Gottlieb OR. 1989. The role of oxygen in phytochemical evolution towards diversity. Phytochemistry. 28, 2359.

Gottlieb OR, and Kaplan MAC. 1993. Phytochemical evolution: the redox theory. Nat. Prod. Lett. 2, 171.

Gottlieb OR, and Borin MR. 1998. Natural products and evolutionary ecology. Pur. Appl. Chem. 70, 299.

Govindrajan VS. 1982. Ginger-Chemistry, Technology and Quality Evaluation. CRC Crit. Rev. Sci. Nutr. 17, 3.

Greeske K, Pohlmann BK. 1996. Horse chestnut seed extract-an effective therapy principle in general practice. Drug therapy of chronic venous insufficiency Fortschritte. der. Medizin. 114, 196–200.

Gribble GW. 1990. Synthesis and antitumour activity of ellipticine alkaloids and related compounds. In: A Brossi (ed) The Alkaloids, Chemistry and Pharmacology Vol. 39. Academic Press, San Diego. pp. 239–352.

Grotewold E. 2005. Plant metabolic diversity: a regulatory perspective. Trends. Plant. Sci. 10, 57–62.

Guha SK, and Chakravorty R. 1952. Effect of sporochin hydro-iodide on surgical and infective cases: a preliminary note. Journal of Indian Medical Association. 22: 97–99.

Guo X-F, and Chem CP. 1980. Isolation and identification of gentianine from Swertia davidi. Chem. Abstr. 94, 7129–3.

Gupta AP, Verma RK, Misra HO and Gupta MM. 1996. Quantitative determination of withaferin. A in different plant parts of *Withania somnifera* by TLC densitometry. J. Med. Arom. Plant Sci. 18,788-790.

Gupta KC, and Chopra IC. 1953. A short note on anti-bacterial properties of chaksine: an alkaloid from *Cassia absus* Linn. Ind. J. Med. Res. 41, 459–460.

Gupta OP, Ali MM, Ray Ghatak BJ, Atal CK. In vitro anthelmintic activity of Embelin and its semi synthetic derivatives. Indian J.Phy. Trans. Med. Phys.Soc. Calcutta., VII1., 85.

Gupta OP, Anand KK, Ghatak BJ, and Atal CK. 1978. Vasicine, alkaloid of *Adhatoda vasica*, a promising uterotonic abortifacient. Ind. J. Exper. Biol. 6, 1075–7.

Gupta OP, Sharma ML, Ghatak BJ, and Atal CK. 1997. Pharmacological investigations of vasicine and vasicinone—the alkaloids of *Adhatoda vasica*. Ind. J. Med. Res. 66, 680–691.

Gupta PN. 1981. Antileprotic action of an extract from 'Anantamul' (*Hemidesmus indicus* R. Br.) Lepr. Ind. 53, 354–359.

Gupta PP, Srimal RC, Srivastava M, Singh KL and Tandon JS. 1995. Antiallergic activity of Arbortristosides from *Nyctanthus arbortristis*. Int. J. Pharmacog. 33, 70–2.

Hamada S, Kataoka T, Woo JT, Yamada A, Yoshida T, Nishimura T, Otake N, and Nagai K. 1997. Immunosuppressive effects of gallic acid and chebulagic acid on CTL-mediated cytotoxicity. Biol. Pharmaceut. Bull. 20, 1017–9.

Hara Y. 1994. Antioxidative action of tea polyphenols: Part 1. Am. Biotech. Lab. 12, 48.

Harrison CMH, Page BM, and Keir HM. 1976. Mescaline as a mitotic spindle inhibitor. Nature. 260, 138–139.

Hartmann T, and Witte L. 1995. Chemistry, biology and chemoecology of the pyrrolizidine alkaloids. In: SW Pelletier (ed) Alkaloids, Chemical and Biological Perspectives. Vol. 9. Wiley, New York. pp. 155–233.

Hartmann T, Witte L, Oprach F and Toppel G. 1986.Reinvestigation of the alkaloid composition of *Atropa belladonna* plants, roots cultures and cell suspension. Planta Med. 53, 390–395.

Hase K, Xiong Q, Basnet P, Namba T, and Kadota S. 1999. Inhibitory effect of tetrahydroswertianolin on tumor necrosis factor-α-dependent hepatic apoptosis in mice. Biochem. Pharmacol. 57, 1431.

He X, Bernart M, Lian L, and Lin L. 1998. High-performance liquid chromatography-electrospray mass spectrometric analysis of pungent constituents of ginger J. *Chromatograp.* 796, 327–334.

Hendriks H, Bos R, Woerdenbag HJ, and Koster AS. 1985. Central Nervous Depressant Activity of Valerenic Acid in the Mouse.Planta. Med. 51, 28–31.

Heptinstall S, Awang DV, Dawson BA, Kindack D, Knight DW and May J. 1992. Parthenolide Content and Bioactivity of Feverfew (*Tanacetum parthenium*). Estimation of Commercial and Authenticated Feverfew Products. J. Pharm. Pharmacol. 1992, 44,391–5.

Hirata K, Ito C, Furukawa H, Itoigawa M, Cosentino LM, and Lee KH. 1999. Substituted 7H-pyrido [4,3-c]carbazoles with potent anti-HIV activity. Biorg. Med. Chem. Lett. 9, 119.

Horie S, Yano S, Aimi N, Sakai S and Watanabe K. 1992. Effects of hirsutine, an antihypertensive indole alkaloid from *Uncaria rhynchophylla,* on intracellular calcium in rat thoracic aorta. Life. Sci. 50, 491–498.

Hsieh TJ, Liu TZ, Chia YC, Chern CL, Lu FJ, Chuang MC, Mau SY, Chen SH, Syu YH, and Chen CH. 2004. Protective effect of methyl gallate from *Toona sinensis* (Meliaceae) against hydrogen peroxide-induced oxidative stress and DNA damage in MDCK cells. Food Chem. Toxicol. 42, 843.

Hsu YL, Kuo PL, and Lin CC. 2004. Acacetin inhibits the proliferation of Hep G2 by blocking cell cycle progression and inducing apoptosis. Biochem, Pharmacol. 67, 823–829.

Hsu YL, Kuo PL, Lin TT, and Lin TC. 2005. Isoliquiritigenin inhibits cell proliferation and nduces apoptosis in human hepatoma cells. Plant Med. 71, 130–134.

Hsu YL, Kuo PL, Lin TC, and Lin CC. 2003. Asiatic acid, a triterpene, induces apoptosis and cell cycle arrest through activation of extracellular signal-regulated kinase and p. 38 mitogen-activated protein kinase pathways in human breast cancer cells. J. Pharmacol. Exp. Ther. 313, 333–334.

Huang YC, Guh JH, and Teng CM. 2004. Induction of mitotic arrest and apoptosis by evodiamine in human leukemic T-lymphocytes. Life Sci. 75, 35.

Huang YL. 1992. Isontetralin: A new lignan from *Phyllanthus niruri.* Planta Med. 58, 473-474.

Jahodar L, Jilek P, Paktova M, and Dvorakova V. 1985. Antimicrobial effect of arbutin and an extract of the leaves of *Arctostaphylos uva-ursi in vitro*. Cesk. Farm. 34, 174–8.

Jain P, and Kulshreshtha DK. 1993. Bacoside A1, a minor saponin from *Bacopa monniera.* Phytochemistry. 33, 449–451.

Jenniskens LHD. 1996. Identification of six taxine alkaloids from *Taxus baccata* needles. J. Nat. Prod. 59, 117–123.

Jiwajinda S, Santisopasri V, Murakami A, Sugiyama H, Gasquet M, Riad E, Balansard G, and Ohigashi H. 2003. *In vitro* anti-tumor promoting and anti-parasitic activities of the quassinoids from *Eurycoma longifolia,* a medicinal plant in Southeast Asia. J. Ethnopharmacol. 85, 173.

Johnson EL, and Emcho SD. 1994. Variation in alkaloid content in *Erythroxylum coca* leaves from leaf bud to leaf drop. Ann. Bot. 73, 645–650.

Johnson IS, Armstrong JG, Gorman M, and Borneet JP. 1983. The vinca alkaloids: a new class of oncolytic agents. Cancer Res. 8, 1390–1427.

Joshi JDS. 1988. Chemistry of Ayurvedic crude drugs: Part VIII: Shatavari: 2. Structure elucidation of bioactive shatavarin I and other glycosides. Ind. J. Chem. 27, 12–16.

Joubert JP, and Schultz RA. 1967. Detection of scilliroside in the preparation of maritime Scille. (*Urginea maritima* Baker) Annals. Pharm. Fr. 42, 17–21.

Kalix P. 1991. The pharmacology of psychoactive alkaloids from *Ephedra* and *Catha*. J. Ethnopharmacol. 32, 201–208.

Kamat VN, Divekar PV, Fernandes F, and Bhatnagar MP. 1958. Studies in medicinal plants. Part 1. Characterization of chakranine, an alkaloid isolated from *Bragantia walichii* R.Br. (N.O. Aristolochiaceae). Ind. J. Med. Res. 46, 418–25.

Kaneda Y, Torii M, and Tanaka T. 1991. *In vitro* effects of berberine sulfate on the growth of Entamoeba histolytica, Giardia lamblia and Tricomonas vaginalis. Ann. Trop. Med. Parasitol. 85, 417–425.

Kashiwada Y, Aoshima A, Ikeshiro Y, Chen YP, Furukawa H, Itoigawa M, Fujioka T, Mihashi K, Cosentino LM, Morris-Natschke S, and Lee KH. 2005. Anti-HIV benzylisoquinoline alkaloids and flavonoids from the leaves of *Nelumbo nucifera*, and structure–activity correlations with related alkaloids. Bioorg. Med. Chem. 13, 443.

Kaur S, Grover IS, Singh M, and Kaur S. 1988. Antimutagenicity of hydrolyzable tannins from *Terminalia chebula* in *Salmonella typhimurium*. Mut. Res. 419, 169–79.

Kaur S, Grover IS, and Kumar S. 1997. Antimutagenic potential of ellagic acid isolated from *Terminalia arjuna*. Ind. J. Exp. Biol. 35, 478–82.

Kavitha D, Shilpa PN, and Niranjali Devaraj S. 2004. Antibacterial and antidiarrhoeal effects of alkaloids of *Holarrhena antidysenterica. Wall.* Ind. J. Exp. Biol. 42, 595–600.

Kawai KY, Nozawa KY, and Ogihara Y. 1977. Biochemical studies on peptide alkaloids: Induction of ion-selective mitochondrial swelling. Experientia. 33, 1454.

Kedzia B, Wrocinski T, Mrugasiewicz K, Gorecki P, and Grzewinska H. 1975. Antibacterial action of urine containing products of arbutin metabolism. Med. Dosw. Mikrobiol. 27, 305–14.

Kelsey RG, and Vance NC. 1992. Taxol and cephalomannine concentrations in the foliage and bark of shade-grown and sunexposed *Taxus baccata* trees. J. Nat. Prod. 55, 912–917.

Kersey RG, and Vance NC. 1996. Taxol and cephaomanine from *Taxus baccata* needles. J. Nat. Prod. 59, 117–123.

Khan MR, Kihara M, and Omoloso AD. 2000. Antimicrobial activity of *Evodia elleryana*. Fitoterapia. 71, 72.

Khanna P, and Bansal R. 1975. Phyllantidine & phyllantine from *Emblica officinalis* Gaertn leaves, fruits, & in vitro tissue cultures. Ind. J. Exper. Biol. 13, 82–83.

Kikuzaki H, Kobayashi M, and Nakatani N. 1991. Diarylheptanoids from rhizomes of *Zingiber officinale*. Phytochemistry. 30, 3647–3651.

Kil KS, Yum YN, and Seo SH. 1996. Antitumor activites of hypericin a protein tyrosine kinase blocker. Pharmacol. Res. 19, 490–496.

Kirmizigül S, Gören N, Yang S W, Cordell GA and Bozok-Johansson C. 1997. Spinonin. A novel glycoside from *Ononis spinosa* subsp, *leiosperma*. J. Nat. Prod. 60, 378–381.

Kitagawa I, Minagawa K, Zhang RS, Hori K, Doi M, Inoue M, Ishida T, Kimura M, Uji T, and Shibuya H. 1993. Dehatrine, an antimalarial bisbenzylisoquinoline alkaloid from the Indonesian medicinal plant *Beilschmiedia madang*, isolated as a mixture of two rotational isomers. Chem.Pharm. Bull. 41, 997.

Kitagawa I, Mahmud T, Yokota K, Nakagawa S, Mayumi T, Kobayashi M, and Shibuya H. 1996. Indonesian medicinal plants. XVII. Characterization of quassinoids from the stems of *Quassia indica*. Chem. Pharm. Bull. 44, 2009.

Kitanaka S, Takido M, Mizoue K, and Nakaike S. 1996. Cytotoxic cardenolides from woods of *Euonymus alata*. Chem. Pharma. Bull. 26, 615–617.

Knabe J, Baldauf J, and Hanke B. 1988. Biological activities of phaeantharine chloride and some synthetic intermediates. Arch. Pharm. 321, 35.

Kovalskaya NE, and Sokolova IV. 2001. Photophysical propeties of furocoumarins. Siberian Physical Technical Institute, Novo-Sobornaya, Tomsk, 634050, Russia.

Krishnamurthy TR. 1959. Some pharmacological actions of evolvine hydrochloride. Curr. Sci. 28, 64–65.

Kulinkovich O, and Masalov N. 1996. New synthetic route to the alkaloid whithasomnine by ring transformation of a functionalized cyclopropanol via the parent pyrrolo (1,2-b) pyrazole. Tetrahedron. Lett. 37, 1095–1096.

Kumar A, and Ali M. 2000. A new steroidal alkaloid from the seeds of *Holarrhena antidysenterica*. Fitoterapia. 71,101–4.

Kumaraguru Atul VS, and Dhananjayan R. 1999. Effect of aegeline and lupeol-two cardio active principles isolated from leavers of *Aegele marmelos* Corr. J. Pharm. Pharmacol. 51, 252.

Kuo PL, and Lin TC. 2003. Green Tea constituent (-) Epigallocatechin-3-Gallate inhibits Hep G2 cell proliferation and nduces apoptosis through p53-dependent and Fas-mediated apoptotic pathway in A549 cells. J. Biomed. Sci. 10, 219–227.

Kuo PL, Hsu YL, Lin TC, and Lin TC. 2005. The antiproliferative activity of prodelphidin B-2'3'-O-gallate from green tea leaf is through cell cycle arrest and Fas-mediated apoptocic pathway in A549 cells. Food Chem. Toxicol. 43, 315–323.

Kuo PL, Cho CY, Hsu YL, Lin TC, and Lin CC. 2005. Putranjivan A form Euphorbia jolkini inhibits proliferation of human breast adenocarcinoma MCF-7 cells via blocking cell cycle progression and inducing apoptosis. Toxicol. App. Pharmacol. 32, 323–325.

Kuo PL, Hsu YL, Lin TC, Lin TC, Chang JK, and Lin CC. 2001. Casuarinin from the bark of *Terminalia arjuna* Linn. Induces apoptosis and cell cycle arrest in human breast adenocarcinoma MCF-7 cells. Plant Med. 71, 237–243.

Kupchan SM, and Baxter RL. 1991. Mezerein: antileukemic principle isolated from *Daphne mezereum* L. Arzneimittelforschung. 41, 72–76.

Kurup PA, and Rao PL. 1954. Antibiotic principle from *Moringa pterygosperma*. 2. Chemical nature of pterygospermin. Ind. J. Med. Res. 42, 85–95.

Kutchan TM. 1995. Alkaloid biosynthesis—the basis for metabolic engineering of medicinal plants. Plant. Cell. 7, 1059–1070.

Lahiri SC, and Dutta NK. 1967. Berberine and chloramphenicol in the treatment of cholera and severe diarrhoea. J. Ind. Med. Assoc. 48, 1.

Lal JB and SJ Dutt. 1993. Constitution of the colouring matter of *Lawsonia alba* (or Indian mehndi). J. Ind. Chem. Soc. 10, 577–582.

Lalla JK, Hamrapurkar PD, and Mukherjee SA. 2005. Isolation and characterization of Saussurine. Ind. Drugs. 42, 209–212.

Leskow P. 1996. Effective treatment with horse chestnut extract (Hoevenol(TM) Capsules) in chronic vein insufficiency. Therapiewoche. 46, 874–877.

Leung AY. 1980. Encyclopedia of Common Natural Ingredients used in Food, Drugs and Cosmetics. 1st. Ed. John Wiley, NY, USA.

Likhitwitayawuid K, Chanmahasathien W, Ruangrungsi N, and Krungkrai J. 1998, Xanthones with antimalarial activity from *Garcinia dulcis*. Planta. Med. 64, 281–282.

Lin CH, Chang GJ, Su MJ, Wu YC, Teng CM, Ko FN. 1994. Pharmacological characteristics of liriodenine, isolated from *Fissistigma glaucescens*, a novel muscarinic receptor antagonist in guinea-pigs. Br. J. Pharmacol. 113, 275.

Liu CX, Xiao PG, and Liu GS. 1991. Studies on plant resources, pharmacology and clinical treatment with berbamine. Phytother. Res. 5, 228–230.

Liu K, Yang S-L, Roberts MF, Elford BC, and Phillipson JD. 1992. Antimalarial activity of *Artemisia annua* flavonoids from whole plants and cell cultures. Planta. Cell. Rep. 11, 637–640.

Liu N, Bonnet F, Delaunay L, Kermarek N, and D'Honneur G. 1993. Partial reversal of the effects of extradural clonidine by oral yohimbine in postoperative patients. British J. Anaesth. 70, 515–8.

Liu YL, Ho DK, and Cassady JM. 1992. Isolation of potential cancer chemopreventive agents from *Eriodictyon californicum*. J. Nat. Prod. 55, 357–363.

Lounasmaa M. 1994. On the structure of the indole alkaloid ajmalicidine. Planta. Med. 60, 480.

Luduena RF, and Roach MC. 1981. Contrasting effects of maytansine and vinblastine on the alkylation of tubulin sulfhydryls. Arch. Biochem. Biophys. 210, 498–504.

Maeda K, and Fukuda M. 1996. Arbutin: mechanism of its depigmenting action in human melanocyte cultures. J. Pharmacol. Exp. Ther. 276, 765–9.

Mahato SB, Garai S, and Chakravarty AK. 2000. Bacopasaponins E and F: two jujubogenin bisdesmosides from *Bacopa monniera*. Phytochemistry. 53, 711–4.

Majinda RR, Waigh RD, and Waterman PG. 1997. Bufadienolides and other constituents of *Urginea sanguinea*. Planta. Med. 63, 188–190.

Mandal S, and Chaterjee A. 1987. Structure of chiratanin-a novel dimeric xanthone. Tetrahedron. Lett. 28, 1309–1310.

Mangathayaru K., Thirumurugan D, Patel PS, Pratap DVV, David DJ and Karthikeyan J. 2006. Isolation and identification of nicotine from *Leucas aspera* (Willd) Link. Ind. J. Pharmaceut, Sci. 68, 88–90.

Manickam M, Ramanathan M, Jahromi MA, Chansouria JP, and Ray AB. 1997. Antihyperglycemic activity of phenolics from Pterocarpus marsupium. J. Nat. Prod. 60, 609–610.

Manske RHF (ed). 1950. The Alkaloids. Vol. 1. Academic Press, New York. pp. 33–206.

Matsuda H, Morikawa T, Oda M, Asao Y, and Yoshikawa M. 2003. Potent anti-metastatic activity of dimeric sesquiterpene thioalkaloids from the rhizome of *Nuphar pumilum* Bioorg. Med. Chem. Lett. 13, 4445.

Meier B, Shao Y, Julkunan-Tiitto R, Bettschart A, and Sticher O. 1992. A chemotaxonomic survey of phenolic compounds in swiss willow species. Planta. Med. 58, 698.

Mohan M, Pant CR, Angra SK, and Mahajan VM. 1982. Berberine in trachoma. Ind. J. Opthalmol. 30, 69–75.

Morimoto I, Nozaka T, Watanbe F, Ishino M, Hirose Y, and Okitsu T. 1983. Mutagenic activites of gentisin and isogentisin from Gentianae radix (Gentianaceae). Mutat. Res. 116: 103–117.

Mulchandani NB, and Hassarajani SA. 1984. 4 methoxy-nor-securinine, a new alkaloid from *Phyllanthus niruri*. Planta. Med. 50, 104–105.

Nadkarni AK, and Nadkarni KM. 1976. Indian Materia Medica. Popular Prakshan, Mumbai.

Nahrstedt A, and Butterweck V. 1997. Biologically active and other chemical constituents of Hypericum perforatum L. Pharmacopsyciat. 30, 129–134.

Nakashima N, Kimura I, Kimura M, and Matsuura H. 1985. Isolation of pseudoprototimosaponin AIII from rhizomes of *Anemarrhena asphodeloides* and its hypoglycemic activity in streptozotocin induced diabetic mice. J. Nat. Prod. 57.

Nakatani N, Ikeda K, Kikuzaki H, Kido M, and Yamaguchi Y. 1988. Diaryldimethylbutane lignans from *Myristica argentea* and their antimicrobial action against *Streptococcus mutans*. Phytochemistry. 27, 3127.

Natarajan PN, Wan AS, and Zaman V. 1974. Antimalarial, antiamoebic and toxicity tests on gentianine. Planta Med. 25, 258–60.

Neerja P, Jain DC, and Bhakuni RS. 2000. Phytochemicals from genus *Swertia* and their biological activities. Ind. J. Chem. 39B, 565.

Neogi NC, Rathor RS, Shrestha AD, and Banerjee DK. 1969. Studies on the anti-inflammatory and antiarthritic activity of achyranthine. Ind. J. Pharmacol. 3, 37–47.

Nikolaeva GG, Glyzin VI, Mladentseva MS, Sheichenko VI and Patudin AV. 2004. Xanthones of *Gentiana lutea*. Chem. Nat. Comp. 19, 106–107.

Nishino H, Kitagawa K, Fujiki H, and Iwashima A. 1986. Berberine sulfate inhibits tumor-promoting activity of teleocidin in two stage carcinogenesis on mouse skin. Oncology. 43, 131–134.

Noble RL. 1990. The discovery of the Vinca alkaloids—chemotherapeutic agents against cancer. Biochem. Cell. Biol. 68, 1344–1351.

Okamoto RS, Sakamoto K, Noguchi K, Kobayashi Y, Suzuki T, and Torii Y. 1976. Effects of ferulic acid on FSH, LH and Prolactin levels in serum and pituitary tisuue of male rats. Proc. Jap. Acad. 52, 264–267.

Okuda J, Miwa I, Inagaki K, Horie T, and Nakayama M. 1982. In hibition of aldose reducatses from rat and bovine lenses by flavonoids. Bioch. Pharmacol. 31, 3807–3822.

Olin J, and Schneider L. 2001. Galantamine of Alzheimer's disease. Coch. Data. Syst. Rev. 1, 1747.

Panda S, and Kar A. 2006. Evaluation of antithyroid, antioxidative and antihyperglycemic activity of scopoletin from *Aegle marmelos* leaves in hyperthyroid rats. Phytother. Res. 20, 1103–1105.

Park S. 1994. The repression of listeriolysin O expression in Listeria monocytogenes by the phenolic beta-D-glucoside, arbutin. Lett. Appl. Microbio. 19, 258–60.

Patil VD, Nayak UR, and Der S. 1972. Chemistry of Ayurvedic Crude Drugs-1. Guggulu (resin from *Commiphora mukul*) 1. Steroidal constituents. Tetrahedron. 28, 2341–2352

Pei YQ. 1983. A review of pharmacology and clinical use of piperine and its derivatives. Epilepsia. 24, 177–82.

Perez-Victoria JM, Chiquero MJ, Conseil G, Dayan G, Di Pietro A, Barron D, Castanys S, and Gamarro F. 1999. Correlation between the affinity of flavonoids binding to the cytosolic site of Leishmania tropica multidrug transporter and their efficiency to revert parasite resistance to daunomycin. Biochemistry. 38, 1736–1743.

Pesonen S, and Ramstad E. 1956. Studies on gentiopicrin. J. Am. Pharm. Assoc. 45, 522–525.

Popik P, and Skolnick P. 1999. Pharmacology of ibogaine and ibogaine-related alkaloids. In: GA Cordell (ed) The Alkaloids, Chemistry and Biology. Vol. 52. Academic Press, San Diego. pp. 197–231.

Prabakan M, Anandan R, and Devaki T. 2000. Protective effect of *Hemidesmus indicus* against rifampicin and isoniazid-induced hepatotoxicity in rats. Fitoterapia. 71, 55–9.

Prakasarao ASC, Bhalla VK, Nayak UR, and Dev S. 1973. Meroterpenoids. II. *Psoralea corylifolia* Linn. 2. Absolute configuration of (+) - bakuchiol. Tetrahedron. 29, 1127–1130.

Rajani M, and Pundarikakshudu K. 1996. A note on seasonal variation of alkaloids in *Adhatoda vasica* Nees. Int. J. Pharmacog. 34, 43.

Rao GS, Sinsheimer JE, and Mcllhenny HM. 1960. Structure of gymnamine, a trace alkaloid from *G. sylverstre* leaves. Chem. Ind. 13, 537.

Rao KN, Bhattacharya RK, and Veanlatchalam SR. 1999. Inhibition of thymidylate synthase by pergularinine, tylophorinidine and deoxytubulosine. Ind. J. Biochem. Biophys. 36, 442–8.

Rastogi S, Pal R, and Kulshreshtha DK. 1994. Bacoside A3—a triterpenoid saponin from *Bacopa monniera*. Phytochemistry. 36, 133–137.

Ray S, Majumdar HK, Chakarvarty AK, Mukhopadhyay S, Gil KR, and Cordell GA. 1996. Amarogentin, a naturally occurring secoiridoid glycoside and a newly recognized inhibitor of topoisomerase 1 from Leishmania donovani. J. Nat. Prod. 59, 27–9.

Ribeiro A, Pilo-Veloso D, Romanha AJ, and Zani CL. 1997. Trypanocidal .avonoids from Trixis vauthieri. Journal of Natural Products. 60, 836–841.

Rizvi SI, Abu ZM, and Suhail M. 1995. Insulin-mimetic effect of (-) epicatechin on osmotic fragility of human erythrocytes. Ind. J. Exper. Biol. 33, 791–2.

Rojas A, Cruz S, Rauch Y, Bye R, and Mata R. 1995. Spasmolytic potential of some plants used in Mexican traditional medicine for the treatment of gastrointestinal disorders. Phytomedicine. 2, 51–55.

Rothkopf M, Vogel G, Lang W, and Leng E. 1977. Animal experiments on the question of the renal toleration of the horse chestnut saponin aescin. Arzneimittel-Forschung. 27, 598–605.

Roussel F, Dalton J, Wissocq JC. 1997. Cytotoxic cardenolides from woods of *Euonymus alata*. Chem. Pharma. Bull. 26: 615–618.

Robinson HM. 1991. Vinca revisited—another happen stance in the discovery of vinblastine. Biochem. Cell Biol. 69, 581–582

Roy MK, Thalang VN, Trakoontivakorn G, and Nakahara K. 2004. Mechanism of mahanine induced apoptosis in human leukemia cells (HL-60). Biochem. Pharmacol. 67, 41.

Roy SK, and et al. 1989. TLC Separation and Quantitative Determination of Guggulsterones. Ind. J. Pharm. Sci. 2, 251–253.

Roy SK, Ali M, Sharma MP, and Ramachandram R. 2001. New pentacyclic triterpenes from the roots of *Hemidesmus indicus*. Pharmazie. 56, 244–6.

Rücker G, Tautges J, Sieck A, Wenzl H and Graf E. 1978. Studies in isolation and pharmacological activity of sesquiterpene Valeranone from *Nardostachys jatamansi*. Arzniem. Forsch. 28, 713.

Sack RB, and Froehlich JL. 1982. Berberine inhibits intestinal secretory response of Vibrio cholerae toxins and Escherichia coli enterotoxins. Infect Immun. 35, 471–475.

Safayhi I, Mack T, Sabieraj J, Anazodo M I, Subramaniam LR, and Ammon HPT. 1992. Boswellic acids: novel, specific, non redox inhibitors of 5-lipoxygenase. J. Pharmacol. Exper. Thera. 261, 1143–1146.

Saksena S, and Dixit VK. 1976. Role of the total alkaloids of *Mucuna prurita* Baker in spermatogenesis in albino rats. Ind. J. Nat. Prod. 3, 3.

Saraswat B, Visen PK, Patnaik GK, and Dhawan BN. 1997. Protective effect of picroliv, active constituent of *Picrorhiza kurrooa*, against oxytetracycline induced hepatic damage. Ind. J. Exper. Biol. 35, 1302–5.

Sarah V, Fuchs K, and Richter E. 2006. Genotoxic effects of myosmine in a human esophageal adenocarcinoma cell line. Toxicology. 222, 71–79.

Sato, and Muro T. 1974. Antiviral activity of scillarenin, a plant bufadienolide. Japan. J. Microbiol. 18, 441–448.

Satyan KS, Jaiswal AK, Ghosal S, and Bhattacharya SK. 1998. Anxiolytic activity of ginkgolic acid conjugates from Indian Ginkgo biloba. Psychopharmacol. 136, 148–52.

Sawant M, Isaac JC, and Naraynan S. 1997. Analgesic studies on total alkaloids and alcohol extracts of *Eclipta alba* (Linn.) Hassk. Phytother. Res. 18, 111–113.

Saxena AM, Murthy PS, and Mukherjee SK. 1996. Mode of action of three structurally different hypoglycemic agents: a comparative study. Ind. J. Exper. Biol. 34, 406–9.

Schmeda-Hirschmann G, Astudillo L, Bastida, J, Codina C, Rojas De Arias A, Ferreira ME, Inchaustti A, and Yaluff G. 2001. Cryptofolione derivatives from *Cryptocarya alba* fruits. J. Pharm. Pharmacol. 53, 563.

Schmeda-Hirschmann G, Astudillo L, Bastida J, Codina C, Rojas De Arias A, Ferreira ME, Inchaustti A, and Yaluff G. 2001. Cryptofolione derivatives from *Cryptocarya alba* fruits. J. Pharm. Pharmacol. 53, 563.

Schmeller T, Sauerwein M, Sporer F, Wink M and Müller WE. 1994. Binding of quinolizidine alkaloids to nicotinic and muscarinic acetylcholine receptors. J. Nat. Prod. 57, 1316–1319.

Schmid B, and Heide L. 1995.The use of Salicis cortex in rheumatic disease: phytotherapie with known mode of action. Planta Medica. 61(Abstracts of 43 rd Ann Congr): 94.

Sekar BC, Mukherjee B, Chakravarti RB, and Mukherjee SK. 1987. Effect of different fractions of *Swertia chirayita* on the blood sugar level of albino rats. J. Ethnopharmacol. 21, 175–81.

Selvam C, and Jachak SM. 2004. A cyclooxygenase (COX) inhibitory biflavonoid from the seeds of *Semecarpus anacardium*. J. Ethnopharmacol. 95, 209.

Shao Y, Poobrasert O, Kennelly EJ, Chin CK, Ho CT, Huang MT, Garrison SA and Cordell GA. 1997. Steroidal saponins from Asparagus officinalis and their cytotoxicity activity. Planta Med. 63, 258–262.

Sharma BR, Rattan RK, and Sharma P. 1981. Marmeline, an alkaloid and other componenets of unripe fruits of *Aegle marmelos*. Phytochemistry. 20, 2606–2607.

Sharma DK, and Hall IH. 1991. Hypolipidemic, anti-inflammatory and antineoplastic activity and cytotoxicity of flavonolignans isolated from *Hydnocarpus wightiana* seeds. Nat. Prod. 54, 1298–302.

Sharma YN, Sharma RC, Zaman A, and Kidwai AR. 1964. Chemical Examination of *Heracleum candicans*-2. Isolation and Structure of a New Furocoumarin Heraclenol. Tetrahedron. 20, 87.

Sharma YN, Zaman A, and Kidwai AR. 1964. Chemical Examination of Heracleum candicans-1. Isolation and Structure of a New Furocoumarin Heraclenin. Tetrahedron. 20, 87–90.

Shelley WB, and Arthur RP. 1955. Mucunin, the active pruritogenic proteinsae of cowhage. Science. 122, 469–470.

Shi Q, Chen K, Li L, Chang JJ, Autry C, Kozuka M, Konoshima T, Estes JR, Lin CM, and Hamel E. 1995. Antitumor agents, 154. Cytotoxic and antimitotic flavonols from *Polanisia dodecandra*. J. Nat. Prod. 58, 1470.

Shukla B, Visen PK, Patnaik GK, and Dhawan BN. 1992. Cholretic effect of Andrographolide in rats and guinea pigs. Planta. Med. 58, 146–149.

Sigler P, Saksena R, Deepak D, and Khare A. 2000. C21 steroidal glycosides from *Hemidesmus indicus*. Phytochemistry. 54, 983–7.

Simonds MA. 1980. Evidence that bicuculline and picrotoxin acts as separate sites to antagonize γ-aminobutryic acid in rat cuneate nucleus. Neuropharmacol. 19, 39–45.

Southwick SM, Morgan CA, Charney DS, and High JR. 1999. Yohimbine use in a natural setting: effects on posttraumatic stress disorder. Biol. Psychiatry. 46, 442–4.

Steinegger E, and Weibel T. 1951. Investigations on the alkaloids of official Gentianaceae. Pharm. Acta. Helv. 26, 259–270.

Stoll B, and Kries W. 1933. Glycesides of Digitalis lanata. Helvetia Chimica Acta 16, 1049–1059.

Subbarao G, and Bramley R. 1971. Hypophyllanthin. Tetrahedron. Lett. 34, 3175.

Sukumar E, Hamsaveni GR, and Bhima RR. 1996. Pharmacological actions of cerpegin, a novel pyridine alkaloid from *Ceropegia juncea*. Fitoterapia. 66, 403–6.

Sun D, Courtney HS, and Beachey EH. 1998. Berberine sulfate blocks adherence of Streptococcus pyogenes to epithelial cells, tibronectin, and hexadecane. Antimicrob. Agents. Chemother. 32, 1370–1374.

Sunthitikawinsakul A, Kongkathip N, Kongkathip B, Phonnakhu S, Daly JW, Spande TF, Nimit Y, and Rochanaruangrai S. 2003. Coumarins and carbazoles from *Clausena excavata* exhibited antimycobacterial and antifungal activities. Planta Med. 69, 155.

Surh YJ, Park KK, Chun KS, Lee LJ, Lee E, and Lee SS. 1999. Anti-tumor-promoting activities of selected pungent phenolic substances present in ginger. J. Environ. Pathol. Toxicol. Oncol. 18, 131–9.

Sutradhar RK, Rahman AM, Ahmad M, Bachar SC, Saha A, and Guha SK. 2006. Bioactive alkaloid from *Sida cordifolia* Linn. with analgesic and anti-inflammatory Activites. Iranian. J. Pharmacol. Thera. 175–78.

Swabb EA, Tai YH, and Jordan L. 1981. Reversal of cholera toxin induced secretion in rat ileum by luminal berberine. Am. J. Physiol. 241, 248–252.

Swaminathan K, Sinha UC, Bhatt RK, Sabata BK, and Tavale SS. 1989. Structure of tinosporide, a diterpenoid furanolactone from *Tinospora cordifolia* Miers. Acta. Cryst. 45, 134–136.

Swant M, Issac JC, and Narayanan S. 2004. Analgesic studies on total alkaloids and alcohol extracts of *Eclipta alba* (Linn.) Hassk. Phytother. Res. 18, 111–3.

Syamasundar KV, Singh B, Thakur RS, Huasin A, Kiso Y and Hikino H. 1985.Antihepatotoxic principles of *Phyllanthus niruri* herbs. J. Ethnopharmacology. 14, 41–4.

Tobin T, Akera T, Brody SL, Ku D, and Brody TM. 1975. Cassine: mechanism of inhibition of Na+ K+ -ATPase and relationship of this inhibition to cardiotonic actions. Eur. J. Pharmacol. 32, 133–145.

Toyomizu M, Okamoto K, Akiba Y, Nakatsu T, and Konishi, T. 2002. Anacardic acid-mediated changes in membrane potential and pH gradient across liposomal membranes. Biochem. Biophys. Acta. Biomembranes. 1558, 54.

Tripathi YB, Sharma M, and Manickam M. 1997. Rubidianin, a new antioxidant from *Rubia cordifolia*. In. J. Biochem. Biophys. 34, 302–6.

Upadhyay RK, Pandey MB, Jha RN, and Pandey VB. 2001. Eclabatin, a triterpene saponin from *Eclipta alba*. J. Asian Nat. Prod. Res. 3, 213–7.

Valsaraj R, Pushpangadan P, Smitt UW, Adsersen A, Christensen SB, Sittie A, Nyman U, Nielsen C and Olsen CE.1997. New Anti HIV-1, antimalarial and antifungal compounds from *Terminalia belerica*. J. Nat. Prod. 60, 739–742.

Vidensek N, Lim P, Campbell A, and Carlson C. 1994. Taxol content in bark, wood, root, leaf, twig and seedling from several Taxus species. J. Nat. Prod. 53, 1609–1610.

Visen B, Shukla G, Patnaik S, Tripathi D, Kulshreshtha R, and Dhawan B. 1992. Hepatoprotective activity of *Ricinus communis* leaves. Int. J. Pharmacog. 30, 241–250.

Visen PK, Shukla B, Patnaik GK, and Dhawan BN. 1993. Andrographolide protects rat hepatocytes against paracetamol-induced damage. J. Ethnopharmacol. 40, 131–136.

Weniger B, Vonthron-Sénécheau C, Arango G J, Kaiser M, Brun R, and Anton R. 2004. A bioactive biflavonoid from *Campnosperma panamense*. Fitoterapia 75, 764.

Wu JH, Chang FR, Hayashi K, Shiraki H, Liaw CC, Nakanishi Y, Bastow KF, Yu D, Chen IS, and Lee, KH. 2003. Antitumor agents. Part 218: Cappamensin A, a new *in vitro* anticancer principle, from *Capparis sikkimensis*. Bioorg. Med. Chem. Lett. 13, 2223.

Xiao JH, Zhang JH, Chen HL, Feng XL, and Wang JL. 2005. Inhibitory effects of isoliensinine on bleomycin-induced pulmonary fibrosis in mice. Planta. Med. 71, 225.

Xie B, Shi H, Chen Q, and Ho CT. 1993. Antioxidant properties of fractions and polyphenol constituents from green, oolong and black teas. Proc. Natl. Sci. Counc. Repub. China. 17, 77–84.

Xu YM, Deng IZ, Ma J, Chen SN, Marshall R, Jones SH, Johnson RK, and Hecht SM. 2003. DNA damaging activity of ellagic acid derivatives. Bioorg. Med. Chem. 11, 1593.

Ya BQ, Nian LC, Li C, and Gen XP. 1999. Protective effect of swerchirin on hematopoiesis in 60 Co-irridiated mice. Phytomedicine. 6, 85–88.

Zavala F, Guenard D, and Potier P. 1978. Interaction of vinblastine analouges with tubulin. Experientia. 34, 1497–1499.

Zhang RX, Dougherty DV, and Rosenblum ML. 1990. Laboratory studies of berberine used alone and in combination with 1,3- bis(2-chloroethyl)-l-nitrosourea to treat malignant brain tumors. Chinese. Med. J. 103, 658–665.

Zhang Y, Zhang QH, Wu LJ, Tashiro S, Onodera S, and Ikejima T. 2004. Atypical apoptosis in L929 cells induced by evodiamine isolated from *Evodia rutaecarpa*. J. Asian Nat. Prod. Res. 6, 19.

Chapter 6

Phytochemicals Research-Emerging Concepts

6.1 Introduction

The active principles in medicinal plants are chemical compounds known as secondary plant products or metabolites. Primary plant metabolites can be considered as those metabolites essential to the life of the plant. Ernst Stahl was the pioneer researcher of secondary metabolites. Many biologically active principles have been discovered from the plant kingdom.

Some of them discourage herbivores, other inhibit bacterial or fungal pathogens. The biological significance of many secondary metabolites is not exactly known. Secondary metabolites of *Rhodiola rosea* are depicted in Fig. 6.1.

Secondary metabolites are produced by routes other than the normal metabolic pathways, mostly after the phase of active growth and under conditions of deficiency. Normally the following pathways are involved in biosynthesis of secondary metabolites:

1. The acetate-malonate pathway
2. Shikiamte pathway
3. Hydroxybenzoate pathway
4. Homegentisate pathway

Primary and secondary compound synthesis in plants is represented in Fig. 6.2.

Biosynthesis of specialty phytochemicals has been represented in the Figs. 6.3 and 6.4.

The active principle or constituent of a drug is usually an alkaloid or glycoside, on the presence of which the characteristic therapeutic action of the substance largely depends. Herbal remedies have been used for centuries but more recently the compounds that are active have

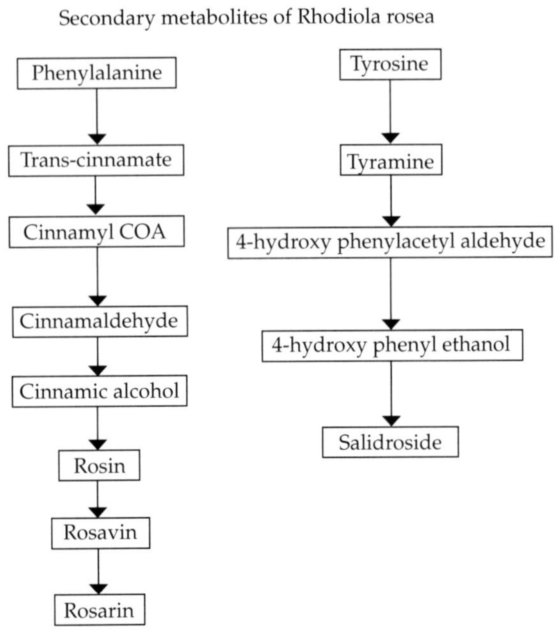

Fig. 6.1 Secondary metabolites of *Rhodiola rosea.*

been identified and this has enabled them to be extracted and purified. Organic chemists have then been able to provide the molecules *in vitro* and so produce them on larger scales. History clearly implies that herbal medicines will continue to play a role in the modern drug discovery and development of new drugs. The principal concepts of this process are shown in Fig. 6.5.

As suggested in the Figure 6.5, not only single active principles, but also active fractions are validated and improved prescriptions can be investigated for new drug discovery. However, issues should be addressed by transferring herbal medicines from folkloric to mainstream international pharmaceutical use, such as supply, quality control, safety, and proven efficacy. High quality, batch-to-batch consistency, and efficacy/safety must be guaranteed using stringent quality control measures including Good Agricultural Practice, Good Manufacturing Practice, and Good Clinical Practice, at each step in the development of standardized new medicines from herbal medicine as shown in Fig. 6.6.

The biological signal for a plant to synthesize initial phytochemicals is maturation of the fruit, vegetable and seed. Different parts of a medicinal plant are picked for use in medicine before the phytochemicals are present. This may be described as the basis for picking the medicinal part of a drug at an appropriate time.

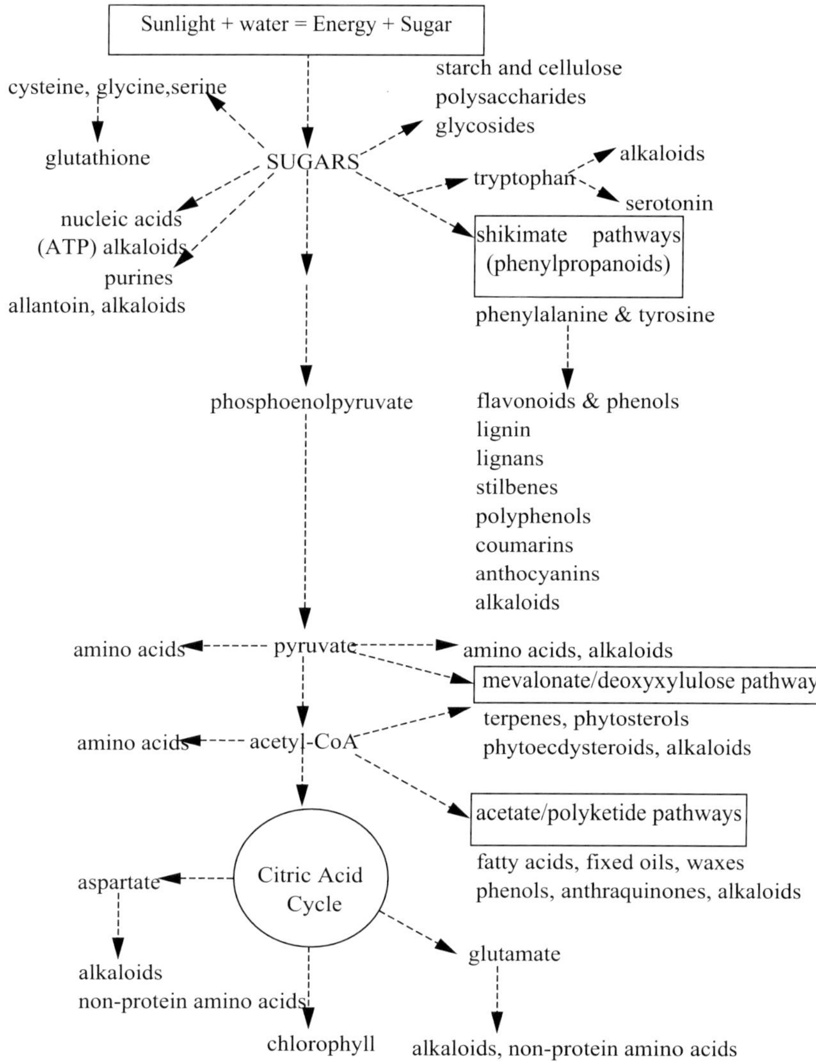

Fig. 6.2 Biosynthesis of primary and secondary metabolites in plants.

Biosynthesis of allicin

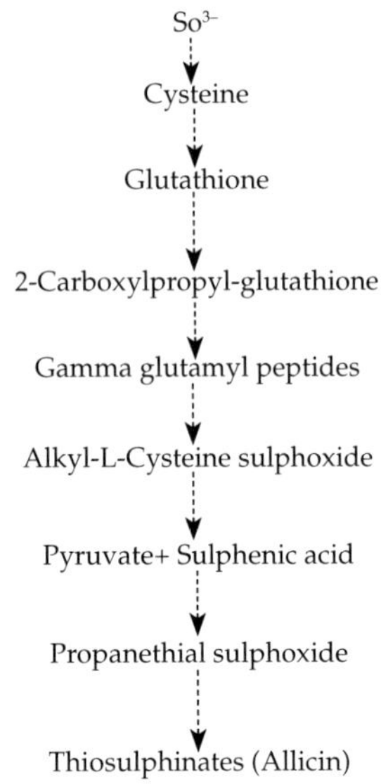

Fig. 6.3 Biosynthesis of allicin.

Biosynthesis of caffeine

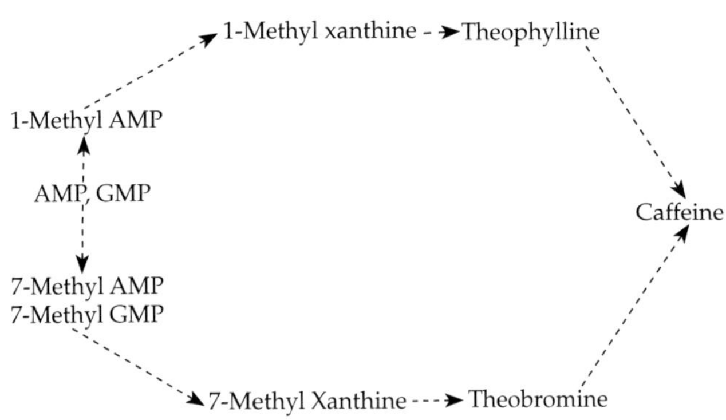

Fig. 6.4 Biosynthesis of caffeine (two pathways).

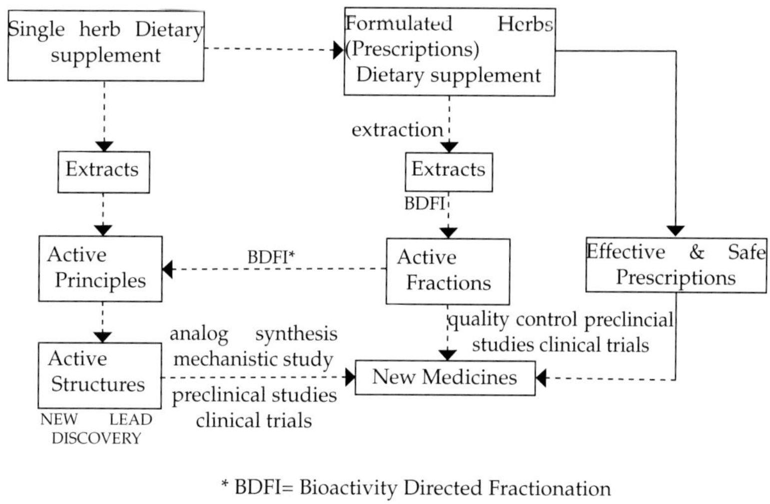

Fig. 6.5 Principal concepts of research on new medicines.

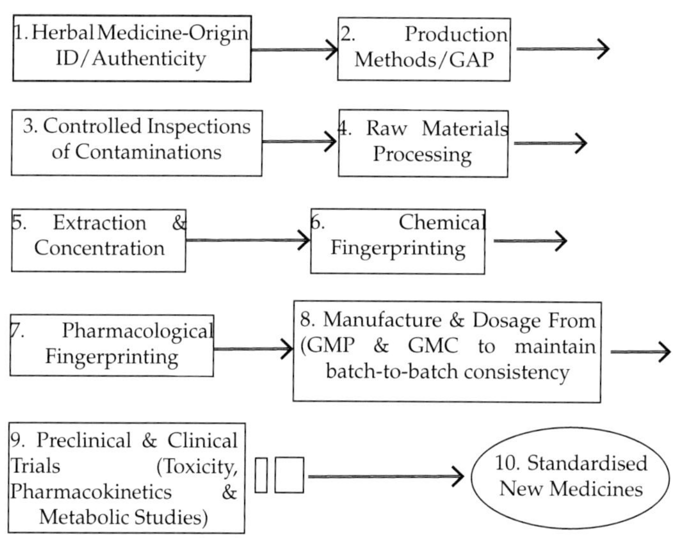

Fig. 6.6 Quality control for standardized new medicines.

Chraka, the great Ayurvedic physician, defined virya as the driving force behind the therapeutic activity of the drug. According to experts, virya is comparable to active constituent of a drug. Ancient scholars developed the virya parameter to study the therapeutic efficacy of the drugs. Other alternative systems of medicine like Unani and Siddha give due importance to virya of a drug. In the Unani system of medicine, virya is known as Tasir.

CAM believes in using the drug as a whole. This does not mean that active principles have no importance in CAM health practice. Using the drug as a whole, means that the chemical constituents are present in natural synergy with each other. Each constituent contributes towards the biological activity of the medicine/formulation.

Let us differentiate between botanical extract and standardized extract. Botanical extracts are prepared by scientifically eliminating the non-beneficial parts of a plant, then continuosly processing it to reach a ratio showing the amount of actual processed vs. the actual amount of beneficial finished product.

Standardized extracts are processed by scientifically isolating the active constituents desired from a plant. The scientific procedure ensures that the finished product reaches a specific percentage of the desired constituent. Standardized extracts should always show the name of the constituent that has been isolated and its percentage. They are standardized to certain limits of markers or active constituents or both. Phytochemicals and standardized herbal extracts are therapeutically potent as compared to fresh or dried material (Fig. 6.7).

During the process of extraction, certain medicinally useful constituents are lost. Like other systems of CAM, Western herbal medicine gives

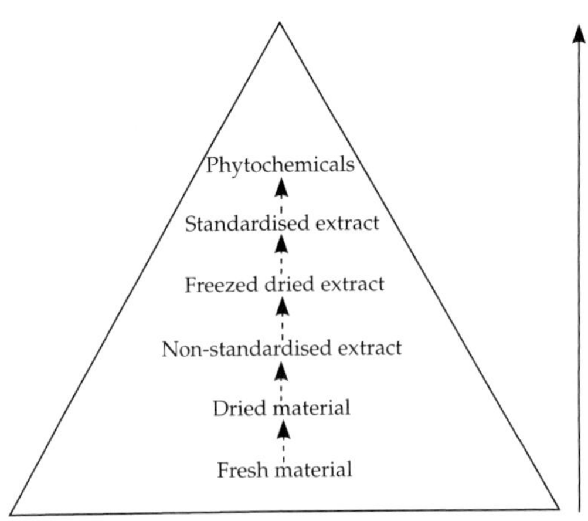

Fig. 6.7 Flow chart for valued added phytochemicals.

due importance to the presence of all chemical constituents rather than a single constituent. According to experts, the standardization of herbals should be in such a way so that a safe and effective dose may be formulated.

6.2 Synergy in relation to the pharmacological action of phytomedicines

Some of the most popular and widely used phytomedicines are those consisting of St. John's wort, Echinacea, ginkgo, garlic, kava, and valerian, which have considerable pharmacological and clinical evidence to support their use, and synergy is generally assumed to play a part in their medicinal effects (Table 5.5). Attempts are rarely made to isolate a single constituent from these extracts. In Traditional Indian Medicine (Ayurveda), Traditional Chinese Medicine, and Western Medical Herbalism, the combination of herbs are fundamental to their philosophy as well as being due to empirical observations and historical usage. As an example, pharmaceuticals containing ephedrine, atropine, and menthol, are rarely considered as phytomedicines, despite being derived from plant sources. Modern herbal medicines are usually found as whole or semi-purified extracts and ideally should be standardized for their active constituents wherever possible to ensure clinical reproducibility.

Table 6.1 *In vitro* and *in vivo* pharmacological evidences for synergy effects.

Ginkgo biloba	**Ginkgolides mixture/Ginkgo extract**
Piper methysticum	Kava lactones/mixture of kava lactones and extract fractions
Glychyrrhiza glabra	Licorice extracts potentiates other susbtances and acts as detoxifier
Cannabis sativa	Cannabis extract/THC
Valeriana officinalis	Valeriana extract/individual constituents
Zingiber officinalis	Zingiber extract/mixture of volatile terpenoids and mixtures
Hypericum perforatum	Hypericum extract/individual constituents

Source: Wagner (2002).

6.2.1 Synergy

Synergy means working together. It occurs when the combined action of constituents is greater than would be expected from a consideration of individual constituents. In herbal medicine, better results are obtained with whole plant extracts rather than with isolated constituents. For example, the side-effects of reseprine are not usually found with the crude drug of *sarpagandha*. If true synergy is occurring in phytomedicines, standardization becomes even more important, since the ratio between synergistic agents is critical, and small changes can make results unpredictable. If combination effects are merely additive, changes in the ratio are less crucial. It may therefore be significant to define synergy using the term 'polyvalent action'.

Synergy is deemed present if the total effect of a combination is greater than would be expected from the sum of effects. This can be expressed as follows:

$$E\ (da, db) = E\ (da) + E\ (db)$$

Polyvalent action includes various effects of multiple active constituents acting in combination, in harmony and possible in synergy. In conventional medicine it is common practice to treat a single complaint as in cancer. This applies more to plant extracts, since combinations are already present within the plant. There are good reasons for not isolating active constituents from herbs like St. John's wort, hawthorn, hops, purple cone flower and black cohosh. In herbals like garlic and valerian, active constituents are unstable and attempts to isolate them would render the extract ineffective.

Example 1: Hypericum perforatum

H. perforatum is traditionally used for the treatment of depression, insomnia, and anxiety. A large body of animal and human clinical research supports its antidepressant effects. Several chemical constituents have been identified in *H. perforatum.* Hypericin, flavanols, and xanthones have been shown to inhibit both monoamine oxidase and/or catechol-o-methyltransferase.

It has been suggested that exacerbated hypothalamic-pituitary-adrenal axis activity has been associated with the incidence of depression. Corticotropin-releasing factor is thought to be major determinant in the regulation of the hypothalamic-pituitary-adrenal activity via activation of CRF (1) receptors. Pseudohypericin (Fig. 6.8) demonstrated antagonistic activity on CRF (1). Thus antidepressant activity of *H. perforatum* is not only due to hypericin (Fig. 6.8) or hyperforin (Fig. 5.204, chapter 5) but pseudohypericin also contributes to the antidepressant mechanism.

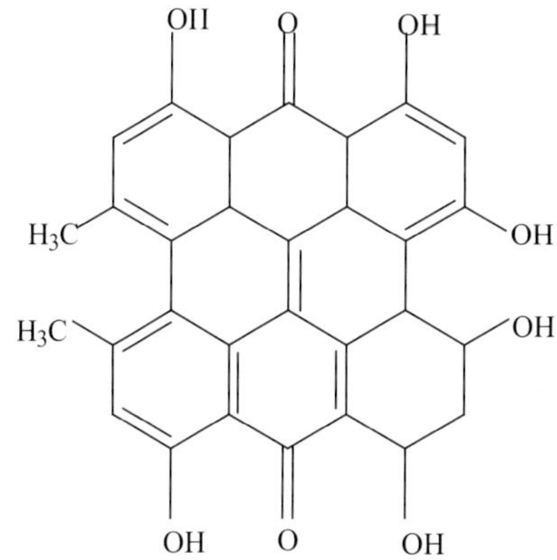

Hypericin

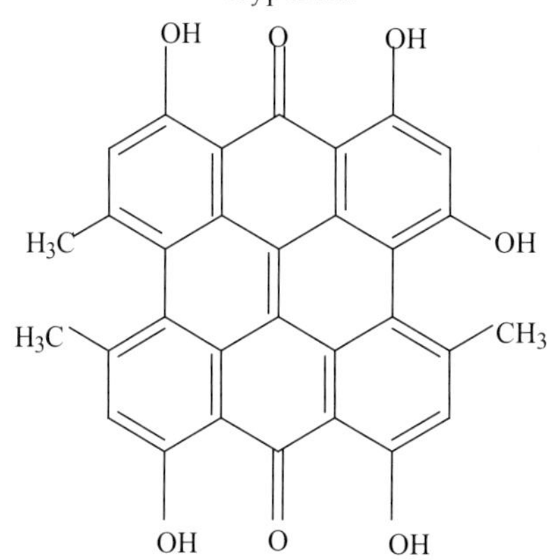

Pseudohypericin

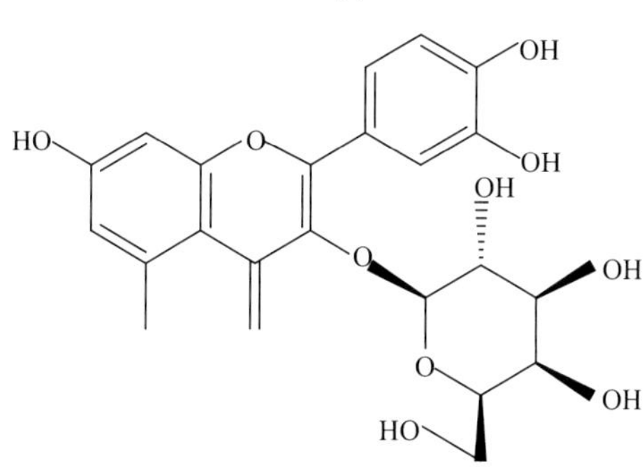

Hyperoside

Fig. 6.8 Possible antidepressant constituents of *H. perforatum*.

Hyperforin is a reuptake inhibitor of serotonin, norepinephrine, dopamine, GABA, and L-glutamate in the nanomolar range (Table 6.2). Radio-labelled hyperforin crosses blood-brain barrier. Four trioxygenated xanthones isolated from the plant and hyperforin-enriched fraction, when combined in suboptimal dosages of 2.5 mg each, caused a significant antidepressant effect which was slightly greater in extent than exhibited by either the hyperforin-enriched- or the xanthone-enriched fraction.

Table 6.2 Inhibitory effects of St. John's wort extracts and pure hyperforin on monoamine oxidase activity and synaptosomal uptake of neurotransmitters. The table shows IC50 values, i.e., the concentration required for 50% inhibition.

	Methanolic extract (1.5% hyperforin)	**CO2-extract (38.8% hyperforin)**	**Hyperforin (100%)**
MAO-A activity	120μg/mL	520μg/mL	-
MAO-B activity	370μg/mL	>500μg/mL	-
Inhibition of serotonin uptake	2.4±0.4 μg/mL	0.26±0.06 μg/mL	0.11±0.02 μg/mL
Inhibition of nor adrenaline uptake	4.5±2.1 μg/mL	0.25±0.08 μg/mL	0.04±0.01 μg/mL
Inhibition of dopamine uptake	0.9±0.1 μg/mL	0.06±0.03 μg/mL	0.06±0.01 μg/mL
Inhibition of GABA uptake	1.1±0.06 μg/mL	0.12±0.04 μg/mL	0.10±0.02 μg/mL
Inhibition of L-glutamate uptake	21.25±10.47 μg/mL	2.83±1.8 μg/mL	0.45±0.37 μg/mL

Source: Bhattacharya, Chakrabarti, Chatterjee (1998).

A flavonoid fraction obtained from crude *H. perforatum* was active in the forced swimming test. The fraction was further separated into two sub fractions. Both the sub fractions were active in the forced swimming test within varied doses. Further work on first sub fraction showed that the main constituents were hyperoside (Fig. 6.8), isoquercitrin, miquelianin and quercetin. The second sub fraction contained small amounts of hyperoside and astiblin. Further testing on these compounds showed that except for quercetin all the rest were active in the forced swimming test.

Pharmacokinetic synergy may occur with *H. perforatum*, where a combination of constituents improves its oral bioavailability. Procyanidin B2 (Fig. 5.224, Chapter 5) and hyperoside of *H. perforatum* extract improves the water solubility of hypericin and thereby increases its pharmacological activity in the forced swimming test of Porsolt. Procyanidin increases the water solubility of hypericin, thus increasing its pharmacokinetic availability.

Example 2: Cannabis sativa (hemp)

The herb has received much attention recently because of the detection of an endogen cannabionoid system in the human brain and the immune system. The endogen cannabionoid system plays a significant role in memory, appetite, lactation, and emesis. Muscle-relaxant, appetite-stimulating, and analgesic effects of (-) trans-delta ^{9}- tetrahydrocannabinol or THC (Fig. 5.205, Chapter 5), are of great interest. A comparative study of cannabis has shown better antispastic activity than THC, as measured in an immunogenic model of multiple sclerosis. The synergistic effect demonstrated with cannabis extract is probably due to presence of cannabidiol or CBD (Fig. 6.9) in the extract, which elevates the level of THC in the brain and at the same time, attenuates the undesired anxiolytic effect of THC. At the same time CBD amplifies the antispastic effect of THC. Amplification of the antispastic effect is not understood but it is assumed that CBD increases the permeation of THC into the muscles.

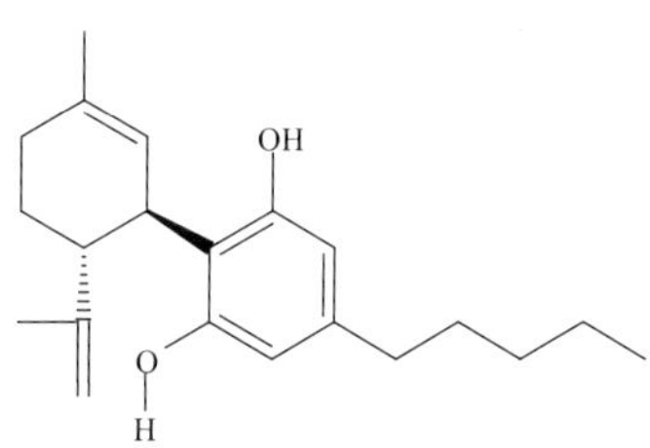

Fig. 6.9 Structure of cannabidiol.

Example 3: Cratageus oxycantha (hawthorn)

Procyanidins and flavone C-glycosides appear to be the main constituents responsible for cardiac tonic effect of the plant. *In vitro* studies have demonstrated the inhibitory effect of hawthorn extract containing procyanidins and flavonoids on angiotensin converting enzyme. As one is aware of the antioxidant, anti-inflammatory and antiplatelet activities of flavonoids, it is necessary to produce hawthorn extract having both procyanidins and flavonoids.

Example 4: Piper methysticum (kava kava)

The plant is well known for anti-anxiety and sedative effects. The active constituents are known as kavalactones or kavapyrones including kavain, dihydrokavain, yongonin, methysticin, and dihydromethysticin. Kavalactones cross the blood-brain barrier and behavorial effects occur at micromolar concentrations. Kavalactones enhance binding to the GABA $_{A}$ receptor in low micromolar concentrations, through a non-benzodiazepine mechanism. They also block voltage gated Na. sup.$^{+}$ and Ca. sup.$^{+}$ channels in micromolar concentrations. Further, kavalactones interact with monoamine oxidase systems by blocking the reuptake of neurotransmitters and inhibiting MAO sub B.

Combined kavalactones, kavain and dihydromethysticin (Fig. 6.10), act in an additive manner to inhibit Ca. sup.$^{+}$ channels. Yongonin given with other kava lactones (i.p.) reaches levels 20 times higher in the brain than when given alone. Similarly, kavain levels in the brain are doubled when given in combination of other kavalactones.

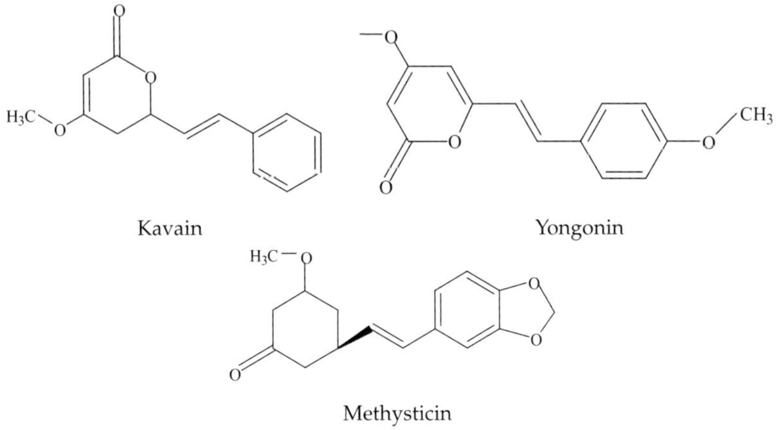

Fig. 6.10 Structure of principal kava lactones.

6.3 Multidrug and multitarget therapy with phytomedicine

In the last decade, two revolutionary shifts occurred in classical traditional medicine:

A. Gradual withdrawal from the dogma of monosubstance therapy and an increasing transition to the treatment of patients with drug combinations.

B. Transition to a new kind of multitarget therapy, which is directed primarily towards the activation of defense, protective and repair mechanisms of the body rather than towards the direct destruction of the damaging agents.

Phytomedicine has long followed and developed the above mentioned strategies by using the monoextracts or extract combinations containing mixtures of bioactive compounds. The strategies are based on therapeutic experiences and the consideration that a complex pathophysiological process can be influenced more effectively and with less side-effects by combination of several low dosage compounds.

6.4 Phytomedicine and cancer

Recently, numerous reviews of plant derived chemo preventive compounds have identified their role in the treatment of cancer. The chemo preventive compounds, precisely known as phytopharmaceuticals, are dietary ingredients, which being food derived, are considered pharmacologically safe. Some of the common chemo- preventive dietary compounds derived from dietary ingredients are depicted in Fig. 6.11.

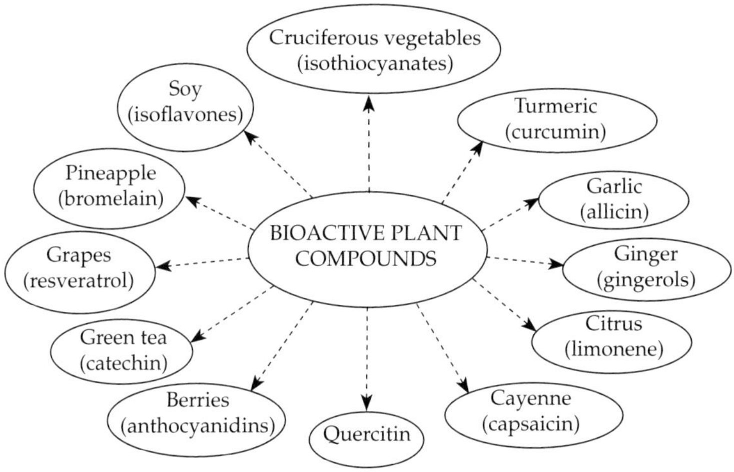

Fig.6.11 Common chemo- preventive dietary compounds.
Source: Jonathan (2005).

Chemo- preventive plant compounds affect all phases of the cancer process, i.e., tumor initiation, promotion and progression. Botanical medicines are complex natural mixtures of pharmacological musltitaskers, simultaneously exerting influence on different levels and via different mechanisms. By contrast, pharmaceutical drugs are classically single synthetic compounds, ideally interfering or disrupting specific mechanism.

6.5 Phytochemicals and bacterial resistance

Recently there has been a growing interest in characterizing new antimicrobial substances from medicinal plants. The interest can be ascribed to three factors:

1. In terms of plant ecology, plants produce antibacterial metabolites as part of chemical defense.
2. There are countless examples of plants, which have been used topically and systemically to treat bacterial infections.
3. Functional group chemistry, chirality and chemical diversity of phytochemicals.

Hyperforin, the acylpholoroglucinol from *Hypericum perforatum, has* been reported to have antibiotic properties. However it has very low minimum inhibitory concentration (MICs) against methicillin resistant *Staphylococcus aureus* (MRSA) and penicillin resistant variants. Sinaicione from *Hypericum sinaicum* and chinensin from *Hypericum chinense* are other plant derived antibacterial.

Filicinic acid derivatives: drummondin D, isodrummondin D, drummondin E, and drummondin F, extracted from *Hypericum drummondii* possesses strong antibiotic activity against the Gram-positive bacteria *S. aureus* and *Bacillus subtilis* and the acid fast bacterium *Mycobacterium smegmatis*. Falcarindiol from roots of the Chinese medicinal plant *Angelica dahurica* have been reported effective against effluxing multidrug-resistant variants of *S.aureus.*

Many clinically relevant bacteria express efflux mechanisms of resistance. By these mechanisms the bacteria are able to prevent penetration of drugs in side cells. Multidrug resistance (MDR) is posing a real problem in treating bacterial infections. Recently phytochemicals have been added with antibiotics to inhibit multi -drug resistance.

Berberine (Fig. 5.5, Chapter 5) and flavonolignan, methoxyhydnocarpin (Fig. 6.12) have been isolated from Berberis sp. Antibacterial activity of berberine is potentiated by methoxyhydnocarpin. This observation has led to the possibility that plants produce both antibacterial compounds and compounds which target bacterial efflux mechanisms to inhibit possible resistance to latent plant antibacterial in bacteria in their environment.

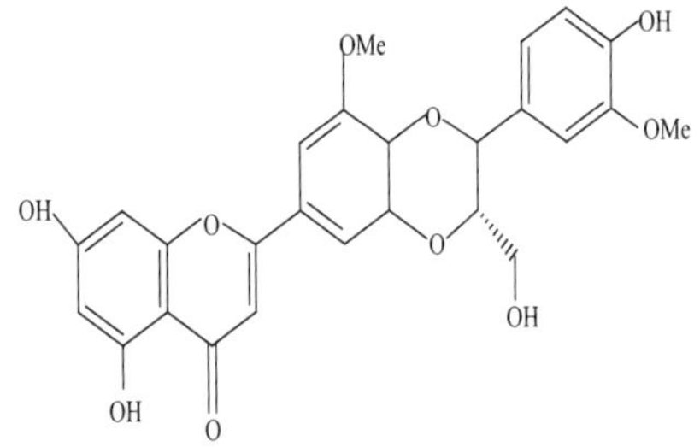

Fig. 6.12 Structure of 5′ methoxyhydnocarpin.

6.6 Role of phytochemicals in healthcare system

Medicinal plants are a significant source of synthetic and herbal drugs in India and China and have been on the forefront when one refers to phyto drugs. The traditional systems of medicines: Ayurveda, Siddha, Unani, Western Herbal Medicine, Traditional Chinese Medicine and Homeopathy have roots in medicinal herbs. A number of distinguished researchers have dealt with herbal medicine and due to its accessibility to traditions; it is still practiced even by lay practitioners.

Sterneur isolated morphine from *Papaver somniferum* (opium poppy) and showed the medical profession that certain phytochemicals produced in plant cells are responsible for pharmacological activity. Later, other alkaloids isolated from opium poppy were investigated for their pharmacological activities. Codeine showed antitussive activity and papaverine antispasmodic activity. The opium based extracts have been utilized for various pharmacological activities, and a number of alkaloids distributed in the plant have different pharmacological activities.

Scientific validation of phytodrugs has been always questioned, but with recent advances and publication of clinical trials, the researchers and the public are viewing herbal products with more of an open mind. In the commercial market, medicinal herbs are used as raw drugs, extracts or tinctures. Isolated active constituents are used for applied research. There has been a dramatic rise in the sale of phyto products like *Allium sativum, Hypericum perforatum, Spirulina, Echinacea angustifolia, Ginkgo biloba* and *Silybum marianum.*

Among phytochemicals distributed in plants, flavonoids sterols and xanthones are of particular interest for a wide range of medicinal activities. Recently several research papers have been published favoring efficacy of these specialty phytochemicals as medicinal agents (see the table below).

Table 6.3 Number of recent publications on flavonoids, sterols and xanthones.

S.no	Class of flavonoids	Number of publications
1.	Epigallocatechin gallate	433,000
2.	Epigallocatechins	1,180
3.	Polyphenols	1.920,000
4.	Proanthocyanidins	446,000
5.	Pycogenol	1,850
6.	Sterols	1,870,000
7.	Xanthones	2,73,000

6.7 Current research on dietary phytochemicals

Dietary phytochemicals are also known as phytonutrients. Phytonutrients are nutrients derived from plant material that have been shown to be necessary for sustaining human life (distinction form phytochemicals). About 800 phytonutrients are known to exist in plant kingdom.

The National Cancer Institute recommends eating at least 5–9 servings of fruits and vegetables per day. Other reputable sources recommend at least 9 servings for men, 7 servings for women, and 5 servings for children.

Research shows that fruits and vegetables are powerful defenders of our health. Research supporting a critical role for fruits and vegetables in good health grows stronger all the time. Scientists now agree that fruits and vegetables should be the foundation of a healthy diet. Phytochemicals in fruits and vegetables can also help reduce your risk of many chronic diseases including cancer, high blood pressure, diabetes, heart disease, stroke and other diseases.

6.7.1 Phytochemicals of Cinnamomum zeylanicum L. *(cinnamon)*

Recent studies have found that cinnamon may have a beneficial effect on blood sugar. Sixty people with type 2 diabetes took 1, 3, or 6 grams of cinnamon in pill form daily, an amount roughly equivalent to one quarter of a teaspoon to 1 teaspoon of cinnamon.

After 40 d, all 3 amounts of cinnamon reduced fasting blood glucose by 18 to 29%, triglycerides by 23 to 30%, LDL cholesterol by 7 to 27%, and total cholesterol by 12 to 26%.

Another study looked at the effect of cinnamon on 79 people with type 2 diabetes who were not on insulin therapy but were taking oral anti-diabetic medications or modifying their diet. They took approximately 3 grams of cinnamon or a placebo 3 times a day for 4 mon. There was a significant reduction in blood glucose in the people taking cinnamon compared to people taking the placebo. Surprisingly, there was no difference in glycosylated hemoglobin (HbA1C) levels, a test that measures how well blood sugar has been controlled during the previous 3 to 4 mon. Coumarin compounds are considered to be active constituents of cinnamon.

6.7.2 Ellagitannins of Punica granatum L. *(pomegranate)*

A recent animal study from University of California reported that ellagitannins present in the fruit of pomegranate have a possible protective role in prostate cancer. Researchers found that ellagitannins accumulate in the prostate and it may be a mode of cancer prevention action. Punicalagin was suggested as a possible anticancer agent.

6.7.3 Flavonoids (phytoestrogens) of Epimedium brevicornum Max *(hoary goat weed)*

A 24-mon randomized, placebo-controlled clinical trial conducted at the Chinese University of Hong Kong and Shanghai University of Chinese Medicine reported the beneficial effect of hoary goat weed in postmenopausal osteoporosis. Eighty five participants were assigned a daily dose of hoary goat weed or placebo. All the women received 300 mg of calcium daily. According to researchers, flavonoids (phytoestrogens) of hoary goat weed may be responsible for increase in bone mineral density at the hip and lumber region thus exerting a useful effect in osteoporosis. The herb contains iicarin and flavonoids genistein and daidezin.

6.7.4 Phytochemicals of *Curcurbita ficifolia* L. (pumpkin)

An experimental study conducted at the East Normal China University reported hypoglycemic and antioxidant activity of fruit extract of pumpkin. According to researchers, the extract may repair damaged β cells in the pancreas thus increasing levels of insulin (insulinotropic effect). D-chiro-inositol was proposed to be active constituent known to mediate insulin activity (Fig. 6.13).

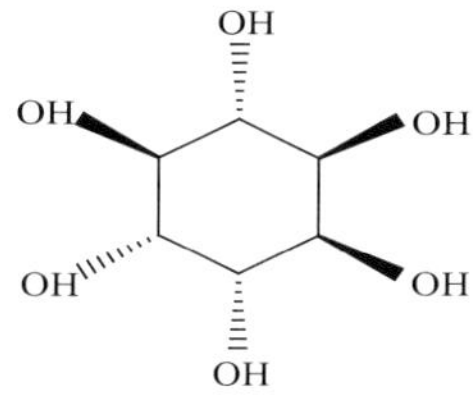

Fig. 6.13 Structure of D-chiro-inositol.

6.7.5 Phytochemicals of *Persea americana* Mill. (Avocado)

A study from the US in 2005 described the inhibitory effect of avocado on the growth of prostate cancer. Recent studies conducted at the Ohio State University reported that phytochemicals found in the fruit can prevent the onset of cancer and even kill some cancer cells. According to experts, phytochemicals extracted from the fruit strike the multiple signaling pathways and prevent cancer by inducing diseased cell death. The phytochemicals have no reaction on healthy cells. The fruit contains proteins (25%) vitamin C, vitamin E, unsaturated fatty acids and sesquiterpenes. The fruit contains zero sodium.

6.7.6 Phytochemical (1-deoxynojirimycin) of *Morus alba* L. (mulberry)

A human study conducted in Japan reported that food grade extract of mulberry is a potent inhibitor of enzyme glucosidase. The enzyme regulates the digestion of carbohydrates. The activity is due to the presence of 1-deoxynojirimycin (DNJ), an antibiotic compound present in mulberry leaves. The study indicated that the single oral administration of 0.8 and

1.2 g of 1-deoxynojirimycin enriched powder significantly suppressed the elevation of postprandial blood glucose and secretion of insulin (Fig. 6.14).

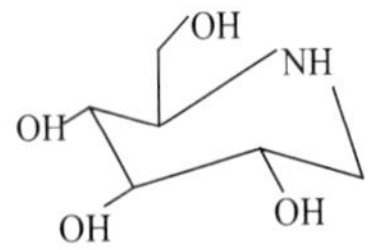

Fig. 6.14 Structure of 1-Deoxynojirimycin.

6.8 Current research on purified phytochemicals

6.8.1 4-(methylthio)-3-butenyl isothiocyanate (MTBITC)

It is a unique chemical found in *Raphanus sativus* and the breakdown product of glucoraphasatin, a glucosinolate present in significant amounts only in radishes. It resembles sulforaphane (found in broccoli) in medicinal properties. The study, conducted by researchers Palmyra, WI, examined that 4-(methylthio)-3-butenyl isothiocyanate is potent inducer of Phase I and Phase II liver detoxification enzymes (Fig. 6.15).

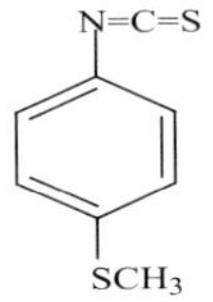

Fig. 6.15 Structure of 4-(methylthio)-3-butenyl isothiocynate.

6.8.2 Brassinolide

It belongs to group of chemical constituents known as brassinosteroids or brassins (Fig. 6.16) and is a plant hormone. They were first discovered in *Brassica napus* pollen. It has an antiviral activity against herpes simplex type 1 and 2 similar to acyclovir.

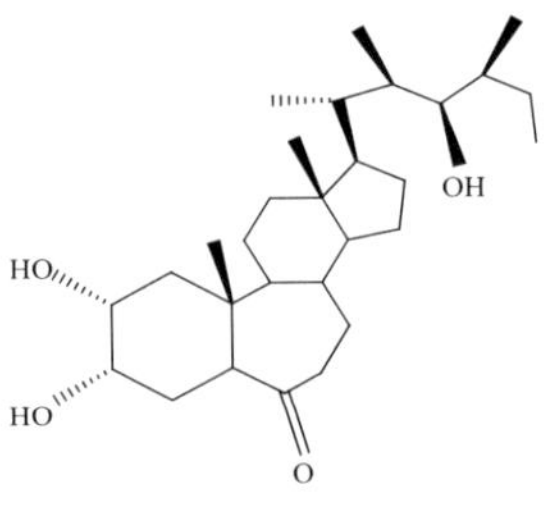

Fig. 6.16 Structure of brassinolide.

6.8.3 Capsaicin

(Fig. 5.205, Chapter 5): It is well-known for counterirritant and rubifacient properties.

Further research

Anti-obesity activity

The researchers, from the National Chung Hsing University in Taiwan, have noted that previous studies suggest that obesity may be reduced by preventing immature fat cells (adipocytes) from developing into mature cells, and other studies have shown that capsaicin can decrease the amount of fat tissue and decrease fat levels in the blood.

6.8.4 Curcumin

Is a coloring matter of turmeric. It is known for its anti-inflammatory property.

Further research

Anticancer activity

Research at the University of Texas has demonstrated that curcumin blocks the activity of a hormone having a link with the development of colorectal cancer. Other studies have demonstrated that curcumin inhibits melanoma cell growth and destroys cancer cells. In addition, animal research conducted at the University of Texas reported that curcumin prevented the spread of breast cancer cells to lungs.

Hypocholesteremic activity

Research from Germany has reported that curcumin has the ability to reduce high levels of blood-cholesterol . In the study, the effect of different doses of curcumin (2–50 micromoles) was investigated on genetics of human hepatocytes.A significant effect was found with 10 micromoles. Furthermore, curcumin was non-hepatotoxic.

Immunomodulator activity

A human-study in isolated macrophages the US reported that curcumin boosts the body's ability to clear the build up of plaques in the brain that are linked to Alzheimer's disease. The isolated macrophages were exposed to a curcumin-derived compound for 24 h and then β-amyloid. It was found that macrophages from three out of six Alzheimer's disease patients showed improved uptake or ingestion of the waste product compared to the patients' macrophages not treated with curcumin.

6.8.5 Glyceollins

Glyceollins are a novel class of soybean phytoalexins with potential cancer-protective antiestrogenic effects. These preliminary findings suggest that soybean glyceollins are natural compounds with potential estrogen-modulating properties in the breast.

6.8.6 Hyperforin

(Fig. 5.192, Chapter 5): It is considered to be the antidepressant constituent of standardized extract of *Hypericum perforatum*.
Further research

Anticancer and anti-angiogenic activity

Hyperforin has been reported to check progress of cancer and leukemia.

6.8.7 Oleocanthal

Is present in extra-virgin olive oil. It is a natural anti-inflammatory compound that has a potency and profile similar to that of ibuprofen. In addition it has an antioxidant activity (Fig. 6.17).

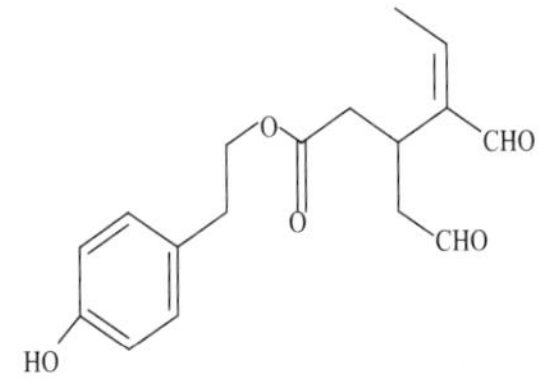

Fig. 6.17 Structure of oleocanthal.

6.8.8 Silymarin (Fig. 5.121)

It is regarded as gold standard in the treatment of liver diseases.

Further research

Anti-psoriatic activity

Slymarin has been reported to be efficacious in psoriasis. The ability of slymarin to inhibit leukotriene synthesis and hepatoprotective action is believed to contribute to anti-psoriatic activity.

Hypoglycemic activity

Researchers at the Institute of Medicinal Plants in Teheran, Iran reported blood sugar lowering activity of slymarin in patients of type 2 diabetes.

6.8.9 Stevioside (Fig. 5.181)

It is a sweet-tasting glycoside isolated from the leaves of *Stevia rebaudiana*.

Further research

Antihypertensive activity

Studies have shown that intravenous injection and powdered drug was able to induce hypotension in hypertensive dogs.

6.8.10 Withaferin-A (Fig. 5.249)

It is a well-defined withanolide present in Indian medicinal plant *Withania somnifera* and known for anticancer and radiomodying effects.

Further research

Research done in America has shown that withaferin-A has the power to inhibit RNA and protein production which can lead to increased cancer death.

6.8.11 Zingerone (Fig.5.194)

An animal study conducted at Taiwan's China Medical University has demonstrated anti diarheal activity of ginger. Ginger extracts and its bioactive components blocked binding of heat-labile eneterotoxin from E.coli to specific receptors on cells in mouse intestine. The binding of the toxin to the receptor results in the case of diarrhea. Biological-activity-guided searching activities lead to zingerone as a possible anti-diarrheal agent.

6.9 Phytochemicals as nutraceuticals or drugs-challenges ahead

There is no doubt that phytochemicals have the potential as pharmaceutical agents or drugs. In the past, several phytochemicals have been used in the medical practice with significant success. With the emergence of points like drug-resistance and cost effectiveness of synthetic drugs, scientists are targeting phytochemicals for developing cheap as well as potent drugs.

According to experts, export of herbal materials and medicines can shoot upto 10,0000 million by 2010 from 4460 million in mid 2003. The international market for herbal remedies is more then US$ 60 billion per year and growing at a rate of 7%. With increased consumer demand for phyto products, the market for phytochemicals is definitely promising. Herbal drug industry has witnessed the emergence of subjects like phytopharmacovigilance. The subject deals with the safety index of herbal drugs. It is defined as "the systematic research of the safety of herbal medicines".

The main challenges in treating phytochemicals as drugs are discussed below:

A. Herbal/phytochemical-drug interactions

Due to the revised interest in herbal medicine, people are favoring nutraceuticals and botanical supplements. As the same time several companies have introduced designed or pharma foods in the market. These are a rich source of phytochemicals having antioxidant, anti-aging and chemo protective effects. Recent studies have reported that phytomedicines have a modulatory effect on drug-metabolizing enzymes, (particularly cytochrome 450) leading to potential drug interactions. Phytochemicals have the tendency to elevate as well as suppress the cytochrome 450 system.

Sales of over-the-counter (OTC) products in 1998 soared to US$46.6 billion, with pharmacies capturing approximately 33%, or US$15.4 billion of the sales. Likewise, the sale of herbal supplements rose to approximately US$5 billion in 2000. Pharmacists routinely advise patients on the use of OTC medicines, and with an increased use of herbal medicines among patients, knowledge of appropriate potential interactions between pharmaceutical drugs, including OTC medications, and herbal supplements is becoming increasingly important.

Some of the best-selling herbal supplements are ginkgo (*Ginkgo biloba*, used for increasing peripheral blood flow and in senile dementia), Asian/Korean ginseng (*Panax ginseng*, used as an 'adaptogen' to help regulate the body's reaction to various stresses), garlic (*Allium sativum*), used for cardiovascular health including hypercholesterolemia), echinacea (*Echinacea purpurea/ angustifolia*), used to support the immune system, especially in the prevention and treatment of colds and influenza), St. John's wort (*Hypericum perforatum)*, used in the management of mild to moderate depression), and saw palmetto (*Serenoa repens)*, used in the management of benign prostatic hyperplasia (Fig. 6.18). Despite the widespread concurrent use of conventional medicines and herbal supplements, documented herb–drug interactions (HDIs) are still sparse and consist mainly of isolated case reports and laboratory studies. For most potential interactions of herbs and drugs that are reported, the theories are based on very high doses of isolated phytochemicals in laboratory models, or conclusions are drawn from similarities in the pharmacology of constituents in the herb and a particular drug. A customer may say, "Herbs have been used safely for thousands of years." While this may be true, the people using those herbs were not on several medications at the same time (Fig. 6.19).

Below is a reference chart listing potential Herb–OTC interactions. Pharmacists should be always aware of the documented or potential interactions between herbal supplements and prescription drugs. Potential interactions are classified by one of the 3 numerals, according to the following key: 1 = Use of these supplements with the listed OTC category may have health benefits. 2 = Recommend caution to individuals taking these herbal supplements in connection with the corresponding OTC

BOTANICAL US Mass Market, 1998 (US Million)

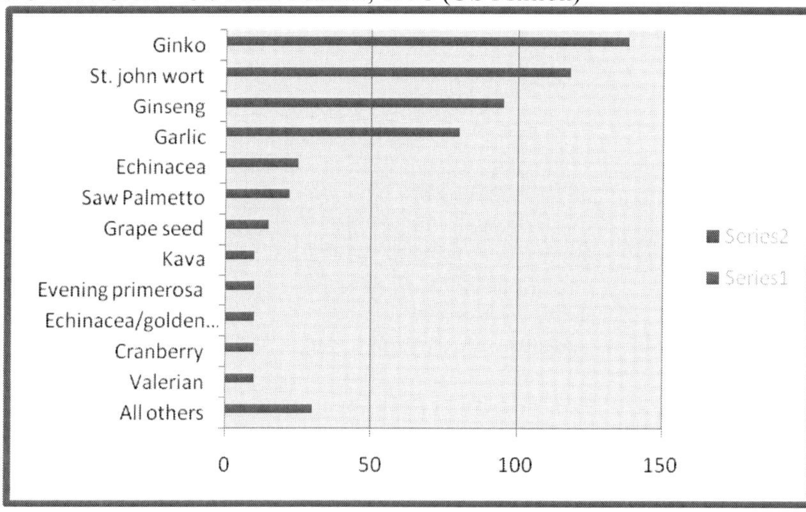

Fig. 6.18 Best selling herbs in US market in 1998.

Color image of this figure appears in the color plate section at the end of the book.

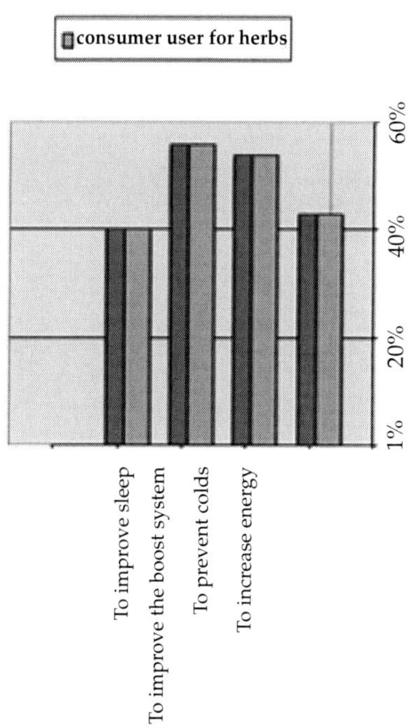

Fig. 6.19 Consumer perception of benefits of herbs.

Color image of this figure appears in the color plate section at the end of the book.

categories.3 = Do not recommend using these herbal supplements with listed OTC categories unless the patient is under the direct supervision of a healthcare professional.

Chart 6.1 Potential Herb–OTC Interaction Chart.

OTC Drug Category	Herbal Category	Interaction Classification
Acetaminophen	Milk thistle (*Silybum marianum*)	1
	Picrorrhiza (*Picrorrhiza kurroa*)	1
	Schisandra (*Schisandra chinensis*)	1
Antihistamines	Belladonna (*Atropa belladonna*)	3
	Bittersweet (*Solanum dulcamara*)	3
	Henbane (*Hyoscyamus niger*)	3
	Hops (*Humulus lupulus*)	2
	Jimson weed (*Datura stramonium*)	3
	Kava kava (*Piper methysticum*)	2
	Passionflower (*Passiflora incarnata*)	2
	Stinging nettle leaf (*Urtica dioica*)	1
	Valerian (*Valeriana officinalis*)	2
Aspirin	Bilberry (*Vaccinium myrtillus*)	2
Aspirin	Cat's claw (*Uncaria tomentosa*)	2
	Coleus (*Coleus forskohlii*)	2
	Cordyceps (*Cordyceps sinensis*)	2
	Dong quai (*Angelica sinensis*)	2
	Evening primrose oil (*Oenothera biennis*)	2
	Feverfew (*Tanacetum parthenium*)	2
	Garlic (*Allium sativum*)	2
	Ginger (*Zingiber officinale*)	2
	Ginkgo (*Ginkgo biloba*)	2
	Ginseng, Asian/Korean (*Panax ginseng*)	2
	Grape seed (*Vitis vinifera*)	2
	Green tea (*Camellia sinensis*)	2
	Horse chestnut (*Aesculus hippocastanum*)	2
	Reishi mushroom (*Ganoderma lucidum*)	2
	St. John's wort (*Hypericum perforatum*)	2
Decongestants	Bitter orange (*Citrus aurantium*)	2
	Cayenne (*Capsicum annuum*)	2
	Coffee (*Coffea arabica*)	2
	Ginseng, Asian/Korean (*Panax ginseng*)	2
	Guarana (*Paullinia cupana*)	2
	Ma huang (*Ephedra sinica*)	3
	Muira puama (*Ptychopetalum* sp.)	2
	Night-blooming cactus (*Selenicereus grandiflorus*)	2
	Yohimbe (*Corynanthe yohimbe*)	3

Chart 6.1 contd...

Chart 6.1 contd...

OTC Drug Category	Herbal Category	Interaction Classification
Diuretics	Celery seed (*Apium graveolens*)	2
	Couch grass rhizome (*Agropyron or Triticum repens*)	2
		2
	Dandelion leaf (*Taraxacum officinale*)	2
	Elder flower (*Sambucus canadensis*)	2
	Horsetail shoots (*Equisetum arvense*)	2
	Juniper berry (*Juniperus communis*)	
Insulin	Aloe (*Aloe vera*)	2
	Bilberry (*Vaccinium myrtillus*)	2
	Bitter melon (*Momordica charantia*)	3
	Fenugreek (*Trigonella foenum-graecum*)	2
	Garcinia (*Garcinia cambogia*)	3
	Garlic (*Allium sativum*)	2
	Ginseng, American (*Panax quinquefolius*)	2
	Gymnema (*Gymnema sylvestre*)	3
NSAIDs	Cayenne (*Capsicum annuum*)	2
	Coleus (*Coleus forskohlii*)	3
	Bilberry (*Vaccinium myrtillus*)	2
	Boswellia (*Boswellia serrata*)	2
	Devil's claw (*Harpagophytum procumbens*)	2
	Garlic (*Allium sativum*)	2
	Grape seed (*Vitis vinifera*)	2
	Green tea (*Camellia sinensis*)	2
	Licorice root (*Glycyrrhiza glabra*)	2
	Milk thistle (*Silybum marianum*)	1
	Turmeric (*Curcuma longa*)	1
		1

Source: Intramedicine, Inc.Cincinnati, Ohio.

Hypericin and hyperforin

As an herbal remedy, *Hypericum perforatum* has not been subjected to the rigorous clinical testing of modern drug candidates. Several recent reports provide evidence that St. John's wort promotes the metabolism of coadministered drugs, including the HIV protease inhibitor indinavir, the immunosuppressant cyclosporin, and oral contraceptives. Because each of these drugs is metabolized by cytochrome P450 (CYP) 3A4, a monooxygenase involved in the metabolism of many xenobiotics, these findings suggested that St. John's wort might induce *CYP3A4* expression. Transcription of *CYP3A4* is known to be induced by a range of xenobiotics, including drugs such as the antibiotic rifampicin, the antimycotic clotrimazole, the insulinsensitizer troglitazone, and the barbiturate phenobarbital.

Hyperforin and *H. perforatum* extracts induce CYP3A4 in hepatocyte cells via the pregnane X nuclear receptor which controls CYP3A4 expression. The crude extracts of *H. perforatum* inhibit CYP2C9, CYP2CI9, CYP2D6 and CYP3A4. The extracts were further fractioned and each fraction was tested for inhibitory action on the cytochrome P450 (CYP) enzyme activities.

Hyperforin, amentoflavone, hypericin, quercetrin and chlorogenic acid, all demonstrated inhibitory affect on cytochrome P450 (CYP) enzyme activities. Amentoflavone was shown to be potent inhibitor of CYP3A4, CYP1A2 and CYP2C9. Hyperforin was shown as potent non-competitive inhibitor of CYP2C9, CYP2D6 and CYP3A4. Hypericin also demonstrated inhibitory activity. The author concluded that *H. perforatum* extracts contains compounds that have potent inhibitory activity on major drug metabolizing enzymes.

A study investigated the effect of *H. perforatum* on cytochrome CYP2D6 and CYP3A4. Seven healthy volunteers were orally administered alprazolam and dextromethorphan along with *H. perforatum*. Urinary concentrations of alprazolam and dextromethorphan and dextromethorphan metabolic ratios were determined. The authors concluded no significant difference in pharmacokinetic profile of alprazolam or dextromethorphan metabolic ratios.

In a human study, 13 healthy volunteers were given 300mg standardized extract *H. perforatum* TID for 14 d. 24 h urinary excretion ratios of 6-betahydroxycortisol/ cortisol were used as an index of CYP3A4 activity. A significant increase, from zero up to around 2.5x over base was found in urinary ratios in 12 subjects, suggestive of 3A4 induction.

According to a study, researchers were able to determine the structure of a significant molecule (PXR) present in the liver and responsible for the metabolism of a majority of the drugs consumed by humans. The molecule is a regulator of the protein known as P450-3A, which is responsible for breaking the drugs. Recently interactions of *H. perforatum* have been suggested with oral contraceptives, antivirals and transplant drugs.

A study reported that hyperforin contributes to the hepatic CYP3A-inducing effect of *H. perforatum* extract in the mouse. A hydro alcoholic extract containing 4.5% hyperforin was given at a dose of 300 mg/kg b.i.d. for 4 and 12 d. Hyperforin, its main phloroglucinol component, was given as dicyclohexylammonium salt (18.1 mg/kg, b.i.d.) on the basis of its content in the extract, to ensure comparable exposure to hyperforin. The extract increased hepatic erythromycin-N-demethylase activity, which is cytochrome P450 enzyme (CYP) 3A-dependent, about 2.2-fold after 4 d of dosing, with only slightly greater effect after 12 d (2.8 times controls).

Hyperforin too increased erythromycin-N-demethylase activity within 4 d to much the same extent as the extract (1.8 times the activity of controls), suggesting that it behaves qualitatively and quantitatively like the extract as regards induction of CYP3A activity. This effect was confirmed by Western blot analysis of hepatic CYP3A expression. Exposure to hyperforin at the end of the four days' treatment was still similar to that with extract of *H. perforatum*, although it was variable and lower than after the first dose in both cases, further suggesting that hyperforin plays a key role in CYP3A induction by the extract of *H. perforatum* in the mouse.

In a subsequent study, it was shown that hyperforin induces the expression of numerous drug metabolism and excretion genes in primary human hepatocytes. The researchers determined the crystal structure of hyperforin in complex with the ligand binding domain of human PXR. Hyperforin induces conformational changes in PXR's ligand binding pocket relative to structures of human PXR elucidated previously and increases the size of the pocket by 250 A. It was found that the mutation of individual aromatic residues within the ligand binding cavity changes PXR's response to particular ligands. Taken together, these results demonstrate that PXR employs structural flexibility to expand the chemical space it samples and that the mutation of specific residues within the ligand binding pocket of PXR tunes the receptor's response to ligands.

Recently it has been reported that *H. perforatum* can reduce the efficacy of chemotherapy drug known as irinotecan. The study was based on previous findings that *H. perforatum* can stimulate cytochrome P450 (CYP3A4), an enzyme system involved in drug metabolism. Specifically, researchers took the premise that hypericin and hyperforin-components of *H. perforatum* seem to enhance CYP3A4 activity and the fact that irinotecan is also partly metabolized by the enzyme system (see Fig. 6.20).

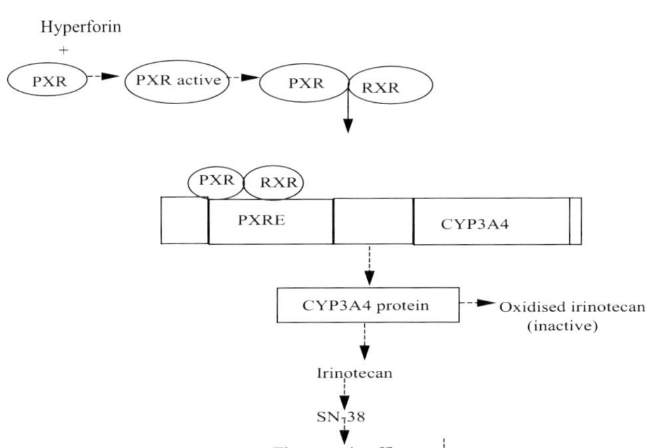

Fig. 6.20 The effect of hyperforin on irinotecan therapy.
Source: Adapted with permission from Meijerman et al., Oncologist 2006.

Hyperforin, binds to and activates X receptor (PXR). Upon activation, PXR forms a heterodimer with the 9-cis retinoic acid receptor (RXR), and this complex binds to the xenobiotic response elements (PXRE) in the CYP3A4 gene. The transcription of the gene is increased, and more CYP3A4 id formed, thereby increasing the metabolism of irinoetcan into an inactive metabolite, oxidized irinotecan. The amount of irinotecan left to be metabolized into SN-38 decreases, leaving less active SN-38 and lower therapeutic efficacy of irinotecan.

Guggulsterones (Fig. 5.205)

The molecular basis for the lipid-lowering action of guggulsterones has been suggested to be an antagonism of the farnesoid X receptor, a member of the nuclear receptor super family of ligand-activated transcription factors. Guggulsterones activate the estrogen receptor, progesterone receptor, and pregnane X receptor with EC_{50} values in low micromolar range. Guggulsterone-mediated activation of pregnane X receptor induces the expression of CYP3A genes both in rodent and human hepatocytes. Pregnane X receptor activation is known to cause herb-drug interactions and it has been suggested that guggulipid therapy should be used cautiously in patients taking prescription medications that are metabolized by the CYP3A family members.

Forskolin (Fig. 5.244)

It is an active ingredient in *Coleus forskohlii* extract. Recent findings have suggested that forskolin and its nonadenyl cyclase-activating analog 1, 9 dideoxyforskolin induce CYP3A gene expression in primary hepatocytes by functioning as agonists of the pregnane X receptor. Activation of the protein kinase-A signaling potentiates pregnane X receptor-mediated induction of CYP3A gene expression in cultured hepatocytes and increases the strength of pregnane X receptor-coactivator protein-protein interaction in cell-based assays.

B. Adverse drug interactions

The use of complementary and adjunctive therapy has been steadily increasing in the U.S. since the 1960s. This can be attributed to several factors, in particular to over 80 million 'baby boomers' embracing alternative medicine as a way to enhance their medical care. The perceived coldness and remoteness of conventional medicine; the entanglement of managed care in red tape; continued emphasis on 'natural' which supposedly connotes ' better', as well as the popularity of vegetarian practices have also contributed to the general population's increasing involvement with alternative medicine. In 1999, 40% of Americans sought alternative medical treatment, outpacing visits to conventional primary care physicians.

Till date, there is scarce or very limited data for adverse drug reaction profile of phytochemicals. This may be attributed to lack of clinical trials with phytochemicals. In phytomedicine, most of the clinical studies have been conducted with herbal extracts standardized to a certain quantity of phytochemical rather than purified phytochemicals.

According to Edwards, World Health Organization Adverse Drug Reaction (WHO ADR) has documented 3.6 million adverse drug reactions. Of these, 41,439 are directly or indirectly related to phytodrugs. It should be noted that most of the adverse drug reactions related to phytodrugs are those of crude preparations or extracts and not of phytochemicals.

Let us concentrate on nephrotxicity of herbs (see Tables 1.1 and 1.2). The consensus of many renal clinicians is that medicinal herbal use, with its known potent pharmacological activity, chemical components (some unknown), microbial content, and ability to interfere with medications, is rendered dangerous and even unadvisable for the renal patient. Herbal preparation-induced acute renal necrosis and chronic renal failure belie the assumed innocuous nature of these preparations. Substantial morbidities such as chronic renal failure, dialysis, or renal transplant and even death have been seen with herbal products. Since these herbal preparations are not single chemical entities, no activity of toxic component can be subjected to standardized tests of efficacy and safety.

Table. 6.4 The five best-selling herbs in the U.S.

Herb	Category/Use	Generally Safe? (In healthy-not renal-population)	Generally Safe? (In healthy-not renal-population)	Comments
St. John's Wort (*Hypericum perforatum*)	Anxiety/ Depression- For mild to moderate depression only	Yes- Not recommended in renal pts.	Yes	Interacts with many drugs- transplant rejection drugs, Not to be taken in pregnancy or with psychoactive Drug/SSRIs.
Echinacea	prevent or improve symptoms of respiratory tract infections	Yes	Yes	Do not take for more than 6 wk. Could worsen autoimmune disease. Type 1 Diabetes.
Garlic (Allium sativum L)	Cardiac/ Cholesterol To decrease serum cholesterol and Triglycerides	Yes	Yes	Effective dose is large; 1-4 cloves Suggest enteric coated Side effects: halitosis, gastritis; diaphoresis, light-headedness.

Tablet 6.4 contd...

Tablet 6.4 contd...

Herb	Category/Use	Generally Safe? (In healthy-not renal-population)	Generally Safe? (In healthy-not renal-population)	Comments
Ginseng (*Panax Ginseng*)	Multiple Categories supposed uses: stress, hypertension, depression, and impaired memory	A matter of debate-Not recommended	NO	Side effects: anxiety, hypertension, insomnia, headaches, and asthma attacks.
Gingko (*Ginkgo biloba*)	Memory/ Concentration-for elderly with Alzheimer's dementia	Yes	Yes- antagonistic effect on platelet activating factorrolc in Immune—mediated renal disease	Side effects: Occasional headache.

Table 6.5 Other popular herbs in the U.S.

Herb	Category/Use	Generally Safe? (In healthy-not renal-population)	Generally Safe? (In healthy-not renal-population)	Comments
Black Cohash (*Cimicifuga racemosa* L)	Health-Menopausal symptoms/ PMS	(Questionable)	(Questionable)	Perspiration.
Saw palmetto (*Serenoa repens*)	Men's Health	Yes	Yes, Clinical evidence getting stronger	High doses headache, diarrhea.
Ma-huang (*Ephedra* sp.)	Weight Loss Anorectic, Bronchodilator Hypertensive crisis; CVA; arrhythmias, death	No	Yes, but potentially lethal even in young healthy.	Hypertensive crisis; CVA; arrhythmias, death.
Feverfew (*Tanacetum parthenium*)	Headaches Vascular migraines	Yes	Yes	May cause mouth ulcers; increased heart rate.
Valerian (*Valeriana officinalis* L.)	Insomnia	Yes-considered safe until cardiac withdrawal syndrome documented	Yes-generally recognized as safe	Headache, excitability, uneasiness, cardiac disturbances.

Source: March 2000 issue of Nephrology News & Issues.

C. Phytochemical variability in commercial samples Example of withaferin-A

Sangwan et al., 2004 reported the variability of withaferin-A (Fig. 5.249, Chapter 5) in commercial preparations of Ashwagandha (*Withania somnifera*). Commercial samples consisting of 250 mg Ashwagandha (as indicated on the label) were investigated for their phytochemical contents. The samples were quantified by High Performance Liquid Chromatography. The amount of withaferin-A estimated in commercial samples is given in Table 5.6. There was significant variation in relative amount of withaferin A. The study reflected a 117-fold higher concentration of withaferin A in the highest (product 1) holding product compared to one with the least (product 10). Further, according to the recommended intake of the product as mentioned on the label, the dose of withaferin A ranged from 0.02 mg to 1.4 mg.

Table 6.6 Withaferin A in selected commercial mono and poly herbal Ashwagandha products.

Products	Withaferin A (mg g-1 of Ashwagandha*)	Percentage of maximum in the sample	Withaferin A (mg) per suggested daily dose**	Dose multiplicity factor for Withaferin A
#001	2.34	100	1.4	70
#002	1.03	44	0.52	26
#003	0.93	39.7	0.37	18.5
#004	0.43	18.4	0.50-1.00	25-50
#005	0.43	18.4	0.86-1.72	43-86
#006	0.41	17.5	0.08-0.16	4.0-8.0
#007	0.32	13.7	0.06	3
#008	0.06	2.6	0.02	1
#009	0.05	2.1	n.a.	n.a.
#010	0.02	0.9	n.a.	n.a.

*Ashwagandha state (herb or herb extract) as specified on the product.

**Daily dose (load) of withaferin A computed from its concentration in the product and suggested intake of product n.a. Not available/applicable.

Source: Adapted with permission from Sangwan et al., Current Science 2004.

The commercial samples were further studied for the relative amount of six phytochemicals (WS-1 and WS-6). Data in Table 5.7 suggests wider variation with regard to amount of WS-1 to WS-6. The example of withaferin-A clearly suggest significant variation in the amount of active constituents in the commercial herbal products.

Table 6.7 Content of unidentified Ashwagandha-specific phytochemicals in selected commercial mono and poly-herbal products.

Products	WS-1 (8.9 min)	WS-2 (25.4 min)	WS-3 (25.7 min)	WS-4 (27.6 min)	WS-5 (29.3 min)	WS-6 (45.1 min)
#001	1.41	0.58	n.d.	n.d.	0.38	1.22
#002	0.21	0.36	0.30	0.02	0.32	0.92
#003	3.10	n.d.	0.68	0.06	0.38	1.14
#004	0.13	0.08	n.d.	n.d.	0.05	0.17
#005	n.d.	n.d.	0.67	n.d.	0.23	0.55
#006	9.18	n.d.	0.45	0.94	0.78	0.17
#007	n.d.	n.d.	n.d.	1.24	n.d.	0.30
#008	27.8	0.03	n.d.	0.11	n.d.	n.d.
#009	4.94	n.d.	0.05	0.79	n.d.	n.d.
#010	n.d.	n.d.	n.d.	n.d.	n.d.	n.d.

*Ashwagandha state (herb or herb extract) as specified on the product n.d., not detected.
Indications in parentheses are according to HPLC retention times of the phytochemicals under the condition of the analysis.
Source: Adapted with permission from Sangwan et al., Current Science 2004.

Example of allicin

The data deals with the evaluation of allicin (Fig. 5.270, Chapter 5): levels in Australian grown garlic. A total of 200 garlic samples were collected from 43 producers in the Australian states of Victoria, Tasmania, Queensland, New South Wales and South Australia. The 200 samples tested recorded allicin yields in the range 0.5 to 9.0 mg/g (by fresh weight) with the majority of samples in the 3.0 to 6.0 mg/g range (Fig. 6.21).

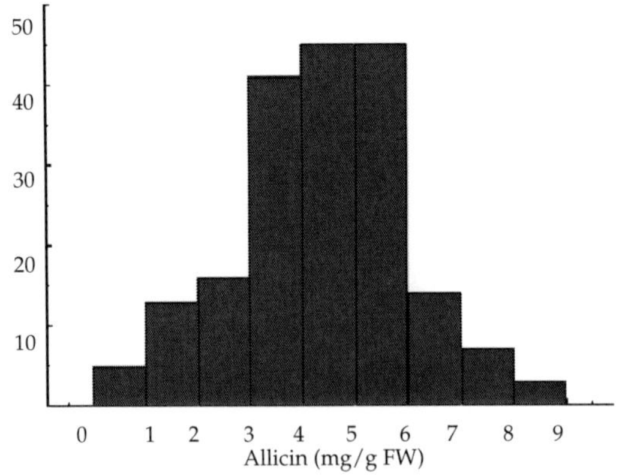

Fig. 6.21 Frequency distribution of allicin levels in 200 garlic samples.
Source: Adapted with permission from Rural Industries Research and Development Corporation, Victoria.

Color image of this figure appears in the color plate section at the end of the book.

Commercial interests determined that an allicin level of 4.5 mg/g (by fresh weight) was required for extraction to be economically viable. Almost 50% (95 out of 200) of the samples assayed had allicin yields over 4.5 mg/g, (i.e., 'pharmaceutical grade') and a further 10% of the samples were in the 4.0–4.5 mg/g range.

As a comparison to allicin levels previously cited in scientific literature, the highest previously recorded allicin level was 7.7 mg/g (by fresh weight) for Romanian Red garlic grown in New York, USA (Table 1.1). Six samples of the Australian grown garlic tested in the current study recorded allicin yields higher than 7.7 mg/g. The next highest allicin yield recorded prior to this study was 6.7 mg/g for Spanish Roja grown in the USA and 6.6 mg/g for Laotouxu garlic obtained from China. Fifteen Australian grown samples from the current study exceeded this level and a number of these samples are listed in Table 5.22.

Table 6.8 Allicin levels (from fresh garlic samples) recorded in Australia and other countries.

Country of origin	Allicin mg/g FW	Country of origin	Allicin mg/g FW
Australia	9.0	Switzerland#	3.6
Australia	8.4	Switzerland#	2.4
Australia	8.3	USA*	7.7
Australia	7.8	USA*	6.7
Australia	7.8	USA*	6.1
Australia	7.7	China^	6.6
Australia	7.6	China^	6.4
Australia	7.4	China^	5.1
Australia	7.3	Japan@	5.3
Australia	7.3	Japan@	5.1
Australia	7.0	Japan@	5.0

#From (7) Ziegler and Sticher, 1989.

*From (5) Koch and Lawson, 1996.

^From (6) Lawson, Wood and Hughes, 1991.

@From (8)Ueda et al., 1991.

D. Toxicology of phytochemicals

This is another gray area where preliminary investigations are to be carried out. LD_{50} of some known specialty phytochemicals are tabulated below:

Table 6.9 Medial lethal dose of some well-known phytochemicals.

S.no	Phytochemical	LD_{50}
1.	Abrin	0.0004 mg/kg (oral in mouse)
2.	Arecoline	100 mg/kg (s.c. in mouse)
3.	Caffeine	192 mg/kg (oral in man)
4.	Chavicol	1230 mg/kg (oral in rat)
5.	Eugenol	1930 mg/kg (oral in mouse)
6.	Harmaline	120 mg/kg (s.c. in rat)
7.	Harmine	200 mg/kg (i.v. in mouse)
8.	Lobeline	40 mg/kg (ipr in mouse)
9.	Mescaline	132 mg/kg (oral in man)
10.	Nicotine	30 mg/kg (oral in man)
11.	Pyridine	891 mg/kg (oral in rat)
12.	Quinine	294 mg/kg (oral in man)
13.	Rotenone	143 mg/kg (oral in man)
14.	Tubocurarine	28 mg/kg (oral in rat)

Source: Adapted from Duke 1998.

E. Cost-effectiveness of isolation methods

Cost-effectiveness of isolation methods of phytochemicals is an important factor in determining the future of phytomedicine. Isolating bioactive compounds is very expensive and complicated. Further a lack of public awareness about health benefits of phytomedicine and phytochemicals is another obstacle.

Conclusion

According to one estimate, only 20% of the plant flora has been screened for drugs. Keeping in view the vast treasure of medicinal herbs, one can expect phytochemicals to play a significant role as modern medical science has limited options for diseases like diabetes mellitus, rheumatoid arthritis, Alzheimer's disease and Parkinson's disease. Work on the identification and isolation of phytochemicals is an ongoing process and phytomedicine is expected to play a critical role in the future of the healthcare system.

References and further reading

Adler LE, Hoffer L, Nagamoto HT, and et al. 1999. Yohimbine impairs Richard IH; Szegethy E; Lichter D et al. Parkinson's disease: a preliminary study of yohimbine challenge in patients with anxiety. Clin. Neuropharmacol. 22, 172–5.

AggarwalBB,TakadaY,andOommenOV.2004.Fromchemopreventiontochemotherapy:common targets and common goals. Exp. Opin. Invest. Drugs. 13, 1327–1338.

Angelina S. 1997. Phytochemicals Taking Their Place. Chemical Market Reporter.

Anonyms. 1992. IUPAC Compendium of Chemical Terminology. 2nd Ed. 64, 165.

Anonyms. 1997. The role of phytochemicals in optimal Health. J. Nat. Acad. Child. Develop. 2, 2.

Anonyms. 2006. Pharmacovigilance of herbal medicines. Pharmaceut. J. 276, 543–545.
Bagchi D, and Preuss H. 2005. Phytopharmaceuticals in Cancer Chemoprevention. CRC Press Boca Raton, FL. USA.
Bell EA, and Charlwood BV. 1980. Secondary plant products. Encyclop. Plant Physiology, Vol. 8. Springer Verlag Berlin-Heidelberg-New York.
Brobst Dan E, Ding X, Creech K, Goodwin B, Kelley B, and Staudinger J. 2004. Guggulsterone Activates Multiple Nuclear Receptors and Induces CYP3A Gene Expression through the Pregnane X Receptor. J. Pharmacol. Exper. Ther. 310, 528–535.
Bruneton J. 1999. Pharmacognosy Phytochemistry Medicinal Plants. 2nd Ed. Lavoisier Publishing, New York.
Chen LJ. 2007. Ginger and Its Bioactive Component Inhibit Enterotoxigenic E.coli Heat-Labile Enterotoxin-Induced Diarrhoea in Mice. J. Agr. Food. Chem. 55(21): 8390–8397.
Dewick P. 2002. Medicinal Natural Products: A Biosynthetic Approach. 2nd Ed. John Wiley & Sons, Ltd. Chichester, UK.
Ding X, and Staudinger J. 2005. Induction of Drug Metabolism by Forskolin: The Role of Pregnane X Receptor and the Protein Kinase X Signal Transduction Pathway. J. Pharmacol. Exper. Ther. 312, 849–856.
Ding H, Chin YW, Kinghorn AD and D'Ambrosio SM. 2007.Chemo preventive characteristics of avocado fruit. Cancer Biol. 17, 386–394.
Ernst E. 2004. Changes for phytopharmacovigilance. Post. Med. 80, 249–250.
Fluck H, and Pharm M. 1995. The influence of climate on the active principles in medicinal plants J. Pharm. Pharmacol. 7, 361–383.
Gear RW, Gordon NC, Heller PH, and Levine JD. 1995. Enhancement of morphine analgesia by the alpha 2-adrenergic antagonist yohimbine. Neuroscience. 66, 5–8.
Gottlieb OR. 1990. Phytochemicals: differentiation and function. Phytochemistry. 29, 171–5.
Gullo VR. (ed). 1994. The Discovery of Natural Products with Therapeutic Potential. Butterworth-Heineman, Boston, USA.
Hanlon PR, Webber DM and Barnes DM. 2007.Aqueous Extract from Spanish Black Radish (Raphanus sativus L. Var. niger) Induces Detoxification Enzymes in the HepG2 Human Hepatoma Cell Line. J. Agr. Food Chem. 55, 6439–6446.
Harborne Jeffrey B, Baxter H, Moss GP. 1999. Phytochemical Dictionary: A Handbook of Bioactive Compounds from Plants. 2nd ed. Taylor & Francis: London, UK.
Hlebowicz J, Darwiche G, Björgell O and Almér LO. 2007. Effect of cinnamon on postprandial blood glucose, gastric emptying and satiety in healthy subjects. Am. J. Clin. Nut. 85,1552–1556.
Ioannides C. 2003. Drug–phytochemical interactions. Inflammopharmacol. 11, 7–42.
James LF, and Painter KE. 1989. Locoweed Poisoning in Livestock. In: L.F. James et al., (eds) Swainsonine and Related Glycosidase Inhibitors. Iowa State University Press, Ames, Iowa.
Jenniskens LHD. 1996. Identification of six taxine alkaloids from *Taxus baccata* needles. J. Nat. Prod. 59, 117–123.
Jonathan T. 2005. Food, Medicine, Poison & Molecular Vitalism. Herbal Bookworm, NY.
Joshi VK, and Sharma PV. 1978. The concept of virya. Sach. Ayur. 31, 1–6.
Kelsey RG, and Vance NC. 1992. Taxol and cephalomannine concentrations in the foliage and bark of shade-grown and sunexposed *Taxus baccata* trees. J. Nat. Prod. 55, 912–917.
Khan A, Safdar M, Ali Khan MM, Khattak KN and Anderson RA. 2003. Cinnamon improves glucose and lipids of people with type 2 diabetes. Diab. Care. 26, 3215–3218.
Kimura T, Nakagawa K, Kubota H, Kojima Y, Goto Y, Yamagishi K, Oita S, Oikawa S and Miyazawa T. 2007. Food-Grade Mulberry Powder Enriched with 1-Deoxynojirimycin Suppresses the Elevation of Postprandial Blood Glucose in Humans. J. Agr. Food Chem. J. Agr. 55, 5869–5874.
Koch H P and Lawson L D. 1996. Garlic. The Science and Therapeutic Application of Allium sativum L. and Related Species (Second Edition). Baltimor: Williams and Wilkins.
Kuo P-Lin, Hsu Ya-Ling, and Lin C. 2005. The Chemopreventive effects of natural products against human cancer cells. Int. J. App. Sci. Engin. 3, 203–214.

Lawson L, Wood S and Hughes B. 1991. HPLC analysis of allicin and other thiosulfinates in garlic clove homogenates. Planta Med. 57,263–270.

Liu YL, Ho DK, and Cassady JM. 1992. Isolation of potential cancer chemopreventive agents from *Eriodictyon californicum*. J. Nat. Prod. 357–363.

Liu N, Bonnet F, Delaunay L, Kermarec N and D'Honneur G. 1993.Partial reversal of the effects of extradural clonidine by oral yohimbine in postoperative patients. Br. J. Anaesth.70, 515–518.

Meijerman I, Beinjen J, and Schellens Jan HM. 2006. Herb-Drug Interactions in Oncology: Focus on Mechanisms of Induction. The Oncol. 2, 742–752.

Obach RS. 2000. Inhibition of human cytochrome P450 (CYP) enzymes by constituents of St. John's Wort, an herbal preparation used in the treatment of depression. J. Pharmacol. Exper. Ther. 294, 88–95.

Pezzuto JM. 1995. Phytochemistry of Medicinal Plants. In: JT Arnason, R Mata and JT Romeo (eds) Recent Advances in Phytochemistry. Vol. 29. Plenum, New York, pp. 19–45.

Raffauf, R.F. 1996. Plant Alkaloids: A Guide to Their Discovery and Distribution. Hawkworth Press, Inc. New York.

Sangwan RS, Chaurasiya ND, Misra LN, Lal P, Uniyal GC, Sharma R, Sangwan NS, Suri KA, Qazi GN and Tuli R. 2004. Phytochemical variability in commercial herbal products and preparations of Withania somnifera (Ashwagandha). Curr. Sci. 86,461–465.

Seeram NP, Aronson WJ, Zhang Y, Henning SM, Moro A, Lee RP, Sartippour M, Harris DM, Rettig M, Suchard MA, Pantuck AJ, Belldegrun A and Heber D. 2007.Pomegranate Ellagitannin-Derived Metabolites Inhibit Prostate Cancer Growth and Localize to the Mouse Prostate Gland. J.Sci. Food Agri. 19; 55,7732–7737.

Singh AP. 2005. Phytochemicals: Their Role in the Modern Era.

Somon G. 2008. Phytochemicals for bacterial resistance–strengths, weaknesses and opportunities. Georg Thieme KG Stuttgart. New York Plenary Lecture Review.

Southwick SM, Krystal JH, Bremner JD, and et al. 1997. Noradrenergic and serotonergic function in posttraumatic stress disorder. Arch. Gen. Psych. 54, 7 49–58.

The Science Advisory Board 2005, Experts Insights from gene to drugs. Arlington, VA 22201

Ueda Y, Kawajiri H, Miyamura N and Miyajima R. 1991. Content of some sulfur-containing components and free amino acids in various strains of garlic. J Japanese Soc Food Sci Tech. 38:429–434.

Vogt HJ, Brandl P, Kockott G, Schimzt JR, Wiegand,MH, Schadrack J and Gierend M. 1997. Double-blind, placebo controlled safety and efficacy trial with yohimbine hydrochloride in the treatment of non-organic erectile dysfunction. Int. J. Impot. 1997,9,155–161.

Wagner H. 2002. Natural products chemistry and phytomedicine research in the new millennium: new developments and challenges. ARKIVOC. 7, 277–284.

Wagner H. 2006. Multitarget therapy-The future of treatment of more than just functional dyspepsia. Phytomedicine. 13, 122–129.

Watkins RE, Maglich JM, Moore LB, Wisely GB, Noble SM, Davis-Searles PR, Lambert MH, Kliewer SA, and Redinbo MR. 2003. A crystal structure of human PXR in complex with the St. John's wort compound hyperforin. Biochemistry. 42, 1430–1438.

Weiner MA, and J.A. Weiner. 1994. Herbs That Heal. Quantum Books, Mill Valley, California.

Xia T, and Wang Q. 2007. Hypoglycemic role of *Cucurbita ficifolia* (Cucurbitaceae) fruit extract in streptozocin-induced diabetic rats. J. Sci. Food Agri. 87, 1753–1757.

Zhang G, Qin L, and Shi Y. 2007. Epimedium-Derived Phytoestrogen Flavonoids Exert Beneficial Effect on Preventing Bone Loss in Late Postmenopausal Women: A 24-Month Randomized, Double-Blind and Placebo-Controlled Trial. J. Bone. Mineral. Res. 22, 1072–1079.

Ziegler SJ and Sticher O. 1989. HPLC of S-alk(en)yl-L-cysteine derivatives in garlic including quantitative determination of (+)S-allyl-L-cysteine sulfoxide (alliin). Planta Med. 55,372–377.

Chapter 7

Therapeutic Utility of Ayurveda

7.1 Introduction

Ayurveda is considered to be the mother of all systems of healing. It is defined as a science of life and teaches us what is harmful and what is useful for health. The subject of Ayurveda is divided into eight distinct branches ranging from medicine to geriatrics. For diagnozing diseases, Ayurveda has its own concept like pulse examination, eye and urine diagnosis. Treatment is based on drugs of herbs, minerals and animal origin. Ayurvedic pharmacy is a vast subject and drugs are formulated as powder, decoction, infusion and fluid extract.

Recent trends have suggested the global acceptance of Ayurveda as a science. Various studies have reported increased patient visits to alternative health care practitioners. World Health Organization (WHO) has also recognized certain alternative systems of healing. Ayurvedic medicinal plants are in high demand and some of these (established in Ayurvedic Pharmacopoeia) have been screened for pre-clinical studies. *Asparagus racemosus* (*shatavari*), *Terminalia arjuna* (*arjuna*), *Terminalia chebula* (*haritaki*), *Tinospora cordifolia* (*guduchi*) and *Withania somnifera* (*ashwagandha*) have been identified as plants of good therapeutic potential.

However there are concerns over the quality of drugs used in Ayurveda. While the process of procurement of the raw material to the finished product appears to be easy, it is not a simple job. The biggest drawback is lack of agricultural and testing standards. These have an ultimate impact on quality control and efficacy of herbal formulations. Recently the Ayurvedic drug market faced a ban on the sale of certain herbo-mineral formulations in the US market.

7.2 Quality standards and Ayurveda

Ayurveda has survived as a science since centuries. Ayurveda was preached by distinguished personalities like Charaka, Sushruta, Madhva and Kashyapa. These people established the fundamental aspects of the science which are are still valid today. Later scholars like Vagabhatta, Chakrapani, Arundutta, and Dalhana did a commendable job in preserving this ancient wisdom. Before screening a medicine from the Ayurvedic point of view, the following parameters are observed:

1. Substance (Dravya).
2. Taste (Rasa).
3. Property (Guna).
4. Potency (Virya).
5. Post-digestion effect (Vipaka).
6. Therapeutics (Prabhva).
7. Pharmacological activity (Karma).

Since Ayurveda has served humanity for centuries, it is but obvious that without internal quality standards it was not possible for the system to survive. Lack of documentation of ancient medical knowledge seems to be the limiting factor for Ayurveda. Oral transmission was the major tool for transferring theoretical as well as practical knowledge. However after 17th century the scholars felt the need of documentation and several 'Nighantu' were written. Nighantu is the Sanskrit word equivalent to Materia Medica (medicinal botany or pharmacognosy).

Just like all health systems, the concept of standardization exists in Ayurveda. Charaka, the great Indian physician, has described methods for collection of plant parts in appropriate seasons. In Ayurveda or other herbal systems of medicine, various plant parts ranging from roots to bark are used in formulations. The theme about collecting plant parts in the appropriate season is related to the presence of the active principle- the curative factor of herbals remedies. Today we have isolated the active principles of the plants and given them names like alkaloids and glycosides. The ancient scholars were well aware of the active principles of plant-based remedies. They described it in terms of virya (potency).

Ayurvedic pharmacy is a highly developed subject. The herbs are processed into various pharmacopoeial preparations like infusions, decoctions, powders, expressed juice and extracts. Herbs are taken in a specific ratio and subjected to various methods like incineration, distillation or extraction. Although these methods apparently seem to be crude but keeping in mind the lack of scientific instruments in the Vedic era it can definitely be said that the science was at its peak. Doses of all the pharmacopoeial preparations and the mode of administration are been well defined in Ayurvedic texts.

When we study the nomenclature of medicinal plants from the Ayurvedic point of view, it can be said that names were assigned on anatomical or physiological basis. Brahami is classical example of this.

The leaf of the plant has a similarity to cerebral hemispheres of the human brain. It may be the basis of extensive use of Brahami in the treatment of brain related diseases.The name of Ashwagandha has been derived from the smell of the root which resembles that of a horse feces. Due to lack of development of subjects like taxonomy the ancient scholars devised names mostly on an organo-leptic basis.

Ayurvedic drugs have given the scientific community a lead to discover molecules from plants. *Cassia tora* (Chakrmurda) is widely used in the treatment of ringworm in Ayurveda. As the name suggests the plant is useful in the treatment of skin disease with round margins. It was on this lead that chrysophanic acid was isolated from this plant. Earlier when standard antipsoriatic medications were not available, chrysophanic acid was the main stay of antipsoriatic treatment. This is not the only case as many plants provide us with lead molecules for respective segments based on description (irrespective of the objective or subjective factor) in ancient texts. On the whole it can be said that quality standards do exist in Ayurveda. Due to a lack of laboratory facilities and proper clinical documentation these standards were never fully developed.

7.3 Controversial Medicinal Plant Identification in Ayurevda

Proper identification of medicinal plants has a significant impact on the finished product and therapeutics. In Ayurveda identification of medicinal plants is a major hurdle in laying pharmacopeial standards for formulations. Astavarga (a group of eight medicinal plants; see Table 5.24) is vital part of Ayurvedic formulations like Chyvanprasha and four plants viz, Riddhi, Vriddhi, Jivaka and Rishbhaka have been discussed as possible members of the family Orchidaceae. Although work has been done on identification of medicinal plants mentioned under Astavarga, much remains to be done to identify the true representatives.

Table 7.1 Eight medicinal plants used in Ashtavarga. The plants marked with stars have been reported to be orchids.

S.No.	Ayurvedic name	Botanical name	Family	Part used
1.	Jivaka*	*Malaxis muscifera* (Lindl.) Kuntze	Orchidaceae	Bulb
2.	Rishbhaka*	*Malaxis acuminta* D.Don	Orchidaceae	Pseudo-bulb
3.	Meda	*Polygonatum verticillatum* (L.) All.	Polygonaceae	Rhizome
4.	Mahameda	*Polygonatum cirrhifolium* (Wall.) Royle	Polygonaceae	Rhizome

Table 7.1 contd...

Table 7.1 contd...

S.No.	Ayurvedic name	Botanical name	Family	Part used
5.	Kakoli	*Roscoea procera* Wall. or *Fritillaria royeli* Hook.f.	Zingiberaceae	Root
6.	Kshira Kakoli	*Lilium polphyllum* D.Don	Liliaceae	Root
7.	Riddhi	*Habenaria intermedia* D.Don	Orchidaceae	Root
8.	Vriddhi	*Habenaria edgeworthii* Hook.f. ex Collett.	Orchidaceae	Root

In Ayurveda pasanbheda is a drug of controversial origin. The plant is widely used in the treatment of kidney ailments particularly kidney stones. Various medicinal plants are used as pasanbheda in different parts of India (see Table 5.25).

Table 7.2 It shows possible representatives of pasanbheda, an Ayurvedic drug of controversial origin.

S.no	Botanical name	Natural order	Parts used
1.	*Aerva lanata* Juss.ex Schult.	Amaranthaceae	Roots
2.	*Bergenia ligulata* (Wall.) Engl.	Saxifragaceae	Seeds
3.	*Bridelia crenulata* L.	Euphorbiaceae	Stem bark
4.	*Bryophyllum calycinum* Salisb.	Crassulaceae	Leaves
5.	*Coleus amboinicus* Lour.	Lamiaceae	Leaves
6.	*Decalepis arayalpatra* Joseph & Chandrasekharan	*Periplocaceae*	
7.	*Didymocarpus pedicellata* R.Br.	Gesneriaceae	Whole plant
8.	*Homonoia riporia* Lour.	Euphorbiaceae	
9.	*Rotula aquatica* Lour.	Boraginaceae	Whole plant

Further work on proper botanical identification of pasanbheda is warranted. In the Indian market several commercial lithontriptic preparations are sold for dissolving kidney stones. Some formulations utilize *Bergenia ligulata* while others *Bryophyllum calycinum.* The situation is quite confusing for practicing physicians. Further work on proper botanical identification of pasanbheda is warranted.

Studies done on plants like *Cratavea nurvala* (varun) have shown that the active constituents of the plant including lupeol possess significant antiurolithiatic activity in animal models. Most of the companies marketing herbal products are dependent on animal findings. Clinical data is very limited with Ayurvedic drugs. In the studies that have been done, the pattern remains objectionable or inconclusive. In case of pasanbheda, studies similar to that of *Cratavea nurvala* should be done for all representatives (mentioned in Table 1.2) so as to understand the mode of action and justify the rationale use of the drug. Botanical identity is of prime importance while creating a formulation.

The list is endless. In the subsequent table we have listed some controversial drugs used in Ayurveda for the benefit of potential readers.

Table 7.3 It Shows Some of the Controversial Medicinal Plant Used in Ayurveda.

S.no	Ayurvedic name	Current representative	Possible representatives
1.	*Adhoguda*	*Euphorbia nivulia* Buch.-Ham.	*Euphorbia acaulis* Roxb., *Argyeria petaloides*
2.	*Agnigurbha*	*Solanum trilobatum* L.	*Solanum xanthocarpum* Schrad et Wendl., *Artemisia vulgaris* Willd.
3.	*Agnimantha*	*Premna integrifolia* L.	*Clerodendron phlomidis* L.
4.	*Ajagandha*	*Gynandropsis gynandra* L. Briq.	*Clemone viscosa* L.
5.	*Alambusa*	*Sphaeranthus indicus* L.	*Mimosa pudica* L., *Biophytum senstivitum* R.Br.
6.	*Amarvalli*	*Cuscuta reflexa* Roxb.	*Cassyatha filiformis* L.
7.	*Amla vetasa*	*Garcinia pedunculata* Roxb.	*Rheum emodi* Wall.
8.	*Ansumati*	*Desmodium gangeticum* DC. var. maculatum (L.) DC. (L.) Baker	Uraria picta (Jacq.) L., Pseudarthria viscida (L.) Wight & Arn.
9.	*Arjuna*	*Terminalia arjuna* Roxb. W. & A.	*Sterculia urens* Roxb., *Terminalia tomentosa* W. & A., *Lagerstroemia flos-regina* Retz
10.	*Arka*	*Calotropis gigentia* (L.) R. Br. ex Ait.	*Calotropis procera* R.Br.
11.	*Asmantaka*	*Ficus rumphii* Blume	*Bauhinia racemosa* Lam.
12.	*Asvaksura*	*Clitoria ternatea* L.	*Medicago sativa* L.
13.	Asoca	*Polyalthia longifolia* Thw.	*Saraca asoca* (Roxb.) De Willd.
14.	*Bala*	*Sida cordifolia* Linn.	*Abutilon indicum* G.Don., *Grewia hirsuta* Vahl., *Sida veronicaefolia* Lam., *Sida rhombifolia* L.
15.	*Bharangi*	*Clerodendron serratum* Linn.	*Clerodendron siphonanthus* (R.Br.) CB Clarke, *Gardenia turgida* Roxb.
16.	*Brahamdandi*	*Tricholepis glaberrima* DC.	*Echinops echinatus* Roxb., *Argemone mexicana* L., *Xanthium strurmarium* Linn., *Voluntarella divaricata* C.B. Clarke
17.	*Ganda durva*	?	?
18.	*Gandira*	*Coleus barbatus* (Andrews) Benth.	*Amaranthus viridis* L.
19.	*Gojihva*	*Elephantopus scaber* L.	*Onosma bracteatum* Wall.
20.	*Granthiparni*	*Angelica glauca* Edgew.	*Clerodendron infortunatum* Gaertn.
21.	*Kākajangha*	*Peristrophe bicalyculata* Nees	*Leea aequata* L.
22.	*Kākanasa*	*Pentatropis microphylla* W. & A.	*Asclepias curassavica* L.
23.	*Karavi*	*Strobilanthes callosus Nees*	*Carum carvi* L.
24.	*Matsyaksi*	*Alternanthera sessilis* R. Br. ex DC.	?

Table 7.3 contd...

Table 7.3 contd...

S.no	Ayurvedic name	Current representative	Possible representatives
25.	Mahamundi	*Sphaeranthus africans* L.	?
26.	*Morata*	*Schrebera swietenioides* Roxb.	*Elaeodendron glaucum* (Rottb.) Pers.
27.	*Murva*	*Marsdenia tenacissima* W.& A.	*Chonemorpha macrophylla* G.Don ., *Clemantis triloba* Heyne ex Roth. *Sansevieria roxburghiana* Schult.f.
28.	*Parpata*	*Fumaria vaillanti* Loisel	*Oldenlandia corymbosa* L., *Rungia repens* Nees, *Mollugo oppositifolia* L. , *Polycarpaea corymbosa* (L.) Lamarck
29.	*Pītachandana*	*Jateorhiza palmata* Miers	*Coscinium fenestartum* (Gaertn) Coleb.
30.	*Rasna*	*Pluchea lanceolata* C.B.Clarke.	*Alpinia galanga* Willd., *Aristolochia indica* L., *Vanda roxburghii* R.Br.
31.	*Rohitaka*	*Aphanamixis polystachya* (Wall.) R. N. Parker	*Tecoma undulata* (Sm.) G. Don
32.	*Rudanti*	*Cressa cretica* L.	*Capparis moonii* Hook. f. Thoms.
33.	*Soma*	*Ephedra gerardiana* Wall.	*Periploca aphylla Decne.* var. *laxiflora* Bornm.
34.	Sprkka	*Melitotus officinalis*	*Anisomeles malabarica* (L.) R.Br.
35.	Tivariya	*Avicennia officinalis* L.	*Hydnocarpus wightiana* Blume.
36.	Tryantika	*Delphinium zalil* Aitch. & Hemsl.	*Gentiana kurroa* Royle
37.	*Uddalaka*	*Paspalum scrobiculatum* L.	*Bauhinia variegata* L.

7.4 Pre-clinical research on medicinal plants used in Ayurveda

In Ayurvedic Materia Medica approximately 800 medicinal plants are actively used in creating formulations. Some of the plants are used as single entity in the form of powder, tablet or extract. There is a clear cut description that certain plants should not be subjected to decoction or infusion as it essentially kills the active principle or constituent. Modern studies have proved that if alkaloid containing drugs are subjected to heat at 60°C, they undergo hydrolysis and get converted into isomeride (a compound having the same kind of atoms but with different stereochemistry). In this case it appears to be coherence between traditional and modern view.

The majority of medicinal plants used in Ayurveda have been subjected to phytochemcial screening and active principles have been identified. Further pre-clinical (pharmacological investigations) have justified the traditional medical claims of Ayurveda. In the subsequent table we have tried to document pre-clinical research on medicinal plants used in Ayurveda with respect to bioactive constituents.

Table 7.4 Pre-clinical research on medicinal plants used in Ayurveda.

S.no	Medicinal Plant	Bioactives	Pre-clinical activity
1.	*Aegle marmelos* Correa	Pyranocoumarin: luvangetin	Gastro-protective
2.	*Andrographis paniculata* Nees.	Diterpene: andrographolide	Chloretic
3.	*Bacopa monnieria* L.	Triterpenoid saponins: Bacoside-A and B	Nootropic
4.	Boswellia serrata	Boswellic acids	Anti-inflammatory
4.	*Butea monosperma* L.	Terpene anhydride: palasonin	Anthelmintic
5.	*Commiphora mukul*	Sesquiterpene ketone: guggulsterones	Hypolipidemic
6.	*Crataeva nurvala*.Buch.-Ham.	Flavonoid: lupeol	Nephroprotective
7.	*Desmodium gangeticum*	Pterocarpenoid: gangetin	Anti-inflammatory
8.	*Eclipta alba* L.	Apigenin, 4-hydroxybenzoic acid and protocatcheuic acid	Hepatoprotective
9.	*Embelia ribes* Burm	Benzoquinone: embelin	Antifertility and anti-implantation
10.	*Glychyrrhiza glabra* L.	Triterpenoid saponin: glycyrrhizin	Antiviral
11.	*Nardostachys jatamansi* DC	Sesquiterpene ketone: jatamansone	Antiarrhythmic, anticonvulsant and tranquillizer
12.	*Nelumbo nucifera* L.	Triterpene: ursolic acid	Anti-inflammatory
13.	*Pergularia pallida* (Roxb.) Wight & Arn.	Phenanthroindolizidine alkaloids: pergularinine and tylophorinidine	Anticancer
14.	*Picrorhiza kurrooa* ex Benth.	Kutkins	Hepatoprotective
15.	*Piper nigrum* L.	Alkaloid: piperine	Anti-diarrheal
16.	*Plumbago rosea* L.	Napthoquinone-plumbagin	Anticancer
17.	*Pongamia pinnata* (L.) Merr.	Flavonoid:pongamol	Anticonvulsant
18.	*Premna integrifolia* L.	Isoxazole alkaloid: premnazole	Anti-inflammatory
19.	*Rubia cordifolia*	Anthraquinone: rubidianin	Antioxidant
20.	*Solanum xanthocarpum* L.	Alkaloid: solasodine	Antifertility
21.	*Swertia chirata* - (Wall.) C.B.Clarke.	Secoiridoid glycosides: amarogentin and xanthone: swerchirin	Topoisomerase inhiniter, central nervous system depressant and hypoglycemic

Table 7.4 contd...

Table 7.4 contd...

S.no	Medicinal Plant	Bioactives	Pre-clinical activity
22.	*Terminalia belerica* sp.	3, 4, 5-trihydroxy benzoic acid (gallic acid)	Hepatoprotective
23.	*Terminalia chebula* Retz	Anthraquinone: chebulin	Antispasmodic
24.	*Tylophora asthmatica* (L. f.) Wight & Arn	Phenanthroindalizidine alkaloid: tylophorine	Anti-inflammatory, anti-anaphylactic and anti-spasmodic
25.	*Withania somnifera* Dunal	Withaferin-A	Anticancer

7.5 Clinical research in Ayurveda

Ayurvedic sciences are expected to play a significant role in the future healthcare system. With the onset of research and development activities in Ayurveda and the herbal system of medicine, there is an acute need for conducting clinical trials with traditional drugs. Most of the work reported in literature is based on pre-clinical studies. Some drugs like *Terminalia arjuna* (for hearts aliments) and *Commiphora mukul* (for obesity and arthritis) have been scientifically validated for ancient claims. The need of the hour is documentation of clinical knowledge on herbs available with us. At the same time clinical trials on ancient formulations are also warranted.

In the subsequent table clinical trials on selected medicinal plants used in Ayurvevda have been elaborated.

Table 7.5 Clinical research on medicinal plants used in Ayurveda.

S.no	Medicinal Plant	Disease area	Part used/ Preparation
1.	*Asparagus officinalis*	Duodenal ulcer	Roots
2.	*Bacopa monnieria*	Anxiety neurosis and epilepsy	Dried herb
3.	Boswellia serrata	Bronchial asthma and rheumatoid arthritis	Gum-resin
4.	*Commiphora mukul*	Acne vulgaris, hyperlipidemia obesity , osteoarthritis and rheumatoid arthritis	Gum-resin
5.	*Dashmula*	Diabetic neuropathy	N/A
6.	*Emblica officinalis*	Dyspepsia, hyperacidity and hyperlipidemia	Whole part
7.	*Enicostemma littorale*	Diabetes mellitus type-2	Pills
8.	*Mucuna pruriens*	Parkinson's disease	Seed powder
9.	*Ocimum sanctum*	Diabetes mellitus type-2	Leaf-powder
10.	*Solanum trilobatum*	Bronchial asthma	Leaf
11.	*Solanum xanthocarpum*	Bronchial asthma	Leaf

Table 7.5 contd...

Table 7.5 contd...

S.no	Medicinal Plant	Disease area	Part used/ Preparation
12.	*Terminalia arjuna*	Congestive cardiac failure	Bark powder and bark extract
13.	*Tylophora asthmatica*	Bronchial asthma	Leaf-powder and alcoholic extract
14.	*Withania somnifera*	Anxiety neurosis and rheumatoid arthritis	Alcoholic extract

7.6 Future trends

Ayurveda is moving from the fringe to mainstream use with a greater number of people seeking remedies free of the side-effects caused by synthesized chemicals. Alternative systems of medicine have become increasingly popular in recent years. The efficacy of some herbal products is beyond doubt, the most recent examples being *Artemisia annua*, *Silymarin marianum* and *Taxus brevifolia*. Several medicinal plants used in Ayurveda (as discussed in this chapter) are being targeted for possible drug development. The future market belongs to standardized Ayurvedic preparations, which have self explanatory advantages over crude drugs.

Further reading

Agrawal P, Rai V, and Singh RB. 1996. Randomized placebo-controlled, single blind trial of the holy basil leaves in patients with non-insulin-dependent diabetes mellitus. Int. J. Clin. Pharmacol. Thera. 34, 406–409.

Anand KK, Singh B, Saxena AK, Chandan BK, Gupta VN, and Bhardwaj V. 1997. 3,4,5-Trihydroxy benzoic acid (gallic acid), the hepatoprotective principle in the fruits of *Terminalia belerica*—bioassay guided activity. Pharmacol. Res. 36, 315–321.

Anand R, Patnaik GK, Kulshreshtha DK, and Dhawan BN. 1994b. Antiurolithiatic activity of lupeol, the active constituent from *Crateva nurvala*. Phytotherapy Research. 8, 417–421.

Anonymous. 1998. Indian Herbal Pharmacopoeia. Vol. 1. Indian Drug Manufacturers Association. Worli, Mumbai.

Arora RB, Singh M, and Arora CK. 1962. Tranquillizing activity of jatamansone, a sesquiterpene from *Nardostachys jatamansi*. Life Science. 6, 225.

Arora RB, Sharma PL, and Kapila K. 1985. Antiarrhythmic and anticonvulsant action of jatamansone. Indian Journal of Medical Research. 46, 782–791.

Authors not listed. 1995. An alternative medicine treatment for Parkinson's disease: results of a multicenter clinical trial. HP-200 in Parkinson's Disease Study Group. Journal of Alternative and Complementary Medicine. 4, 68.

Authors not listed. 1998. Flexible dose open trial of Vijayasar in cases of newly-diagnosed non-insulin-dependent diabetes mellitus. Ind. J. Med. Res.108, 24–9.

Babbar OP, Chhatwal VK, Ray IB, and Mehra MK. 1982. Effect of berberine chloride eye drops on clinically positive trachoma patients. Ind. J. Med. Res. 76, 83–88.

Bahl CP, and Seshadri TR. (eds). 1978. Pashanbhedi: drugs for urinary calculus, K.N. Udupa. 77–98.

Bajad S, Bedi KL, Singla AK, and Johri RK. 2001. Antidiarrheal activity of piperine in mice. Plant. Med. 67, 284–287.

Barik BR, Bhowmik T, Dey AK, and Patra A. 1976. Premnazole, an isoxazole alkaloid of *Premna integrifolia* L. and *Gmelina arborea* L. with anti-inflammatory activity. Fitoterapia. 639, 395.

Basu SP, Mandal JK, and Mehdi NS. 1994. Anticonvulsant effect of pongamol. Ind. J. Pharmaceuti. Sci. 56–163.

Bharani A, Ganguly A, and Bhargava KD. 1995. Salutary effect of *Terminalia arjuna* in patients with severe refractory heart failure. Int. J. Cardiol. 49, 191–199.

Bhargava SK, Dixit VP, and Khanna P. 1984 Anti-fertility effect of embelin in female rats. Fitoterapia. 55, 302–304.

Bhattacharya SK, Reddy PK, Ghosal S, Singh AK, and Sharma PV. 1976. Chemical constituents of Gentianaceae XIX: CNS-depressant effects of swertiamarin. J. Pharmaceut. Res. 65, 1547–9.

Bhavan BV. 1992. Selected Medicinal Plants of India. Tata Press, Bombay, India. pp. 333–336.

Chawla YK, Dubey P, Singh R, Nundy S, and Tandon BN. 1987.Treatment of dyspepsia with amalaki (*Emblica officinalis*), an Ayurvedic drug. Vagbhata. 5, 24–26.

Dash B. 1994. Materia Medica of Ayurveda. South Asia Books, Delhi, India.

Devi PU. 1996. *Withania somnifera* Dunal (Ashwagandha): potential plant source of a promising drug for cancer chemotherapy and radio sensitization. Ind. J. Exper. Biol. 34, 927–932.

Devi PU, Akagi K, Ostapenko V, Tanaka Y, and Sugahara T. 1996. Withaferin-A: A new radio sensitizer from the Indian medicinal plant *Withania somnifera*. International Journal of Radiation Biology 69, 193–197.

Devi U, Solmon FE, and Sharda AC. 1999. Plumbagin- A Plant Napthoquinone with Antitumour and Radimodifying Properties. Pharmaceut. Biol. 37, 231–236.

Dixit VP. 1986. Antifertility effects of solasodine obtained from *Solanum xanthocarpum* berries in male rats and dogs. J. Steroid Biochem. 25, 27.

Dwivedi KN. 1961. Sushruta's dashamula and its application in diabetic neuropathy. Sach. Ayur. 23, 121–127.

Dwivedi S, Aggarwal A, Agarwal MP, and Rajpal S. 2005. Role of *Terminalia arjuna* in ischaemic mitral regurgitation. Int. J. Cardiol. 100, 507–8.

Ghosh D, and Anadakumar A. 1981. Anti-inflammatory and analgesic activities of Gangetin–a pterocarpenoid from *Desmodium gangeticum*. Ind. J. Pharmacol. 15, 391–402.

Goel RK, Maiti RN, Manickam M, and Ray AB. 1997. Antiulcer activity of naturally occurring pyrano-coumarin and iso-coumarins and their effect on prostanoid synthesis using human colonic mucosa. Ind. J. Exper. Biol. 35, 1080–1083.

Gopalakrishnan C, Shankaranarayan D, Kameswaran L, and Natarajan S. Pharmacological investigations of tylophorine, the major alkaloid of *Tylophora indica*. Ind. J. Med. Res. 69, 513–20.

Gore KV, Rao AK, and Guruswamy MN. 1980. Physiological studies with *Tylophora asthmatica* in bronchial asthma. Ind. J. Med. Res. 71, 144–8.

Govindan S, and et al. 1999. A pilot study on the clinical efficacy of *Solanum xanthocarpum* and *Solanum trilobatum* in bronchial asthma. J. Ethnopharmacol. 66, 205–10.

Gupta I, Gupta V, Parihar A, Gupta S, Lüdtke R, Safayhi H, and Ammon HP. 1998. Effects of *Boswellia serrata* gum resin in patients with bronchial asthma: results of a double-blind, placebo-controlled, 6-week clinical study. Eur. J. Med. Res.3, 511–514.

Gupta I, Parihar A, Malhotra P, Singh GB, Lüdtke R, Safayhi H, and Ammon HP. 1997. Effect of *Boswellia serrata* gum resin in patients with ulcerative colitis. Eur. J. Med. Res.2, 37–43.

Jacob A, Pandey M, Kapoor S, and Saroja R. 1998. Effect of the Indian gooseberry (amla) on serum cholesterol levels in men aged 35-55 years. Eur. J. Clin. Nut.42, 939–944.

Kidd PM. 1999. A review of nutrients and botanicals in the integrative management of cognitive dysfunction. Alter. Med. Rev. 4, 144–161.

Kishore P, Pandey P N, Pandey S N, and Dash S.1980. Treatment of duodenal ulcer with *Asparagus racemosus* Linn. J. Res. Ind. Med. Yoga. Homeopat. 15, 409–15.

Krishnamurthy A. 1969. The Wealth of India. Vol. VIII. Publication and Information Directorate, Council of Scientific and Industrial Research. New Delhi.

Kumar D, Mishra SK, Tandan SK, and Tripathi HC. 1995. Possible mechanism of anthelmintic action of palasonin on Ascaridia galli. Indian Journal of Pharmacology. 27, 161–8.

Kuppurajan K, Rajagopalan SS, Koteswara RT, and Sitaraman R1978. Effect of guggulu (*Commiphora mukul* Engl.) on serum lipids in obese, hypercholesterolemic and hyperlipidemic cases. J. Assoc. Phys. Ind. 26, 367–73.

Malhotra SC, Ahuja MM, and Sundaram KR. 1997. Long term clinical studies on the hypolipidemic effect of Commiphora mukul and clofibrate. Ind. J. Med. Res. 65, 390–5.

Mathew KK, and Shivpuri DN. 1974. Treatment of asthma with alkaloids of *Tylophora indica*: a double-blind study. Asp. All. App. Immunol. 7, 179–7.

Mukherjee PK, Saha K, Pal M, and Saha BP. 1997. Studies on the anti-inflammatory activity of rhizomes of *Nelumbo nucifera* (letter). Plant Med. 63, 67–9.

Nadkarni K. 1998. Indian Materia Medica. Vol. 1. Popular Prakashan, Bombay, India.

Rao KN, Bhattacharya RK, and Venkatachalam SR. 1997. Inhibition of thymidine synthase and cell growth by the phenanthroindolizidine alkaloids pergularinine and tylophorinidine. Chem. Biol. Interact. 106, 201–12.

Ray S, Majumder HK, Chakravarty AK, Mukhopadhyay S, Gil RR, and Cordell GA. 1996. Amarogentin, a naturally occurring secoiridoid glycoside and a newly recognized inhibitor of topoisomerase 1 from Leishmania donovani. J. Nat. Prod.59, 27–9.

Saraswat B, Visen PK, Patnaik GK, and Dhawan BN. 1997. Protective effect of picroliv, active constituent of *Picrorhiza kurrooa*, against oxytetracycline induced hepatic damage. Ind. J. Exp. Biol. 35, 302–5.

Saxena AK, Singh B, and Anand KK. 1993. Hepatoprotective effects of *Eclipta alba* on sub cellular levels in rats. J. Ethnopharmacol 40, 155–61.

Saxena AM, Murthy PS, and Mukherjee SK. 1996. Mode of action of three structurally different hypoglycemic agents: a comparative study. Ind. J. Exper. Biol. 34, 406–9.

Shah CS, Shah N, and Mody KD. 1972. Pharmacognostic study of pashanbhed: I-III: *Bergenia ciliata* and *Didymocarpus pedicellata*. Quar. J. Crude Drug Res. 12, 182–193.

Shivpuri DN, Singhal SC, and Parkash D. 1972. Treatment of asthma with an alcoholic extract of *Tylophora indica*: a cross-over, double-blind study. Ann. Aller. 30, 407–12.

Shukla B, Visen PK, Patnaik GK, and Dhawan BN. 1992. Cholretic effect of Andrographolide in rats and guinea pigs. Plant. Med. 58, 146–149.

Singh AP. 2005. Dravyaguna Vijnana. Chaukhambha Orientalia, New Delhi.

Singh AP, and Sandhu AS. 2005. A Dictionary of Medicinal Plants. Sundeep Publishers, New Delhi.

Singh AP, Malhotra S, and Munjal G. 2002. Hepatoprotective Potential of Phytomedicines. Gastroenterol. Today 6, 76–77.

Singh HK, and Dhawan BN. 1982. Effect of *Bacopa monniera* Linn. (Brahmi) extract on avoidance responses in rat. J. Ethnopharmacol. 5, 205–214.

Singh HK, Rastogi RP, Srimal RC, and Dhawan BN. 1988. Effect of bacosides A and B on avoidance responses in rats. Phytother. Res. 2, 70–75.

Singh RB, Niaz MA, and Ghosh S. 1994. Hypolipidemic and antioxidant effects of *Commiphora mukul* as an adjunct to dietary therapy in patients with hypercholesterolemia. Cardiovas. Drugs Thera. 8, 659–664.

Singh V. 2003. *Cultivation, processing and quality evaluation of medicinal & aromatic plants.* Institute of Himalayan Bio resource Technology (CSIR) 17.

Singh BB, Mishra LC, Vinjamury SP, Aquilina N, Singh VJ, and Shepard N. 2003. The effectiveness of *Commiphora mukul* for osteoarthritis of the knee: an outcomes study. Alter. Ther. Heal. Med9, 74–79.

Singh BN, and Sharma PV. 1971. Effect of amalaki on amlapitta. Journal of Research in Indian Medicine. 5, 223–230.

Thappa DM, and Dogra J. 1994. Nodulocystic acne: oral gugulipid versus tetracycline. J. Dermatol. 21, 729–731.

Thiruvengadam KV, Haranath K, Sudarsan S, Sekar TS, Rajagopal KR, Zacharian MG, and Devarajan TV.1978. *Tylophora indica* in bronchial asthma: a controlled comparison with a standard anti-asthmatic drug. J. Ind. Med. Assoc.71, 172–176.

Trivedi V, Nesamany S, and Sharma VK. 1982. A clinical study of the antitussive and antiasthmatic effect of *Vibhitakphal churana* (*Terminalia bellerica* Roxb.) in the cases of Kasa-swasa. J. Res. Ayur. Sidd. 3, 1–8.

Uma DP, Akagi K, and et al. 1996. Withaferin-A: A new radio sensitizer from the Indian medicinal plant *Withania somnifera*. International Journal of Radiation Biology. 69, 193–197.

Uniyal MR, and Issar RK. 1996. Botanical studies of *Astavarga*. Nagarjun. 9, 276–284.

Upadhay UM, and Goyal RK. 2004. Efficacy of *Enicostemma littorale* in type 2 diabetic patients. Phytother. Res. 18, 233–235.

Vyas SN. 1987. A clinical study on the effect of shudha guggulu in rheumatoid arthritis. Rheumatism. 23, 15–26.

Chapter 8

Amchi System of Medicine

8.1 Introduction

The Amchi System of medicine (ASM) or Tibetan Medical System (TMS) is one of the practicing systems of complementary and alternative medicine (CAM), also known as Sowa Rigpa (classical Tibetan), which means 'science of life'. Amchi signifies doctor or physician. Fundamental aspects of the Amchi system of medicine are similar to traditional Indian Medicine (TIM) or Ayurveda.

It is estimated that the Amchi System of medicine is the outcome of aspects of medicine that were transmitted from India to Tibet between the 7th and 12th centuries. ASM however, did not make advances as Traditional Indian Medicine, Traditional Chinese Medicine (TCM), Homeopathy and Western Medicinal Herbalism.

8.2 Basic principles

Historically, the amchi would begin their medical training at an early age. Their knowledge and skills have been transferred from teacher to student, often from father to son. Thus, lineages of amchi families exist throughout the Tibetan cultural world. After learning how to read and write classical Tibetan and studying relevant religious texts, students would learn their vocation by apprenticing with an elder amchi and studying the Gyud Shi, or the Four Root Tantra texts of Tibetan medicine. Young amchi would also learn how to identify and collect medicinal plants, make medicine, remove poisonous qualities of certain ingredients, diagnoze disease using pulse and urine analysis, and provide prescriptions for patients. They would also be trained in moxibustion, cauterization, and other healing techniques. Likewise, the amchi would receive training in astrology, which is an essential component of diagnosis and treatment within the Tibetan medical tradition.

8.3 Range of drugs used

The present study deals with the Tibetan Medical System and the ingredients used in preparing various ethno medicines to cure several ailments by amchis inhabiting Ladakh and the Lahaul-Spiti region of the Indian trans-Himalaya. A total of 337 plant species, 38 species of animals and six minerals were documented during the survey period. Among 83 amchis interviewed, 36% had disciples or students, primarily their own sons and daughters. The study reflects that the Tibetan system of medicine is declining in the study area due to the shift in socio-economic patterns and unwillingness of the younger generation to adopt amchi as a profession.

Cold deserts are found in the interior of Asia and in the inter mountain zone of North America. In India, cold desert comes under the trans-Himalayan zones, which are confined to Ladakh in Jammu and Kashmir, Lahaul and Spiti in Himachal Pradesh. The Indian cold desert flora is dominated by a few dwarf bushes. Like other parts of the Himalayas, these areas are considered as a treasure of medicinal plants. Keeping in view the above, the Field Research Laboratory (FRL) DRDO carried out extensive ethno-medico- botanical surveys of different parts of the Indian cold desert and has collected 425 plant species, which are used by amchis (herbal doctor) for various ailments. The local tribals have a strong belief in the amchi system of medicine. The important medicinal taxa are *Podophyllum hexandrum, Aconitum heterophyllum, Saussurea obvallata, Gentiana aligida, Artemisia* sp., *Oxytropis microphylla*, and *Cremanthodium ellisii.*

Detailed phytochemical and pharmacological studies are required for positive exploitation and wider application of these Amchi-medico-taxa. The majority of medicinal flora of the Indian cold desert are rare, endangered and threatened due to unscientific exploitation, overgrazing, uprooting for fuel and natural disasters. Therefore, there is a need to conserve the rare and precious cold desert medicinal plants by opening nature reserves and botanical gardens and by developing practices to popularize their commercial cultivation.

Amchis may also create confidential mixtures of medicines for a patient. This confidentiality was maintained and no data regarding those mixtures were recorded. When permission was granted, the plants species has been indicated (for example with the confidential nature of the mixture noted). Many plant species are used by amchis to treat a broad range of ailments, while some plant species are found to treat only one aliment. For example, *Allium oreoprasum, Aster diplostephioides,* and *Neopicrorhiza scrophulariiflora* treat a broad range of ailments, while *Maharanga bicolor, Maharanga emodi,* etc. are used to treat only one ailment (ear pain). In total, 91 medicinal plants were found to be used to treat 93 ailments. The most commonly treated ailments include gastritis, cough and cold, edema, stomachache, diarrhea, dysentery, jaundice, rheumatism, and numbness of limbs.

8.4 Major medical plants used in amchi system of medicine

Cicerbita macrorhiza (Royle) P Beauv (Asteraceae)

The plant is used as febrifuge.

Rubus foliolosus (Rubiaceae)

It is given in fever, dyspepsia, cough, cold, vertigo. The plant from Japan is reported to contain labdene type diterpene glycosides.

Aster diplostephioides (DC) CB Clarke. (Asteraceae)

The flowers are used in Tibetan medicine, they are said to have a bitter taste and a cooling potency. Antidote, febrifuge, haemostatic and tonic, are used in the treatment of infectious fevers, influenza, nose bleeds, poisoning, sores from environmental poisoning and an inability to stretch or contract the limbs.

Delphinium brunonianum Royle (Ranunculaceae)

Amchi name: Bcha-rog-spos. It is commonly known as Musk Larkspur. In the Himalayas it is only used to destroy ticks on animals.

Rhododendron anthopogon D.Don (Ericaceae)

In Nepal, the leaves are boiled and the vapor inhaled to treat coughs and colds.

Rhododendron lepidotum Wall ex D Don (Ericaceae)

Two new coumarin glycosides, named rhodonin and rhodonetin were isolated from the aerial parts of *R. lepidotum*. The plant is reported to have antioxidant and antiviral activities.

Neopicrorhiza scrophulariiflora (Pennell) DY Hong (Scrophulariaceae)

This plant is found in the inventory of medicinal plants used in the Amchi system of medicine. It is among rare plants used in the amchi system of medicine, which has been investigated phytochemically. Phenylpropanoid glycosides: scrophulosides A and B, androsin and picroside- I have been reported. Three iridoid glycosides, picrorosides A, B and C, and a cucurbitacin glycoside, scrophoside A, have been isolated from the rhizomes, along with two known iridoid glycosides, picrosides I and II, and three known cucurbitacin glycosides.

Four new non-glycosidic iridoids, piscrocins D, E, F, and G , as well as two new iridoid glycosides, piscrosides A and B, were isolated from the roots, together with seven known iridoids. The compounds demonstrated hepatoprotective activity on CCl4-induced hepatocytes damage *in vitro,* and the structure-activity relationships was established.

Maharanga bicolor (Boraginaceae)

It is used in Nepalese ethnomedicine for the treatment of several diseases. In the course of screening investigations the dichloromethane extract of the roots of Maharanga bicolor was found to inhibit the growth of gram positive bacteria. Bio-assay directed fractionation led to the isolation of five active naphthazarins, deoxyalkannin, alkannin, acetylalkannin, alkannin 3-hydroxyisovalerate and alkannin ß-acetoxyisovalerate. Alkannin, acetylalkannin, alkannin 3-hydroxyisovalerate and alkannin ß-acetoxyisovalerate showed antibacterial activity against multi resistant human pathogenic Staphylococcus and Enterococcus species. Deoxyalkannin, alkannin 3-hydroxyisovalerate and alkannin ß-acetoxyisovalerate, showed antiviral activity, against herpes simplex virus type.

Cremanthodium ellisii Hook. f.) Kitam. ex Kitam. and Gould (Asteraceae)

The plant is reported to contain esters of sesquiterpene polyalcohols, phenylpropanoids and lignans. These compounds are reported to have antibacterial and anticancer activities.

Further reading

Ballabh B, and Chaurasia OP. 2007. Traditional medicinal plants of cold desert Ladakh-used in treatment of cold, cough and fever. J. Ethnopharmacol. 2, 341–9.

Chen H, Zhu Y, Shen X, and Jia Z. 1996. Four new Sesquiterpene Polyol Esters from *Cremanthodium ellisii*. Nat. Prod. 59, 1117–20.

Chopra RN, Nayar SL, and Chopra IC. 1986. Glossary of Indian medicinal plants (including the supplement). Council Scientific Industrial Research, New Delhi.

Gartoulla RP. 1992. Alternative medication practices. Dissertation. North Bengal Univ. Darjeeling, India.

Kala CP. 2005. Health traditions of Buddhist community and role of amchis in trans-Himalayan region of India. Curr. Sci. 89, 25.

Khan R, Shawl AS, Tantray M, and Alam MS. 2008. New coumarin glycosides from *Rhododendron lepidotum*. Fitoterapia. 79, 232–233.

Kim IH, Kaneko N, Uchiyama N, Lee JE, Takeya K, Kawahara N and Goda. 2006. Two Phenylpropanoid Glycosides from *Neopicrorhiza scrophulariiflora*. Chem. Pharma. Bull. 54, 22.

Kim IH, Uchiyama N, Kawahara N and Goda Y. Iridoid glycosides and cucurbitacin glycoside from *Neopicrorhiza scrophulariiflora*. Phytochemistry 2006,67, 2691–6.

Manandhar NP. 2002. Plants and people of Nepal. Timber Press, Portland, OR.

Ohtani K, Chong-Ren Y, Miyajima C, Zhou J, and Tanaka O. 1991. Labdane-type diterpene glycosides from fruits of *Rubus foliolosus*. Chem. Pharmaceut. Bull. 39, 2443–5.

Prakash DG, Upadhyay BN, Singh R, Dhakarey, Kumar S and Singh KK. 2007 Free-radical scavenging activities of Himalayan rhododendrons. Curr. Sci.92, 526–531.

Rahbhandari M, Mentel R, Jha PK, Chaudhary RP, Bhattarai S and Gewali MB. 2009. Antiviral activity of some plants used in Nepalese traditional medicine. Evidence based Complement Alter Med. 6, 517–22

Rajbhandari NM, Schoepke T, Mentel R, and Lindequist U. 2007. Antibacterial and antiviral naphthazarins from *Maharanga bicolor*. Pharmazie. 62, 633–5.

Singh B, Chaurasia OP. 2000.Medical flora of Indian cold desert. Acta Hort. 523, 65–72.

ISHS Acta Horticulturae 523, XXV Int Horticultural Congr, Part 13.

Tamang TY. 2005. A glance to the Tibetan medical tradition. Reflections millennium batch 2000. Pokhara. 2, 62–3.

Wang AX, Zhang Q, and Jia ZJ. 2004. Phenylpropanosids, lignans and other constituents from *Cremanthodium ellisii*. Pharmazie. 59, 889–92.

Wang H, Wu FH, Xiong F, Wu JJ, Zhang LY, Ye WC, Li P and Zhao SX.2006. Iridoids from *Neopicrorhiza scrophulariiflora* and their hepatoprotective activities in vitro. Chem. Pharma. Bull.54, 1144–9.

Chapter 9

Ethnopharmacology of Algae

9.1 Introduction

The oldest account known about the use of algae comes from the Chinese King, Shen Nang 2,700 years B.C. Data on the usage of algae in Traditional Indian Medicine (TIM) or Ayurveda is scarce with the exception of *Ceratophyllum demersum* Linn. (Ceratophyllaceae). The paste of *C. demersum* is used in bleeding hemorrhoids. It is also used in polydipsia, diarrhea, fever, bleeding diathesis, and biliousness.

In Traditional Chinese Medicine (TCM), several sea-weeds have been used. *Nostoc muscorum* Ag. and *N. sphaeroides* Kütz (Nostocaceae) are used in the treatment of night-blindness and burns. *Saragassum fusiforme* (Harvey) Setchell (Sargassaceae) has been used as an antifungal and anticoagulant in TCM since 8th century A.D. Kunbu (*Laminaria* and *Ecklonia*), Haizao (*Sargassum*) and Zicai (*Porphyra*) are integral parts of TCM. The *Sargassum* decoction is used in the treatment of obesity, breast and ovarian tumors and lymphadenitis. Kunbu and haizao are used in promoting diuresis. *Laminaria japonica* Aresch. (Laminariaceae) and *Undaria pinnatifida* (Harvey) Suringar (Phaeophyceae) are used traditionally in Korea to promote maternal health.

Carrageenan is obtained from Irish moss (*Chondrus crispus*); because of its mucus forming properties it has been used in lung diseases and to improve the bitter drug taste. Carrageenan has also been used in cases of digestive tract irritations and in diarrhea and dysentery. In France and Great Britain, carrageenan has been used to treat stomach ulcers due to its mucous properties.

Agar-agar obtained from *Gelidium* and *Gracilaria* sp. is used in constipation and stomach prolapse. *Fucus vesiculosus* Linn. (Fucoideae) has been used for centuries in the treatment of thyroid, obesity, arteriosclerosis and digestive diseases. The recorded use of *Fucus vesiculosus* dates back to at least the time period of the Eclectic Physicians of the 19th century. *Chlorella vulgaris* yields chlorellin, an antibiotic compound.

Chondrus crispus Stackh. (Gigartinaceae) is used in bronchitis and diarrhea. Marine algae *Gelidiopsis* sp. (Gaisellaceae) is consumed in Indonesia by coastal communities for anti-aging property. Interestingly, the chloroform extract of *Gelidiopsis* sp. has an antioxidant activity. In Vietnam, seaweeds are used for treatment of cough, asthma, hemorrhoid, boils, goiter, urinary diseases, tumors, ulcers, and headaches.

9.2 Phytochemical and pharmacological investigations

Significant phytochemicals reported from algae

Plastocyanin and ferredoxin have been reported from *Ceratophyllum demersum*. A diterpenoid (dictyotetraene) has been isolated from *Dictyota* sp. from North Brittany Sea. Four diterpenoids (dischototetroal, dichotopentanol, dichotone and dichotodione) have been reported from *Dictyota dichotoma* (Hudson) Lamouroux (Dictyotaceae) growing in Pakistan. Several sterols, terpenes, polyols and certain metabolites have been isolated from these seaweeds. Some significant phytochemicals include dicdichane from *Dictyota dichotoma* and hauckiosterol from *Dictyota hauckiana* (Dictyotaceae) and *Spatoglossum variabile* Figari et De Notaris (Dictyoaceae); dictinol, dictindiol, dictintriol, amijiol and sargasterol from *Dictyota indica* Sonder (Dictyotaceae), padinolide from *Colpomernia sinuosa* (Mertens ex Roth) Derbes et Solier (Scytosiphonaceae) and *Padina tetrastromatica* Hauck (Dictyotaceae).

Sterols and volatiles have been reported from *Cystoseira barbata* J.Ag. (Phaeophaceae) collected from the Black Sea. A diterpenoid (stypolactone) has been isolated from *Stypopodium zonale* (J.V. Lamouroux) Papenfuss (Dyctiotaceae). *Spirulina platensis* Linn. is rich in vitamins, proteins, pigments and polyunsaturated fatty acids.

Diterpenes (isopachydictyolal from *D. dichotoma* and 4α-acetyldictyodial from *D. linearis*) were isolated from plants growing in Greece. *Dunaliella salina* UTEX is a single celled, salt-water micro-alga that accumulates massive amounts of carotenoids (α-Carotene, and β-Carotene) and xanthophylls (cryptoxanthin, lutein, and zeaxanthin). It is also reported to contain volatile compounds like β-Cyclocitral, α- and β-ionone, neophytadiene, and phytol.

Anticoagulant activity

It has reported that aqueous extract of carrageenan, even in great dilution, acts as blood anticoagulant.

Antimicrobial activity

Antiviral

Polysaccharides and polyphenols obtained from *Fucus vesiculosus* have anti-HIV activity. It has been postulated that adhesion is the first step in HIV infection; the active compounds from *Fucus vesiculosus* block adhesion which prevents transmission of HIV-1 (Beress et al., 1993). A novel reverse transcriptase inhibitor has been isolated from *Fucus vesiculosus*. It inhibits HIV reverse transcriptase by competing with the nucleic acid substrate.

Polysaccharides of *Gracilaria corticata* J. Agardh. (Gracilariaceae) showed selective antiviral activity against herpes simplex virus types 1 and 2. The galatofurans isolated from *Adenocystis utricularis* (Bory de Saint-Vincet) Skottsberg (Adenocystaceae) have inhibitory activity against herpes simplex virus type 1 and type 2.

A sulfated polysaccharide fraction isolated from hot water extract of *Caulerpa racemosa* (Forsskal) J. Agardh. (Caulerpaceae) was reported to be active against herpes simplex virus type 1 (HSV-1) and type 2 (HSV-2) in vitro cells (Ray, 2004). Galactofucan, a sulfated polysaccharide isolated from aqueous extract of *Undaria pinnatifida* (Harvey) Suringar (Phaeophyaceae) has significant anti viral activity against several strains of herpes simplex virus.

Ditepenoids isolated from Northeast Brazil algae, *Dictyota pfaffii* Schnetter (Dictyotaceae) showed strong anti-HSV-1 activity *in vitro* but only one compound inhibited the reverse transcriptase enzyme of HIV-1. Meroditerpenes (maric acid, epitaondiol and the peroxylactone of 5′a-desmethyl-5′-acetylatomaric acid) isolated from Brazilian algae, *Stypopodium zonale* showed strong anti-HSV-1 activity *in vitro*. It also inhibited the transcriptase reverse enzyme of HIV-1.

Antibacterial and antifungal

Crustose germlings of *Chondrus crispus* have antibacterial activity against benthic diatoms. The methanolic and ethyl acetate extracts of *Gelidiopsis* sp. from Indonesia have antibacterial activity against *Escherichia coli, Pseudomonas aeruginosa, Vibrio damsela, Vibrio harveyi* and *Staphylococcus aureus*. Fractions of the hexane, ethyl ether and dichloromethane extract of *Cystoseira tamariscifolia* Hudson (Cystoseriaceae) has significant antimicrobial and cytotoxic activities.

Methanolic extracts of *Nitella tenuissima* (Desv.) Kuetzing. (Characeae), *Microchaete tenera* Thuret ex Bornet (Microchaetaceae) and *Sphaeroplea annulina* (Roth) Ag. (Sphaeropleaceae) have been investigated for *in vitro* antimicrobial activity against *Proteus vulgaris, Bacillus cereus, Escherichia coli, Pseudomonas aeruginosa, Aspergillus niger, Aspergillus flavus* and *Rhizopus nigricans* using agar cup-plate method. Blue-green alga, namely, *Microchaete tenera*; and green algae, namely, *Nitella tenuissima* and *Sphaeroplea annulina* showed significant antibacterial activity against *Pseudomonas aeruginosa*. *Microchaete tenera* showed good antimicrobial activity against *Proteus vulgaris* and *Aspergillus niger*. *Sphaeroplea annulina* showed feeble antifungal activity against *Aspergillus flavus*.

Different pressurized liquid extracts obtained from *Dunaliella salina* has significant antimicrobial activity against *coli*, *Staphylococcus aureus*, *Candida albicans*, and *Aspergillus niger* (Miguel et al., 2006). Fatty acids were identiifed as possible antimicrobial agents. Methanolic extracts of *Corallina officinalis* Linn. (Rhodophyaceae), *Cladostephus spongiosis* (Hudson) C Agardh, *Cystoseira barbata* J.Ag., *Dictyota dichotoma* (Hudson) Lamouroux and *Halopteris filicina* (Grateloup) Kützing) (Phaeophyaceae) and *Ulva rigida* C.Ag. (Cholorophyaceae) demonstrated antibacterial activity against *Staphylococcus aureus* and *Enterobacter aerogens*.

The four tricyclic diterpenoids isolated from *Stochospermum marginatum* C. Agardh Kutzing have also been tested against three gram positive and six gram negative bacteria as well as six pathogenic fungi, and were found to exhibit strong antimicrobial activity.

Antifilarial

The crude extract of Indian algae, *Botryocladia leptopoda* (J.Agardh) Kylin (Rhodymeniaceae) and its hexane fraction has significant antifilarial activity against the animal filarial species included *Litomosoides sigmodontis* and *Acanthocheilonema viteae*. There was a marked reduction in the peripheral microfilarial level in both of the rodent filarial parasites.

Antioxidant activity

The chloroform extract of *Gelidiopsis* sp. from Indonesia has been reported to have significant antioxidant activity. Water soluble extract of *Sargassum thunbergii* Mertl O. Kuntze and purified proteases have significant antioxidant activity. The antioxidant activity heightened with increased concentrations of the extracts. The extract has a significant protective effect against DNA damage. Methanolic extracts of *kaki*, *Laminaria japonica* and *Undaria pinnatifida* have free radical scavenging effect against 1, 1-diphenyl-2-picrylhydrazyl in animal models.

Antitumor activity

A phycocolloid (carbohydrates) developed from *Chondracanthus chamissoi* (J. Agardh) Kützing ex *Gigartina chamissoi* (C. Ag.) J. Agardh (endemic to temperate southern Pacific coasts) has been used in the treatment of oncologic patients, producing a good recuperation in 68% of 162 patients.

Pahayokolide A, an active metabolite of *Lyngbya* sp. isolated from the Florida Everglades, was shown to inhibit a number of cancer cell lines.

Neutral and acidic polysaccharides and their protein complexes fractionated and purified from *Sargassum thunbergii* Mertl O. Kuntze demonstrated antitumor activity in mice with Ehrlich carcinoma.

Hypoglycemic and lipolytic activity

In some *Chondrus* species, a genus related with *Gigartina*, a series of sterols have been found, all of which have hypocholesterolemic activity. These sterols can also depress blood pressure in human atherosclerosis as well as low cholesterol levels in rabbits and rats.

A polysaccharide extracted from *Fucus vesiculosus* was evaluated for hypolipidemic activity in normal and hyperlipiemic rats. The alga significantly lowered total cholesterol and low-density-lipoprotein-cholesterol levels, and significantly increased the antiatherogenic index.

Compounds obtained from extracts of *Cystoseira barbata* have shown hypoglycemic and lipolytic activity. A new phloroglucinol derivative isolated from brown alga *Eisenia bicyclis* (Kjellman) Setchell (Phaeophyceae) from Japan has been reported to be effective in the treatment of diabetic complications. Polysaccharides from *Laminaria japonica* Aresch. reduced the concentration of cholesterol, triglycerides, and high density lipoprotein in quails.

Miscellaneous activity

A polyhalogenated monoterpene, plocamadiene A isolated from red marine algae, *Plocamium corallorhiza* Turner J. Hooker & Harvey (Plocamiaceae) releases histamine from peritoneal mast cells in the rats and guinea pigs. *Gracilaria corticata* J. Agardh. (Gracilariaceae) has been reported to have significant cholinesterase activity.

9.3 Spirulina, Aphanizomenon and Chlorella sp.

Spirulina platensis Linn., Spirulina fusiformis Voronichin (Oscillatoriaceae)

These algal genera deserve special mention as regards to medicine . Animal research has reported antiviral, antioxidant, neuroprotective, and cardio protective activities of *S. platensis. Spirulina fusiformis* Arch. has been reported to be nephroprotective in animal studies. C-phycocyanin from *S. platensis* has hepatoprotective, anti-inflammatory and antioxidant activities. C-phycocyanin has been reported to affect the stem cells found in bone marrow.

Clinical studies suggest that *Spirulina* is effective in melanosis and keratosis due to chronic arsenic posioning. It improves hemoglobin levels in malnourshied children and is effective in allergenic rhinitis or hay fever. A clinical trial from India, suggested a possible role of *Spirulina fusiformis* in chemo prevention for oral cancer. A recent clinical trial, reported hypolipidemic and antihypertensive activities of 4.5 G of Spirulina per day for 6 wk.

Aphanizomenon flos aquae (Linn.) Ralfs ex Bornet & Flahault (Nostocaceae)

This alga is reported to contain more protein and chlorophyll than any other food source. Animal trials have reported immunomodulator and hypolipidemic activities of the alga. A clinical trial, reported positive result of *Aphanizomenon flos aquae* in Alzheimer's disease.

Chlorella vulgaris Whittaker & Margulis, Chlorella pyrenoidosa Whittaker & Margulis and Chlorella kessleri Fott & Novakova (Chlorellaceae)

Animal trials in India, have reported anti-tumor, radio protective, and gastroprotecive activities of *Chlorella vulgaris and Chlorella kessleri*. Human trials have reported the usefulness of *Chlorella pyrenoidosa* in hyperlipidemia, hypertension and fibromylagia.

Conclusion

One of the reasons for exploring biological compounds in algae is the potential for medical use. Algae contain numerous bioactive constituents as shown in investigational studies. However, much work remains to link medical effects with specific algae species. Traditional medical claims need to be justified by pre-clinical and clinical research.

References and further reading

Abourrchine A, Charrouf M, Berrada M, Bennamara A, Chaib N, and Francisco C. 1999. Antimicrobial activities and cytotoxicity of the brown algae *Cystoseira tamariscifolia*. Fitoterapia. 70, 611–614.

Anyinam C. 1995. Ecology and Ethnomedicine: Exploring Links between Current Environmental Crisis and Indigenous Medical Practices. Soc. Sci. Med. 40, 321–329.

Ayehunie S, Belay A, Baba TW, and Ruprecht RM. 1998. Inhibition of HIV-1 Replication by an Aqueous Extract of *Spirulina platensis* (*Arthrospira platensis*). J. Acq. Imm. Def. Syn. Human, Retrovirolo. 18, 7–12.

Babadzhanov AS, Abdusamatova N, Yusupova FM, Faizullaeva N, Mezhlumyan LG, and Malikova MK. 2004. Chemical Composition of *Spirulina platensis* Cultivated in Uzbekistan. Chem. Nat. Comp. 40, 3.

Bannerman RHO, Burton J, and Ch'en W-C. 1983. Traditional Medicine and Health Care Coverage: A Reader for Health Administrators and Practitioners. World Health Organization. Geneva.

Barbosa JP, Pereira RC, Abrantes JL, Cirne-dos-Santos CC, Rebello MA, Frugulhetti IC, and Texeira VL. 2004. *In vitro* antiviral diterpenes from the Brazilian brown alga *Dictyota pfaffii*. Plant Med. 70, 856–60.

Belay A, Ota Y, Miyakawa K, and Shimamatsu H. 2006. Current knowledge on potential health benefits of *Spirulina*. J. Appl. Physiol. 5, 235–241.

Bensky D, and Barolet, R. 1990. Chinese Herbal Medicine: Formulas and Strategies Rev. Ed. Eastland Press Seattle, WA.

Béress A, Wassermann O, Tahhan S, Bruhn T, Béress L, Kraiselburd EN, Gonzalez LV, de Motta GE and Chavez PI.1993.A new procedure for isolation of anti-HIV compounds (polysaccharides and polyphenols) from the marine alga *Fucus vesiculosis.* J. Nat. Prod. 56, 478–488.

Berry JP, Gantar M, Gawley RE, Wang M, and Rein KS. 2004. Pharmacology and toxicology of pahayokolide A, a bioactive metabolite from a freshwater species of *Lyngbya* isolated from the Florida Everglades. Comp. Biochem. Physiol. Part C: Toxicol. Pharmacol. 139, 231–238.

Bhakuni DS, and Silva M. 1974. Biodynamic substances from marine flora. Bot. Mar. 17, 40–51.

Chapman VJ, and Chapman D J. 1980. Seaweeds and their uses. 3ª Ed. Chapman and Hall Ltd. Gran Bretaña.

Christopher S, and Michael J. 1981. The Biology of Seaweeds. Blackwell Scientific Publications. Gran Bretaña.

Claudio F, and Stendardo B. 1996. Contributo clinico sperimentale sull'uso un. fitocolloide in oncologia. Min. Med. 367, 3617–3622.

Cousens G. 1985. Report of Treatment of Alzheimer's disease with *Alphanae Klamathomenon Flos-Aqua.* Orthomedicine. 8, 17.

Demattè B, Guerriero A, and Pietra, F. 1995. Dictyotetraene, a new diterpenoid from a *Dictyota* sp. (Chromophycota, Dictyotaceae) of the North Brittany sea. J. Chem. Soc. Chemi. Comm. 391–393.

Dorta E, Mercedes C, Az-Marrero Ana R, and Darias J. 2003. Stypolactone, an Interesting Diterpenoid from the Brown Alga *Stypopodium zonale.* Tetrahed. Lett. 43, 9043–9046.

Fabricant DS, and Farnsworth NR. 2001. The value of plants used in traditional medicine for drug discovery. Envir. Heal. Pros. 109, 69–75.

Gemma C, Mesches MH, Sepesi B, Choo K, Holmes DB, and Bickford PC. 2002. Diets enriched in foods with high antioxidant activity reverse age-induced decreases in cerebellar beta-adrenergic function and increases in proinflammatory cytokines. Exper. Neurol. 22, 6114–20.

Gupta A, Vijayaraghavan MR, and Gupta R. 1998. The presence of cholinesterase in marine algae. Phyochemistry. 7, 875–1877.

Güven KC, Aktin E, Koyuncuoglu H, and Bergisadi N. 1974. Lipolytic and hypoglycemic compounds obtained from *Cystoseira barbata* J.Ag. Arzneimittelforschung. 24, 144–7.

Han J, Kang S, Choue R, Kim H, Leem K, Chung S, Kim C, and Chung J. 2000. Free radical scavenging effect of *Diospyros kaki, Laminaria japonica* and *Undaria pinnatifida.* Fitoterapia. 73, 710–712.

Herrero M, Ibáñez E, Cifuentes A, Reglero G and Santoyo S. 2006. Dunaliella salina Microalga Pressurized Liquid Extracts as Potential Antimicrobials. J. Food. Prot. 69, 2471–2477.

Hong DD, and Hoang TMH. 2004. Nutritional analysis of Vietnamese seaweeds for food and medicine. Biofactors. 22, 323–325.

Jan LBZH, Paulsen OBS, and Michaelsen EHCT. 2000. Studies of polysaccharides from three edible species of *Nostoc* (cyanobacteria) with different colony morphologies: structural characterization and effect of the complement system of polysaccharides from *Nostoc commune.* J. Phycol. 35, 871–888.

Jensen GS, Ginsberg DI, Huerta P, Citton M, and Drapeau MS. 2000. Consumption of *Aphanizomenon flos aquae* has rapid effects on the circulation and function of immune cells in humans. A novel approach to nutritional mobilization of the immune system. JANA. 2, 50–58.

Khafji AK, and Boney AD. 1979. Antibiotic Effects of Crustose Germlings of the Red Alga *Chondrus crispus* Stackh. on Benthic Diatoms. Ann. Bot. 43, 231–232.

Khan M, Shobha JC, Mohan IK, Naidu MU, Sundaram C, Singh S, Kuppusamy P, and Kutala VK. 2005. Protective effect of *Spirulina* against doxorubicin-induced cardiotoxicity. Phytother. Res. 19, 1030–1037.

Konishi F, Tanaka K, Himeno K, Taniguchi K, and Nomoto K. 1985. Antitumor effect induced by a hot water extract of *Chlorella vulgaris* (CE): resistance to Meth-A tumor growth mediated by CE-induced polymorphonuclear leukocytes. Can. Immunol. Immunother. 19, 73–78.

Kuhad A, Tirkey N, Pilkhwal S, and Chopra K. 2006. Renoprotective effect of *Spirulina fusiformis* on cisplatin-induced oxidative stress and renal dysfunction in rats. Renal Fail. 28, 247–254.

Kushak RI, Drapeau C, Van Cof EM, and Winter HH. 2002. Favorable effects of blue-green algae *Aphanizomenon flos-aquae* on rat plasma lipids. JANA. 2, 59–65.

Lakshmi V, Kumar R, Gupta P, Varshney V, Srivastava M, Dikshit M, Murthy P, Misra and Bhattacharaya S. 2004. The antifilarial activity of a marine red alga, *Botryocladia leptopoda*, against experimental infections with animal and human filariae. Parasitol. Res. 93, 468–474.

Li, C, Gao Y, Li M, Shi W, and Liu Z. 2005. Effect of *Laminaria japonica* polysaccharides on lowering serum lipid and anti-atherosclerosis in hyperlipemia quails. Zhong, Yao, Cai. 28, 676–679.

Linawati H. 1999. Bioactive Compound of Marin Organisms Used as Traditional Medicine in Indonesia: A Preliminary Investigation. Bandung Institute of Technology, Bandung, Indonesia.

Mao TK, Van-de-Water J, and Gershwin ME. 2005. Effects of a Spirulina-based dietary supplement on cytokine production from allergic rhinitis patients. J. Med. Food. 8, 27–30.

Mathew B, Sankarannarayanan R, Nair PP, Varghese C, Somanthan T, Amma BP, Amma NS, and Nair MK. 1995. Evaluation of chemoprevention of oral cancer with *Spirulina fusiformis*. Nut. Can. 24, 197–202.

Mazumdar S, Ghosal PK, Pujol CA, Carlucci MJ, Damonte EB, and Ray B. 2002. Isolation, chemical investigation and antiviral activity of polysaccharides from *Gracilaria corticata* (Gracilariaceae, Rhodophyta). Int. J. Biol. Macromol. 31, 87–95.

Merchant RE, and Andre CA. 2001. A review of recent clinical trials of the nutritional supplement *Chlorella pyrenoidosa* in the treatment of fibromyalgia, hypertension, and ulcerative colitis. Alt. Ther. Health. Med. 7, 79–91.

Merchant RE, Carmack CA, and Wise CM. 2000. Nutritional supplementation with *Chlorella pyrenoidosa* for patients with fibromyalgia syndrome: a pilot study. Phytother. Res. 14, 167–173.

Miguel H, and et al. 2006. *Dunaliella salina* Microalga Pressurized Liquid Extracts as Potential Antimicrobials. J. Food. Prot. 69, 2471–2477.

Milkova T, Talev G, Christov R, Dimitrova-Konaklieva S, and Popov S. 1997. Sterols and volatiles in *Cystoseira barbata* and *Cystoseira crinita* from the Black Sea. Phtyochemistry. 45, 93–95.

Misbahuddin M, Islam AZMM, Khandker S, Mahmud IA and Anjumanara IM. 2006. Efficacy of spirulina extract plus zinc in patients of chronic arsenic poisoning: a randomized placebo-controlled study. (Risk factors). J. Toxicol. Clin. Toxicol.44, 135–7.

Moen LK, and Clark GF. 1993. A novel reverse transcriptase inhibitor from *Fucus vesiculosus*. Int. Conf. AIDS. 9: 145 (abstract no. PO-A03-0061).

Oganezova EP, and Nalbandian RM. 1976. Purification and properties of plastocyanin and ferredoxin from *Ceratophyllum demersum* L. Biokhimiia. 41, 794–800.

Okada Y, Ishimaru A, Suzuki R, and Okuyama T. 2004. A new phloroglucinol derivative from the brown alga *Eisenia bicyclis*: potential for the effective treatment of diabetic complications. J. Nat. Prod. 67, 103–5.

Parl P-J, Park EJ, Kim SK, Byun HG, Jeon BT, and Jeon, YJ. 2005. Reactive Oxygen Scavenging Effect of Enzymatic Extracts from *Sargassum thunbergii*. J. Agr. Food. Chem. 53, 666–6672.

Ponce MA, Carlos A, Pujol EB, Maria LF, and Stortz CA. 2003. Fucoidans from the brown seaweed *Adenocystis utricularis*: extraction methods, antiviral activity and structural studies. Carb. Res. 338, 153–165.

Prashantkumar P, Angadi SB, and Vidyasagar GM. 2006. Antimicrobial activity of blue-green and green algae. Ind. J. Pharmaceut. Sci. 68, 647–648.

Ray B. 2004. *In vitro* anti-herpetic activity of sulfated polysaccharide fractions from *Caulerpa racemosa*. Phytochemistry. 65, 3151–3157.

Reddy CM, Bhat VB, Kiranmai G, Reddy MN, Reddanna P, and Madyastha KM. 2000. Selective Inhibition of Cyclooxygenase-2 by C-Phycocyanin, a Biliprotein from *Spirulina platensis*. Biochem. Biophys. Res. Comm. 277, 599–603.

Siamopoulou P, Iliopoulou D, Vagias C, Cos P, Vanden Berghe D, and Roussis V. 2004. Diterpenes from the brown algae *Dictyota dichotoma* and *Dictyota linearis*. Phytochemistry. 65, 2025–2030.

Thomas SG, and Beveridge AA. 1993. The novel marine natural product plocamadiene A causes histamine release from mast cells of the guinea-pig and rat *in vitro*. Clin. Exper. Pharmacol. Physiol. 20, 223–9.

Wazir S. 1992. Taxonomic and phytochemical studies of certain brown algae from the coast of Karachi. Ph. D. thesis, University of Karachi, Karachi.

Chapter 10

Ethnopharmacology of Lichens

10.1 Introduction

The Lichen division is comprised of at least 8 orders, 45 families, and 6,000 species. Information on the edible and medicinal uses of lichens is scattered. Many lichens are known to have potent antibiotic properties, and many are edible. However, some lichens do contain toxic substances, so one should not graze randomly on them.

10.2 Lichens in folk and traditional medicine

'DoctrineofSignatures' writtenin the15thcenturystated "Aplantcouldtreat a disease it most looked like". This formed the basis of phytotherapeutics in traditional systems of medicines like Traditional Indian Medicine (TIM) or Ayurveda, Traditional Chinese Medicine (TCM), and Western Medical Herbalism. Interestingly , the word lichen is derived from the Greek word 'Leprous' and refers to use of lichens in treating skin diseases due to their peeling-skin appearance. Lichen like *Lobaria pulmonaria* (L.) Hoffm. (Stictacea) and *Parmelia sulcata* Taylor (Parmeliaceae) have been used in the treatment of pulmonary and cranial diseases, respectively. Similarly, *Xanthoria parietina* (L.) Th. Fries (Lobariaceae), being yellow, was used to cure jaundice.

Medicinal uses of lichens are linked with folklore. The medicinal use of lichens can be traced back to the 18th dynasty (1700–1800 BC) when *Evernia furfuracea* (L.) Mann or (Parmeliaceae) was first used as a drug. Some lichens were claimed to be good for coughs, jaundice, rabies and restoring loss of hair. Herbal medicinal texts included an account of several species of lichens including *Cladonia, Evernia, Lobaria, Parmelia, Peltigera, Pertusaria, Physica, Rocella, Usnea* and *Xanthoria* . During the Middle Ages, lichens figured prominently in herbals used by practitioners.

Peltigera sp., mixed with 2 drachms of black pepper, for 4-d, in half-pint of warm milk, was used for rabies or hydrophobia. Linnaeus named *Peltigera,* in 1753 as Lichen caninus (the Dog Lichen) as it has rhizines on the lower surface that resembles teeth, hence a specific epithet of canina. In traditional medicine, *Peltigera canina* (L.) Willd is used as liver tonic and laxative. In Ireland, *Peltigera aphthosa* (L.) Wiild. was used as a vermifuge.

People of northern California used *Letharia vulpina* (L.) Hue. (Parmeliaceae) in stomach diseases. A novel species of *Dictyonema* was used by the Waorani as a hallucinogen. In Arabian medicine, *Alectoria usneoides* was used in the treatment of splenomegaly (enlarged spleen).

In China, *Usnea diffracta* Vain (Parmeliaceae) was used in medicine around 500 A.D. (Strickmann,). Hippocrates prescribed *Usnea barbata* for uterine ailments. In Sweden, *Parmelia saxatilis* (L.) Ach. is used to treat warts. In China, *Lethariella cashmeriana* Krog-Wei, *L. sernanderi* (Motyka) Obermayer, *L. sinensis* J. C. Wei & Jiang (Parmeliaceae) and *Thamnolia vermicularis* Hytter i Norden Danske (Icmadophilaceae) are used as medicated teas.

Usnea sp. are used in Traditional Chinese Medicine (TCM), homeopathic system of medicine and traditional medicine in Pacific Islands and New Zealand. *Usnea* sp. are valued for demulcent properties and finds use for mild inflammation of the oral and pharyngeal mucosa. *Usnea filipendula* Stirt was used in the former Soviet Union for cuts and wounds.

Spanish folk medicine has documented the use of lichens in various medical aliments. Decoction of *Pseudoevernia furfuracea* (L.) Zopf. (Parmeliaceae) is used in Alfacar and Viznar in respiratory ailments. *Ramalina bourgeana* Mont. ex Nyl. (Ramalinaceae) is consumed for diuretic and stone–dissolving (lithontriptic) properties (González -Tejero, 1995). *Xanthoparmelia scabrosa* (Taylor) Hale (Parmeliaceae) is an ingredient for various aphrodisiac formulations sold in the international market. Tea prepared from *Flavocetraria nivalis* (L.) Kärnefelt & Thell (Parmeliaceae) is used in the treatment of motion-sickness and heart attacks by the natives of Qollahuaya Andeans in Poland.

Cetraria islandica (L.) Ach. (Parmeliaceae) commonly known as Irish moss, is a valued folk remedy for lung diseases, kidney and bladder complaints as well as externally for poorly healed wounds (Bown, 2001). Traditionally it is used for mild inflammation of oral and pharyngeal mucosa. It is also valued in dyspepsia and loss of appetite. In European folk medicine, it is used in the treatment of cancer.

Reindeer lichens are important medicinal agents. *Cladonia rangiferina* (L.) F. H. Wigg. syn. *Cladina rangiferina* (L.) Nyl. (Cladoniaceae) is the commonly studied reindeer lichen. Northern native people used reindeer lichen in medicinal teas to treat colds, arthritis, fevers and other problems. Reindeer lichens were also used as a poultice to relieve the ache of arthritic joints. Reindeer lichens have been taken to treat fever, jaundice constipation, convulsions, coughs, and tuberculosis. *Cladonia pyxidata* (L.) Hoffm. is a useful remedy for whooping cough.

Three *Parmelia* sp. (*Parmelia chinense* (Osbeck) Hale & Ahti., syn *P. perlata* (Huds.) Ach. *P. sancti-angeli* (Lynge) Hale and *P. peforatum* (Jacq.) A. Massal. are used for the Indian drug chharila, which is used as an aphrodisiac. In India *Parmelia chinense* is used as a diuretic and as a liniment for headaches and powder to help wounds heal. *Parmelia sancti-angeli* is used in Central India to treat Tinea (ringworm) disease. Ash of the lichen, mixed with mustard or linseed oil, is applied to the affected area. *Parmelia peforatum* is medically recognized in Afghanistan.

Parmelia nepalense (Talyor) Hale ex Sipman is used in Nepal in the treatment of toothache and sore throat. In the western Himalayas *Thamnolia vermicularis* (Schwartz) Ach. (Icmadophilaceae) is used as an antiseptic. In Sikkim, *Heterodermia diademata* (Talyor) D.D. Awas. (Physciaceae) is used for cuts and wounds.

10.3 Phytochemical and pharmacological investigations

Lichen metabolites exert a wide variety of biological actions including antibiotic, antimycobacterial, antiviral, anti-inflammatory, analgesic, antipyretic, antiproliferative and cytotoxic effects. Even though these manifold activities of lichen metabolites have now been recognized, their therapeutic potential has not yet been fully explored and thus remains pharmaceutically unexploited. The utility of lichens is due to the of range of secondary compounds produced by them.

Antimicrobial activity

Antibacterial

Antibiotic properties of lichens are of special interest to scientists. According to one estimate, 50% of all lichens have antibiotic properties. Burkholder and co workers (1944) were pioneers initiating research on lichens as antibacterial agents. They tested 42 lichens for antibiotic properties and 27 were reported to inhibit growth of bacteria.

Parmelia physodes (L.) Ach. was reported to be an antibiotic. Usnic acid is a wide-spectrum antibiotic characterized from lichens. Vulpinic acid has mild antibiotic property. Atranorin has been found to be much less biologically active than usnic and vulpinic acids. Usnic , evernic and vulpinic acids inhibited the growth of gram positive bacteria *Staphylococcus aureus, Bacillus subtilis,* and *Bacillus megaterium*, but the acids had no effect on the gram negative bacteria *Escherichia coli* or *Pseudomonas aeruginosa*.

Usnic acid isolated from lichens from the south of Spain has high activity against Gram-positive bacteria. Acetone, diethyl ether and ethanol extracts of the lichen *Cetraria aculeata* (Schreber) Fr. and its active constituent protolichesterinic acid were tested positive against *Escherichia coli, Staphylococcus aureus, Aeromonas hydrophila, Proteus vulgaris, Streptococcus faecalis, Bacillus cereus, Bacillus subtilis, Pseudomonas aeruginosa,* and *Listeria monocytogenes.*

Alectosarmentin, (-)-usnic acid, physodic acid and 8′-O-ethyl-beta-alectoronic acid isolated from the alcoholic extract of the lichen *Alectoria sarmentosa* (Ach.) Ach. (Alectoriaceae) showed antimicrobial activity. *Hypogymnia apinnata* (atranorin), *Letharia columbiana* (vulpinic acid), *Lobaria pulmonaria* (Stictic acid, constictic acid, and norstictic acid) and *Usnea filipendula* (Usnic acid and salazinic acid) have been reported to have significant antibiotic activity against *Micrococcus luteus, Staphylococcus aureus, Salmonella gallinarum* and *Serratia marcescens* respectively.

The acetone and methanol extracts of *Lasallia pustulata* (L.) Mérel. (Umbilicariaceae), *Parmelia sulcata* Taylor, and *Umbilicaria crustulosa* (Ach.) Frey (Umbilicariaceae), manifested antibacterial activity against the majority of species of bacteria tested, in addition to selective antifungal activity. Acetone, chloroform, diethyl ether, methanol, and petroleum ether extracts of *Parmelia sulcata* and its constituent (salazinic acid) demonstrated antibacterial activity against *Aeromonas hydrophila, Bacillus cereus, Bacillus subtilis, Listeria monocytogenes, Proteus vulgaris, Yersinia enterocolitica, Staphylococcus aureus, Streptococcus faecalis, Candida albicans, Candida glabrata, Aspergillus niger, Aspergillus fumigatus,* and *Penicillium notatum.*

3-hydroxyphysodic acid isolated from *Hypogymnia tubulosa* (Schaerer) Havaas (Parmeliaceae) showed antimicrobial activity against *Aeromonas hydrophila, Bacillus cereus, Bacillus subtilis, Escherichia coli, Klebsiella pneumoniae, Listeria monocytogenes, Proteus vulgaris, Salmonella typhimurium, Staphylococcus aureus, Streptococcus faecalis, and Candida albicans.*

Extracts of *Xanthoparmelia pokornyi* (Vainio) Hale (Parmeliaceae) and its constituents (gyrophoric and stenosporic acid) have been reported to be potential antimicrobials. Extracts of *Pseudevernia furfuracea* and its constituents (physodic acid, chloroatranorin, atranorin, and olivetoric acid) have significant antimicrobial activity. Hirtusneanoside isolated from *Usnea hirta* (L.) Wigg. (Parmeliaceae) showed growth inhibitory activities against Gram-positive bacteria.

Antifungal

Parietin, anthraquinone isolated from methanol extract of *Caloplaca cerina* (Ehrh. ex Hedwig) Th.Fr. (Teloschistaceae) has been reported to have significant antifungal activity. Extracts of Andean lichens *Protousnea poeppigii* (Nees & Flot.) Vain. (Parmeliaceae) and *Usnea florida* var. *rigida* Acharius demonstrated antimicrobial activity against the pathogenic fungi *Microsporum gypseum, Trichophyton mentagrophytes* and *T. rubrum.* Isodivaricatic acid, 5-propylresorcinol, divaricatinic acid and usnic acid were identified as antifungal agents.

Antiviral

Usnic acid isolated from *Teloschistes chrysophthalmus* (L.) Th. Fr. (Teloschistaceae) and parietin isolated from *Ramalina celastri* demonstrated antiviral activity against the arena viruses Junin and Tacaribe.

Antioxidant activity

Phenolic constituents from the lichen *Parmotrema stuppeum* (Nyl.) Hale (Parmeliaceae) including methyl orsenillate, orsenillic acid, atranorin and lecanoric acid showed moderate antioxidant activity. An animal study reported antioxidant activity of the lichen *Cetraria islandica.* Stictic acid derivatives from the lichen *Usnea articulata* (Ach.) Motyka were reported to have significant antioxidant activity.

Antitumor activity

Pannarin inhibited cell growth and induces cell death in human prostate carcinoma DU-145 cells.The orcinol derivatives tenuiorin and methyl orsellinate present in the extract of *Peltigera leucophlebia* (Nyl.) Gyeln. (Peltigeraceae) exhibited in vitro inhibitory activity against 15-lipoxygenase from soybeans. A correlation has been observed between 5-lipoxygenase inhibition and antiproliferative effects for related lichen metabolites. On this account, tenuiorin and methyl orsellinate were further tested for antiproliferative activity on cultured human breast, pancreatic and colon cancer cell lines. Methyl orsellinate lacked antiproliferative activity but tenuiorin depicted moderate activity. Bianthraquinone glycosides, colleflaccinosides isolated from *Collema flaccidum* (Ach.) Ach. (Collemataceae) collected in Israel and Russia, were reported to have antitumor activity.

Immunomodulator activity

Heteroglycans and a beta-glucan isolated from *Thamnolia vermicularis* var. *subuliformis* were tested for *in vitro* immunomodulating activity and reported to have various influences on the immune system.

Tyrosinase-inhibitory activity

Methanol extracts of *Graphina glaucorufa* (Vain.) Zahlbr., *Graphina multistriata* Müll. Arg., *Graphina salacinilabiata* Patw. & CR Kulk., *Graphis assamensis* Nagarkar & Patw., *Graphis nakanishiana* Patw. & CR Kulk, and *Phaeographopsis indica* (Patw. & Nagarkar) Sipman & Aptroot (Graphidaceae) have significant tyrosinase-inhibitory activity. Methanolic extracts of the edible and medicinal lichens, *Umbilicaria* (*Gyrophora*) *esculenta* and *Usnea longissima* manifested *in vitro* melanogenesis inhibitory. It was concluded that lichen extracts affected the activity of tyrosinase via the inhibition of tyrosinase glycosylation.

10.4 Recent phytochemical investigations on lichens

Two tridepsides (2,4-Di-O-methylgyrophoric acid and 2,4,5-tri-O-methylhiascic acid) have been isolated from *Parmelia damaziana* Zahlbr. Lasallic acid, a tridepside has been reported from *Lasallia asiae-orientalis* Whittaker & Margulis (Umbilicariaceae). γ-butyrolactone acid, (-)-isomuronic acid and gyrophoric acid have been isolated from *Punctelia microsticta* (Mull. Arg.) Krog (Parmeliaceae). Dasypogalactone, a lactone has been reported from *Usnea dasygopa* Rohl growing in Indonesia.

A new red anthraquinone, draculone, has been isolated from the corticolous tropical lichen *Melanotheca cruenta* (Mont.) Miill. (Trypetheliaceae) together with anthraquinone pigment; haematommone. Beta-orcinol metabolites viz hypotrachynic acid, deoxystictic acid, cryptostictinolide and 8′-methylconstictic acid have been reported from the lichen *Hypotrachyna revoluta* (Flörke) Hale. (Parmeliaceae).

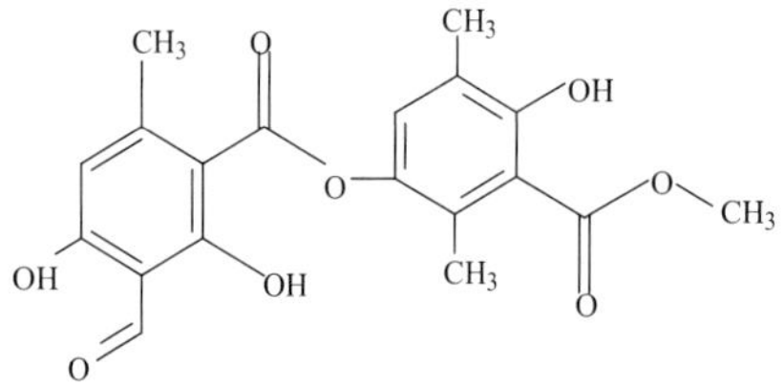

Atranorin

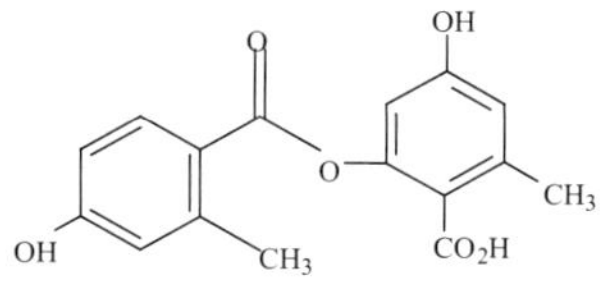

Lecoanoric acid

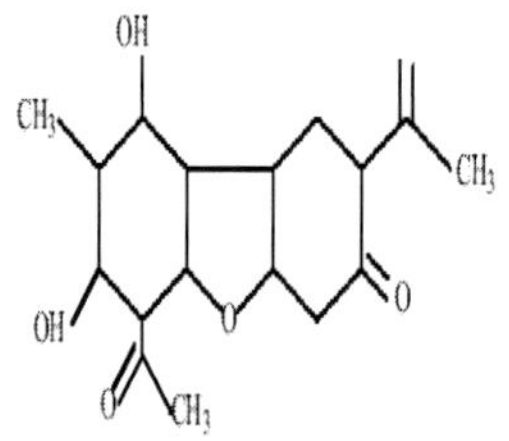

Usnic acid

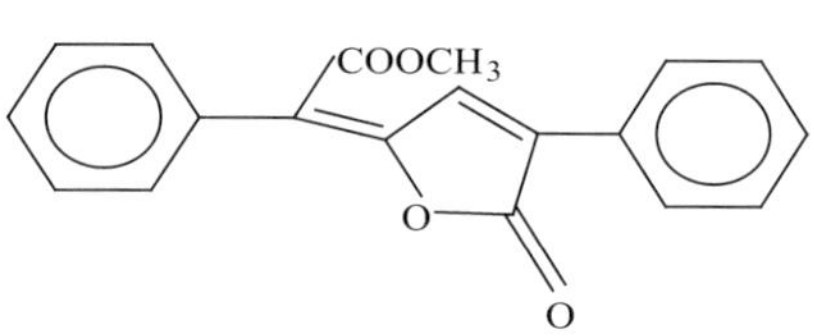

Vulpinic acid

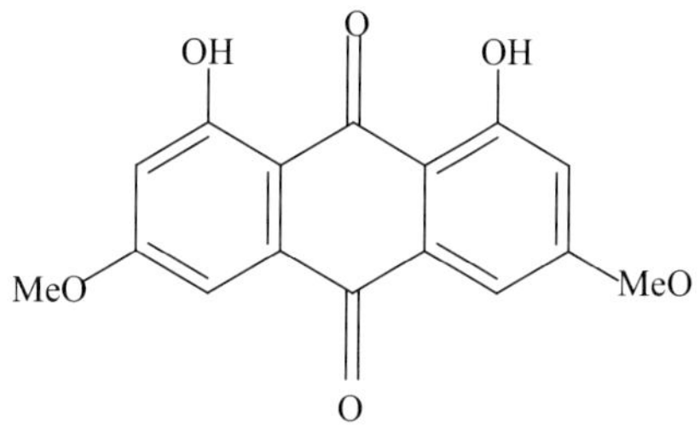

Parietin

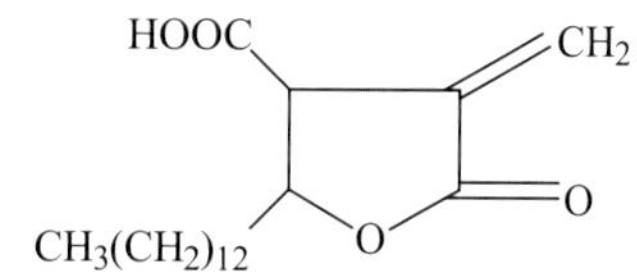

Protolichensterinic acid

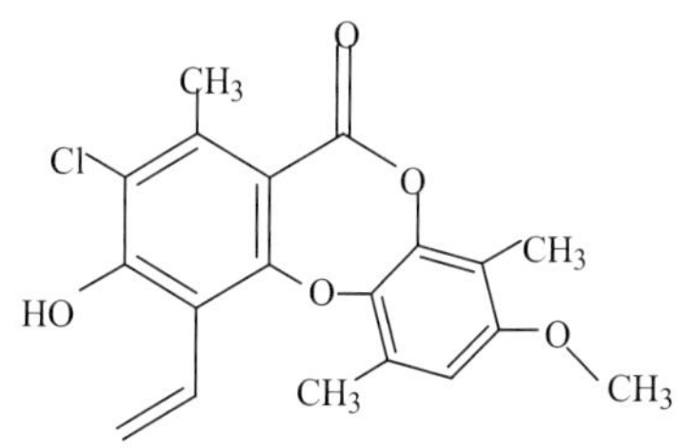

Pannarin

Fig. 10.1 Significant phytochemicals reported from lichens.

References and further reading

Bastien JW. 1983. Pharmacopoeia of the Qollahuaya Andeans. J. Ethnopharmacol. 8, 97–111.

Behera BC, Adawadkar B, Makhija U. 2006. Tyrosinase-inhibitory activity in some species of the lichen family Graphidaceae. J. Herb. Pharmacother. 6, 55–69.

Boustie J, Grube M. 2005. Lichens—a promising source of bioactive secondary metabolites. Plant. Gene. Resou. 3, 273–278.

Bown D. 2001. Encyclopedia of Herbs and their Uses. Dorling Kindersley, London.

Burkholder PR, Evans AE, McVeigh I and Thorton HK. 1994. Antibiotic activity of lichens. Proceedings of the National Academy of Sciences of the USA. 30, 250–255.

Bylicka H. 1952. Antibiotics of lichens Acta. Microbiol. Pol. 1, 185.

Candan M, Yilmaz M, Tay T, Kivanç M and Türk H. 2006. Antimicrobial activity of extracts of the lichen Xanthoparmelia pokornyi and its gyrophoric and stenosporic acid constituents. Zeit. Für. Naturfors. 61, 319–23.

Candan M, Yilmaz M, Tay T, Erdem M and Türk AO. 2007. Antimicrobial activity of extracts of the lichen Parmelia sulcata and its salazinic acid constituent. Zeit. Für. Naturfors. 62, 619–21.

Chandra S, Singh A. 1971. A lichen crude drug (chharila) from India. J Res. Ind. Med. 6, 209–215.

Chevallier A. 1996. The Encyclopedia of Medicinal Plants Dorling Kindersley. London.

Cocchietto M, Skert N, and Nimis PL. 2002. A review on usnic acid, an interesting natural compound. Naturwissenschaften. 89, 137–146.

Crockett M. 2003. Antibacterial Properties of four Pacific Northeast Lichens. Botany 465 Lichenology at Oregon State University.

Davis EW, Yost JA. 2001. Novel hallucinogens from Eastern Ecuador. Bot. Mus. Leaf. 29, 291-295.

Dobrescu D, Tănăsescu M, Mezdrea A, Ivan C, Ordosch E, Neagoe F, Rizeanu A, Trifu L and Enescu V. 1993. Contributions to the complex study of some lichens--Usnea genus. Pharmacological studies on *Usnea barbata* and *Usnea hirta* species. Rom. J. Physiol. 30, 101–107.

Elix JA, Jayanthi VK, and Leznoff CC. 1981. 2, 4-Di-O-methylgyrophoric acid and 2, 4, 5-tri-O-methylhiascic acid. New tridepsides from *Parmelia damaziana*. Aust. J. Chem. 34, 1757–1761.

Fazio AT, and et al., 2007. Lichen secondary metabolites from the cultured lichen mycobionts of *Teloschistes chrysophthalmus* and *Ramalina celastri* and their antiviral activities. Zeit. Für. Naturfors. 62: 543–9.

Gollapudi SR, Telikepalli H, Jampani HB, Mirhom YW, Drake SD, Bhattiprolu KR, Vander Velde D and Mitscher LA. 1994. Alectosarmentin, a new antimicrobial dibenzofuranoid lactol from the lichen, *Alectoria sarmentosa*. J. Nat. Prod. 57, 934–8.

Gonza´lez-Tejero MR, Martı´nez-Lirola MJ, Casares-Porce M and Molero-Mesa J. 1995.Three lichen used in popular medicine in Eastern Andalucia (Spain). Econ. Bot. 49, 96–98.

Gülçin I, Oktay M, Küfrevioğlu OI and Aslan A. 2002. Determination of antioxidant activity of lichen *Cetraria islandica* (L) Ach. J. Ethnopharmacol.79, 325–329

Ingólfsdóttir K, Gudmundsdóttir GF, Ogmundsdóttir HM, Paulus K, Haraldsdóttir S, Kristinsson H and Bauer R. 2002. Effects of tenuiorin and methyl orsellinate from the lichen Peltigera *leucophlebia* on 5-/15-lipoxygenases and proliferation of malignant cell lines in vitro. Phytomedicine.9, 654–8.

Jayapraksha GK, Rao LJ. 2000. Phenolic constituents from the lichen *Parmotrema stuppeum* (Nyl.) Hale and their antioxidant activity. Zeit. Für. Naturfors. 55, 1018–22.

Kim MS, Cho HB. 2007. Melanogenesis Inhibitory Effects of Methanolic Extracts of *Umbilicaria* esculenta and *Usnea longissima*. J. Microbiol. 45, 578–82.

Kumar K, Upreti DK. 2001. *Parmelia* sp. (lichens) in ancient medicinal plant lore of India. Eco. Bot. 55, 458–459.

Kumar S, Banskota AH, Manandhar MD. 1996 Isolation and identification of some chemical constituents of *Parmelia nepalensis*. Plant. Medi. 62, 93–94.

Lal BM, Upreti DK. 1996. Ethno botanical notes on three Indian lichens. Lichenologist. 27, 77–99.

Lauterwein M, Oethinger M, Belsner K, Peters T and Marre R. 1995. *In vitro* activities of the lichen secondary metabolites vulpinic acid, (+)-usnic acid, and (-)-usnic acid against aerobic and anaerobic microorganisms. Antimicrob. Agents. Chemother.39, 2541–2543.

Lawrey JD. 2002. Biological role of lichen substances. The Bryologist 1986;89:111–122. Cocchietto M, Skert N, Nimis PL. A review on usnic acid, an interesting natural compound. Naturwissenschaften. 89, 137–146.

Lohézic-Le Dévéhat F, Tomasi S, Elix JA, Bernard A, Rouaud I, Uriac P and Boustie J. 2001. Stictic acid derivatives from the lichen Usnea articulata and their antioxidant activities. J. Nat. Prod.70, 1218–20.

Maier MS 1999. Revised Structure for (-)-Dihydropertusaric Acid, a -Butyrolactone Acid from the Lichen *Punctelia microsticta*. J. Nat. Prod. 62, 1565–1567.

Manojlovic NT, Solujic S, Sukdolak S and Milosev M. 2005. Antifungal activity of *Rubia tinctorum, Rhamnus frangula* and *Caloplaca cerina*. Fitoterapia.76, 244–6.

Mathey A, Spiteller P, Steglich W. 2002. Draculone, a new anthraquinone pigment from the tropical lichen *Melanotheca cruenta*. Zeit. Für. Naturfors. 57, 565–7.

Mordraksi M. 1956. Antibiotics from lichens of *Parmelia physodes* species. Archi.Immunol. Thera. Exepr. Immunol. 4, 299–334.

Müller K. 2000. Pharmaceutically relevant metabolites from lichens. App. Microbiol. Biotec. 56, 9–16.

Narui T, Sawada K, Okuyama T, Takatsuki S and Culberson CF. 1996. Lasallic acid, a tridepside from the lichen, *Lasallia asiae-orientalis*. Phytochemistry42, 839–842.

Negi HR, Kareem A. 1996. Lichens: the unsung heroes. Amruth. 1, 3–6.

Omarsdottir S, Freysdottir J, Olafsdottir ES. 2007. Immunomodulating polysaccharides from the lichen *Thamnolia vermicularis* var. *subuliformis*. Phytomedicine. 14, 179–84.

Papadopoulou PP, Tzakou O, Vagias C, Kefalas P and Roussis V. 2007. Beta-orcinol metabolites from the lichen *Hypotrachyna revoluta*. Molecules. 12, 997–1005.

Pereira J. The Elements of Material Medica and Therapeutics, Vol. 2, 3th American ed. 1853.

Perez- Llano GA. 1994. Lichens. Their biological and economic significance. Bot. Rev. 10, 1.

Perez- Llano GA. 1994b. Economic uses of lichens. Eco. Bot. 2, 15–45.

Rankovic B, Misic M. Sukdolak S. 2007. Evaluation of antimicrobial activity of the lichens *Lasallia pustulata, Parmelia sulcata, Umbilicaria crustulosa* and *Umbilicaria cylindrica*. Mikrobiologya. 76, 817–821.

Rezanka T, Dembitsky VM. 2006. The colleflaccinosides, two chiral bianthraquinone glycosides with antitumor activity from the lichen *Collema flaccidum* collected in Israel and Russia. Natu. Prod. Res. 20, 969–980.

Renzaka T, Sigler K. 2007. Hirtusneanoside, an unsymmetrical dimeric tetrahydroxanthone from the lichen *Usnea hirta*. J. Natu. Prod. 70, 1487–1491.

Rizzini, CT. 1999. The uses of lichens in medicine. Bras. Med., e 1952, 66, 589–96. Some lichen products have antimicrobial activity. Zeit. Für. Naturfors.54, 605–9.

Rowe G, and et al., 1999. Some lichen products have antimicrobial activity. Zeit. Für. Naturfors. 54, 605–9.

Rowe JG, Saenz MT, Garcia MD and Gil AM. 1991.Additional contribution to the study of the antimicrobial activity and identification of lichenic substances in some lichens from Southern Spain. Ann. Pharmacol.49, 278–285.

Russo A, Piovano M, Lombardo L, Vanella L, Cardile V and Garbarono J. 2006. Pannarin inhibits cell growth and induces cell death in human prostate carcinoma DU-145 cells. Antican. Drugs.2006, 17:1163–9.

Saklani A, Upreti DK. 1992. Folk uses of some lichens of Sikkim. J. Ethnopharmacol. 37, 229–233.

Schmeda-Hirschmann G, Tapia A, Lima B, Pertino M and Sortino M. 2007. A new antifungal and antiprotozoal depside from the Andean lichen *Protousnea poeppigii*. Phytother Res. 22, 349–55.

Shibamoto T, Wei CI. 1984. Mutagenicity of lichen constituents. Envir. Muta. 6, 757–762.

Suwarso WP, Gani, RL, Krohn K and John, M. 1999. Dasypogalactone , a new C3-Symmetric Macrolactone from the Indonesian Lichen *Usnea dasypoga* Rohl. Eur.J. Org. Chem. 7, 1719–21.

Türk AO, Yilmaz M, Kivanç M, Türk H. 2003. The antimicrobial activity of extracts of the lichen *Cetraria aculeata* and its protolichesterinic acid constituent. Zeit. Für. Naturfors., 2003, 58, 850–4.

Türk AO, Yilmaz M, Kivanç M, Türk H. 2006.Antimicrobial activity of extracts of chemical races of the lichen *Pseudevernia furfuracea* and their physodic acid, chloroatranorin, atranorin, and olivetoric acid constituents. Zeit. Für. Naturfors., 2006, 61, 499–507.

Vartia KO. 1973. Antibiotics in lichens. The Lichens (eds. V. Ahmadjian and M. E. Hale), Academic Press, New York, pp. 547–561.

Wang LS. 2001. Ethnic Uses of Lichens in Yunnan, China. Lichenologist, 104, 345–349.

Yilmaz M, Tay T, Kivanç M and Türk H. 2005.The antimicrobial activity of extracts of the lichen *Hypogymnia tubulosa* and its 3-hydroxyphysodic acid constituent. Zeit. Für. Naturfors. 60, 35–8.

Chapter 11

Ethnopharmacology of Bryophytes

11.1 Introduction

Liverworts and mosses are small, low-growing plants, which constitute the phylum bryophyta. Bryophytes, which are phylogenetically placed between vascular plants and algae, form a unique division in the plant kingdom. There are more than 22,000 members of mosses (Bryophyta) in the world. About 2850 bryophytes are reported to have medicinal value. They are now increasingly used for diverse purposes including pollution control and as new sources of pharmaceuticals.

11.2 Bryophytes in traditional systems of medicine

There is no historical record, explaining the use of liverworts and mosses in Ayurveda, the Indian system of medicine. One research claims that mosses from the species of *Barbula, Fissidenc, Minium, Thuidium* and species of liverworts like *Asterella, Dumortiera, Marchantia, Pellia, Plagiochasma* and *Stephenrencella-Anthoceros* were present in the vicinity of Shilajit (*Asphaltum punjabianum*) exuding rocks and these bryophytes are responsible for the formation of Shilajit. Bryophytes reveal the occurrence of minerals and metals in their tissue such as copper, silver, zinc, iron, lead etc., which are similar to the elements present in Shilajit.

Chinese Traditional Medicine named 40 kinds of bryophytes used to treat cardiovascular diseases, tonsillitis, bronchitis, cystitis, and skin infections. Some species of *Fissidens* and *Polytrichum* were utilized as diuretic and hair growth stimulating drugs in China more than 400 years ago. *Marchantia polymorpha* is used in the treatment of liver diseases like jaundice and hepatitis. In China, *Rhodobryum giganteum* (Schwaegr.) Par and *R. roseum* (Hedw) Limpr are used in the treatment of heart ailments.

North American Indians used *Bryum, Mnium, Philonotis* spp., and *Polytrichum juniperinum* to heal burns, bruises, and wounds. The Seminole people in North America were reported to use *Barbula unguiculata*

Hedwig, *Bryum capillare* T. Bickley and *Octoblepharum albidum* (L.) Hedw as a febrifuge and anodyne. Willow moss, *Fontinalis antipyretica* Hedw was once reputed as a febrifuge (documented in the Journal of Linnaeus). Dried *Sphagnum* was used extensively as a surgical dressing during the First World War. In France, *Marchantia polymorpha* was used to promote diuresis. In the Himalayas, *Riccia* sp. was used in the treatment of Tinea (ringworm) infestations.

The use of bryophytes as antibacterial or disinfectant agents deserves special mention. *Sphagnum teres* (Schimp.) Ångstr is used in ophthalmologic diseases. In China and Bolivia, *Fissidens osmundoides* Hedwig is used an antibacterial agent to treat inflammatory conditions of the pharynx and larynx. *Haplocladium microphyllum* (Hedw) Broth is used as a demulcent medicine in inflammatory conditions like bronchitis, cystitis, tonsillitis and tympanitis. *Philonotis fontana* (Hedw) Brid is used by Go suite native people as a soothing preparation for healing burns.

11.3 Phyochemistry of bryophytes

Extracts of many species of bryophytes are known to contain phenolic compounds. Sterols, isoflavonoids, flavonoids, and bioflavonoids have been reported. Terpenoids, phenolics and volatile constituents have also been investigated in some bryophytes. *Rhodobryum giganteum* is reported to contain volatile oil, lactone and amino acid. Essential oil of *Plagiomnium undulatum* (Hedw) T Kop is reported to contain butenolides. Quercetin, luteolin and apigenin have been identified as major flavonoids in *Marchantia convoluta* Gao et KC Zhang.

Marchantia polymorpha Linn from Japan has been reported to accumulate eicosapentaenoic acid. Diterpenoids have been reported from the New Zealand liverwort *Jungermannia* species. *Isotachis japonica* Steph is reported to contain aromatic esters. *Anoectangium bicolor* Renauld & Cardot has been reported to contain the highest amount of neurotransmitter, acetylcholine, among bryophytes.

11.4 Pharmacological investigations

The antifungal and antifeedant activity of bryophytes is widely known, but mainly from *in vitro* studies. Recent exploratory studies done in Germany, Peru and Bolivia have demonstrated that extracts derived from native bryophyte species have significant effects on human pathogenic fungi and may cure skin diseases. Several Indian bryophytes have been identified as a potential source of antimycobacterial agents. As pointed out, *Anoectangium bicolor* is a rare bryophyte with anti cholinesterase activity. A brief account of pharmacological investigations carried out on bryophytes is summarized below:

Antifungal activity

Fractionation of the ethanolic extract of *Homalia trichomanoides* (Hedw) B.S.G. lead to isolation of 3 β-methoxyserrat-14-en-21 β-ol, atranorin (Fig. 10.1), and methyl 2, 4-dihydroxy-3, 6-dimethylbenzoate as antifungal compounds active against *Candida albicans*.

Pallavicinia lyellii (Hook) Carruth is known to have antifungal activity against *Aspergillus fumigatus* under *in vitro* conditions. Further investigations led to the discovery of steroid as the antifungal compound from *P. lyellii*. The compound was found to be effective against aspergillosis-induced mortality in immuno-compromised mice.

Hypnum cupressiforme Hedwig has significant antifungal effects. *Herbertus aduncus* subsp. *hutchinsiae* (Gottsche) R.M.Schust. has antifungal activity against *Botrytis cinerea, Pythium debaryanum* and *Rhizooctonia solani*. (-) α-herbertenol, (-) β-herbertenol, and (-) α-formylherbertenol were identified as antifungal constituents (Fig. 11.1).

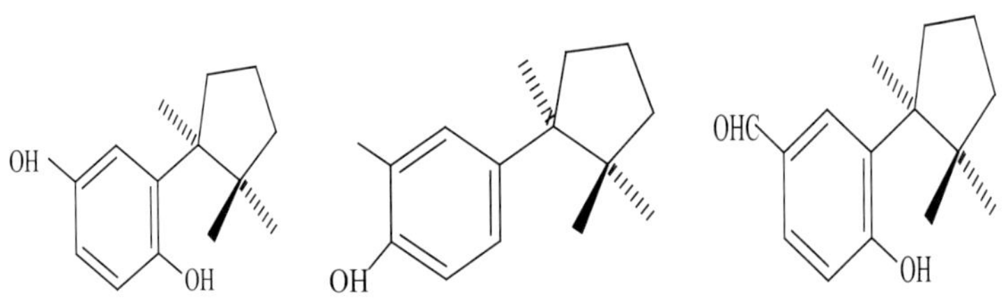

(-) α-herbertenol (-) β-herbertenol (-) α-formylherbertenol

Fig. 11.1 Antifungal constituents of *Herbertus aduncus* subsp. *hutchinsiae*.

Bioassay-directed fractionation of the MeCOEt extract of the American liverwort, *Porella cordeana* (Hueben.) Evans yielded drimenin and aristolone which were moderately toxic towards DNA-repair-deficient mutant of *Saccharomyces cerevisiae* (Fig. 11.2).

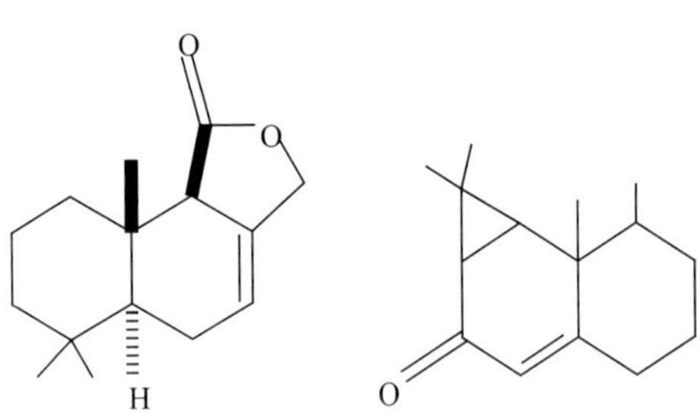

Drimenin Aristolone

Fig. 11.2 Antifungal constituents of *Porella cordeana*.

Antibacterial activity

An acetone extract of the Chinese moss, *Pleurochaete squarrosa* (Brid.) Lindb. was reported to be active against some gram-negative strains. The acetone and methanol extracts of *Palustriella commutata* (Hedw.) Ochyra (native to Turkey) were studied for antibacterial activity. The acetone extract had a potential activity against gram-negative and gram-positive bacteria. Acetone extracts of the moss, *Rhynchostegium riparioides* (Hedw) C Jens has significant antibacterial activity against gram-negative bacteria.

Sanionins A and B isolated from the moss *Sanionia georgico-uncinata* Ochyra, (collected on the Antarctic Livingston Island) showed activity against important Gram-positive pathogens, such as mycobacteria, multiresistant staphylococci, and vancomycin resistant enterococci. Sanionins A and B showed anti-inflammatory activity and low cytotoxicity along with antibacterial activity. The extract of Stringy Moss, *Leptodictyum riparium* (Hedw.) Warnst. inhibited activity against conventional antibiotic-resistant *Pseudomonas aeruginosa. Dicranum scoparium* Hedw. has strong antibacterial activity against gram negative bacteria.

Anti-cancer activity

Extracts of *Polytrichum juniperum* Hedw were reported to have anticancer activity against Sarcoma 37. Diplophyllin (Fig. 11.3), an ent-eudesmanolide isolated from *Diplophyllum albicans* (L) Dum and *D taxifolium* (Wahlenb) Dumort showed significant anticancer activity against human epidermoid carcinoma.

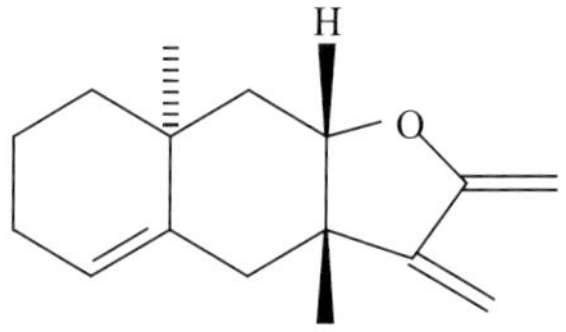

Fig. 11.3 Structure of Diplophyllin.

Sesquiterpenoids, costunolide and tulipinolide (Fig. 11.4), isolated from *Conocephalum supradecomositum* (Lindb.) Steph, *Frullania monocera* (Hook f & Tayl) Gottsche, Lindenb & Nees, *F. tamarisci* (L) Dumortier, *Machantia polymorpha, Porella japonica* (Sande Lac) Mitt. and *Wiesnerella denudata* (Mitt) Steph, *Lepidozia vitrea* Steph and *Plagiochila semidecurrens* Lehm et Lindenb demonstrated anticancer activity against carcinoma of the nasopharynx.

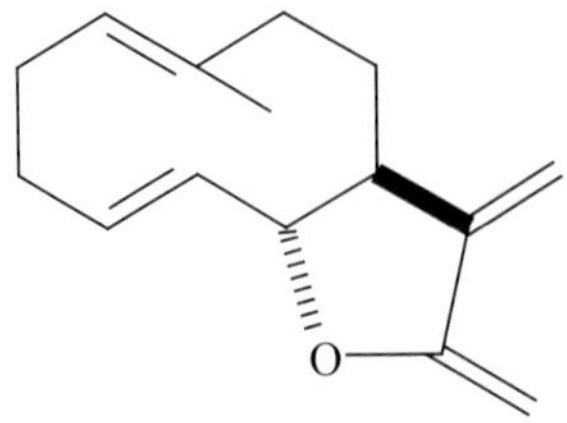

Costunolide

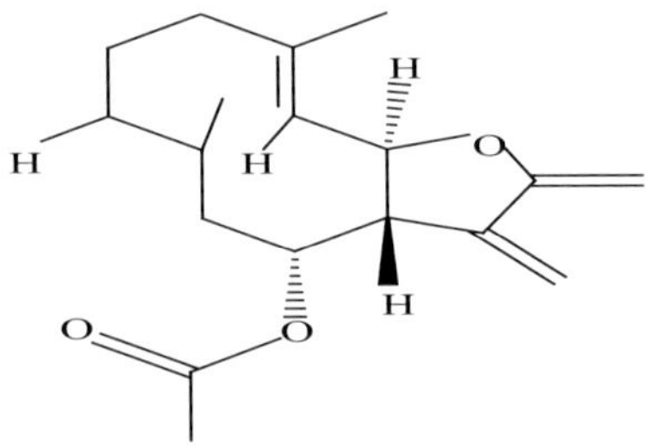

Tulipinolide

Fig. 11.4 Sesquiterpenoids of *Conocephalum supradecomositum*.

Marchantin A (Fig. 11.5) from *Marchantia palacea* Bertal, *M polymorpha* L and *M tosana* Steph demonstrated anticancer activity against KB cells.

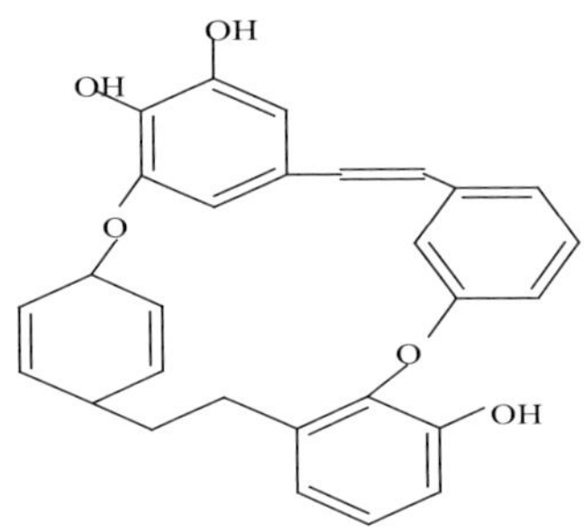

Fig. 11.5 Structure of Marchantin A.

Perrottetin E isolated from *Radula perrottetti* and riccardin (Fig.11.6) isolated from *Riccardia multifida* (L) Gray showed anticancer activity similar to marchantin.

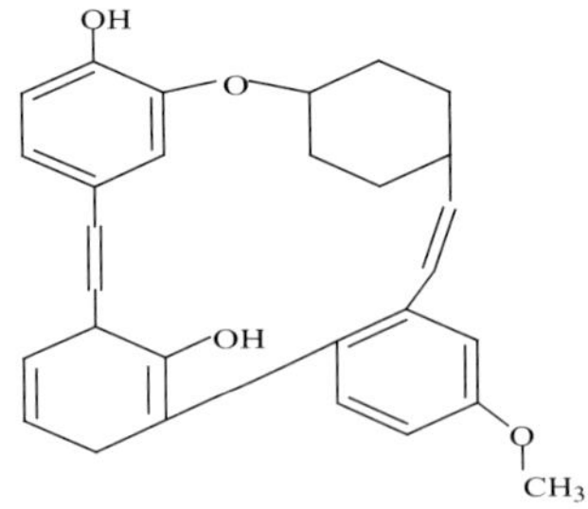

Fig. 11.6 Structure of Riccardin.

Benzonaphthoxanthenones (ohioensin A- E) isolated from *Polytrichum ohioense* Ren et Card, following bioassay-directed fractionation and three novel benzonaphthoxanthenones (1-O-methylohioensin B, 1-O-methyldihydroohioensin B and 1,14-di-O-methyldihydroohioensin B) and two novel cinnamoyl bibenzyls (pallidisetin A and B) isolated from the ethanol extract of *Polytrichum pallidisetum* Funck showed cytotoxicity against 9PS and several human tumor cell lines.

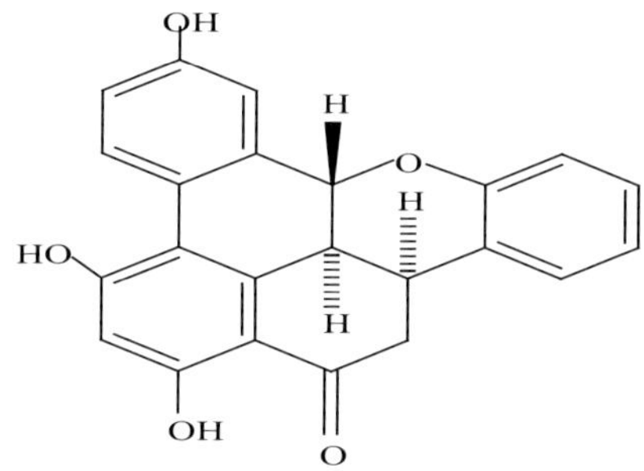

Fig. 11.7 Structure of Ohioensin A.

Crude methanolic extracts of mosses, *Isothecium subdiversiforme* Broth. and *Thamnobryrum sandei* (Besch.) Iwatsuki seen commonly in South Japan demonstrated *in vitro* cytotoxicity against P-388 lymphocytic leukemia cells. Mayatansinoids are active cytotoxic principles of the two mosses.

Claopodium crispifolium Hook) Ren & Card and *Plagiomnium venustum* (Mitt) T Kop were found to have P-388 activity against P-388 lymphocytic leukemia cells. *Bazzania trilobata* (L) S Gray, a liverwort from New Hampshire, showed KB activity and is reported to contain lignans. *Atrichum undulatum* (Hedw) PBeauv has broad spectrum antibacterial activity. *Barbula* and *Timmiella* species have been reported to have significant antibacterial activity.

Cardio protective activity

An ether extract of *Rhodobryum giganteum* (Schwaegr) Par. was reported to cure angina (chest-pain) and reduce the oxygen resistance by increasing the rate of flow in the aorta by over 30%.

Miscellaneous activity

In Britain, sphagnol, a derivative of *Sphagum* is used to relieve the itch of insect bites. Sphagnol is recognized as a useful application in eczema, psoriasis, pruritus, hemorrhoids, chilblains, scabies and acne.

One of the reasons for exploring biological compounds in bryophytes is the potential for medical use. Bryophytes contain numerous bioactive constituents as shown in investigational studies. However, much work remains to link medical effects with specific bryophyte species. Traditional medical claims need to be justified by pre-clinical and clinical research.

Further reading

Akerele O. 1993. Nature's medicinal bounty: don't throw it away. World, Health, For. 14, 390–5.

Askawa Y. 1981. Biologically active substances obtained from bryophytes. J. Hattor. Bot. Lab. 50, 123–142.

Askawa Y. 1982. Chemical constituents of the Hepaticae. Prog. Chem. Org. Nat. Prod. 42, 1–285.

Banejree RD. 1974. Studies on the antibiotic activities of bryophytes and pteridophytes. Thesis, Uni Kalyani, Kalyani, India.

Basile A, Giordano S, López-Sáez JA and Cobianchi RC. 1999.Antibacterial activity of pure flavonoids isolated from mosses. Phytochemistry. 52, 1479–82.

Basile A, Sorbo S, Giordano S, Lavitola A and Castaldo Cobianchi R. 1997. Antibacterial activity in *Pleurochaete squarrosa* extracts (Bryophyta). Int. J. Antimicrob. Agents. 10, 169–72.

Basile A, Vuotto ML, Ielpo MTL, Moscatiello V, Ricciardi L, Giordano S and Cobianchi RC. 2001. Antibacterial activity in *Rhynchostegium riparioides* (Hedw) Card. extract (bryophyta). Phytother. Res.12, 146–8.

Belkin M, Fitzgerald DB, and Felix MD. 1952. Tumor-damaging capacity of plant materials. Vol. 2. Plants used as diuretics. J. Nat. Can. Inst. 53, 13: 741–4.

Castaldo Cobianchi R, Giordano S, Basile A and Violante U. 1988. Occurrence antibiotic activity in *Conocephalum conicum, Mnium umdulatum* and *Leptodictyum riparium* (Bryophyta). Giorn. Bot. Ital., 122: 303–311.

Chen XQ, and Xiao JB. 2005. RP-HPLC-DAD determination of flavonoids: separation of quercetin, luteolin and apigenin in *Marchantia convoluta*. Ira. J. Pharmaceut. Res. 4, 175–81.

Cheng WP, and Yu J. 1998. The medicinal uses of bryophytes. Act. Bot. Yun. 25, 1–55.

Crum H. 1998. A focus on Peatlands and Pear mosses. Univ Michigan Press, An Arbor, MI.

Ding H. 1982. Medicinal spore-bearing plants of China. Shanghai. p. 409.

Flowers S. 1957. Ethnobryology of the Gosuite Indians of Utah. Bryologist. 60, 11–4.

Frahm JN. 1957. Recent developments of commercial products from bryophytes. Bryologist. 107, 277–283.

Gautam R, Saklani A, and Jachak SM. 2007. Indian medicinal plants as a source of antimycobacterial agents. J. Ethnopharmacol. 110, 200–234.

Gupta KG, and Singh B. 1971. Occurrence of antibacterial activity in moss extracts. Res.Bullt. Punjab Univ. 22, 237–239.

Hahn H, Seeger T, Geiger H, Zinsmeister, HD and Markham KR and Wong H. 1995. The first biaurone, a triflavone and biflavonoids from two Aulacomnium species. Phytochemistry.40, 573–6.

Harrigan GG, Ahmad A, Baj N, Glass TE, Gunatilaka AA and Kingston DG. 1993. Bioactive and other sesquiterpenoids from Porella cordeana. J. Nat. Prod. 56, 921–5.

Ivanova V, Kolarova M, Aleksieva K, Dornberger KJ, Haertl A, Moellmann U, Dahse HM and Chipev N. 2007. Sanionins: anti-inflammatory and antibacterial agents with weak cytotoxicity from the Antarctic moss *Sanionia georgico-uncinata*. Biochem. Biotechonol., 37, 343–352.

Jorgnersteen JH. Selection of antimicrobial agents for routine testing in clinical microbiology laboratory: diagnosis of microbial infection. Pharmacology., 1998, 16, 245–249.

Joshi GC, Tiwari KC, Pande NK and Pande G. 1994. Bryophytes, the source of the origin of Shilajit – a new hypothesis. BMEBR.15, 106–11.

Matsuo A, and Sato A. 1991. Sterols of mosses. Phytochemistry. 30, 2305–6.

Matsuo A, Nakayama A, and Hayashi S. 1971. Aromatic esters from the liverwort Isotachis japonica. Zeit. Natur. 26, 1023–35.

Matsuo A, Kubota N, Uto S, Nozaki H, Nakayama M and Hayashi S. 1980. Structure of three novel sesquiterpene aldehydes, (-)-isobicyclogermacernal, (-)-lepidozenal, and (+)-vitrenal, displaying plant growth inhibitory effect from the liverwort *Lepidozia vitrea*. 23rd Symposium of Chemistry Natural Products. Symposium papers.: 420–7.

Matsuo A, Atsumi K, Nakayama M and Hayashi S. 1981a.Structure of ent-2, 3-seco-alloaromadendrane sesquiterpenoids having plant growth inhibitory activity from *Plagiochila semidecurrens* (liverwort). J. Chem. Soc. Perkin. Trans.1, 2816–24.

Matsuo A, Kubota N, Nakayama M and Hayashi S. 1981b. Lepidozenal, a sesquiterpenoid with a novel trans-fused bicyclo [8.1.0] undecane system from the liverwort *Lepidozia* vitrea. Chem. Lett. 34, 1097–1100.

Matsuo A, Yuki S, Nakayama M and Hayashi S. 1982. Structure and biological activity of several sesquiterpenoids having a novel herbetane skeleton from the liverwort *Herberta adunca*. 25th Symp Chemistry Natural Products, Symposium papers. 242–9.

Matsuo A, Nozaki A, Kubota N, Uto S and Nakayama M. 1984.Structures and conformation of (-)-isobicyclogermacrenal and 9-lepidozenal, two key sesquiterpenoids of the cis- and trans-10, 3-bicyclic ring system, from the liverwort *Lepidozia vitrea*. J Chem Soc Perkin Trans. 1, 203–214.

Miller NG, Miller H. Make ye the bryophytes. Horticulture., 1979, 57, 40–47.

Nagashima F, Kasai W, Kondoh M, Fujii M, Watanabe Y, Braggins JE, Asakawa Y.New ent-kaurene-type diterpenoids possessing cytotoxicity from the New Zealand liverwort *Jungermannia* species. Pharmaceut. Bull., 1983, 51, 1189–92.

Ohta Y, Andersen NH, and Liu CB. 1977. Sesquiterpene constituents of two liverworts of genus *Diplophyllum*. Novel eudesmenaolides and cytotoxicity studies for enantiomeric methylene lactones. Tetrahedron. 33, 617–628.

Pant G, and Tewari SD. 1989. Bryophytes and mankind. Ethnobotany. 2, 97–103.

Richardson DHS. 1981. The biology of mosses. Blackwell, London.

Sakai K, Ichikawa T, Yamada K, Yamashita M, Tanimoto M, Hikita A, Ijuin Y, Kondo K.Antitumor principles in mosses: The first isolation and identification of maytansinoids, including a novel 15-methoxyansamitocin-P-3. J. Nat. Prod., 1998, 51, 845–850.

Saritas Y, Sonwa MM, Iznaguen H, König WA, Muhle H, Mues R. Volatile constituents in mosses (Musci). Phytochemistry., 2001;57: 443–57.

Saruwatari M, Takio S, and Ono K. 1999. Low temperature-induced accumulation of eicosapentaenoic acids in *Marchantia polymorpha* cells. Phytochemistry. 52, 367–72.

Scher JM, Zapp J, and Becker H. 2003. Lignan derivatives from the liverwort *Bazzania trilobata*. Phytochemistry. 62, 769–777.

Spjut RW, Suffness M, Cragg GM and Norris.Mosses DH. 1986. Liverworts and hornworts screened for antitumor agents. Eco. Bot. 40, 310–338.

Sturtevant W. 1954. The Mikauski Seminole: Medical beliefs and practices. Dissertation: Yale Univ, Yale.

Subhisha S, and Subramoniam A. 2005. Antifungal activities of a steroid from Pallavicinia lyellii, a liverwort. Ind. J. Pharmacol. 37, 304–308.

Subhisha S, and Subramoniam A. 2006. *In vivo* efficacy of an antifungal fraction from *Pallavicinia lyellii*, a liverwort. Ind. J. Pharmacol. 38, 211–212.

Ülhani S, Savaroulu F, Olak F, En C and Erdemgul FZ. 2006. Antimicrobial Activity of *Palustriella commutata* (Hedw) Ochyra Extracts (Bryophyta). Turk. J. Biol. 30, 149–52.

Wang X, Yu W, and Lou H. 2001. Antifungal Constituents from the Chinese Moss Homalia trichomanoides. Chem. Biodiver. 2, 139–145.

Wu PC. 1982. Some uses of mosses in China. Byrol. Times. 13, 5.

Zheng G. 1991. Cytotoxic agents from the mosses *Polytrichum ohioense* and *Polytrichum pallidisetum*. Part 1. Design and synthesis of inhibitors of myristoyl coa: protein N-myristoyltransferase as potential anticancer drugs. Part 2. Diss. Abstr. Int. 52, 182.

Chapter 12

Ethnopharmacology of Anthelmintic Ferns

12.1 Introduction

Pteridophytes constitute the primitive vascular plant groups which are found scattered all over the world. While not much consideration has been given towards the utility of pteridophytes; though they possess equal economic and medicinal importance . Ferns are represented by about 305 genera, comprising more than 10,000 species all over the world. The medicinal value of pteridophytes is known to man for more than 2000 years.

12.2 Ferns in traditional medicine

About 300 B.C., Theophrastus recommended oil extracted from ferns to expel internal parasites. Rhizomes of various shield fern species (*Polystichum* and *Dryopteris*) have been used since the 18th century as a cure for intestinal worms. *Cyathea manniana* Hook. (Cyatheaceae) from East Africa has been used by the Chagga and by German troops in the First World War as an anthelmintic. Pteridophytes find use in Homoeopathic, Ayurvedic, Tribal and Unani prescriptions for worm infestation. Aromatic compounds, glycosides and α- and γ-pyrones are responsible for the anthelmintic, antibacterial, mutagenic and antifeedant effects of ferns.

12.3 Materia Medica and phytochemistry of anthelmintic ferns

Actiniopteris radiata L. (Adiantaceae)

It is commonly found in India and possesses anthelmintic activity. Powdered rhizome is used as an anthelmintic in homeopathic medicine. It contains rutin and sterols.

Athyrium filix-femina (L.) Roth. ex Mert. (Dryopteridaceae)

It is native to Britain and popularly known as lady fern. A liquid extract of the root is an effective anthelmintic, though it is less powerful than the male fern, *D. filix-mas.*

Cheilanthes hirta Sw. (Adiantaceae)

This fern is found in Swaziland. Powdered rhizome is used as an anthelmintic for tapeworm.

Cibotium barometz (L.) J. Smith (Dicksoniaceae)

In India, it is used as a remedy for tape worm. In veterinary practice it is used as a vermifuge.

Cyathea medullaris G. Forst. Sw. (Cyatheaceae)

It is native to New Zealand. The gum obtained from the fern is used as a vermifuge.

Cystopteris fragilis (L.) Bernh. (Dryopteridaceae)

It is commonly known as brittle bladder fern. It is cosmopolitan in distribution. A decoction of the roots has been used as an anthelmintic enema.

Cyrtomium falcatum (L.f.) C. Presl. (Dryopteridaceae)

It is known as holly fern and found from East Asia to the Himalayas. It is reported to be naturalized in parts of Britain. The rhizome is used as an anthelmintic, chiefly for expulsion of tapeworm.

Cyrtomium fortunei-J.Sm (Dryopteridaceae)

This fern is found in East Asia. A decoction is used in the treatment of worm infestations including hookworm, tapeworm, ascariasis, and filiariasis.

Dryopteris crassirhizoma Nakai (Dryopteridaceae)

It is found in Korea. The rhizome is used against tapeworm and hookworm. It is reported to contain triterpenes and phenols.

Dryopteris expansa (C.Presl.) Fraser-Jenkins & Jermy (Dryopteridaceae)

It is found in the northern temperate zone, including Britain. The roots are rich in filicin. It is one of the most effective treatments known for tapeworms—its use should be immediately followed by a non-oily purgative such as magnesium sulfate in order to expel the worms from the body. An oily purge, such as caster oil, increases the absorption of the fern root and can be dangerous.

Dryopteris Filix-mas (L.) Schott (Dryopteridaceae)

It is grown as an ornamental fern in gardens. The oldest known anthelmintic now in use, as established historically, appears to be a male fern. It is sometimes referred to in ancient literature as Worm Fern. Many fern constituents display various types of bioactivities. The early physician, Theophrastus, recognized the value of the fern for treating tinea (ringworm) infections. Classical use of the plant as an anthelmintic included 5 to 8 g doses of the oleoresin extract. The fern has been used in traditional medicine for the treatment of worm infections.

The chemical composition of oleoresin of the rhizome of *D. filix mas* composition was ascribed to a group of related compounds commonly known as 'phenol fraction of the oleoresin' which were derivatives of 2-acylcyclol~exane-1, 3-dione.

Böhm showed the medical world that phloroglucinol group of compounds were responsible for anthelmintic activity of *D. Filix-mas.* Major pholoroglucinol of *D. Filix-mas* is filicin or filicic acid (1.5%) and is considered to be the active principle. According to pharmacopoeia, the rhizome of *D. Filix-mas* should not contain less than 1.5% of filicin. Liquid extract of *D. Filix-mas* contains 25% w/w of filicin. Other pholoroglucinols present in the rhizome of D.Filix-mas includes aspidin, desaspidin, paraspidin, margaspidin,aspidine, filixic acid, flavaspidic acid and filmarone.

Aspidin

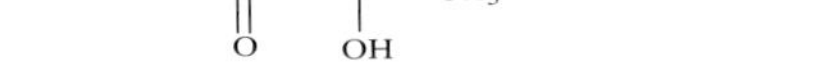

Desaspidin

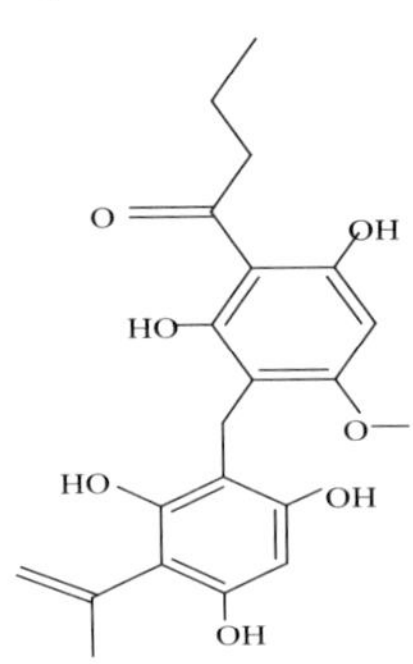

Margaspidin

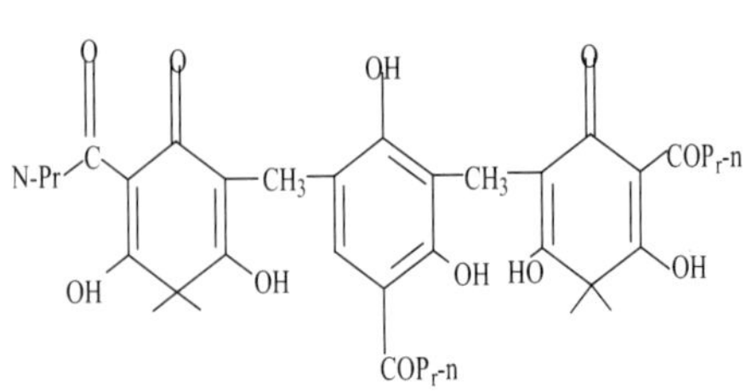

Filixic acid

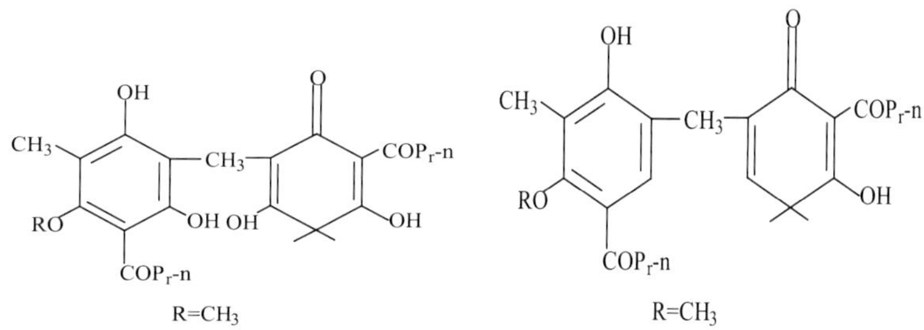

Filicic acid

Flavaspidic acid

Fig. 12.1 Major phloroglucinols of *Dryopteris Filix-mas.*

Penittlä and Sundma (1963) reported trisaspidin, trisdesaspidin and trisflavaspidic acid as antelmnitic principles of genus *Dryopteris*. Aspidin and desapidin were reported to be active among phloroglucinols by Bowden et al., in 1965. Murakami et al., 1984 holds aspidinol and desaspidinol responsible for the anthelmintic effects detected in extracts from various species of ferns of the genus *Dryopteris*. Aspidinol and desaspidinol have been reported from petroleum ether extract of the leaves and flowers *Leucosidea sericea* (Rosaceae).

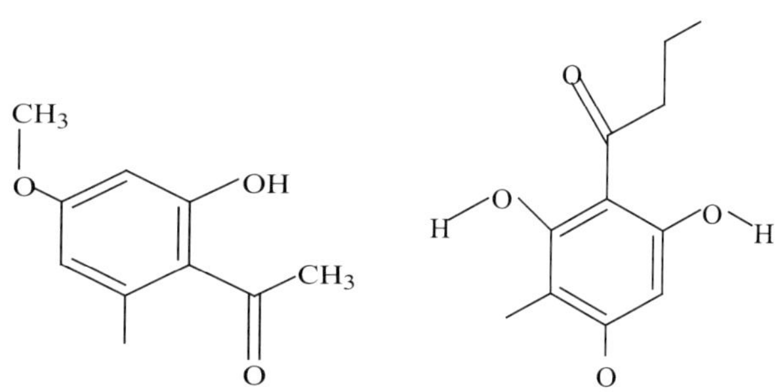

Desapidinol

Aspidinol

Fig. 12.2 Other hloroglucinols of *Dryopteris Filix-mas.*

Filicin and filmarone are active vermifuges and are particularly toxic to tapeworms. Following ingestion of the drugs, tinea are expelled within hours; however, a purgative typically is ingested concomitantly with the vermifuge to aid expulsion. The oleoresin paralyzes intestinal voluntary muscle and the analogous muscles of the tapeworm, which is then readily eliminated by the action of the purgative.

Preliminary preparation of *D. Filix-mas* with a milk diet and alba mixture for 2 d is followed by the drug in 3 divided doses in succession on an empty stomach in the morning followed by magnesium sulfate purgative. The head of the tapeworm is searched for in the stools that are passed. Castor oil must not be used because the oil aids the absorption of filicic acid which is toxic.*D. Filix-mas* is contraindicated in alcoholism, pregnancy and in advanced diseases of the heart, liver and kidneys.

The rhizome of *D. Filix-mas* is strongly cytotoxic against band worms and liver flukes, although round worm (*Ascaris lumricoides*) and pin worm (*Oxyuris vermicalulis*) are resistant. In combination with other anthelmintics, male fern extract has been reported to be effective against *Hymenolepis nana,* a small intestinal tapeworm. The components of the plant have been used as veterinary vermifuges. *D. Filix-mas* should be used with caution as in large doses it is poisonous. It is reported to be hepatotoxic and nephrotoxic.

The rhizomes of *Asplenium Filix-foemina* (Bernh.), *Aspidium oreopteris* (Sw.), and *A. spinulosum* (Sw.) resemble those of the male fern and have often been found mixed with it when imported. *D. juxtaposita* Christ is used as a substitute for the European male fern.

Dryopteris lewalleana Pic.Serm. (Dryopteridaceae)

In folk medicine of Swaziland, the Zulu and Xhosa, rhizome of the fern is used as anthelmintic.

Dryopteris oreades Fomin. *Polypodium vulgare*

It is commonly known as Mountain male fern. It is found in Britain. Its anthelmnitic use is similar to *D. expansa.*

Polypodium vulgare L. (Polypodiaceae)

It is found in Europe, the Mediterranean, temperate Asia and eastern N. America. A tea or syrup of the whole plant is anthelmintic.

References and further reading

Bosman AA, Combrinck S, Roux-van der Merwe R, Botha, BM, McCrindle RI. 1965. Isolation of an anthelmintic compound from *Leucosidea sericea*. S. Afri. J. Bot. 70, 509–511.

Bowden JL, Broadbent JL,and Ross WJ. 1965. Some simple anthelmintics. Br. J. Pharmacol. Chemother. 24, 714–724.

Bown. D. 1995. Encyclopaedia of Herbs and their Uses. Dorling Kindersley, London.

Chopra RN, Nayar SL, and Chopra IC. 1986. Glossary of Indian Medicinal Plants (Including the Supplement). Council of Scientific and Industrial Research, New Delhi.

Dixit RD. 1974. Ferns—a much neglected group of medicinal plants. J. Res. Ind. Med. 9, 59–68.

Duke JA. 1985. Handbook of Medicinal Herbs. CRC Press, Boca Raton, FL.

Foster S, and Duke JA. 1990. A Field Guide to Medicinal Plants. Eastern and Central N. America. Houghton Mifflin Co.

Fowst G. 1955. The toxicology of anthelmintic ferns. I. Boll. Chim. Farm. 94, 306–24.

Fowst G. 1955. The toxicology of anthelmintic ferns. II. Boll. Chim. Farm. 94, 346–57.

Grieve A. 1984. A Modern Herbal. Penguin.

Houtzagers JJ. 1965. Renal insufficiency after *D. filix Mas* administration. Ned. Tijdschr. Geneeskd. 23, 178–9.

Hsu H-Y. 1986. Oriental Materia Medica. Oriental Healing Art Institute, Long Beach, CA. p. 743.

Krotov AI. 1964. Experimental study of the effectiveness of the combinations of aminocrichine, imozen and male fern extract. 1. Effect of the imaginary stage of *Hymenolepis nana* in white mice. Med. Parazitol. 33, 215–21.

Kumaraswamy R, and Chelladurai V. 1974. Pharmacognostical studies on the Siddha drug `iruvi' (*Dryopteris filix mas*). J. Res. Ind. Med. 9, 28–34.

Mannan MM, Maridass M, and Victor B. 2008. A Review on the Potential Uses of Ferns. Ethnobot. l Leaf. 12, 281–285.

Mehra PN, and Mittal TC. 1960. Indian substitutes of male fern in pharmacognosy. Nature. 186, 722–723.

Morton CV, and Benedict RC. 1953. Recent Fern Literature. Am. Fern. J. 43, 84–91.

Murakami T, and Tanaka N. 1988. Progress in the Chemistry of Organic Natural Products, 54, Springer, Vienna.

Palva IP, Rantanen I, and Salmenpera A. 1965. Explusion of the fish tapeworm using aspidin and desaspidin. Ann. Med. Intern. 54, 137–8.

Parihar P, and Parihar L. 2006. Some pteridophytes of medicinal importance from Rajasthan. Nat. Prod. Rad. 4, 297–301.

Penittlä A, and Sundma J. 1963. Trisaspidin, trisdesaspidin and trisflavaspidic acid, three new three ring phloroglucinol derivatives from *Dryopteris austriaca*. Acta. Chem. Scand. 17, 2361–2369.

Puri HS. 1970. Indian Pteridophytes Used in Folk Medicine. Am. Fern. J. 60, 137–143.

Roux JP.2003. Swaziland Ferns and Fern Allies, Southern African Botanical Diversity Network Report No. 19. Keats Publishing, New Canaan, CT.

Takki S. 1967. Anthelmintic effect of fern preparations and their action on some liver function tests in man. Ann. Med. Ex. p. Biol. Fenn. 45, 341–51.

Taneja SC, and Tiwari HP. 1974. Chemical Constituents of *Actiniopteris Radiata* (SW.) Link. Curr. Scie. 43, 749–750.

Thomas TL, and Taurins A. 1962. Constituents of the oleoresin of the male fern. Part 1. Alcohol and sterol fractions. Can. J. Chem. 40, 302–10.

Widĕn CJ, Sarvela J, and Britton OM. 1983. On the location and distribution of phloroglucinols (filicins) in Ferns. Ann. Bot. Fennici. 20, 407.

Chapter 13

Ethnopharmacology of Medicinal Orchids

13.1 Introduction

Orchidaceae is one of the largest families among angiosperms. According to one estimate the family includes 800 genera and 25,000 species. Orchids are well known for their economic importance and are widely cultivated for ornamental purposes. Orchids are cosmopolitan in distribution. *Vanilla planifolia* is a commercially important orchid as it is source of vanillin used as a foodstuff flavoring.

13.2 Historical aspects

The term orchid was coined by Theophrastus as the anatomy of the plants resembled testicles. The Greek word 'orchid' literally means testicles. This may account for the use of orchids as aphrodisiacs in ancient civilizations. Among the history of ancient alternative systems of medicine—Ayurveda and Traditional Chinese Medicine (TCM) are on the forefront.

Asthavarga is an important ingredient of various classical Ayurvedic formulations like Chavyanprasa. Out of eight constituents of Ashtavarga, four have been reported to be orchids (see Table 7.1).

Traditional Chinese medicine widely utilizes orchids in medicines. A few of them have been subjected to phytochemical and pharmacological studies. In India work has been carried out on the chemical analysis of some medicinally useful orchids. *Eulophia campestris, Orchis latifolia, Vanda roxburgii* are among some of the important plants. *Dendrobium macraei* is another orchid of value from the Ayurvedic point of view as it is reported to be source of Jivanti. *Cypripedium parviflora* is widely used as an aphrodisiac and nervine tonic in Western Herbalism.

Many medicinal orchids are reported to contain alkaloids. Antimicrobial activities of some orchids have been suggested although detailed investigations are still warranted. Recent studies have reported isolation of anthocyanins, stilebnoids and triterpenoids from orchids. Orchinol, hircinol, cypripedin, jibantine, nidemin and loroglossin are some important phytochemicals reported from orchids. Some of the phytochemicals isolated from orchids along with their biological source have been tabulated below:

Table 13.1 Phytochemicals isolated from orchids along with the biological source.

S.No	Name of phytochemical	Phytochemical class	Source
1.	Aeridin	Phenanthropyran	*Aerides crispum*
2.	Agrostophyllinol	Triterpenoid	*Agrostophyllum brevipes*
3.	Agrostophyllinone	Triterpenoid	*Agrostophyllum brevipes* and *Agrostophyllum callosum*
4.	Isoagrostophyllol	Triterpenoid	*Agrostophyllum callosum*
5.	Orchinol, 6-methoxycoelonin, imbricatin, flaccidin, oxoflaccidin, isooxoflaccidin, flaccidinin, agrostophyllin, callosin, callosinin, callosumin, callosuminin and callosumidin.	Stilbenoids	*Agrostophyllum callosum*
6.	Arundinan	Stilbenoid	*Arundina graminifolia*
7.	Cypripedin	1-4 phenanthrenequinone	*Cypripedium calceolus pubescens*
8.	Loroglossin	Glucoside	*Orchis latifolia*
9.	Jebantine	Alkaloid	*Dendrobium macraei*
10.	Gigantol	Bibenzyl	*Dendrobium nobile*
11.	Moscatilin	Bibenzyl	*Dendrobium nobile*
12.	Dendrobine	Alkaloid	*Dendrobium nobile*
13.		Anthocyanins	*Dracula chimaera*(21)
14.	Nudol	Phenanthrene	*Eulophia nuda*
15.	Melianin	Glycoside	*Vanda roxburghii*
16.	Nidemin	Triterpenoid	*Nidema boothi* and *Scaphyglottis*
17.	Kinsenoside	Glycoside	*Anoectochilus formosanus*
18.	Rotundatin and moscatin	Phenanthrene	*Dendrobium moscatum*
19.	Gymopusin	Phenanthrene	*Bulbophyllum gymopus*

13.3 Materia Medica of medicinally important orchids

Aerides crispum Lindl.

Origin: India

Phytochemistry: Phenanthropyran: aeridin.

Part used: Tubers

Aerides multiflorum Roxb.

Syn: *Aerides affine* Lindl.

Distribution: The Himalayas (Garhwal to Sikkim), Assam, India and Myanmar.

Part used: Tubers.

Pre-clinical studies: Antibacterial.

Agrostophyllum brevipes Ridley

Distribution: E. Himalayas to Indo-China.

Phytochemistry: Triterpenoids: agrostophyllinol and agrostophyllinone. (9).

Part used: Tubers.

Agrostophyllum callosum Rchb. f.

Distribution: The Himalayas (Nepal to Bhutan), Assam, Myanmar, Thailand, Malaysia. (10)

Botany: *Agrostophyllum callosum* is 30–60 or higher plant. Stalks are creeping rhizomes. Rhizome 3–4 mm and stem erect. Leaves 8–13 cm wide and inflorescence 1–2 cm in diameter with short pedicles. Flowers pink or white colored. (10)

Phytochemistry: Triterpenoids: agrostophyllinone and isoagrostophyllol, stilbenoids:orchinol,6-methoxycoelonin,imbricatin,flaccidin,oxoflaccidin, isooxoflaccidin, flaccidinin, agrostophyllin, callosin, callosinin, callosumin, callosuminin and callosumidin.

Part used: Tubers.

Anoectochilus formosanus Hayata

Distribution: Taiwan.

Botany: *Agrostophyllum callosum* is 30–60 or higher plant. Stalks are creeping rhizomes. Rhizome 3–4 mm and stem erect. Leaves 8-13 cm wide and inflorescence 1–2 cm in diameter with short pedicles. Flowers pink or white colored.

Phytochemistry: Glycoside: kinsenoside, and polysaccharide.

Part used: Tubers.

Action: Anticancer.

Therapeutics: Hepatitis, hypertension and cancer.

Pre-clinical studies: Antioxidant, antihperglycemic, hepatoprotective: kinsenoside, and immuno-modulating: polysaccharide.

Arundina graminifolia **(D. Don) Hochr.**

Syn: *Arundina bambusifolia* Lindl., *Cymbidium bambusifolium* Roxb.

Common name: Bamboo Orchid.

Distribution: The Himalayas of Nepal, Sri Lanka, Thailand, Laos, Cambodia, Vietnam, southern China, Japan, Taiwan and south to Malaya and Java.

Botany: *Arundina graminifolia* is a large terrestrial plant with erect stems that are 1.5–2.5 cm tall and up to 1.5 cm in diameter. The leaves are borne in two ranks and are narrowly oblong and grass-like, 12–30 cm long and 1.6–2.5 cm. The simple, terminal inflorescence may be branched and is 15-30 cm long. The large cattleya-like flowers are purple-red, flesh-colored or white and are up to 10 cm across. The lip is darker than the sepals and petals, often veined dark purple and has a yellow to orange-yellow patch at the base. The short-lived, scented flowers last for about 3 d and usually several open at the same time.

Phytochemistry: Benzyldihydrophenanthrene: arundinaol, stilbenoid: arundinan and phenanthrene constituents.

Part used: Rhizome.

Pre-clinical studies: Antibacterial.

Bletilla striata **(Thunb.) Rchb.f.**

Syn: *Bletilla hyacinthine* (Sm.) R.Br.

Common name: Hyacinthina orchid, urn orchid.

Distribution: East Asia: China and Japan.

Botany: *Bletilla striata* is a deciduous terrestrial orchid. The tuberous rhizomes go up to 60 cm, papery, thin leaves. Light green leaves are plicate and are about 7.5 cm wide.

Phytochemistry: Polysaccharide.

Parts used: Pseudo bulbs.

Actions: Antibacterial, anti-inflammatory, antiphlogistic, demulcent, pectoral, skin, styptic and vulnerary.

Therapeutics: Internal hemorrhage.

Human study: Vascular embolizing agent in interventional treatment of primary hepatic carcinoma.

Cypripedium calceolus pubescens **(Willd.) Correll**

Syn: *Cypripedium pubescens* Willd., *Cypripedium parviflorum pubescens* (Willd.) Knight.

Common name: Lady's Slipper orchid.

Distribution: N. America to E. Asia - Japan.

Botany: Plants erect, 70–700 cm. Flowers: sepals greenish or yellowish (often obscured by darker markings); dorsal sepal suborbiculate or ovate to ovate-lance-acuminate, 19–80 × 7–40 mm; lateral sepals connate; synsepal 11–80 × 5–34 mm; petals horizontal to strongly descending, same color as sepals, commonly spirally twisted or undulate, sometimes flat, linear-lanceolate to lance-ovate or oblong, 24–97 × 3–12 mm; lip rather pale to deep yellow, very rarely white, rarely with reddish spots or suffusion on adaxial external surface, 15–54 mm; orifice basal; staminode cordiform-ovoid, deltoid, lance-ovoid, or ovoid-oblong .

Parts used: Roots.

Phytochemistry: The active constituents are soluble in alcoholic extract of the plant and is known as cyprepedin. The plant is reported to contain 1-4 phenanthrenequinone known as cypripedin.

Actions: Antispasmodic, diaphoretic, hypnotic, nervine, sedative, tonic. The plant is used as substitute for *Valeriana officinalis* L. although it is inferior.

Therapeutics: Diabetes, diarrhea, dysentery, paralysis, convalescence, impotence and malnutrition.

Dactylorhiza hatagirea **(D.Don) Soo.**

Syn: *Orchis latifolia* L.

Common name: Salampanja, Marsh orchis, salep orchid.

Ayurvedic name: Munjataka.

Distribution: Western Himalayas, Afghanistan and Iran.

Botany: *Dactylorhiza hatagirea* is a terrestrial orchid with fleshy tuberous roots. Tubers are slightly flattened, palmately lobed. Stem is usually 30–50 cm tall, leafy and with few sheathing scales in the lower portion. Leaves are erect, oblong-lanceolate, 7–15 cm long, obtuse and with a sheathing base. Flowers are pink-purple, crowded in terminal, spicate racemes.

Parts used: Roots.

Phytochemistry: Mucilage, starch, glucoside: loroglossin, albumen, volatile oil and ash. Five new compounds known as dactylorhins A-E and two natural compounds known as dactyloses A-B have been reported from plants growing in Nepal.

Actions: Aphrodisiac, expectorant and nervine tonic.

Therapeutics: Diabetes, diarrhea, dysentery, paralysis, convalescence, impotence and malnutrition.

Dendrobium macraei **Auct**

Syn: *Ephemerantha macraei* (Lindl.) Hunt et Sunmeh, *Flickingeria nodosa* (Dalz.) Seiden f.

Ayurvedic name: Jivanti, Jeva jevaniya, saka shreshtha, yasasvini, jiva bhadra.

Distribution: The Himalayas.

Botany: An air plant, growing on jabmul tree, much branching, stems long, pendulous and knotty, with many oblong pseudo bulbs, leaf one, red sessile and long. The flowers are white, with a yellow lip 3 or 4 inches in diameter and fragrant.

Parts used: Tubers.

Phytochemistry: α and β jibantic acid and alkaloid: jebantine.

Actions: Tonic.

Therapeutics: General debility.

Dendrobium nobile **Lindl.**

Syn: *Dendrobium lindleyanum, Dendrobium coerulescens.*

Distribution: The Himalayas and China.

Phytochemistry: mucilage, alkaloid: dendrobine, 1–4: phenanthrenequinone: denbinobine. Recently gigantol has been reported from methanolic extract of the plant growing in Japan. A bibenzyl compound, moscatilin has been isolated from the storage stem of the plant.

Actions: Antiphlogistic, pectoral, sialogogue, stomachic and tonic.

Therapeutics: In Vietnam the plant is used in the treatment of pulmonary tuberculosis, general debility, flatulence, dyspepsia, reduced salivation, parched and thirsty mouth, night sweats, fever and anorexia.

Pre-clinical studies: Anti-mutagenic.

Eulophia nuda **Lindl.**

Syn: *Eulophia dabia* (D.Don) Hochr.

Common name: Whitton root, Salep.

Ayurvedic name: Mankand.

Distribution: The Himalayas.

Botany: The tubers, conical, surrounded with circular marks. The remains of leaflets are yellowish white or of a green color.

Parts used: Tubers.

Phytochemistry: Phenanthrenes: chief is nudol.

Actions: Demulcent and anthemnintic.

Therapeutics: Worm infestation and scrofula.

Eulophia campestris **Wall. Ex Stapf**

Syn: *Eulophia dabia* (D.Don) Hochr.

Distribution: The Himalayas.

Botany: The tubers, conical, surrounded with circular marks. The remains of leaflets are yellowish white or of a green color.

Parts used: Tubers.

Phytochemistry: Mucilage.

Actions: Demulcent and anthemnintic.

Therapeutics: Worm infestation and scrofula.

Habenaria edgeworthii **Hook.f. ex Collett.**

Syn: *Habenaria acuminata* Lindl. syn *Platanthera edgeworthii* (Hook.f. ex Collett) R.K. Gupta)

Ayurvedic names: Riddhi, Laksmi, Mangala, Rathanga, Risisrista, Saravajanpriya, Siddhi, Sukha, Vasu and Yuga.

Distribution: E. Asia—The Himalayas. In dry grassy slopes, field borders upto 2800m (26).

Botany: Stem: 30 to 60 cm. high, leafy, stout. Leaves-Ovate, oblong-lanceolate, 4–10 cm long acute, acuminate thick, upper leaves gradually smaller, nerves 5–7, base sheathing. Flowering spike- 7 cm to 25 cm long bearing many flowers. Flowers- Yellow-green 1 to 1.5 cm across with lanceolate acute bracts , the lower shorter, the upper longer; than the ovary sepals green, pubescent, the margins slightly fringed; petals yellow thick, erect; lip yellow longer than the sepals concave narrowing to a long strap shaped limb, spur about twice the length of ovary, yellowish-green curving upwards with tip curved down.

Ayurvedic dynamics: Sweet in taste and pacifies *vata* and *pitta* but aggravates *kapha.*

Actions: Cooling and spermopiotic.

Therapeutics: Diseases of the blood.

Parts used: Leaves and roots.

Substitute: *Pueraria tuberosa* DC.

Habenaria intermedia **D.Don**

Syn: *Habenaria arietina* H.f.

English name: Wild orchid.

Ayurvedic names: Riddhi, Laksmi, Mangala, Rathanga, Risisrista, Saravajanpriya, Siddhi, Sukha, Vasu and Yuga.

Distribution: E. Asia—The Himalayas. In dry grassy slopes, field borders upto 2600 m. Botany: Erect, 25-60 cm high, terete, robust leafy. Leaves-Scattered usually 5, nerved ovate-lanceolate acuminate, cordate at the base. Inflorescence: 2–6 flowered. Flowers: 5 cm across white or greenish-white few, distant. Bracts leaf like lanceolate, acuminate, equal or more than ovary. Sepals persistent, 20–25 mm long, green, spreading tips reflexed, upper one white inside. Petals white, 5-nerved. Lip 3–lobed, longer than sepals, green spur 5–6 cm stout, longer than ovary more or less curved. Side lobes deeply fringed.

Ayurvedic dynamics: Sweet in taste and pacifies *vata* and *pitta* but aggravates *kapha*.

Actions: Cooling and spermopiotic.

Therapeutics: Diseases of the blood.

Parts used: Leaves and roots.

Substitute: *Pueraria tuberosa* DC.

Habenaria pectinata D.Don

Distribution: The Himalayas.

Common name: Safed musli.

Therapeutics: The leaves are crushed and applied in snake bites. Tubers mixed with condiments are used in arthritis.

Malaxis muscifera (Lindl.) Kuntze

Ayurvedic names: Jivaka, Chiranjivi, Dirghayu, Harsanga, Ksveda, Kurchasira, Pranda, Sringaka and Svadu.

Distribution: The Himalayas 1850 m to 2300 m from Himachal Pradesh to Arunachal Pradesh.

Botany: *Microstylis muscifera* is a terrestrial, robust herb, up to 25 cm high. Stem tending to be psuedobulbous at base. Leaves –usually 3 may be more, 5–10 cm ovate-lanceolate, acute with prominent veins and light green. Flowers- shortly stalked about 10 mm in diameter, yellowish-green with purple centre. Sepals oblong, 2 lateral shorter than the dorsal, margins recurved. Petals linear longer than sepals, margin recurved. Lip-slightly convex, tip notched or bilobulate, auricles straight and slightly over lapping.

Phytochemistry: No information.

Ayurvedic dynamics: Sweet in taste, cold in potency, pacifies *vāta* and aggravates *kapha*.

Actions: Cooling, febrifuge and spermopiotic.

Therapeutics: Bleeding diathesis, burning sensation, fever and phthisis.

Part used: Bulb.

Substitute: *Pueraria tuberosa* DC.

Malaxis acuminta D.Don

Syn: *Microstylis wallichii* Lindl., *Malaxis wallichii* Deb.

Ayurvedic names: Rishbhaka, Bandhura, Dhira, Durdhara, Gopati, Indraksa, Kakuda, Matrika, Visani, Vrisa and Vrisnabha.

Distribution: The Himalayas 1800 m to 3500 m eastwards to Sikkim.

Botany: Stem- 3 to 25 cm high with ovoid pseudo bulbs. Leaves- One or two (unequal) 3–10 x 2–4 cm sessile, ovate to ovate-lanceolate obtuse, narrowed at base to sheathing petiole. Infloresence-10 to 25 cm long. Flowers- 3–4 mm long, pale-yellow-green, bracts lanceolate shorter than ovary sepals broadly lanceolate. Petals liner shorter than sepals. Lip ovate abruptly pointed, margins thickened. Flowering time- July-August. Tuber- round, shining bearing stem giving the shape of a bullock horn having a similar curvature. The taste is slightly bitter with a fat like substance.

Phytochemistry: No information.

Ayurvedic dynamics: Sweet in taste, cold in potency, pacifies *vata* and aggravates *kapha*.

Actions: Cooling, febrifuge and spermopiotic.

Therapeutics: Bleeding diathesis, burning sensation, fever and phthisis.

Part used: Pseudo bulb.

Substitute: *Pueraria tuberosa* DC.

Orchis laxiflora Lam.

Syn: *Orchis ensifolia* Vill.

Common name: Oriental Salep, Marsh Orchis.

Distribution: South Europe, North Africa and West Asia.

Botany: *Orchis laxiflora* is a terrestrial orchid with fleshy tuberous roots. Tubers are slightly flattened. Stem contains sheathing scales in the lower portion. Leaves are erect and oblong-lanceolate. Flowers are dark-purple in spicate racemes.

Phytochemistry: Mucilage.

Actions: Astringent and expectorant.

Therapeutics: Diarrhea, bronchitis and convalescence.

Part used: Bulb.

Vanda spathulata (L.) Spreng.

Distribution: Peninsular India and Sri Lanka.

Therapeutics: Powdered flowers are used in the treatment of consumption, asthma and mania.

Vanda tessellata (Roxb.) Hook. Ex Don

Syn: *Vanda roxburghii* R.Br.

Common name: Vanda.

Ayurvedic names: Atirasa and Rasna.

Distribution: India, Sri Lanka and Myanmar.

Botany: *Vanda tessellata* is an epiphytic orchid, 30–60 cm high, with leafy stem. Leaves are thickly coriaceous, recurved, plicate, obtuse keeled. Flowers are greenish yellow, mottled with brown on the mid lobe of lip with purple caruncles.

Phytochemistry: Alkaloid, glucoside, bitter principle, tannins, resin, saponin, sitosterols and coloring matter. A glycoside (melianin) and a complex withanolide have been reported from plants growing in Pakistan.

Actions: Aphrodisiac, analgesic and nervine tonic.

Therapeutics: Paste of leaves is used as application in fevers. It is ingredient of *Rasna Panchaka Quatha*, the Ayurvedic formulation used in the treatment of arthritis and rheumatism. Expressed juice of the leaves is used in the treatment of otitis media.The root is used as an antidote against scorpion stings and as a remedy for bronchitis.

Parts used: Whole plant.

Pre-clinical studies: Aphrodisiac, anti-inflammatory, anti-arthritic, antimicrobial, and wound-healing.

Further reading

Ahmed F, Sayeed A, Islam A, Salam SMA, Sadik G, Satter MA and Astaq Mohal Khan GRM. 2001. Characterization and *in vitro* antimicrobial activity of 17-β-hydroxy-14, 20-epoxy-1-oxo-[22R]-3β-[O-β-d-glucopyranosyl]-5, 24-withadienolide from *Vanda Roxburghii* Br. The Sciences 1, 324–326.

Ahmed F, Sayeed A, Islam A, Salam SMA, Sadik G, Satter MA and Astaq Mohal Khan GRM. 2002. Antimicrobial Activity of Extracts and a Glycoside from *Vanda roxburghii* R. Br. Pak. J. Biol. Sci. 5, 189–191.

Anonymous. 1973. The Wealth of India, Vol. X. pp. 432.

Anuradha V, and Prakash NS. 1998. Aeridin: A phenanthropyran from *Aerides crispum*. Phytochemistry 48, 85–186.

Basu KB, Dasgupta SK, Bhattacharya R and Das PK. 1971.Anti-inflammatory principles of *Vanda roxburghii*. Curr. Sci. 21, 84–86.

Bechtel H, Cribb P, and Launert E. 1992. The Manual of Cultivated Orchid Species. 3rd Ed. Blandford Press, London.

Bhandari SR, Kapadi AH, Mujumder PL. Joardar M and Shoolery JN. Nudol, a phenanthrene of the orchids Eulophia nuda, *Eria carinata* and *Eria stricta*. Phytochemistry. 24, 801–4.

Chauhan NS. 1999. Medicinal and Aromatic Plants of Himachal Pradesh. Indus Publishing Company, New Delhi.

Fossen T, and Ovstedal DO. 2003. Anthocyanins from flowers of orchids *Dracula chimaera* and *Dracula cordobae*. Phytochemistry. 63, 783–7.

Ghanaksh A, and Kaushik P. 1999. Antibacterial effect of *Aerides multiflora* Roxb. A study *in vitro*. Journal of Orchid Society of India. 13, 65–68.

Kasera PK, and Shukla JK. 2001. Bio-medicinal properties and cultivation of *Leptadaenia reticualta* (Jivanti)-an endangered plant of the Thar Desert, India. Current Science. 84, 877–8.

Khasin SM, and Mohana Rao PR. 1999. Medicinal Importance of Orchids. The Botanica. 49, 86–91.

Khory N. 1982. Materia Medica of India and their Therapeutics. Neeraj Publishing House.

Kizu H, Kaneko E-I, and Tomimori T. 1999. Studies on Nepalese crude drugs. XXVI. Chemical constituents of Panch aunle, the roots of *Dactylorhiza hatagirea* D. Don. Chemistry Pharmaceutical Bulletin 47, 1618–1625.

Kumar S, Subramoniam A, and Pushpangadan P. 2000. Aphrodisiac activity of *Vanda tessellata* extract in mice. Indian Journal of Pharmacology. 32, 300–304.

Lin CC, Huang PC, and Lin JM. 1992. Antioxidant and hepatoprotective effects of *Anoectochilus formosanus* and *Gynostemma pentaphyllum*. American Journal of Chinese Medicine. 28, 87–96.

Liu MF, Han Y, Xing DM, Shi Y, Xu LZ, Du LJ and Ding Y.2004.A new stilbenoid from *Arundina graminifolia*. Asian J Nat Prod Res. 2004, 6, 229–232.

Liu MF, Han Y, Xing DM, Wang W, Xu LZ, Du LJ and Ding Y. 2005.One new benzyldihydrophenanthrene from *Arundina graminifolia*. Asian J Nat Prod Res. 7, 767–770.

Liu MF, Ding Y, and Zhang DM. Phenanthrene constituents from rhizome of *Arundina graminifolia*. Zhongguo Zhong Yao Za Zhi 200, 30, 353–6.

Lu TJ. 2005. Water-soluble polysaccharides with potential Immuno-modulating activities of *Anoectochilus formosanus* Hayata. Fifth International Conference of Food Science and Technology, Wuxi. China.

Majumder PL, and Banerjee S. 1987. Revised structure of gymopusin a ring-B oxygenated phenanthrene derivative isolated from *Bulbophyllum gymopus* (J.). Indian Journal of Chemistry Sec. B 28, 1085–88.

Majumder PL and Banerjee S. 1987. Structure of moscatibn- a new derivative from Orchid *Dendrobium moscatunm*. Ind. J. Chem. Sec B, 26, 18.

Majumder PL and Pal P. 1992. Rotundatin, a new 9, 10-dihydrophenanthrene derivative from *Dendrobium roturdatum*. Phytochemistry. 31, 3225.

Majumder PL, Banerjee S, Maiti DC and Sen S. 1995. Stilbenoids from the orchids *Agrostophyllum callosum* and *Coelogyne flaccida*. Phytochemistry. 39, 649–653.

Majumder PL, Banerjee S, and Sen, S. 1996. Three stilbenoids from the orchid.

Majumder PL, Sen S, and Banerjee S. 1999. Agrostophyllol and Isoagrostophyllol, Two Novel Diastereomeric 9,10-Dihydrophenanthropyran dderivatives from the Orchid *Agrostophyllum callosum*. Tetrahedron. 55, 13.

Majumdar PL, Majumder S. and Sen S.2003. Triperpenoids from the orchids *Agrostophyllum breirps* and *Agrostophyllum callosum*. Phytochemistry. 62, 591.

Miyazawa M, Shimamura H, Nakamura S and Kameoka H. 1997.Antimutagenic Activity of Gigantol from *Dendrobium nobile*. Agricul Food Chem. 45, 2849–2853.

Miyazawa M, Shimamura H, Nakamura S, Sugiura W, Kosaka H and Kameoka H. 1999. Moscatilin from *Dendrobium nobile*, a naturally occurring bibenzyl compound with potential antimutagenic activity. Agricul Food Chem.47, 2163–2167.

Nayak BS, Suresh R, Rao AV, Pillai GK, Davis EM, Ramkissoon V and McRae A. 2005. Evaluation of Wound Healing Activity of *Vanda roxburghii* R. Br. (Orchidaceae): A Preclinical Study in a Rat Model. Int. J. Lower Extremity Wounds. 4, 200–204.

Patoomratana T, and et al., 1988. Phenanthrenes of the *Eulophia nuda*. Phytochemistry. 27, 3267–3271. 32. S.R.

Prasad DN, and Achari G. 1966. A study of anti-arthritic action of *Vanda roxburghii* in albino rats. Journal of Indian Medical Association. 46, 234–7.

Shih CC, Wu YW, and Lin WC. 2002. Antihyperglycemic and antioxidant properties of *Anoectochilus formosanus* in diabetic rats. Clinical and Experimental Pharmacology and Physiology. 29, 684–688.

Singh AP, and Sandhu AS. 2005. A Dictionary of Medicinal Plants Sundeep Publishers, New Delhi.

Singh AP. 2005. *Dravyaguna Vijnana*. Chaukhambha Orientalia, New Delhi.

Singh AP. 2006. *Dhanwantri Nighantu*. Chaukhambha Orientalia, New Delhi.

Singh AP. 2006. *Raj Nighantu*. Chaukhambha Orientalia, New Delhi.

Stewart J, and Griffiths M. 1995. Manual of Orchids. Timber Press, Portland, Oregon.

Stuart GA. 1984. Chinese Materia Medica. A translation of an ancient Chinese herbal. Southern Materials Centre, Taipei.

Tuchinda P, Udchachon J, Khumtaveeporn K, Taylor WC, Engelhardt LM and White AH.1988. Phenanthrenes of the Eulophia nuda. Phytochemistry. 27, 3267–3271. 32.

Uniyal MR. 1975. Astavarga *Sandigdha Vanaushadhi*. Dhanwantri Partrika. Sri Jwala Ayurevd Bhawan Aligarh.

Uniyal MR, and Issar RK. 1966. Botanical studies of Astavarga. Nagarjun. 9, 276–284.

Uniyal MR, and Uniyal M. 1977. Medicinal Plants and Minerals of Uttarakhand. Himalaya. Baidyanath, Ayurved Shodh Sanstha, Patna.

Uniyal MR, and Joshi GC. 1993. Historical view of the basic principles of the identification of controversial drugs, problems and suggestion. Sachitra Ayurved. 45, 531–536.

Van Dan N, and Nhu D. 1989. Medicinal Plants in Vietnam. World Health Organisation.

Wang C, Sun J, Luo Yi, Xue W, Diao H, Dong L. Chen J and Zhang J. 2005. A Polysaccharide Isolated from the Medicinal Herb Bletilla striata Induces Endothelial Cells Proliferation and Vascular Endothelial Growth Factor Expression *in vitro*. Biotech. Lett. 28, 539–543.

Wu JB, Lin WL, Hsieh CC, Ho HY, Tsay HS and Lin WC.2006.The hepatoprotective activity of kinsenoside from *Anoectochilus formosanus*. Phytotherapy Research 2006, 21, 58–61.

Yeung Him-Che. 1985. Handbook of Chinese Herbs and Formulas. Institute of Chinese Medicine, Los Angeles.

Zheng C, Feng G, and Liang H. 2000. *Bletilla striata* as a vascular embolizing agent in interventional treatment of primary hepatic carcinoma. Chinese Medicine Journal. 111, 1060–3.

Chapter 14

Salacia sp. Potential Hypoglycemic Plants

14.1 Introduction

Salacia species (Celasteraceae) including *S. chinensis*, *S. reticulata* and *S. oblonga*, are used as antidiabetic agents in traditional systems of medicine like Ayurveda and Unani. The plants have been used in India, Japan and Korea to treat high blood glucose levels. Pre-clinical investigations have demonstrated the anti-obesity activity of Salacia. Large-scale clinical-studies are warranted for the potential use of *Salacia* species in treating diabetes, while pre-clinical research and isolated clinical trials have, thus far, supported this claim. This chapter summarizes the pharmacological investigations carried out on various species of *Salacia* with respect to diabetes.

14.2 *Salacia chinensis Linn.* Syn: *S. prinoides* DC.

S. chinensis is commonly studied among plants of the genus *Salacia*. It is an erect or straggling tree or a woody climbing shrub found along seashores and river banks as well as in forests. The plants bear flowers in clusters and small single seeded, red fruits having an edible pulp.

Proanthocyanidins has been reported from *S. chinensis*. Two new friedelane-type triterpenes: salasones D and E (Fig. 14.1), a new norfriedelane-type triterpene: salaquinone B (Fig. 14.2), and a new polyacylated eudesmane-type sesquiterpene: salasol B (Fig. 14.3), were isolated from the stems of *S. chinensis*. Some norfriedelane-type triterpene, lignan, and catechin constituents were found to show an antioxidant activity.

Three new friedelane-type triterpenes called salasones A, B, and C (Fig. 14.4), a new norfriedelane-type triterpene, salaquinone A (Fig. 14.5), and a new acylated eudesmane-type sesquiterpene, salasol A (Fig. 14.6), were reported from 80% aqueous methanolic extract of the stems of *S.*

chinensis collected in Thailand. 3 β, 22 β -dihydroxyolean-12-en-29-oic acid, tingenone, tingenine B, regeol A, triptocalline A, and mangiferin, showing an inhibitory effect on rat lens aldose reductase were also isolated.

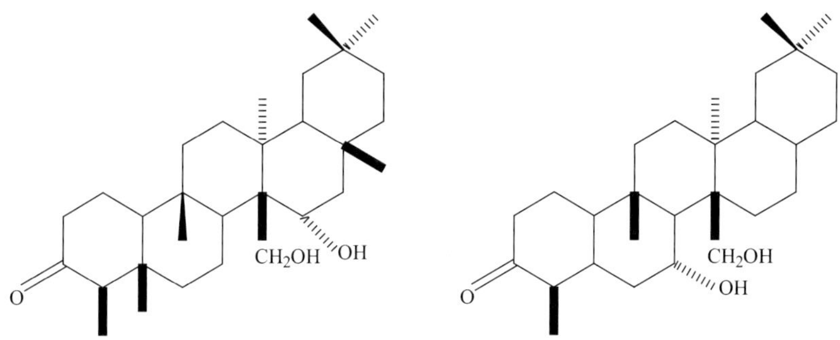

Fig. 14.1 Structure of Salasone D and Salasone E.

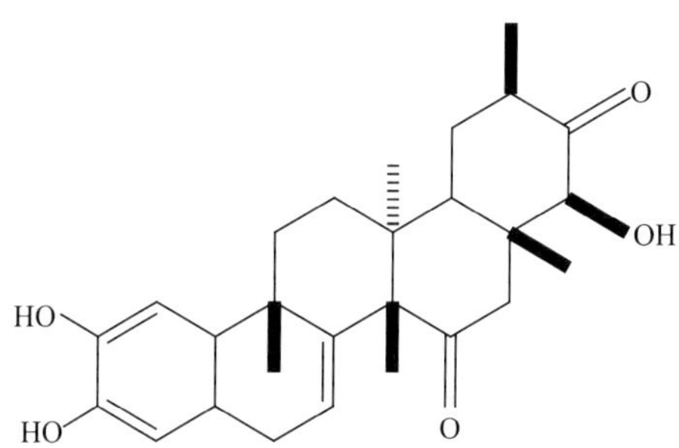

Fig. 14.2 Structure of Salaquinone B.

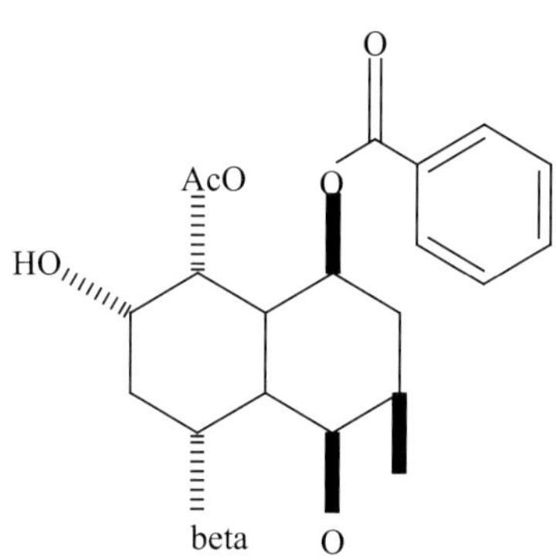

Fig. 14.3 Structure of Salasol B.

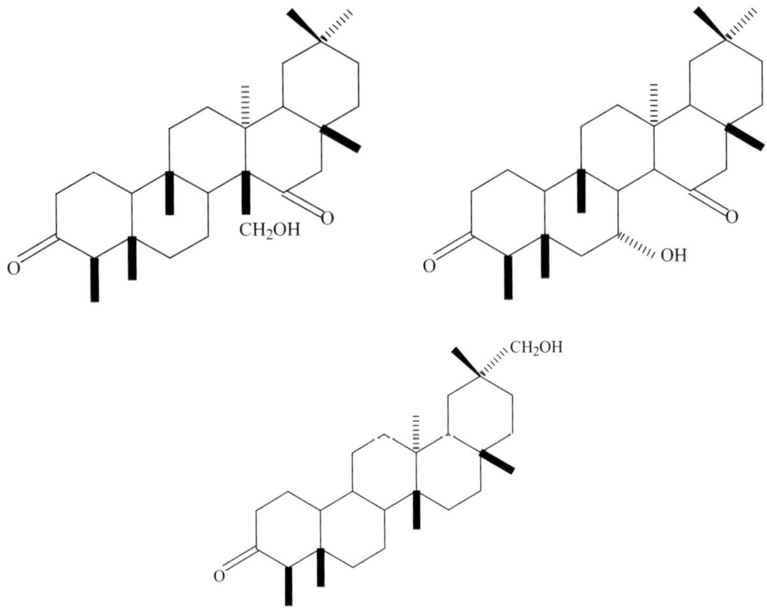

Fig. 14.4 Structure of Salasones A, B, and C.

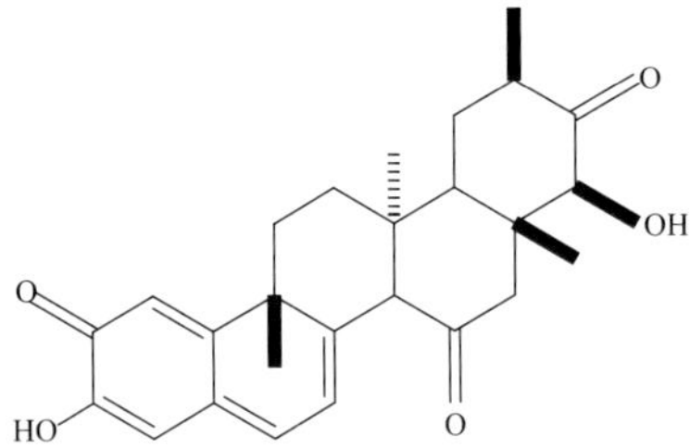

Fig. 14.5 Structure of Salaquinone A.

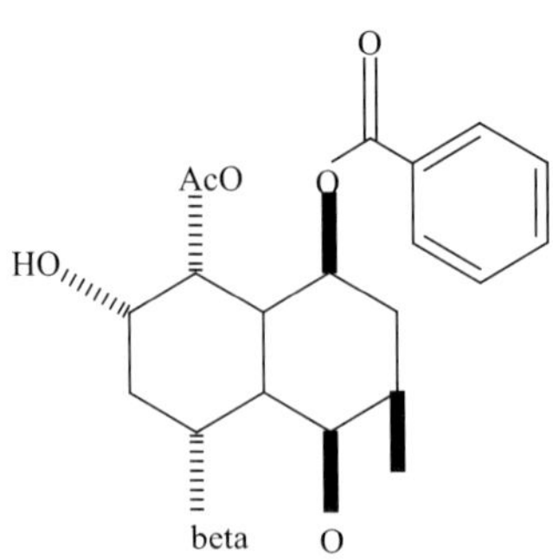

Fig. 14.6 Structure of Salasol A.

Recently megastigmane glycosides have been reported from the leaves of *S. chinensis.* Seven new megastigmane glycosides: foliasalaciosides E 1-3, F, G, H, and I, were recently isolated from the leaves of *S. chinensis* Linn. collected from Thailand. New triterpene constituents, foliasalacins A (1)-A (4), B (1)-B (3), and C, have been isolated from the leaves of *S. chinensis.*

In Ayurveda, the plant is known as Saptchakra. In addition to Ayurveda, the plant is used as a remedy for type 2 diabetes in the Siddha system of medicine. A clinical study in 1984 reported the usefulness of *S. chinensis* based formulation in diabetes mellitus. Kadal, a proprietary Siddha medicine comprising of roots and bark of *S. chinensis* and Triphala were given to 25 type 2 diabetes mellitus patients. Kadal was administered in a dose of 500 mg twice a day and Triphala was administered at 2.5 g thrice a day with water for a period of 4 mon. They reported a hypoglycemic effect of the formulation.

An animal study reported potent anti-hyperglycemic effect of methanolic extract from the stems of *S. chinensis,* in oral sucrose or maltose-loaded rats. The extract also demonstrated inhibition of alpha-glucosidase and rat lens aldose reductase, as well as nitric oxide production from lipopolysaccharide-activated mouse peritoneal macrophage, and radical scavenging activities. Salacinol was identified as the alpha-glucosidase inhibitory principle of the extracts.

14.3 *Salacia reticulata* Wight

S. reticulata is a climbing perennial woody plant growing wild in South India and in North Sri Lanka. Its roots and stems are been used in Ayurveda to relieve dry mouth. Mangiferin has been isolated from the root bark of *S. reticulata.* A novel nortriterpenoid aldehyde called salacenonal (Fig. 14.7) has been reported from *S. reticulate.* Kotalanol having potential natural alpha-glucosidase inhibiting activity was isolated from the roots and stems of *S. reticulata* through bioassay-guided separation. Kotalanol was found to show more potent inhibitory activity against sucrose than salacinol and acarbose.

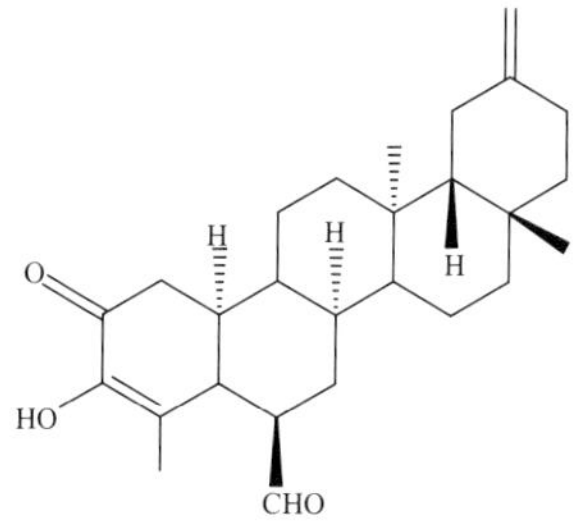

Fig. 14.7 Structure of Salacenonal.

A study reported that lipase inhibitory and lipolytic activities have mild antiobesity effects of hot water-soluble extract from the roots of *S. reticulata* and its polyphenolic constituents in rats. A randomized single center double blind crossover trial, studied the efficacy of an herbal tea containing *S. reticulata* (Kothala Himbutu tea) in patients with type 2 diabetes mellitus. Fifty-one patients with type 2 diabetes mellitus for longer than 6 mon and with evidence of stable glycemic control over the preceding 6 mon (as assessed by HbA1C) participated in the study. They were randomized to receive a standard preparation of Kothala Himbutu tea for 3 mon followed by placebo in similar tea bags for a further 3 mon (n = 28) or in reverse order (n = 23). All patients received detailed advice on diet, exercise and lifestyle modification. HbA1C was measured at recruitment, at 3 mon and on completion of the study at 6 mon. Liver and renal functions were assessed biochemically at baseline, at 3 and 6 mon and adverse events were recorded. There were no significant differences between the two groups in age, body mass index, male/female ratio, and glycemic control and baseline laboratory tests. All patients completed both arms of the trial. The HbA1C at the end of drug treatment was significantly lower than after treatment with placebo.

A mixture of the aqueous extract of *S. reticulata* and cyclodextrin has been reported to reduce the accumulation of visceral fat mass in mice and male Sprague-Dawley rats with high-fat diet-induced obesity.

Recently a polyhydroxylated cyclic 13-membered sulfoxide (Fig. 14.8) was isolated from an aqueous extract of *S. reticulata*. The alpha-glucosidase inhibitory activity was much greater than the inhibitory activity of salacinol and kotalanol, previously isolated from *S. reticulata*. The same authors reported isolation and alpha-glucosidase inhibitory activity of a novel 13-membered ring thiocyclitol from an aqueous extract of *S. reticulata*. The inhibitory activity was investigated by maltose- and sucrose-loading on Wistar rats.

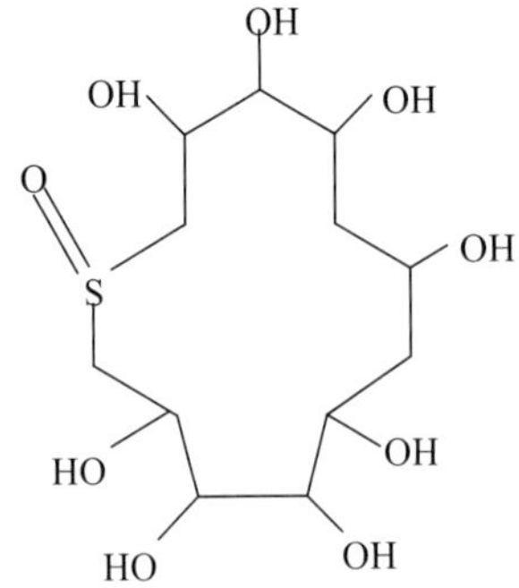

Fig. 14.8 Structure of polyhydroxylated cyclic 13-membered sulfoxide.

14.4 *Salacia oblonga* Wall.

In Japan *S. oblonga* has been sold as a food supplement for several years. *S. oblonga* plants grow in limited regions of India and Sri Lanka, and it is not yet well known in the U.S.

A study reported two biologically active principles from petroleum ether extract of the root bark of *S. oblonga*. Two principles demonstrated about 60% and 76% hypoglycemic potency of an equal dose of tolbutamide (250 mg/kg) in albino rats.

It was demonstrated that the aqueous methanolic extract of the roots of *S. oblonga* inhibited the increase in serum glucose level in sucrose- and maltose-loaded rats. The water-soluble and ethyl acetate-soluble portions from the aqueous methanolic extract showed inhibitory activities on alpha-glucosidase and aldose reductase, respectively. From the water-soluble portion, potent alpha-glucosidase inhibitors, salacinol and kotalanol, were isolated, together with nine sugar related components, while a new friedelane-type triterpene, kotalagenin 16-acetate, was isolated from the ethyl acetate-soluble portion along with known diterpenes and triterpenes. The structure of kotalagenin 16-acetate was elucidated on the basis of physicochemical evidence. The principal components from this natural medicine were examined in terms of inhibitory activity on aldose reductase, and the diterpene and triterpene constituents, including the new kotalagenin 16-acetate, were found to be responsible components for the inhibitory activity on aldose reductase.

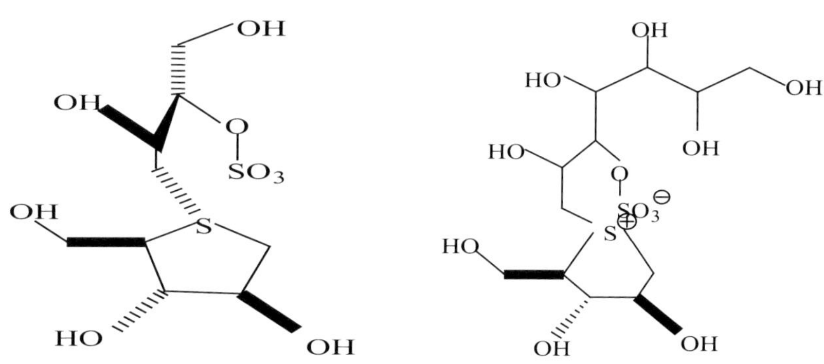

Salacinol Kotalanol

Fig. 14.9 Structure of possible hypoglycemic principles of *Salacia* sp.

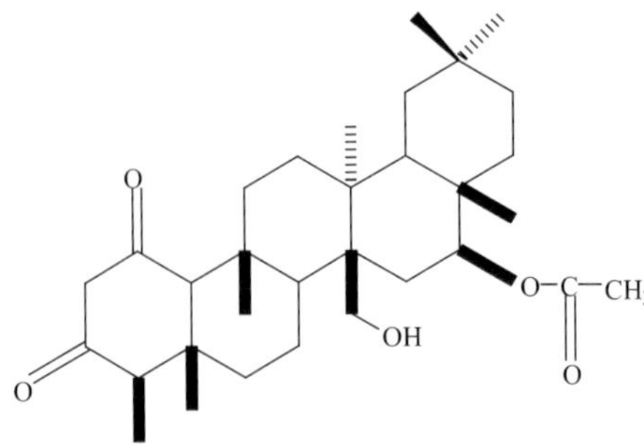

Fig. 14.10 Structure of kotalagenin 16-acetate.

A clinical study evaluated the effect of an herbal extract of *S. oblonga* on postprandial glycemia and insulinemia in patients with type 2 diabetes after ingestion of a high-carbohydrate meal. Sixty-six patients with diabetes were studied in this randomized, double-blinded crossover study. During a fasting stage, subjects consumed 1 of the following 3 meals: a standard liquid control meal, a control meal + 240 mg *S. oblonga* extract, and a control meal + 480 mg *S. oblonga* extract. Both doses of the Salacia extract significantly lowered the postprandial positive area under the glucose curve (14% for the 240 mg extract and 22% for the 480 mg extract) and the adjusted peak glucose response (19% for the lower dose and 27% for the higher dose of extract) to the control meal. In addition, both doses of the salacia extract significantly decreased the postprandial insulin response, lowering both the positive area under the insulin curve and the adjusted peak insulin response (14% and 9%, respectively, for the 240 mg extract; 19% and 12%, respectively, for the 480 mg extract) in comparison with the control meal.

Scientists investigated the effect of the water extract of *S. oblonga* on obesity and diabetes-associated cardiac hypertrophy and discussed the role of modulation of cardiac angiotensin II type 1 receptor (AT(1)) expression in the effect. *S. oblonga* 100 mg/kg) was given orally to male Zucker diabetic fatty rats for 7 wk. At the end-point of the treatment, the hearts and left ventricles were weighed, cardiomyocyte cross-sectional areas were measured, and cardiac gene profiles were analyzed. On the other hand, angiotensin II-stimulated embryonic rat heart-derived H9c2 cells and neonatal rat cardiac fibroblasts were pretreated with water extract of *S. oblonga* and one of its prominent components mangiferin, respectively. Atrial natriuretic peptide, mRNA expression and protein synthesis and [(3) H] thymidine incorporation were determined.

S. oblonga treated Zucker diabetic fatty rats showed less cardiac hypertrophy (decrease in weights of the hearts and left ventricles and reduced cardiomyocyte cross-sectional areas). *S. oblonga* treatment suppressed cardiac over expression of Atrial natriuretic peptide, brain natriuretic peptide and AT (1) mRNAs and AT (1) protein in Zucker diabetic fatty rats. Aqueous extract of *S. oblonga* (50–100 microg/ml) and mangiferin (25 micromol) suppressed angiotensin II-induced ANP mRNA over expression and protein synthesis in H9c2 cells. They also inhibited angiotensin II-stimulated [(3) H] thymidine incorporation by cardiac fibroblasts. The findings demonstrate that *S. oblonga* decreases cardiac hypertrophy in Zucker diabetic fatty rats, at least in part by inhibiting cardiac AT(1) over expression.

14.5 Salacia as peroxisome proliferator-activated receptor-alpha activator

In a present study, water extract of *S. oblonga* (100, 300 and 900mg/kg/day/p.o., over a 28 d period) elicited dose-related increases in liver weight (LW) by 1.6%, 13.4% and 42.5%, respectively, and in the ratio of LW to body weight by 8.8%, 16.7% and 40.2%, respectively, in male rats. These effects were less pronounced in females. Water extract of *S. oblonga* selectively increased liver mass in male rats but Sudan red staining was not different, which indicates that hepatic lipid accumulation was similar in both genders. However, water extract of *S. oblonga* even at the highest dosage did not influence serum ALT and AST activities in male or female rats. Moreover, water extract of *S. oblonga* was found to activate peroxisome proliferator-activated receptor-alpha and acylCoA oxidase mRNA expression. Thus, water extract of *S. oblonga* dependent peroxisome proliferator-activated receptor-alpha activation may precede the development of the gender difference in hepatic hypertrophy.

14.6 Toxicity

Reproductive toxicity

A study determined the effects of the *S. reticulata* root extract on the reproductive outcome of normal Wistar rats (250–260 g) when administered orally (10 g/kg) during early (d 1–7) and mid- (d 7–14) pregnancy. The root extract significantly ($P<0.05$) enhanced post-implantation losses (control *vs* treatment: early pregnancy, 4.7 ± 2.4 *vs* 49.3 ± 13%; mid-pregnancy, 4.7 ± 2.4 *vs* 41.7 ± 16.1%). Gestational length was unaltered but the pups born had a low birth weight ($P<0.05$) (early pregnancy, 6.8 ± 0.1 *vs* 5.3 ± 0.1 g; mid-pregnancy, 6.8 ± 0.1 *vs* 5.0 ± 0.1 g) and low birth index ($P<0.05$) (early pregnancy, 95.2 ± 2.4 *vs* 50.7 ± 12.9%; mid-pregnancy, 95.2 ± 2.4 *vs* 58.3 ± 16.1%), fetal survival ratio ($P<0.05$) (early pregnancy, 95.2 ± 2.4 *vs* 50.7 ± 12.9; mid-pregnancy, 95.2 ± 2.4 *vs* 58.3 ± 16.1), and viability index ($P<0.05$)

(early pregnancy, 94.9 ± 2.6 *vs* 49.5 ± 12.5%; mid-pregnancy, 94.9 ± 2.6 *vs* 57.1 ± 16.1%). However, the root extract was non-teratogenic.

Genotoxicity and subchronic toxicity

Another study investigated the genotoxicity of a *S. oblonga* root extract using the standard battery of tests (reverse mutation assay; chromosomal aberrations assay; mouse micronucleus assay) recommended by US Food and Drug Administration (FDA) for food ingredients. *S. oblonga* was determined not to be genotoxic under the conditions of the reverse mutation assay and mouse micronucleus assay, and weakly positive for the chromosomal aberrations assay. Continuing further studies, Flammang et al., 2007, investigated the toxicity of a *S. oblonga* root extract in a subchronic 90-d feeding study in rats. An *in vivo-in vitro* rat peripheral blood lymphocyte chromosomal aberrations assay was added at termination of the subchronic rat study to examine cultured lymphocytes for possible chromosomal aberration induction. The present study results indicate that aqueous extract of *S. oblonga* was negative for the induction of chromosomal aberrations in cultured rat peripheral blood lymphocytes after 90 consecutive days of treatment. The no observable adverse effect level was determined to be 2,500 mg/kg/day, p.o. following subchronic administrations to rats.

14.7 Doses of Salacia based preparations

Although no authentic information is available regarding the dosage, a typical dose of *Salacia* based preparations is 2.5 to 5.0 grams daily of the whole herb, or a comparable amount as extract.

All the three species of Salacia have demonstrated alpha-glucosidase inhibiting activity like acarbose, with salacinol and kotalanol, as possible active principles. These studies have provided insights into potential protective and anti-obesity roles of Salacia species also. Some animal studies have demonstrated that Salacia might have antidiabetic action like conventional PPAR-gamma activators. Clinical trials have also reported efficacy of Salacia species in the treatment of diabetes mellitus. Toxicity data shows this herb is devoid of genotoxic and teratogenic effect but it should be avoided during pregnancy.

References and further reading

Augusti KT, Joseph P, Babu TD. 1995. Biologically active principles isolated from *Salacia oblonga* Wall. Ind. J. Physiol Pharmacol. 39, 415–417.

Flammang AM, Erexson GL, Mecchi MS, and Murli H. 2006. Genotoxicity testing of a *Salacia oblonga* extract. Food Chem. Toxicol. 44, 1868–1874.

Flammang AM, Erexson GL, Mirwald JM, and Henwood SM. 2007. Toxicological and cytogenetic assessment of a *Salacia oblonga* extract in a rat subchronic study. Food Chem. Toxicol 45,1954–1962.

Huang TH, He L, Qin Q, Yang Q, Peng G, Harada M, Qi Y, Yamahara J, Roufogalis BD, and Li Y. 2008. *Salacia oblonga* root decreases cardiac hypertrophy in Zucker diabetic fatty rats: inhibition of cardiac expression of angiotensin II type 1 receptor. Diabetes, Obes, Metab. 10, 574–585.

Jayawardena MH, de Alwis NM, Hettigoda V, and Fernando DJ. 2005. A double blind randomized placebo controlled cross over study of an herbal preparation containing *Salacia reticulata* in the treatment of type 2 diabetes. J. Ethnopharmacol. 97, 215–218.

Jennifer A Williams, Yong S Choe, Michael J Noss, Carl J Baumgartner, and Vikkie A Mustad. 2007. Extract of *Salacia oblonga* lowers acute glycemia in patients with type 2 diabetes. Am.J. Clin. Nut. 86, 124–130.

Karunanayake EH, and Sirimanne SR. 1985. Mangiferin from the root bark of *Salacia reticulata*. J. Ethnopharmacol. 13, 227–228.

Kishi A, Morikawa T, Matsuda H, and Yoshikawa M. 2003. Structures of new friedelane- and norfriedelane-type triterpenes and polyacylated eudesmane-type sesquiterpene from *Salacia chinensis* Linn. (*S. prinoides* DC., Hippocrateaceae) and radical scavenging activities of principal constituents. Chem. Pharm. Bull. 51, 1051–1055.

Kishino E, Ito T, Fujita K, and Kiuchi Y. 2006. A mixture of the *Salacia reticulata* (Kotala himbutu) aqueous extract and cyclodextrin reduces the accumulation of visceral fat mass in mice and rats with high-fat diet-induced obesity. J. Nut. 136, 433–439.

Krishnaa V, and Rangaswami S. 1967. Proanthocyanidins of *Salacia chinensis* Linn. Tetrahedron. 26, 2441–2446.

Matsuda H, Murakami T, Yashiro K, Yamahara J, and Yoshikawa M. 1999. Antidiabetic principles of natural medicines. IV. Aldose reductase and qlpha-glucosidase inhibitors from the roots of *Salacia oblonga* Wall. (Celastraceae): structure of a new friedelane-type triterpene, kotalagenin 16-acetate. Chemi. Pharmaceuti, Bull. Tokyo. 47, 1725–1729.

Morikawa T, Kishi A, Pongpiriyadacha Y, Matsuda H, and Yoshikawa M. 2003. Structures of new friedelane-type triterpenes and eudesmane-type sesquiterpene and aldose reductase inhibitors from *Salacia chinensis*. J. Nat. Prod. 66, 1191–1196.

Nakamura S, Yi Zhang, Pongpiriyadacha Y, and et al., 2008. Megastigmane Glycosides from the Leaves of *Salacia chinensis*. ChemInform. Published Online: 24 Apr.

Oe H, and Ozaki S. 1964. Hypoglycemic effect of 13-membered ring thiocyclitol, a novel alpha-glucosidase inhibitor from Kothala-himbutu (*Salacia reticulata*). Bio. sci. Biotech. Biochem. 72, 1962–1964.

Ozaki S, Oe H, and Kitamura S. 2008. Alpha-glucosidase inhibitor from Kothala-himbutu (*Salacia reticulata* Wight). J. Nat. Prod. 71, 981–4.

Ratnasooriya WD, Jayakody JRAC, and Premakumara GAS. 2003. Adverse pregnancy outcome in rats following exposure to a *Salacia reticulata* (Celastraceae) root extract. Brazi. J. Med. Biol. Res. 36, 931–935.

Rong X, Kim MS, Su N, Wen S, Matsuo Y, Yamahara J, Murray M, and Li Y. 2008. An aqueous extract of *S. oblonga* root, an herb-derived peroxisome prolifertor—activated receptor-alpha factor, by oral gavage over 28 days inducers gender-dependent hepatic hypertrophy in rats. Food. Chem. Toxicol. 46, 2165–72.

Sivaprakasam K, Rao KK, Yashoda R and Veluchamy G. 1984. Siddha remedy for diabetes mellitus. J. Res. Ayur. Sidd., 1984, 5, 25.Yoshikawa M, Murakami T, Yashiro K, and Matsuda H. 1998. Kotalanol, a potent alpha-glucosidase inhibitor with thiosugar sulfonium sulfate structure, from antidiabetic ayurvedic medicine *Salacia reticulata*. Chem. Pharmaceut. Bull. 46, 1339–1340.

Yoshikawa M, Shimoda H, Nishida N, Takada M, and Matsuda H. 2002. *Salacia reticulata* and its polyphenolic constituents with lipase inhibitory and lipolytic activities have mild antiobesity effects in rats. J. Nut. 132, 1819–24.

Yoshikawa M, Pongpiriyadacha Y, Kishi A, Kageura T, Wang T, Morikawa T, and Matsuda H. 2003. Biological activities of *Salacia chinensis* originating in Thailand: the quality evaluation guided by alpha-glucosidase inhibitory activity. Yakugaku. Zasshi. 123, 871–880.

Yoshikawa M, Zhang Y, Wang T, Nakamura S, and Matsuda H. 2008. New triterpene constituents, foliasalacins A (1)-A (4), B (1)-B (3), and C, from the leaves of *Salacia chinensis*. Chem. Pharmaceut., Bull. 56, 915–20.

Zhang Y, Nakamura S, Pongpiriyadacha Y, Matsuda H, and Yoshikawa M. 2008. Absolute structures of new megastigmane glycosides, foliasalaciosides E(1), E(2), E(3), F, G, H, and I from the leaves of *Salacia chinensis*. Chem. Pharmaceut. Bull. 56, 547–53.

Chapter 15

Ethnopharmacology of *Taraxacum officinale* Weber

15.1 Introduction

Complementary and Alternative Systems of Medicine (CAM) viz. Ayurveda, Siddha, and Traditional Chinese Medicine, have become more popular in recent years. Medicinal herbs and extracts prepared from them are widely used in the treatment of liver diseases like hepatitis, cirrhosis, and loss of appetite. A number of reviews have focused on the therapeutic utility and adverse effects of herbal products, particularly hepatotoxicity.

Silymarin (flavanolignan mixture from *Silybum marianum* Linn.), daphnoretin (coumarin from *Wilkstroemia indica),* kutkin or picrosides and kutkosides (iridoid glycosides from *Picrrorhiza kurroa* Royle*), gly*cyrrhizin (triterpene saponin from *Glycyrrhiza glabra* Linn.), schizandrins (lignans from *Schizandra chinensis*), andrographolide (diterpene lactone from *Andrographis paniculata Nees),* and lignans (phyllanthin from *Phyllanthus niruri*), are proven hepatoprotective phyto-constituents, as they have shown genuine utility in experimental and clinical studies.

More than 700 mono and polyherbal preparations in the form of decoction, tincture, tablets and capsules from more than 100 plants are in clinical use. Keeping in mind the non-availability of a standard hepatoprotective agent in synthetic medicine and the discovery of silymarin as hepatoprotective agent from natural source, we need to initiate screening for herbals with documented effect on the liver. Among several plants used for liver diseases, *T. ofiicinale* is well recognized in Complementary and Alternative Medicine (CAM). The folk medicines of China, India and Russia have recognized *T.officinale* as a liver tonic. Traditional Chinese Medicine combines *T.officinale* with other herbs to treat hepatitis.

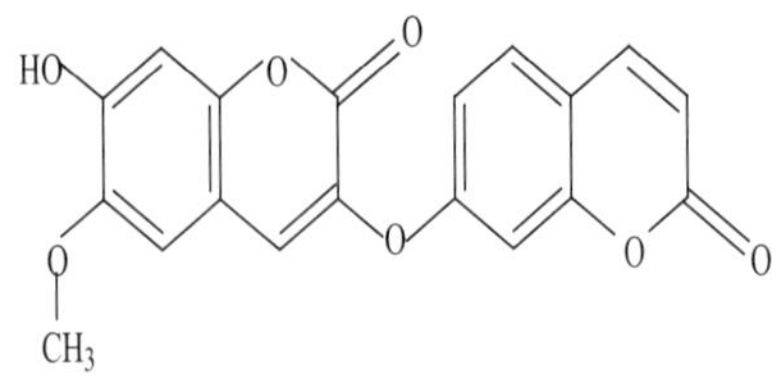

Fig. 15.1 Daphnoretin-hepatoprotective principle.

15.2 Dandelion and traditional medicine

The genus name *Taraxacum* is derived from the Greek word 'taraxos', meaning 'disorder' and 'akos' meaning 'remedy'. As a medicinal plant, *T.officinale* has been considered to be an aperient or mild laxative, diuretic, stimulant, stomachic, tonic, and detoxicant. Tea prepared from *T.officinale* has been used against fever, insomnia, jaundice, rheumatism, eczema and other skin diseases, and constipation. *T.officinale* and other *Taraxacum* species have also been used against warts, cancers, and tumors.

15.3 Phytochemistry

Carotenoids: Lutein (Fig. 15.2) and violaxanthin (Fig. 15.3).

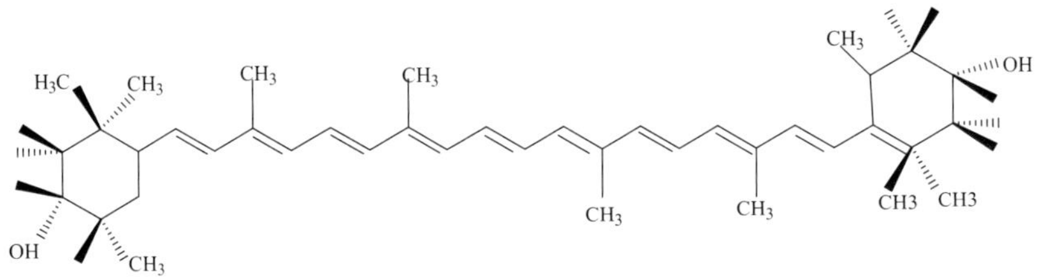

Fig. 15.2 Lutein.

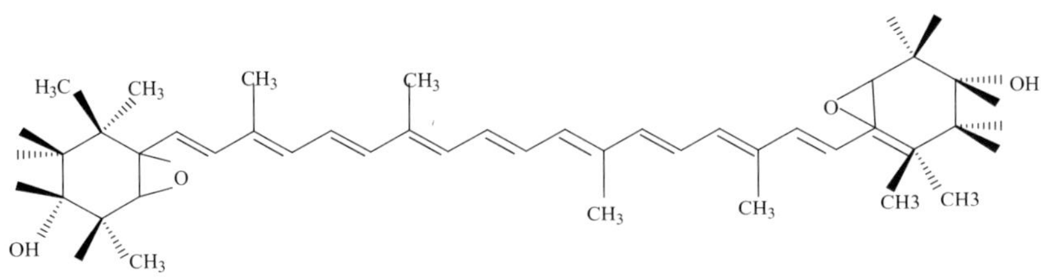

Fig. 15.3 Violaxanthin.

Coumarins: Esculin (Fig. 15.4) and scopoletin (Fig. 15.5).

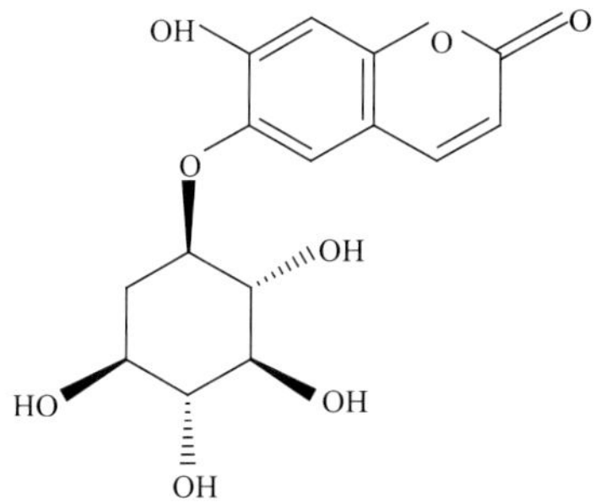

Fig. 15.4 Esculin.

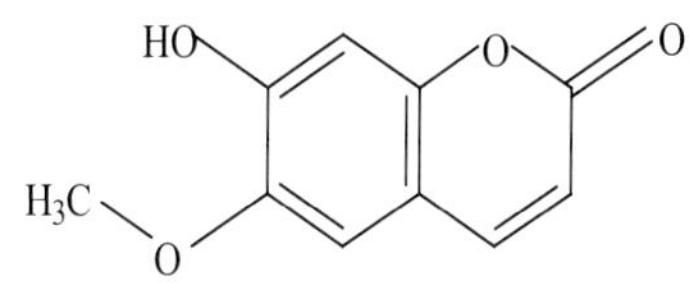

Fig. 15.5 Scopoletin.

Flavonoids: Apigenin-7-glucoside (Fig. 15.6), luteolin-7-glucoside (Fig. 15.7), isorhamnetin 3-glucoside (Fig. 15.8), luteolin-7-diglucoside (Fig. 15.9), quercetin-7-glucoside (Fig. 15.10), quercetin (Fig. 15.11), luteolin (Fig. 15.12), rutin (Fig. 15.13) and chrysoeriol (Fig. 15.14).

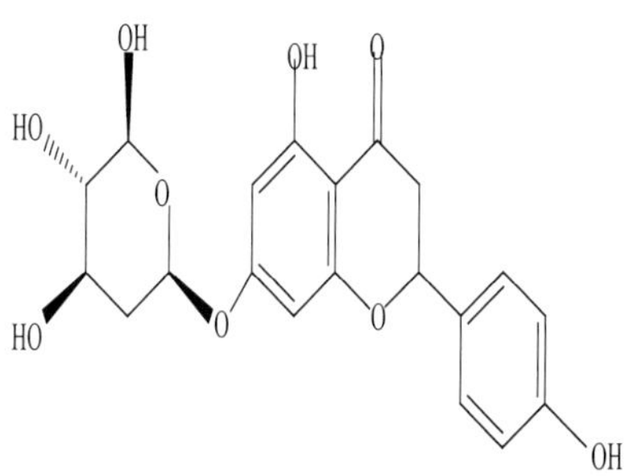

Fig. 15.6 Apigenin-7-glucoside.

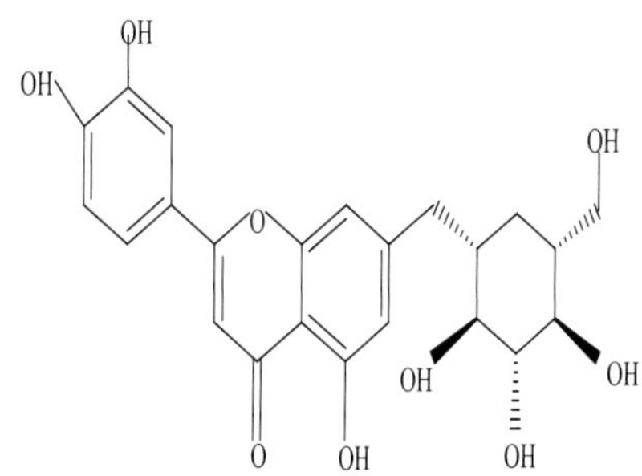

Fig. 15.7 Luteolin-7-glucoside.

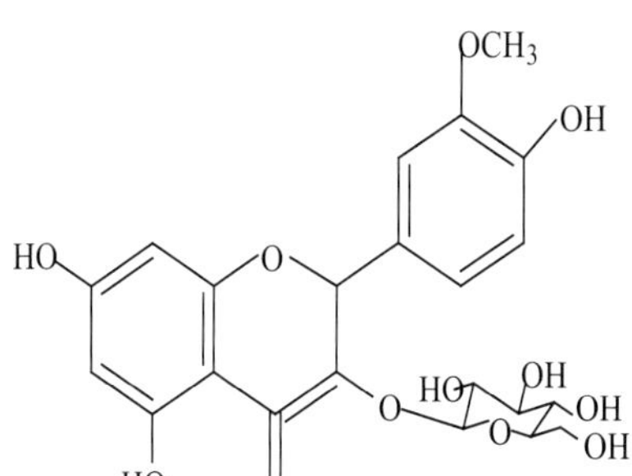

Fig. 15.8 Isorhamnetin 3-glucoside.

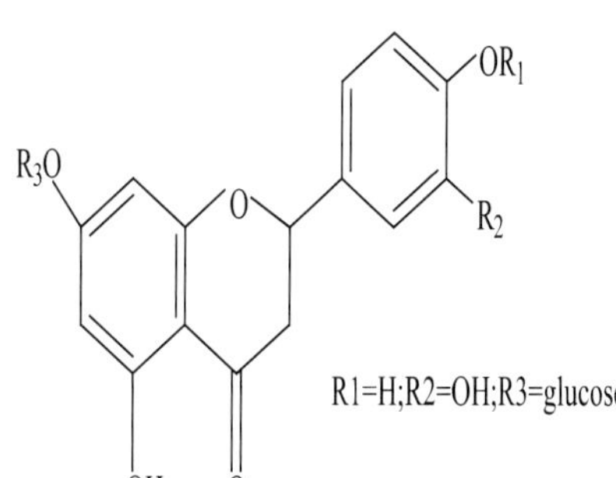

Fig. 15.9 Luteolin-7-diglucoside.

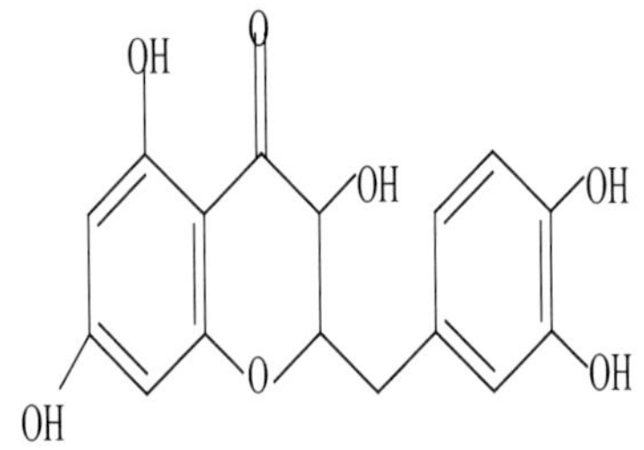

Fig. 15.10 Quercetin-7-glucoside.

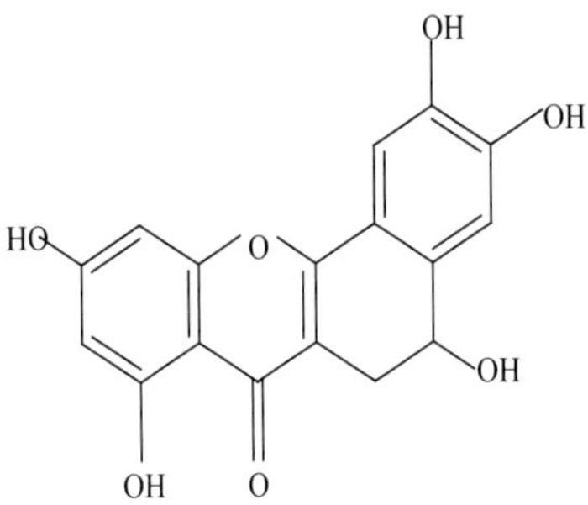

Fig. 15.11 Quercetin.

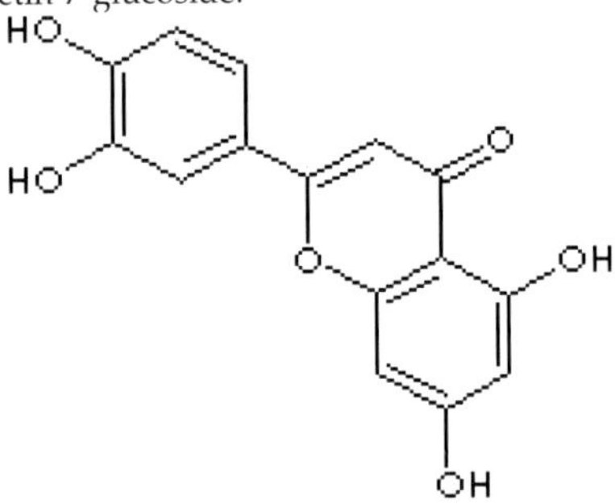

Fig. 15.12 Luteolin.

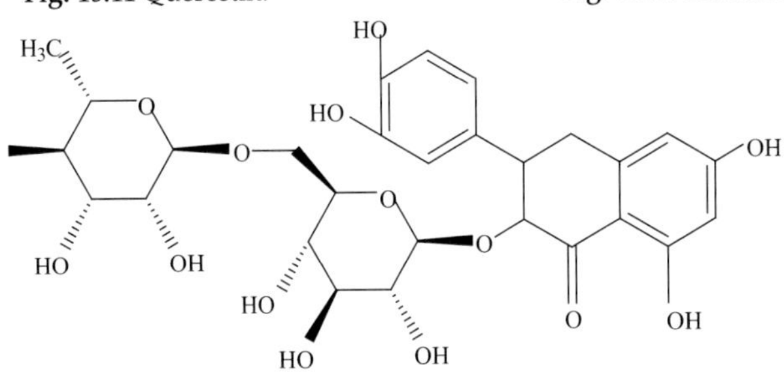

Fig. 15.13 Rutin.

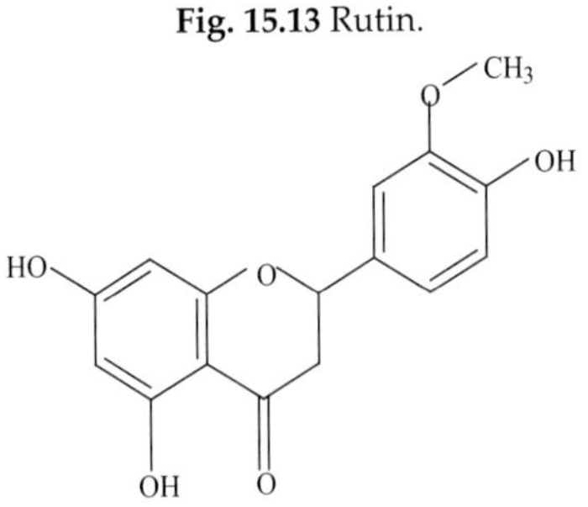

Fig. 15.14 Chrysoeriol.

Phenolic acids: Caffeic, chlorogenic, chicoric (Fig. 5.193), dicaffeoyltartaric acid and ρ-hydroxyphenylacetic acids.

Polysaccharides: Glucans and mannans and inulin.

Sesquiterpene lactones: The plant is a rich source of sesquiterpene lactones or bitter principles. Taraxacin or taraxinic acid or lactucopicrin (Fig. 15.15), lactucin (Fig. 15.16) and cichorin (Fig. 15.17) are chief bitter principles and belong to the guaianolide class. Taraxacin in concentrated solutions, forms precipitates with a number of alkaloidal reagents. The plant contains a crystalline substance, taraxacerine or taraxaceron, which is reported to be bitter resin. Taraxacoside, a type of acylated gamma-butyrolactone glycoside has been reported from the plant. Other sesquiterpene lactones are of the germacranolide type including 11β, 13-dihydrolactucin (Fig. 15.18), ixerin D (Fig. 15.19), ainslioside taraxinic acid β-glucopyranosyl, taraxinic acid 1 '-glucosyl ester, and 11, 13-dihydrotaraxinic acid l'-glucoside. Eudesmanolides including tetrahydroridentin-B (Fig. 15.20) and taraxacolide-O-β-glucopyranoside are reported. Recently, a cyanogenic glycoside, prunasin has been reported from the extract of the plant. Phenylpropanoid glycosides: dihydroconiferin (Fig. 15.21), syringin (Fig. 15.22) dihydrosyningin (Fig. 15.23) have been reported.

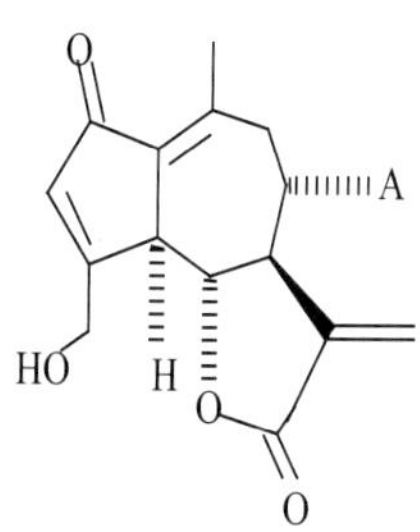

Fig. 15.15 Lactucopicrin.

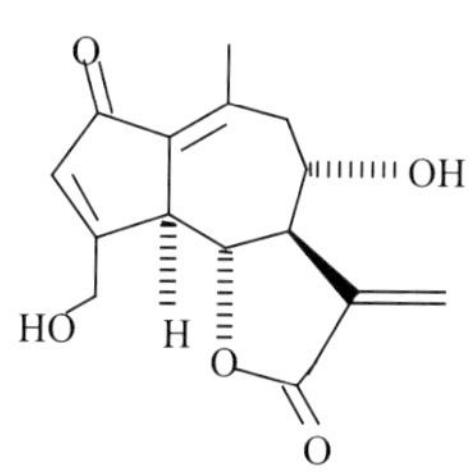

Fig. 15.16 Lactucin.

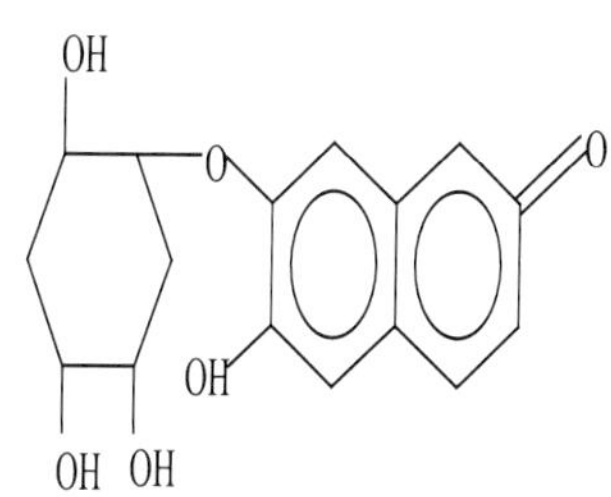

Fig. 15.17 Cichorin.

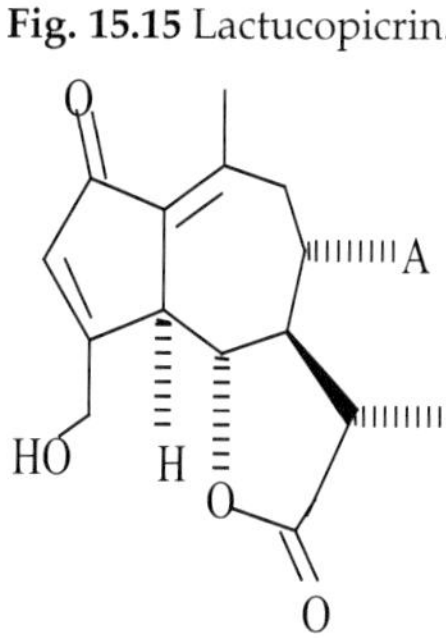

Fig. 15.18 11β,13-dihydrolactucin.

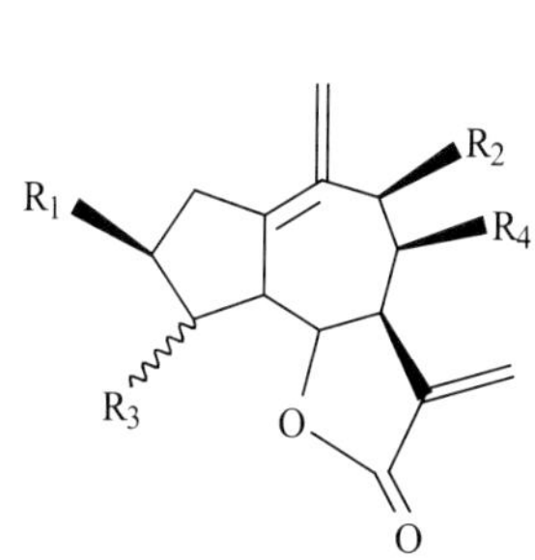

Fig. 15.19 Ixerin D.

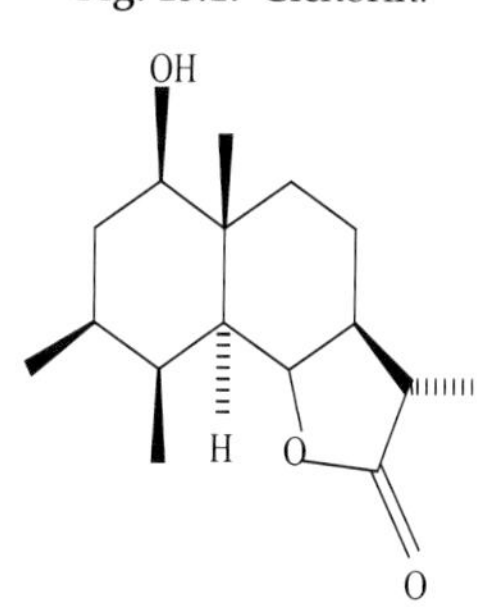

Fig. 15.20 Tetrahydroridentin.

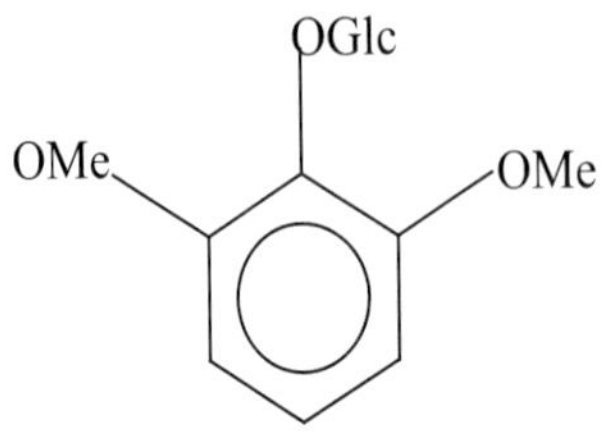

Fig. 15.21 Dihydroconiferin.

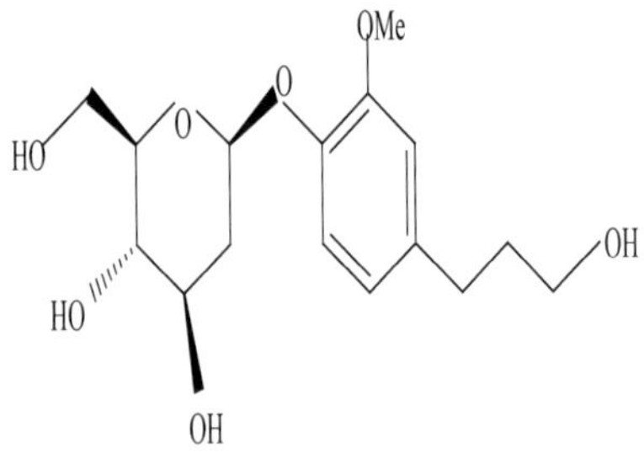

Fig. 15.22 Syringin.

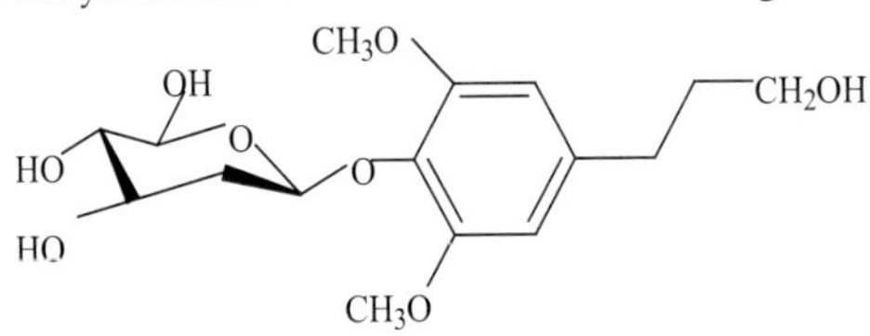

Fig. 15.23 Dihydrosyringin.

***Sterols*:** Taraxasterol (Fig. 15.24), ψ-taraxasterol, (Fig. 15.25) and homo-taraxasterol, β-sitosterol (Fig. 15.26), stigmatsterol (Fig. 15.27), campesterol (Fig. 15.28).

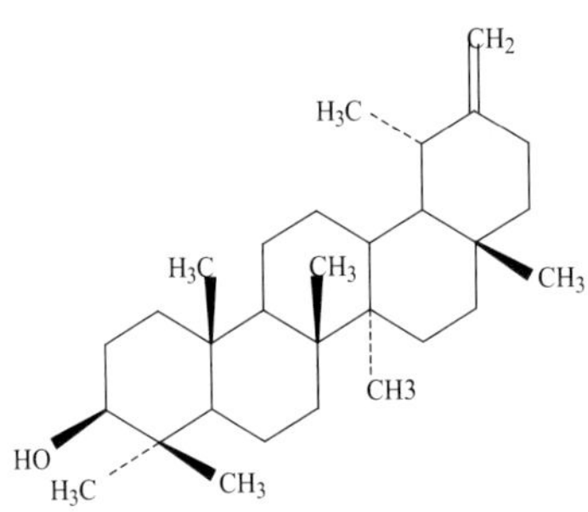

Fig. 15.24 Taraxasterol.

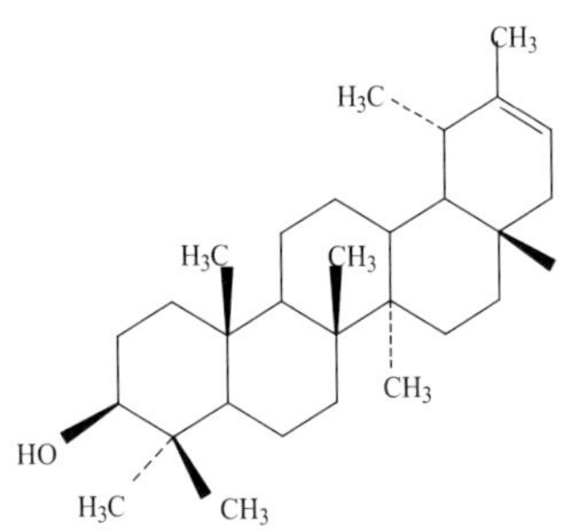

Fig. 15.25 ψ-taraxasterol.

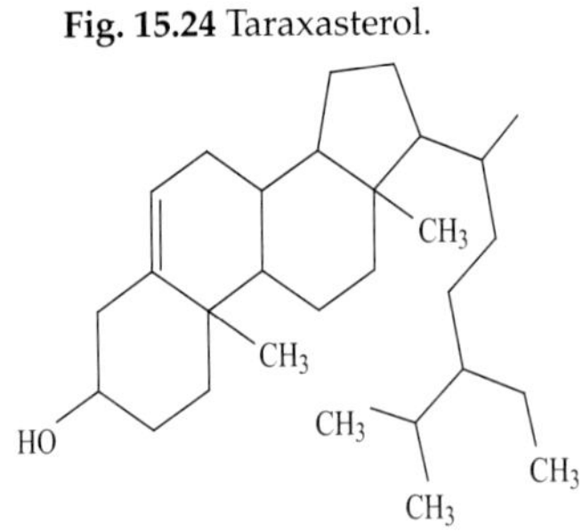

Fig. 15.26 β-sitosterol.

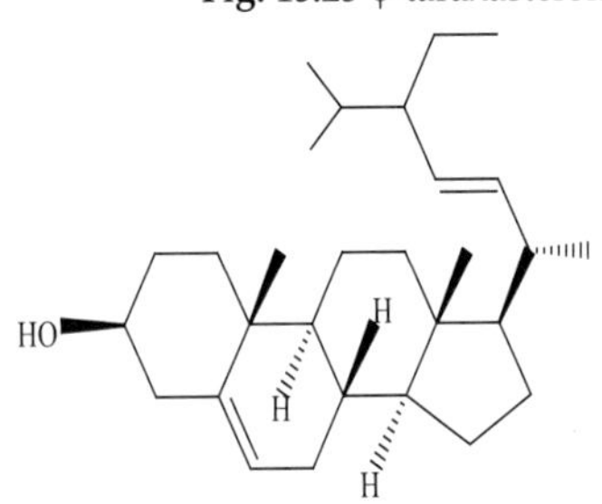

Fig. 15.27 Stigmatsterol.

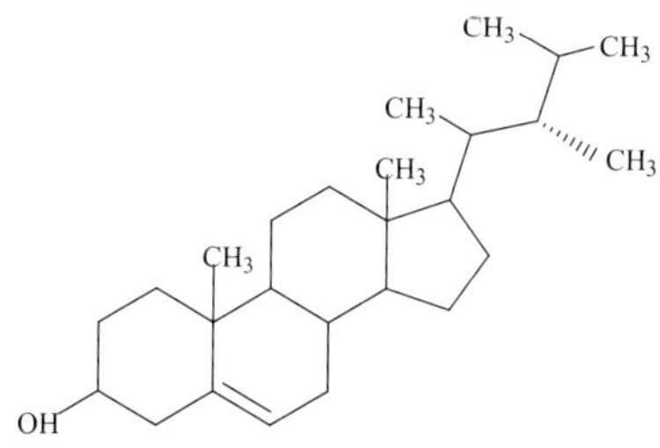

Fig. 15.28 Campesterol.

Triterpenes: α-amyrin (Fig. 15.29), β-amyrin (Fig. 15.30), lupeol (Fig. 15.31), taraxol, taraxaserol, and cycloartenol (Fig. 15.32) are present in the roots. 3β-hydroxylup-18(19)-ene-21-one (Fig. 15.33) has been reported from fresh roots of the plant. Arnidiol (Fig. 15.34) and faradiol (Fig. 15.35) have been reported.

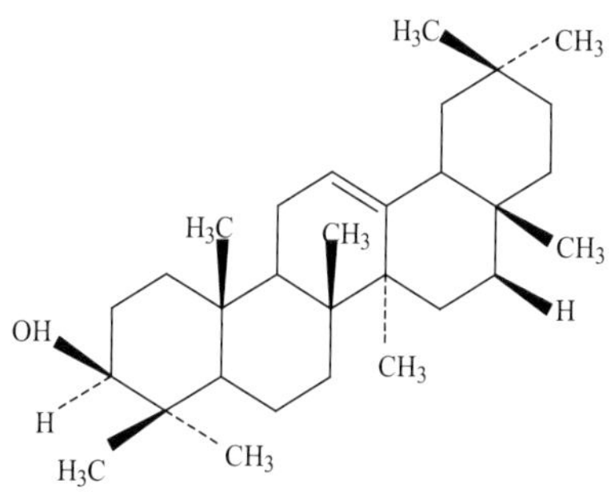

Fig. 15.29 α-amyrin.

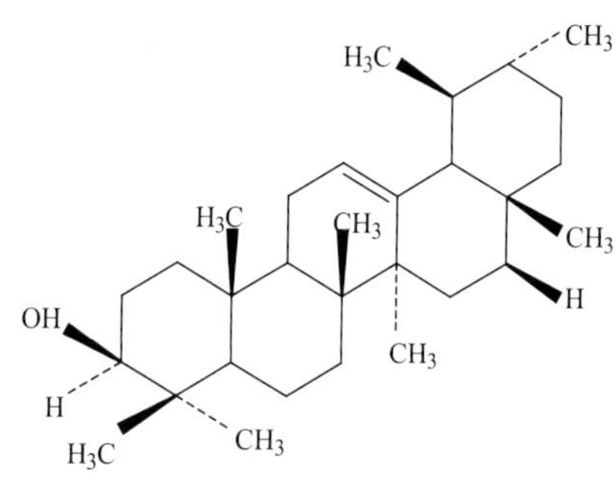

Fig. 15.30 β-amyrin.

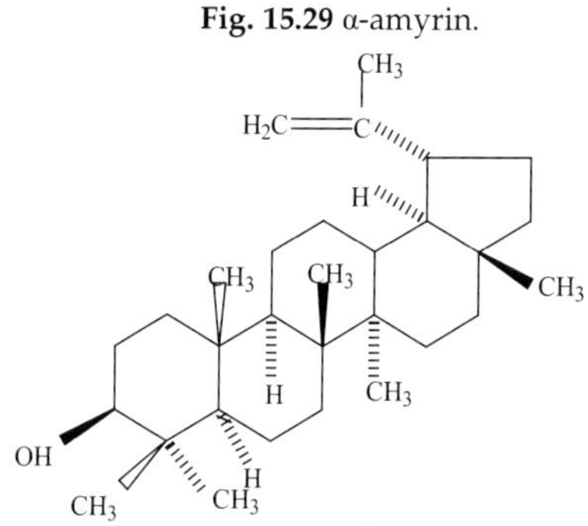

Fig. 15.31 Lupeol.

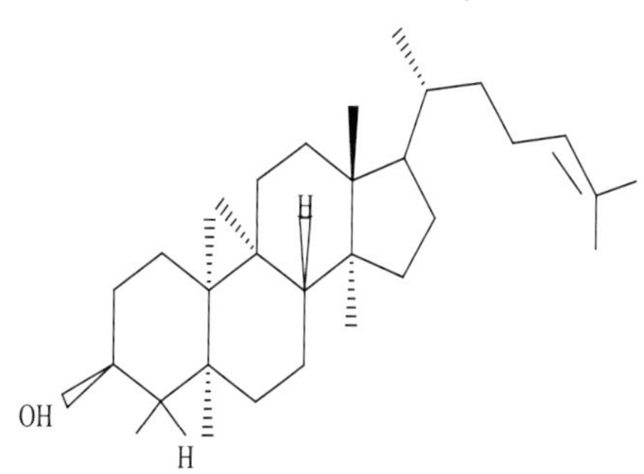

Fig. 15.32 Cycloartenol.

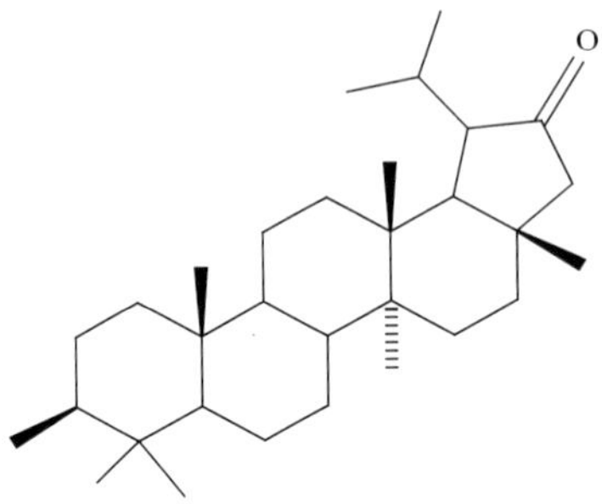

Fig. 15.33 3β-hydroxylup-18(19)-ene-21-one.

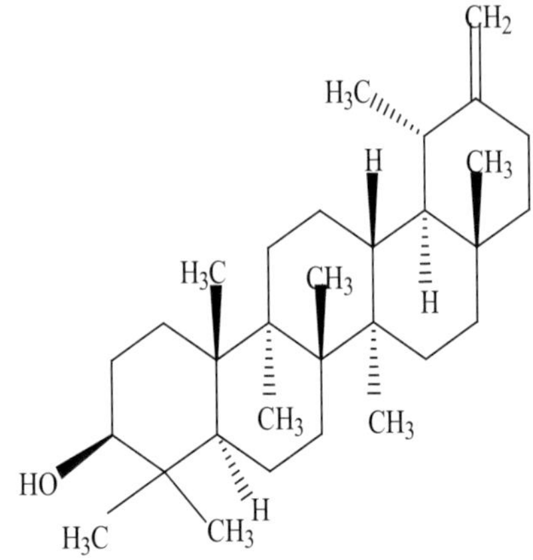

Fig. 15.34 Arnidiol.

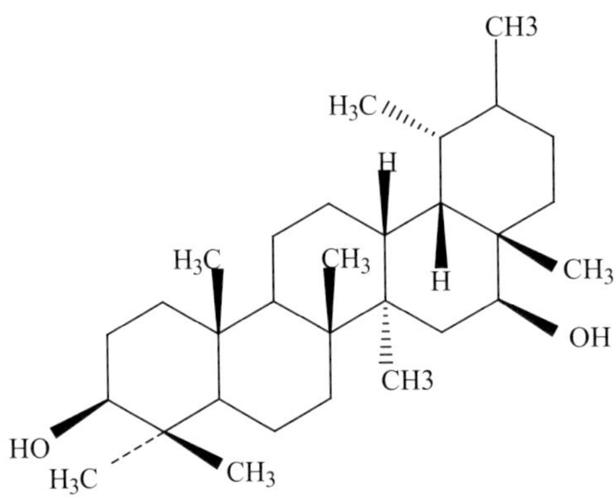

Fig. 15.35 Faradiol.

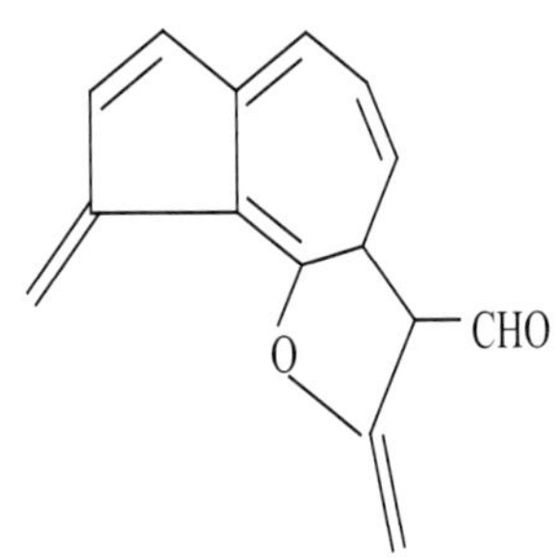

Fig. 15.36 Lettucenin A.

Other: Lettucenin A (Fig. 15.36), a serine proteinase: taraxalisin, amino acids, choline, mucilage, and pectin.

15.4 Pharmacology

The bitter compounds in the leaves and root help stimulate digestion and are mild laxatives. These bitter principles also increase bile production in the gallbladder and bile flow from the liver. Oral administration of extracts from the roots of *T. officinale* has been shown to act as a cholagogue, increasing the flow of bile. In German studies, dandelion leaf extracts increased bile secretion by 40% in rats. In French studies, giving dogs a decoction of fresh dandelion root doubled their bile output. In two Chinese studies of animals with gastric ulcers, gastric metaplasia and hyperplasia, dandelion-containing herbal combinations led to significant histologic improvement.

This chapter reviews the effect of *T. officinale* on the hepatobiliary apparatus and gastrointestinal system. The extracts of *T. officinale* have demonstrated antitumor, hypoglycemic, diuretic, antibacterial and nitric oxide regeneration activity. We need to focus on other pharmacological investigations particularly antioxidant activity and anti-inflammatory activities, which might explain the use of *T.officinale* in liver diseases.

In vivo

Anti-inflammatory activity

Extracts of *T. officinale* have an inhibitory effect on tumor necrosis factor-alpha production from rat astrocytes. Sesquiterpene glucosides isolated from fractionation of the extract of *T. officinale* have anti-leukotriene activity. Luteolin and luteolin-7-O-glucoside from the flower of *T. officinale* has a suppressive effect on iNOS and COX-2 in RAW264.7 cells. *T. officinale* has been reported to have a protective effect against cholecystokinin-induced acute pancreatitis in rats.

Antioxidant

Hydro-alcoholic acid of *T. officinale* roots demonstrated antioxidant activity in rats. Extract of *T. officinale*, in the dose of 100 mg/kg, p.o., improved the superoxide dimutase, catalase, glutathione, and peroxidase levels decreased by $CCl4$ treatment.

Cholretic

Bile secretion was doubled in dogs by a decoction of fresh roots (equivalent to 5 g dried plant); a similar activity has been observed for rats.

In vitro

Antioxidant

Water and ethyl acetate fractions of *T.officinale* flower extract showed antioxidant activities in a stable 2, 2-diphenyl-1-picrylhydrazyl radical model and reduced the breakage of super coiled DNA strand induced by both non-site-specific and site-specific hydroxyl radical. Oxidation of structured phosphatidylcholine liposome induced by peroxyl radical was reduced in the presence of both water and ethyl acetate fractions of the extract. Luteolin and luteolin 7-glucoside were identified as potential antioxidant agents.

Flower extract of *T. officinale* resulted in oxidation of linoleic acid emulsion and suppressed superoxide and hydroxyl radical. The antioxidant activity of flower extract of *T. officinale* against Diphenylpicryl-hydrazyl and a synergistic effect with α-tocopherol were attributed to the reducing activity of flavonoids and coumaric acid derivatives present in the extract. The extract further inhibited peroxyl-radical-induced intracellular oxidation of RAW264.7 cells with range of concentrations.

15.5 Clinical studies

Chronic colitis

A study reported the efficacy of polyherbal formulation containing *T.officinale* in chronic pain associated with colitis. As multiple herbs were used, and this study was not well-designed or reported, the effects of dandelion are not clear.

Hepatitis B

One human study, reported improved liver function in people with hepatitis B after taking a combination herbal preparation containing *T.officinale, Artemisia capillaries* Thunb (Asteraceae), *Taraxacum mongolicum* Hand-Mazz. (Asteraceae), *Plantago ovata* Forssk. (Plantaginaceae), *Cephalanoplos segetum* (Bunge) Kitam (Asteraceae), *Hedyotis diffusa* Willd. (Rubiaceae), *Chrysanthemum indicum* Linn. (Asteraceae), *Smilax glabra* Roxb. (Smilaceae), *Astragalus membranaceus* (Fisch.ex Link.) Bunge (Fabaceae), *Salvia miltiorrhizae* Bunge. (Lamiaceae), *Polygonum orientalis* L. (Polygalaceae), *Paeonia alba* L. (Ranunculaceae), *and Polygonatum sibiricum* F. Daelaroche. (Convallariaceae). Since polyherbal preparation was used in the study, the possible benefits of *T.officinale* need to be explored.

Studies have indicated that chicoric acid inhibits the penetration of viruses in cells. Chicoric acid also acts as an antioxidant by preventing the oxidation of collagen and cells. Chlorogenic acid is cholagogue; its regular ingestion helps the flow of bile and thus reduces the adverse effects of bile stagnation. While chlorogenic acids are not the only compounds that serve well as cholagogues, the evidence for their effectiveness is by far the strongest. Both chicoric and chlorogenic have an action on the heaptobiliary apparatus and therefore the role in extracts of *T. officinale* needs further exploration.

Further reading

Akashi T, Furuno T, Takahashi T, and Ayabe S-I. 1994. Biosynthesis of triterpenoids in cultured cells, and regenerated and wild plant organs of *Taraxacum officinale*. Phytochemistry. 36, 303–308.

Akhtar MS, Khan QM, and Khaliq T. 1985. Effects of *Portulaca oleracae* (Kulfa) and *Taraxacum officinale* (Dhudhal) in normoglycaemic and alloxan-treated hyperglycaemic rabbits. J. Pak. Med. Assoc. 35, 207–210.

Anonyms: ESCOP Monographs. 2003. The Scientific Foundation for Herbal Medicinal Products. 2nd Ed. ESCOP & Thieme, Exeter, UK and Stuttgart, Germany. pp. 499–504.

Baba K, Abe S, and Mizuno D. 1981. Antitumor activity of hot water extract of dandelion, *T. ofiicinale*-correlation between antitumor activity and timing of administration. Yakugaku. Zasshi. 101. 538–543.

Barnes J, and Anderson LA. 2003. Dandelion. Herbal Medicines. 2nd Ed. Pharmaceutical Press, London, UK. pp. 171–173.

Böhm K. 1959. Choleretic action of some medicinal plants. Arzneimittelforsch. 9, 376–8

Budzianowski J. 1997. Coumarins, caffeoyltartaric acids and their artifactual methyl esters from *Taraxacum officinale* leaves. Plant Med. 63, 288.

Chabrol E, Charonnat R, Maximin M, Waitz R, and Porin J. 1931. L'action choleretique des Composees. C R Soc. Biol. 108, 1100–1102.

Chakurski I, Matev M, and Koichev A. 1981. Treatment of chronic colitis with an herbal combination of *Taraxacum officinale, Hypericum perforatum, Melissa officinalis, Calendula officinalis* and *Foeniculum vulgare*. Vutreshni. bolesti. 20, 51–54.

Chen Z. 1990. Clinical study of 96 cases with chronic hepatitis B treated with jiedu yanggan gao by a double-blind method. Zhong. Xi.Yi. Jie. He. Za. Zhi. 10, 71–74.

Cho S-Y, Park J-Y, Park E-M, Choi M-S, Lee M-K, and Jeon S-M. 2002. Alternation of hepatic antioxidant enzyme activities and lipid profile in streptozotocin-induced diabetic rats by supplementation of dandelion water extract. Clin. Chim. Acta. 317, 109–117.

Cordatos E. 1992. *Taraxacum officinale*. In: M Murray, and J Pizzorno (eds) A Textbook of Natural Medicine. Bastyr University Press, Seattle.

Cupp MJ. 1999. Herbal remedies: adverse effects and drug interactions. Am. Fam. Phys. 59, 1239–1244.

Eisenberg DM, Kessler RC, Foster C, Norlock FE, Calkins DR, and Delbanco TL. 1993. Unconventional medicine in the United States – Prevalence, costs and patterns of use. New. Elect. J. Medi. 32, 246–252.

Elferink RO, Yin ZH. 2004. Huang and other plant-derived preparations: where herbal and molecular medicines meet. J. Hepatol. 41, 691–693.

Fang J. 1991. Effect of fu-zheng qu-xie on gastric disease infected with Campylobacter pyloridis. Chin. J. Mod. Dev. Trad. Med. 11, 150–152.

Hansel R, Kartarahardja M, Huang J, and Bohlmann F. 1980. Sesquiterpene lactone-beta-D-glucopyranosidee sowie ein neues Eudesmanolid aus *Taraxacum officinale*. Phytochemistry. 19, 857–61.

Hausen BM. 1982. (Taraxinic acid 1'-O-beta-D-glucopyranoside, the contact sensitizer of dandelion (*Taraxacum officinale* Wiggers). Derm. Beruf. Umwelt. 30, 51–53.

Hook I, McGee A, and Henman M. 1993. Evaluation of dandelion for diuretic activity and variation in potassium content. Int. J. Pharmacog. 31, 29–34.

Hu C, and Kitts DD. 2003. Antioxidant, prooxidant, and cytotoxic activities of solvent-fractionated dandelion (*Taraxacum officinale*) flower extracts *in vitro*. J.Agr. Food Chem. 51, 301–310.

Hu C, and Kitts DD. 2005. Dandelion (*Taraxacum officinale*) flower extract suppresses both reactive oxygen species and nitric oxide and prevents lipid oxidation in vitro. Phytomedicine. 12, 588–97.

Ingber A. 2000. Seasonal allergic contact dermatitis from *Taraxacum officinale* (dandelion) in an Israeli florist. Cont. Derma. 43, 49.

Jovanovic M, Poljacki M, Mimica-Dukic N, Boz P, Vujanovic L, and Duran V. 2004. Sesquiterpene lactone mix patch testing supplemented with dandelion extract in patients with allergic contact dermatitis, atopic dermatitis and non-allergic chronic inflammatory skin diseases. Cont. Derma. 51, 101–10.

Kashiwada Y, Takanaka K, Tsukada H, Miwa Y, Taga T, and Tanaka S. 2001. Sesquiterpene glucosides from anti-leukotriene B4 release fraction of *Taraxacum officinale*. J. Asian. Nat. Prod. Res. 3, 191–7.

Kim HM, Lee EH, Shin TY, Lee KN, and Lee JS. 1998. *Taraxacum officinale* restores inhibition of nitric oxide production by cadmium in mouse peritoneal macrophages. *Immunopharmacol*. Immunotoxicol. 20, 283–97.

Kim HM, Oh C-H, and Chung C-K. 1999. Activation of inducible nitric oxide synthase by *Taraxacum Officinale* in mouse peritoneal macrophages. Gen. Pharmcol. 32, 683–8.

Kim HM, Shin HY, Lim KH, Ryu ST, Shin TY, and Chae HJ. 2000. *Taraxacum officinale* inhibits tumor necrosis factor-alpha production from rat astrocytes. Immunopharmacol. Immunotoxicol. 22, 19–30.

Kisiel W, and Barszcz B. 2000. Further sesquiterpenoids and phenolics from *Taraxacum officinale*. Fitoterapia. 71, 269–73.

Kisiel W, Barszcz B, and Szneler E. 2000. Pol. J. Chem. 74, 281.

Kitts DD, and Hu C. 2004. Luteolin and luteolin-7-O-glucoside from dandelion flower suppress iNOS and COX-2 in RAW264.7 cells. Mol. Cell. Biochem. 265, 107–13.

Kristo TS, Szoke E, Kery A, Terdy PP, Selmeczi B, and Simandi B. 2001. Production and characerisation of Taraxacum officinale extracts prepared by supercritical fluid and solvent extractions. ISHS Acta Horticulturae 597. International Conference on Medicinal and Aromatic Plants (Part 2).

Kuusi T, Pyylaso H, and Autio K. 1985. The bitterness properties of dandelion. II. Chemical investigations. Lebensm-Wiss Technol. 18,347–9

Liu X, Han W, and Sun D. 1992. Treatment of intestinal metaplasia and atypical hyperplasia of gastric mucosa with xiao wei yan powder. Chung. Kuo. Chung. Hsi. Chieh. Ho. Tsa. Chih. 12, 602–3.

Malhotra S, Singh AP, and Munjal G. 2001. Hepatotoxic potential of commonly used herbal products. Gastroenterol. Today. 5, 110–111.

Malhotra S, Singh AP, and Munjal G. 2002. Hepatoprotective Potential of Phytomedicines. Gastroenterol. Today. 6, 76–77.

Mizushina Y, Yoshida H, Matsumoto H, and Iida A. 2003. Cyanogenic glycoside from *Taraxacum officinale*. Nat. Med. 57, 36.

Petlevski R, Hadzija M, Slijepcevic M, and Juretic D. 2001. Amino acids in nettle root, juniper berries and dandelion root (Urticae radix, Juniperi fructus et Taraxaci radix). Farmaceutski. Glasnik. 57, 139–147.

Racz-Kotilla E, Racz G, and Solomon A. 1974. The action of *Taraxacum officinale* extracts on the body weight and diuresis of laboratory animals. Plant Med. 26, 212–7.

Ram VJ. 2001. Herbal preparations as a source of hepatoprotective agents. Drug. News. Persp. 14, 353–63.

Rana AC, and Sumanth M. 2006. *In vivo* anti-oxidant activity of hydroalcoholic acid of *Taraxacum officinale* root in rats. Ind. J. .Pharmacol. 38, 54–5.

Rauwald HW, and Huang JT. 1985. Taraxacoside, a type of acylated gamma-butyrolactone glycoside from *Taraxacum officinale*. Phytochemistry. 24, 1557–1559.

Rudenskaya G.1998. Taraxalisin—a serine proteinase from *Taraxacum officinale* Webb. FEBS. Lett. 437, 237–240.

Sang-Wan Seo H-NK, Hyo-Jin An, Kang-Beom Kwon, Byung-Cheal Lim, Eun-A Seo, Do-Gon Ryu, Goo Moon,, Hong-Yeoul Kim H-MK, and Seung-Heon Hong. 2005. *Taraxacum officinale* protects against cholecystokinin-induced acute pancreatitis in rats. World Journal of Gastroenterology. 11, 597–599.

Saraswat B, Visen PK, Patnaik GK, and Dhawan BN. 1997. Protective effect of picroliv, active constituent of *Picrorhiza kurrooa* against oxytetracycline induced hepatic damage. Ind. J. Expér. Biol. 35, 1302–1305.

Schütz K, Kammerer DR, Carle R, and Schieber A. 2004. Characterization of phenolic acids and flavonoids in dandelion (*Taraxacum officinale* WEB. ex WIGG.) root and herb by high-performance liquid chromatography/electrospray ionization mass spectrometry. Rapid Commun. Mass. Spectrosco. 19, 179–186.

Sweeney B, Vora M, Ulbricht C, and Basch E. 2005. Evidence-based systematic review of dandelion (*Taraxacum officinale*) by natural standard research collaboration. J. Herb. Pharmacother. 5, 79–93.

Trojanova I, Rada V, Kokoska L, and Vlkova E. 2004. The bifidogenic effect of *Taraxacum officinale* root. Fitoterapia. 75, 760–763.

Vogel G. 1997. Natural substances with effects on the liver. In: P Wagner, and P Wolff (eds) New Natural Products and Plant Drugs with Pharmacological, Biological or Therapeutic Activity. Springer-Verlag, Heidelberg.

Westermann L, and Roddick JG. 1981. Annual Variation in Sterol levels in Leaves of *Taraxacum officinale* Weber. Plant. Physiol. 68, 872–875.

Williams C, Goldstone F, and Greenham J. 1996. Flavonoids, cinnamic acids and coumarins from different tissues and medicinal preparations of *Taraxacum officinale*. Phytochemistry. 42, 121–127.

Chapter 16

Ethnopharmacology of *Acanthus ilicifolius* Linn.

16.1 Introduction

Acanthus ilicifolius (sea holly) is found in tropical Asia and Africa, from Malaya to Polynesia. It is a viny shrub or tall herb, upto 1.5 m high, scarcely woody, bushy, with very dense growth. Shallow tap roots, but occasionally stilt roots are conspicuous. Leaf simple, opposite, decussate, cauline, exstipulate, petiole short, flattened, glabrous, pulvinous to sheathing base. Flower bisexual, typically zygomorphic, complete, erect, sessile, hypogynous. Fruit 1 cm green and 2.5–2.0 cm long, kidney shaped 4 seed drupe, Seed 0.5–1.0 cm long.

16.2 Phytochemistry

Alkaloid, acanthicifoline has been reported. Two new cyclolignan glycosides, (+)-lyoniresinol 3a-O-b-D-galactopyranosyl-(1→6)- b -D-glucopyranoside and (+)-lyoniresinol 2a-*O*-b-D-galactopyranosyl-3a-*O*-b -D-glucopyranoside have been reported from aerial parts of *A. ilicifolius*. A phenylethanoid glycoside (ilicifolioside A) and an aliphatic alcohol glycoside (ilicifolioside B) have been isolated from the aerial parts.

Two lignan glucosides, (+)-lyoniresino l 3a-[2-(3, 5-dimethoxy-4-hydroxy)-benzoyl]-O-beta-glucopyranoside, and dihydroxymethyl-bis (3, 5-dimethoxy-4-hydroxyphenyl) tetrahydrofuran-9(or 9')-O-beta-glucopyranoside have been isolated from the aerial parts. 11-epoxymegastigmane glucoside and megastigmane glucosides (roseoside) have been reported from *A. ilicifolius* growing in China.

2-benzoxazolinone and blepharin have been reported from plants growing in Vietnam. A new coumaric acid derivative acancifoliuside, acteoside, isoacteoside, acanthaminoside, (+)-lyoniresinol 3a-O-beta-glucopyranoside, (-)-lyoniresinol, and alpha-amyrin, have been isolated from the methanolic extract of the leaves of *A. ilicifolius*.

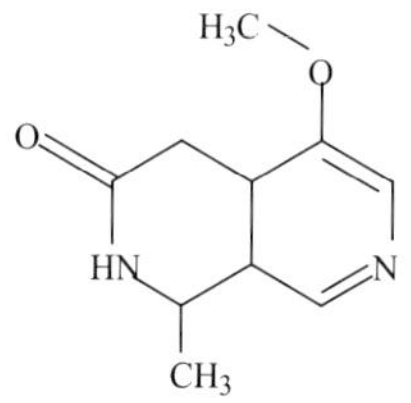

Fig. 16.1 Structure of Acanthicifoline.

16.3 Traditional Medicinal Uses

Malaysia: The leaves of *A. ilicifolius* are used to treat rheumatism, neuralgia and poison arrow wounds. It is widely believed by mangrove dwellers that chewing the leaves protects against snake bite.

Malay: The pounded seeds of *A. ilicifolius* and *A. ebracteatus* are used to treat boils, and the juice of the leaves to prevent alopecia. Both species are also used to treat urolithiasis.

India: In Ayurveda, the plant is known as Sahachara. In India the drug is considered to be astringent and makes a good nervine tonic, expectorant, and stimulant. He says that the root is expectorant, and is used in coughs and asthma. The root, boiled in milk, is largely used in leucorrhoea and general debility. The Siamese and Indo-Chinese consider the roots to be cordial and attenuant, and useful in paralysis and asthma. The tender shoots and leaves are used in India for bites. In Goa, the leaves, which abound in mucilage, are used as an emollient fomentation in rheumatism and neuralgia.

Thailand: Water extracted from the bark is used to treat colds and dermatitis. Ground fresh bark is used as an antiseptic. Tea brewed from the leaves relieves pain and purifies the blood.

16.4 Pharmacology

Anti-inflammatory

The methanolic fraction of *A. ilicifolius* leaf extract produced significant inhibition of rat paw oedema, when administered both prior to and after carrageenan administration, in a manner similar to BW755C a synthetic cyclooxygenase and lipoxygenase inhibitor. The extract decreased protein exudation and leukocyte migration in the peritoneal fluid, thereby indicating its effectiveness towards inhibiting peritoneal inflammation. It also produced significant inhibition of cyclooxygenase (1 and 2) and lipoxygenase activity. Preincubation of the extract inhibited the production of proinflammatory cytokines in lipopolysaccharide stimulated peripheral blood mononuclear cells.

The methanolic fraction of the extract was also found to possess significant free radical scavenging activity. The extract on intraperitoneal administration augmented the endogenous antioxidant status, as evident from the significant increase of ferric reducing ability of plasma and total peroxyl radical trapping activity of plasma.

Anti-osteoporotic activity

The effects of the compounds isolated from *A. ilicifolius* on the function of osteoblastic MC3T3-E1 cells were tested. Acteoside, isoacteoside, and (+)-lyoniresinol 3a-O-beta-glucopyranoside (30 microM) increased the growth and differentiation of osteoblasts significantly ($P < 0.05$), indicating that *A. ilicifolius* leaves may help prevent osteoporosis.

Hepatoprotective

The alcoholic extract of *A. ilicifolius* leaves inhibited the formation of oxygen derived free radicals *in vitro* with IC (50) of 550 microg/ml, 2750 microg/ml, 670 microg/ml and 600 microg/ml (Fe (2+)/ascorbate system), 980 microg/ml (Fe (3+)/ADP/ascorbate system) for superoxide radical production, hydroxyl radical generation, nitric oxide radical formation and lipid peroxide formation, respectively. The oral administration of the extract (250 and 500 mg/kg) significantly reduced CCl4 induced hepatotoxicity in rats, as judged from the serum and tissue activity of marker enzymes; glutamate oxaloacetate transaminase, glutamate pyruvate transaminase and alkaline phosphatase. The results were comparable with those obtained with curcumin {100 mg/kg, p.o.}.

Chemo preventive

A. To investigate the chemo preventive efficacy of *A. ilicifolius* in a transplantable Ehrlich ascites carcinoma (EAC)-bearing murine model, male Swiss albino mice were divided into four groups: Group A was the untreated normal control; Group B was the Ehrlich ascites carcinoma control mice group that received serial, intraperitoneal (ip) inoculations of rapidly proliferating 2 x 10(5) viable Ehrlich ascites carcinoma cells in 0.2 mL of sterile phosfate buffered saline; Group C was the plant extract-treated group that received the aqueous leaf extract of *A. ilicifolius* at a dose of 2.5 mg/kg body weight by single ip injections, once daily for 10, 20 and 30 consecutive days following tumor inoculation (aqueous leaf extract of *A. ilicifolius*); and Group D was the Ehrlich ascites carcinoma + aqueous leaf extract of *A. ilicifolius* treatment group.

The chemo preventive potential of the aqueous leaf extract of *A. ilicifolius* was evaluated in a murine model by studying various biological parameters and genotoxic markers, such as tumor cell count, mean survival of the animals, haematological indices, hepatocellular histology, and immuno-histochemical expression of liver metallothionein protein, sister-chromatid exchanges, and DNA alterations. Treatment of the Ehrlich ascites carcinoma -bearing mice with the aqueous leaf extract of *A. ilicifolius* significantly ($P < 0.001$) reduced viable tumor cell count by 68.34% (228.7 x 10(6) +/–0.53) when compared to Ehrlich ascites carcinoma control mice (72.4 x 10(6) +/–0.49), and restored body and organ weights almost to the normal values. Aqueous leaf extract of *A. ilicifolius* administration also increased ($P < 0.001$) mean survival of the hosts from 35 +/–3.46 d in Ehrlich ascites carcinoma control mice to 83 +/–2.69 d in Ehrlich ascites carcinoma + aqueous leaf extract of *A. ilicifolius* treated mice.

B. Alcoholic extract of *A. ilicifolius* (250, 500 mg/kg b wt) was found to be effective against tumor progression and carcinogen induced skin papilloma formation in mice. The extract was found to be cytotoxic towards lung fibroblast (L-929) cells in 72 h MTT assay and the concentration required for 50% cell death was 18 μg/ml. Oral administration of the extract (500 mg/kg b wt) reduced the tumor volume and administration of the same concentration increased the life span by 75% in ascites tumor harboring animals. The extract also significantly delayed the onset of dimethylbenzanthrazene/Croton oil induced skin papilloma in mice in a dose dependent manner.

References and further reading

Babu BH, Shylesh BS, and Padikkala J. 2001. Antioxidant and hepatoprotective effect of *Acanthus ilicifolius*. Fitoterapia. 72, 272–277.

Babu BH, Shylesh BS, and Padikkala J. 2002. Tumour reducing and anticarcinogenic activity of *Acanthus ilicifolius* in mice. J. Ethnopharmacol. 79, 27–33.

Chakraborty TD, Bhuniya M, Chatterjee M, Rahaman and Singha D. 2007 *Acanthus ilicifolius* plant extract prevents DNA alterations in a transplantable Ehrlich ascites carcinoma-bearing murine model. World. J.Gastroenrerol. 13: 6538–6548.

Huo C, Liang H, Tu G, Zhao Y, and Lin WA. 2008. New 5, 11-epoxymegastigmane glucoside from *Acanthus ilicifolius.* Nat. Prod. Res. 22, 896–900.

Kanchanapoom T, Kamel MS, Kasai R, Yamasaki K, Picheansoonthon C, and Hiraga Y. 2001. Lignan glucosides from *Acanthus ilicifolius*. Phytochemistry. 56, 369–72.

Kanchanapoom T, Kasai R, and Yamasaki K. 2002. Flavonoid Glycosides from *Acanthus ilicifolius* L. Nat. Med. 56, 122.

Mani Senthil Kumar KT, Gorain B, Roy DK, Zothanpuia, Samanta SK, Pal M, Biswas P, Roy A, Adhikari D, Karmakar S and Sen T. 2008. Anti-inflammatory activity of *Acanthus ilicifolius*. J. Ethnopharmacol., 120, 7–12.

Mastaller M. 1997. Mangroves: The Forgotten Forest between Land and Sea. Tropical Press, p. 97.

Nagao T, Otsuka H, Kohda H, Sato T, and Yamasaki K. 2005. Benzoxazinones from *Acanthus ilicifolius*. Biochem. Syst. Ecol. 33, 643–645.

Van Kiem P, Quang TH, Huong TT, Nhung le TH, Cuong NX, Van Minh C, Choi EM and Kim YH.2008. Chemical constituents of *Acanthus ilicifolius* L. and effect on osteoblastic MC3T3E1 cells. Arch. Pharm. Res. 31: 823–829.

Wu J, Zhang S, Xiao Q, Li Q, Huang J, Long L and Huang L. 2003. Phenylethanoid and aliphatic alcohol glycosides from *Acanthus ilicifolius*. Phytochemistry. 63: 491–495.

Xie LS, Liao YK, Huang QF, and Huang MC. 2005. Pharmacognostic studies on mangrove *Acanthus ilicifolius*.. Zhongguo. Zhong. Yao Za. Zhi. 30, 1501–1503.

Zhang W, Li Q, Huang J, Xiao Z, and Long L. 2004. Two New Cyclolignan Glycosides from *Acanthus ilicifolius*. Z. Naturforsch. 59, 341–344.

Chapter 17

Ethnopharmacology of *Alstonia macrophylla* ex A. DC

Introduction

Alstonia is a widespread genus of evergreen trees and shrubs from Apocynaceae (dogbane family). *Alstonia* (devil tree) consisting of about 40–60 species, native to tropical and subtropical Africa, Central America, Southeast Asia and Australia (Sidiyasa, 1998). About 200 alkaloids have been reported from the genus, many of which are pharmacologically active.

Alstonia macrophylla Wall. ex A. DC. is native of Malaysia and frequently planted as an ornamental. The decoctions of *A. macrophylla* leaves and stem bark are widely used by the Onge and Nicobarese of Little Andaman Islands, India, to treat stomachache, skin diseases and urinary infections. Leaves are reported to have anticholeretic and vulnerary effect, and are greased with hot coconut oil for sprains, bruises and dislocated joints as poultice and used as a febrifuge.

Phytochemistry: *A. macrophylla* is a rich source of monomeric and dimeric indole alkaloids. The total alkaloid content is the highest in the stem bark and lowest in leaves. Alstophylline, macralstonine, macralstonidine, macrocasalhine, macrosalhine, macrocapamine (Fig. 17.1) and villalstonine (Fig. 17.2) have been reported from the stem bark.

Stem bark of the Indian plant yields villalstonine (0.8–1.0%) and macralstonine (0.06%). 11-methoxy-geissaschizine (Fig. 17.3), 10-11-dimethoxy-geissaschizine (Fig. 17.4) and 16-hydroxy-Nb—demethylalstophylline (Fig. 17.5) have been reported from plants growing in Sri Lanka.

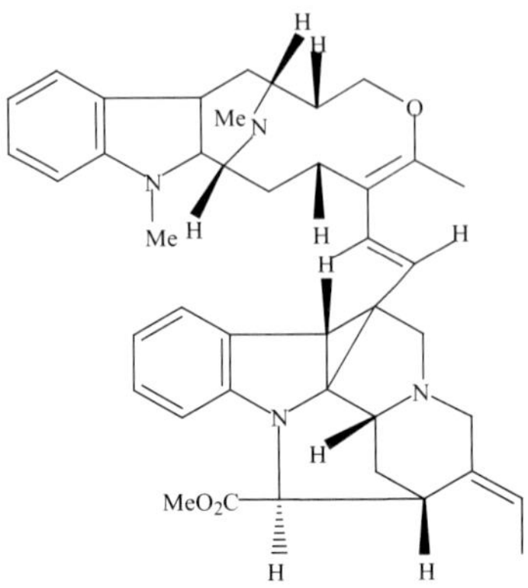

Fig. 17.1 Structure of macrocapamine.

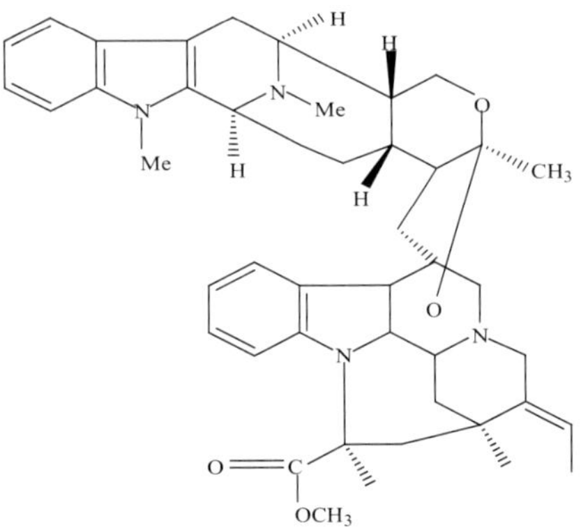

Fig. 17.2 Structure of villalstonine.

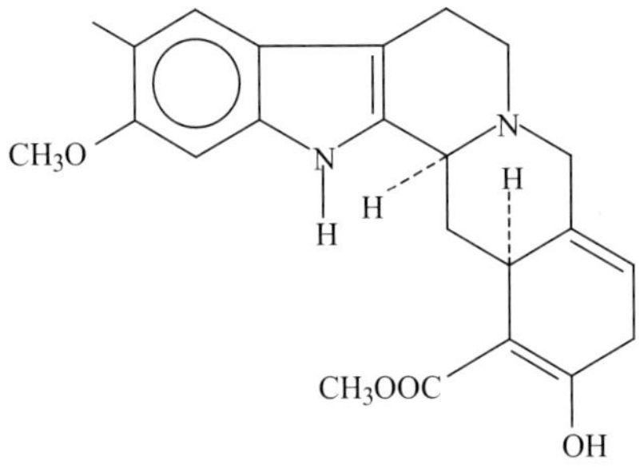

Fig. 17.3 Structure of 11-methoxy-geissaschizine.

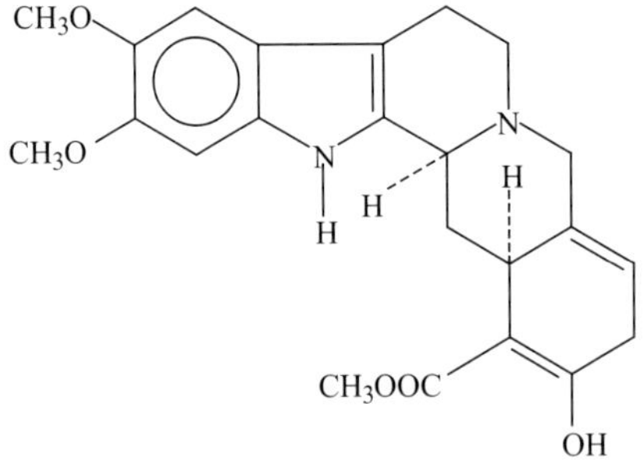

Fig. 17.4 Structure of 10, 11-dimethoxy-geissaschizine.

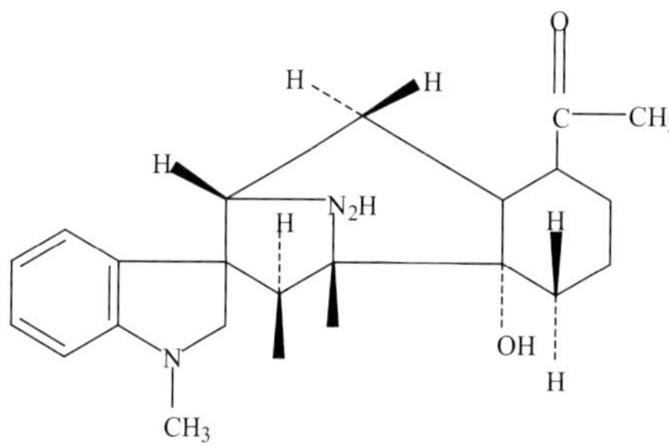

Fig. 17.5 Structure of 16-hydroxy-Nb-demethylalstophylline.

Alstozine N-oxide (Fig. 17.6) has been isolated from the leaves of *A.macrophylla*.

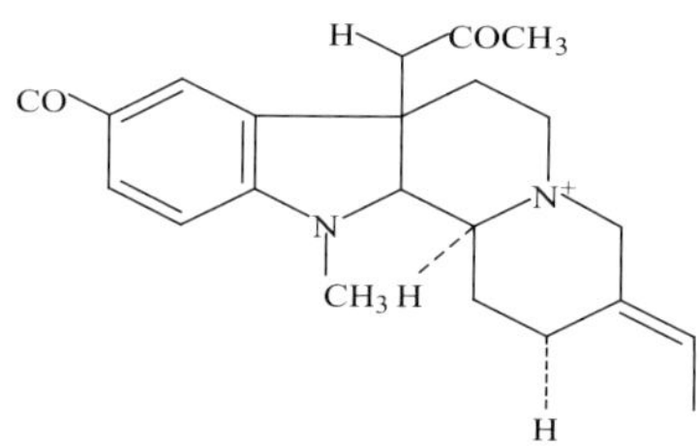

Fig. 17.6 Structure of Alstozine N-oxide.

Oxindole alkaloids, Nb-demethylalstophyllal oxindole and alstonal have been isolated. Activity-directed fractionation led to the isolation of a novel indole alkaloid, *O*-methylmacralstonine, from the most active fraction of *A. macrophylla* along with four known alkaloids, talcarpine, villalstonine, pleiocarpamine (Fig. 17.7), and macralstonine.

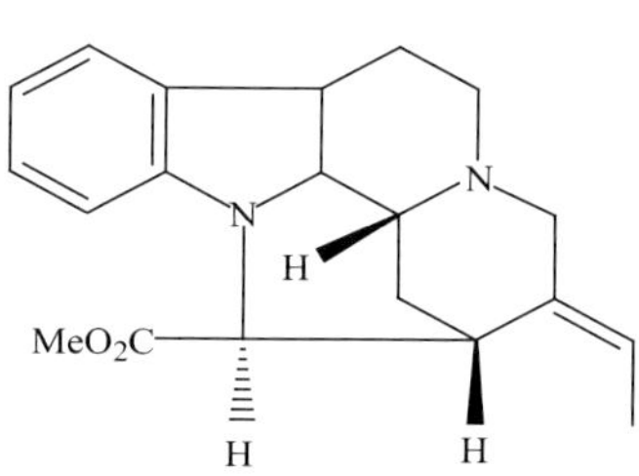

Fig. 17.7 Structure of Pleiocarpamine.

Three new indole alkaloids, viz. 10-methoxyaffinisine, 10-methoxycathafoline and alstonerinal, in addition to alstonerine, alstonisine (Fig. 17.8), alstonal, alstophylline, vincamajine (Fig. 17.9), lochnerine and cathafoline have been isolated from the stem-bark extract.

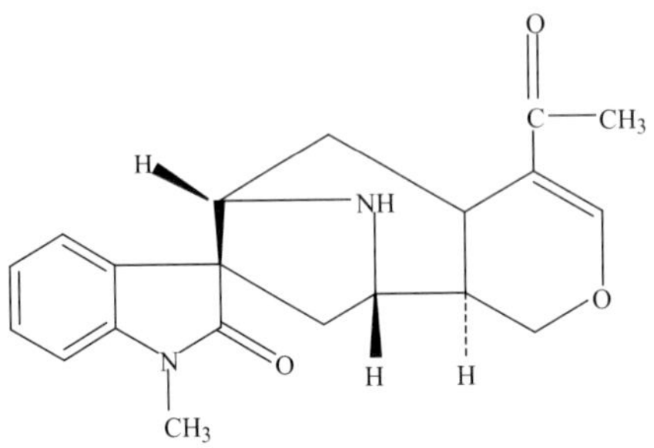

Fig. 17.8 Structure of Alstonisine.

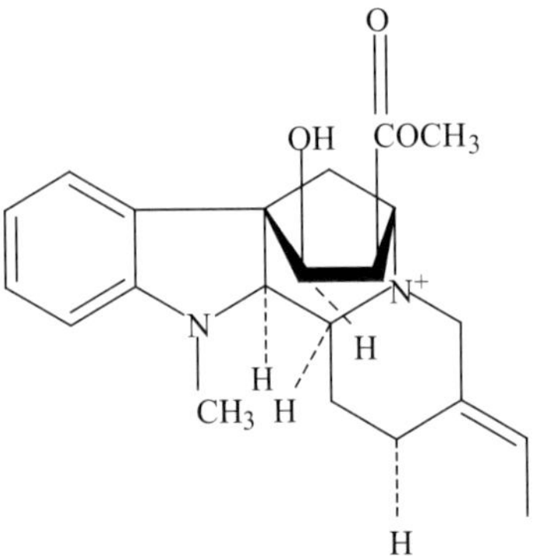

Fig. 17.9 Structure of vincamajine.

Alstoctazine (Fig. 17.10), a novel bisindole alkaloid has been isolated from plants growing in Malaysia.

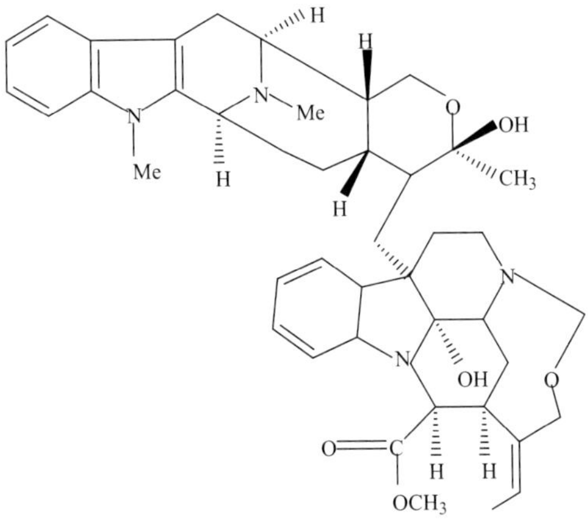

Fig. 17.10 Structure of alstoctazine.

Ten new indole alkaloids, alstomaline, 10,11-dimethoxynareline, alstohentine, alstomicine, 16-hydroxyalstonisine, 16-hydroxyalstona, 16-hydroxy-N(4)-demethylalstophyllal oxindole, alstophyllal, 6-oxoalstophylline, and 6-oxoalstophyllal, in addition 21 other known ones, were obtained from the leaf extract of the Malayan A. macrophylla. Recently alstiphyllanines A–D (Fig. 17.11–17.14) have been reported.

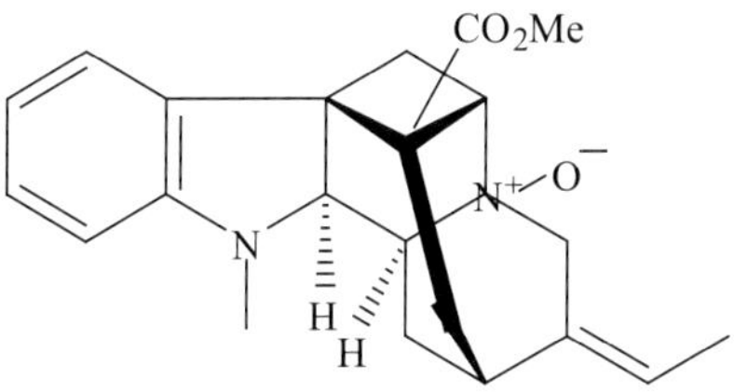

Fig. 17.11 Structure of Alstiphyllanine A.

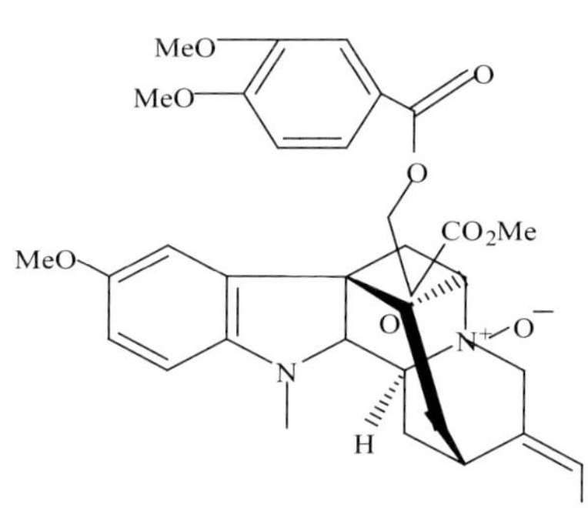

Fig. 17.12 Structure of Alstiphyllanine B.

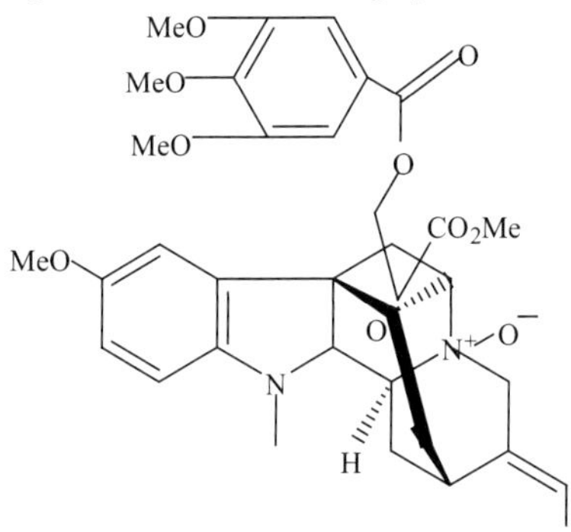

Fig. 17.13 Structure of Alstiphyllanine C.

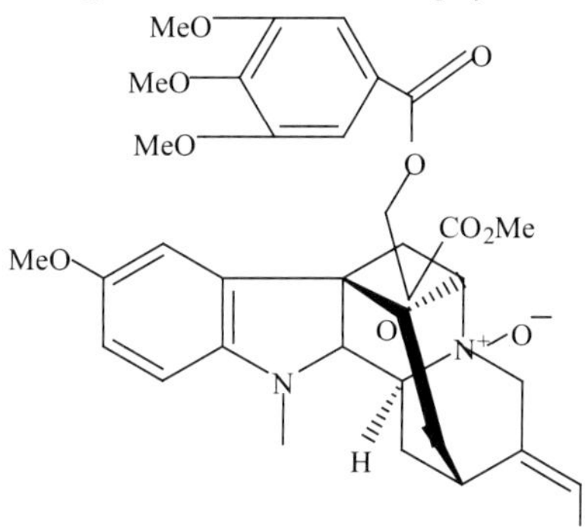

Fig. 17.14 Structure of Alstiphyllanine D.

Other alkaloids reported in literature include alstonidine (Fig. 17.15), alstomacrocine (Fig. 17.16), alstomacroline (Fig. 17.17), alstopicralamine (Fig. 17.18), alstoumerine (Fig. 17.19), macroxine (Fig. 17.20), quebrachidine (Fig. 17.21), 10-Hydroxystrictamine (Fig. 17.22), and Nb-demethylalstophyllinoxindole (Fig. 17.23).

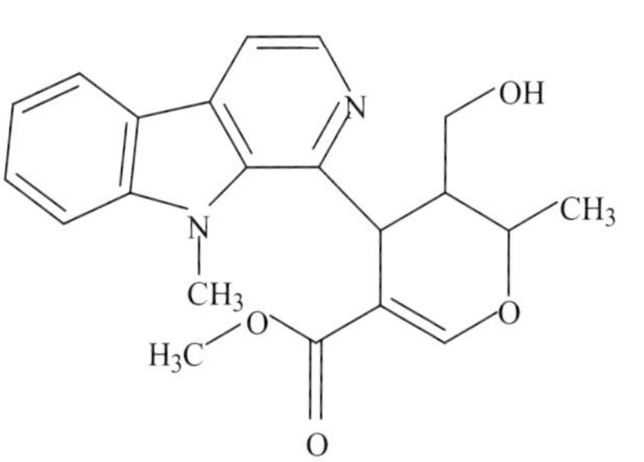

Fig. 17.15 Structure of Alstonidine.

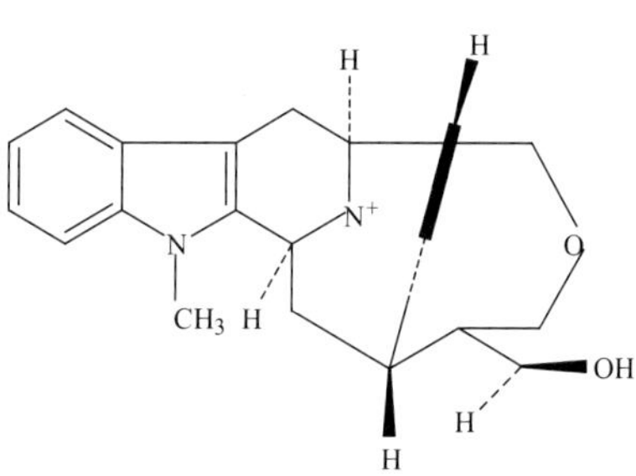

Fig. 17.16 Structure of Alstomacrocine.

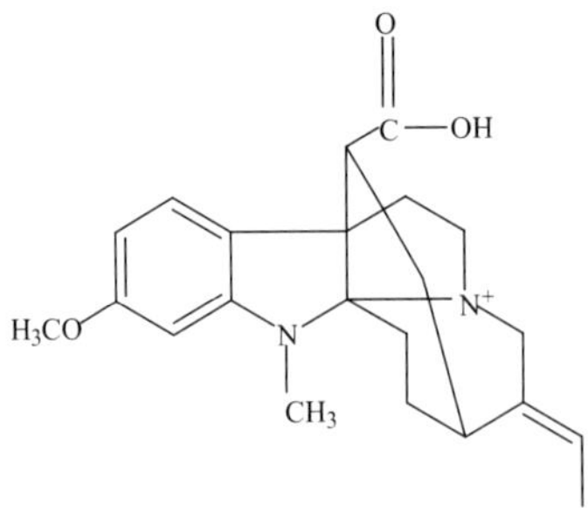

Fig. 17.17 Structure of Alstomacroline.

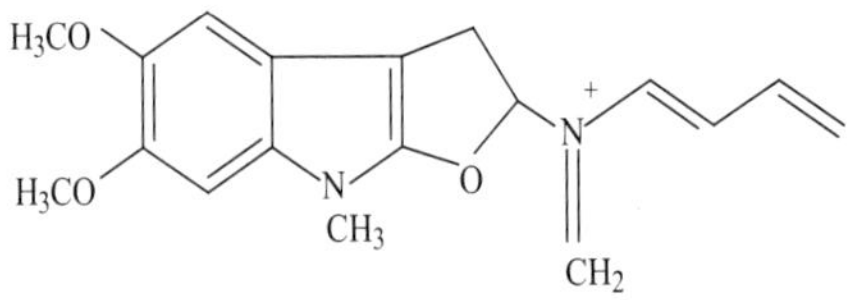

Fig. 17.18 Structure of Alstopicralamine.

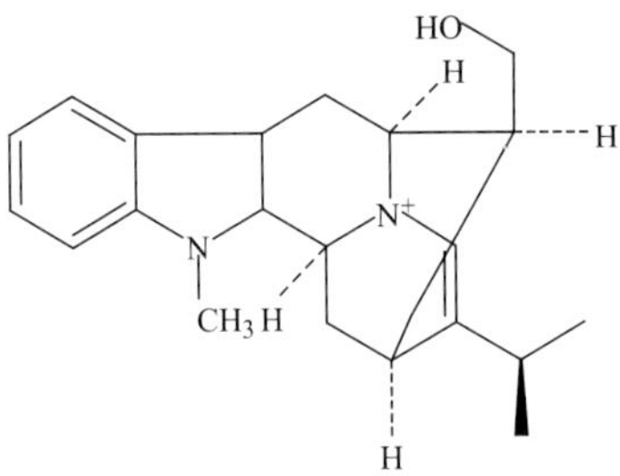

Fig. 17.19 Structure of Alstoumerine.

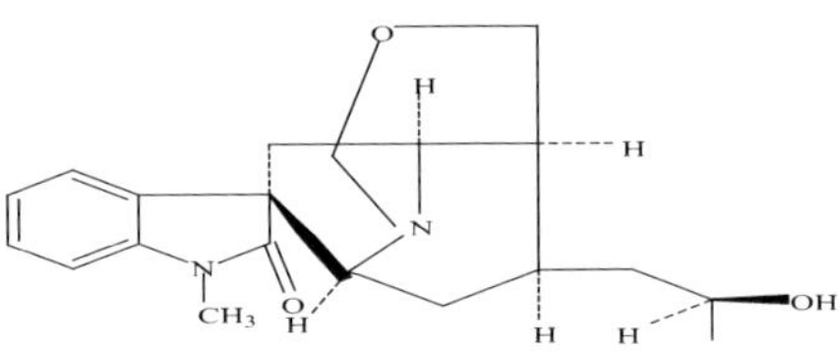

Fig. 17.20 Structure of Macroxine.

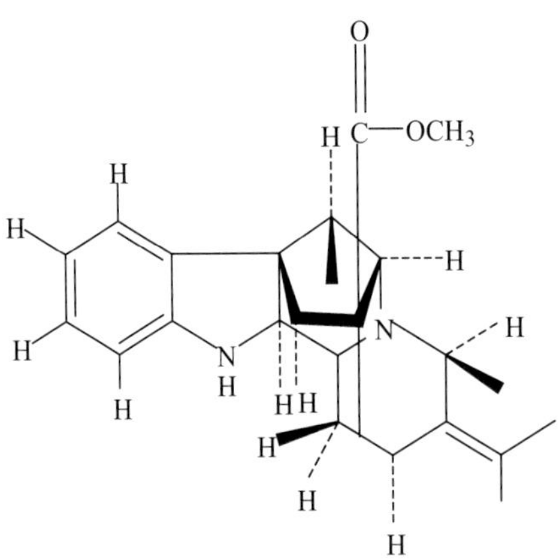

Fig. 17.21 Structure of quebrachidine.

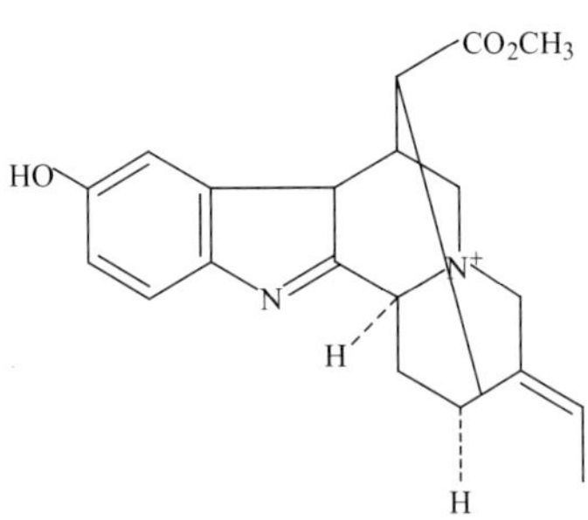

Fig. 17.22 Structure of 10-Hydroxystrictamine.

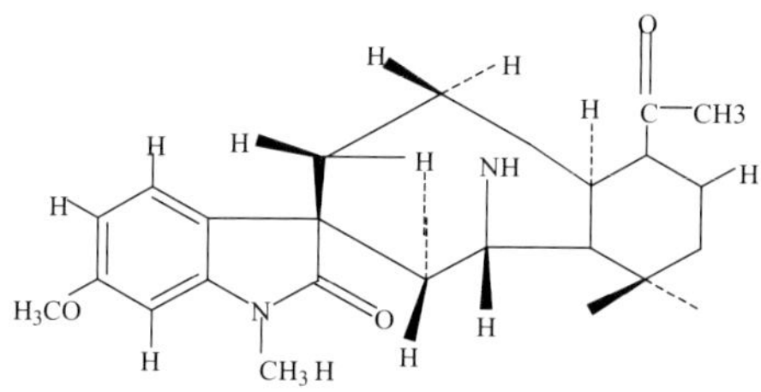

Fig. 17.23 Structure of Nb-demethylalstophyllinoxindole.

Pharmacology

A. macrophylla extracts

Antimicrobial activity: The methanolic crude and methanol-aqueous extract of *A. macrophylla* leaves and n-butanol part of the crude extract showed antimicrobial activity against various strains of *Staphylococcus aureus, Staphylococcus saprophyticus, Streptococcus faecalis, Escherichia coli, Proteus mirabilis, Trichophyton rubrum, Trichophyton mentagrophytes var. mentagrophytes* and *Microsporum gypseum*. The minimum inhibitory concentration (MIC) values ranges from 64 to 1000 microg/ml for bacteria and 32–128 mg/ml for dermatophytes. The stem- bark extract prepared similarly was found to be less active when compared to the leaves.

Anti-inflammatory activity: Methanolic extract of dried leaves of *A. macrophylla* and its fractions were investigated for its anti-inflammatory activity. The extract at a concentration of 200 mg [kg.sup.-1] and 400 mg [kg.sup.-1], p.o. and its fractions at 25 mg [kg.sup.-1] and 50 mg [kg.sup.-1], p.o. showed the significant dose dependent anti-inflammatory activity in carrageenan and dextran-induced rats hind paw edema (acute models) as well as in cotton pellet-induced granuloma (chronic model) in rats.

Antipyretic activity: The methanol extract of *A. macrophylla* and its fractions were tested on normal body temperature and yeast-induced pyrexia in Wistar Albino rats. The leaf extract at oral doses of 200 and 300 mg/kg, and the *n*-butanol fractions of the extract at 50 mg/kg showed significant reduction in normal body temperature and yeast-provoked elevated temperature in a dose-dependent manner comparable to that of standard antipyretic drug paracetamol. The antipyretic effect was started at 1h and extended for at least 5h after the drug administration.

CNS activity: Methanol extract at 100–200 mg/kg p.o. and major non-polar fraction B at 50 mg/kg of *A. macrophylla* leaves caused a significant reduction in spontaneous activity, remarkable decrease in exploratory behavioral pattern, a reduction in muscle relaxant activity and also significantly potentiated phenobarbitone sodium-induced sleeping time.

Alkaloids of A. macrophylla

Antihypertensive effect: Root alkaloidal mixture is reported to be antihypertensive in animals and humans with no side-effects. Total alkaloid mixture is reported to decrease 10–60% blood pressure in dogs.

Intravenous administration of macroxine is reported to decrease in arterial blood pressure and slight bradycardia in pentothal sodium anaesthetized rats. Four doses viz. 50μg/kg, 100μg/kg, 250μg/kg and 1000μg/kg of macroxine were injected. The effects of macroxine on % fall on mean arterial blood pressure were studied. The dose of 100 μg/kg showed 24.03% fall in the mean arterial blood pressure and resulted in 6.6% bradycardia (Table 1.1).

Table 17.1 Effect of macroxine on mean arterial blood pressure.

Dose (μg/kg i.v.)	Mean arterial blood pressure (% fall)
50	17.31
100	24.03
250	16.96
1000	10.0

Alstopicralamine demonstrated 55.36% fall in blood pressure with a dose of 30μg/kg given intravenous in an anaesthetized rat (Table 1.2).

Table 17.2 Effect of alstopicralamine on mean arterial blood pressure.

Dose (μg/kg i.v.)	Mean arterial blood pressure (% fall)
20	42.50
30	55.36

Macralstonine has marked antihypertensive effect. Villastonine, in a dose of 2mg/kg body wt of an anaesthetized cat caused fall in blood pressure which was unaffected by atropine. Villastonine is reported to be neuroleptic; 20 mg/kg i.p. dose caused depletion of 5-OH-tryptamine on rat brain after 15 min.

Cytotoxic activity: Significant cytotoxic activity was exhibited by the extract of *A. macrophylla* on two human lung cancer cell lines, MOR-P (adenocarcinoma) and COR-L23 (large cell carcinoma). Activity-directed fractionation led to the isolation of *O*-methylmacralstonine, talcarpine, villalstonine, pleiocarpamine, and macraistonine. Villalstonine was found to possess pronounced activity on both cell lines with an IC_{50} value less than 5 μM.

Thirteen indole alkaloids isolated from the root bark of A. *macrophylla* and a semisynthetic bisindole O-acetylmacralstonine were assessed for cytotoxic activity against two human lung cancer cell lines, MOR-P (adenocarcinoma) and COR-L23 (large cell carcinoma), using the SRB assay. Pronounced cytotoxic activity was exhibited by the bisindoles on both cell lines.

The potent alkaloids were further tested against a human normal cell line (breast fibroblasts) and other human cancer cell lines including StMl1 la (melanoma), Caki-2 (renal cell carcinoma), MCF7 (breast adenocarcinoma), and LS174T (colon adenocarcinoma). O-acetylmacralstonine, villalstonine and macrocarpamine were found to possess pronounced activity against cancer cell lines with IC_{50} values in the range of 2–10 μM, with no discernible cell-type selectivity. However, O-acetylmacralstonine displayed discernibly less toxicity against the normal breast fibroblasts.

Mutagenic activity: The Ames Salmonella typhimurium microsomal screening system was standardized and pleiocarpamine was screened for mutagenic potential. The results indicated that pleiocarpamine exhibited mutagenic activity in TA98 tester strain.

References and further reading

Arunachalam G, Chattopadhyay D, Chatterjee S, Mandal AB, Sur TK, and Mandal SC. 2002. Evaluation of anti-inflammatory activity of *Alstonia macrophylla* Wall ex A. DC. leaf extract. Phytomedicine. 4, 77–81.

Asolkar LV, Kakkar KK, and Chakre OJ. 1992. Second supplement to Glossary of Indian Medicinal Plants with active principles, Part 1, Publications & Information Directorate, CSIR, New Delhi, India. pp. 51–52.

Bhargava N. 1983. Ethnobotanical studies of the tribes of Andaman and Nicobar Islands, India. I. Onge. Eco. Bot. 37, 110–119.

Chattopadhyay D, Maiti K, Kundu AP, Chakraborty MS, Bhandra R, Mandal SC, and Mandal AB. 2001. Antimicrobial activity of *Alstonia macrophylla*: folklore of bay islands. J. Ethnopharmacol. 77, 49–55.

Chattopadhyay D, Arunachalam G, Ghosh L, and Mandal AB. 2004. CNS activity of *Alstonia macrophylla* leaf extracts: an ethno medicine of Onge of Bay Islands. Fitoterapia. 75, 673–682.

Chattopadhyay D, Arunachalam G, Ghosh L, Rajendran K, Mandal AB, and Bhattacharya SK. 2005. Antipyretic activity of *Alstonia macrophylla* Wall ex A. DC: An ethnomedicine of Andaman Islands. J. Pharm. Pharmaceut. Sci. 8, 558–564.

Chauh, C.H. 1997b. Alstoctazine, a novel bisindole alkaloid from *Alstonia macrophylla*. Mal. J. Chem. 6, 1–3.

Kam TS, Iek IH, and Choo YM. 1999. Alkaloids from the stem-bark of *Alstonia macrophylla*. Phytochemistry. 51, 839–844.

Kam TS, Iek IH, and Choo YM. 2004. New Indole Alkaloids from *Alstonia macrophylla*. J. Nat. Prod. 67, 47–552.

Keawpradub N, Eno-Amooquae E, Burke PJ, and Houghton PJ. 1999. Cytotoxic activity of indole alkaloids from Alstonia macrophylla. Planta. Med. 65, 311–315.

Keawprdub N, Houghton PJ, and Eno-Amooquaye E. 1997. Activity of Extracts and Alkaloids of Thai *Alstonia* Species against Human Lung Cancer Cell Lines. Planta. Med. 63, 97–101.

Kishi T, Hesse M, Gemenden CW, Taylor WI, and Schmid H. 1965. Alstophylline, a new indole alkaloid from Alstonia macrophylla Wall. Helv. Chim. Acta. 48, 1349–62.

Rahman A. 1984. Isolation and structural studies on new natural products of potential biological importance. Pure. App. Chem. 61, 453–456.

Rahman A, and Qureshi MM. 1999. Applications of modern 2D-NMR techniques in structure elucidation of natural products. Pure. App. Chem. 62, 1385–1388.

Rassol SA, and Mushtaq R. 1991. Genetic activity of certain chemicals in Salmonella –microsomal screening system. J. Islamic. Acad. Sci. 44, 301–304.

Sidiyasa K. 1998. Taxonomy, phylogeny, and wood anatomy of Alstonia (Apocynaceae) Suppl. 11, Blumea, New York.

Wong WH, Lim PB, and Chauh CH. 1996. Nb-demethylalstophyllal oxindole and alstonal, Oxindole alkaloids from *Alstonia macrophylla*. Phytochemistry. 41, 313–315.

Chapter 18

Ethnopharmacology of *Alstonia venenata* R. Br

Introduction

Alstonia venenata R. Br, a member of the family Apocynaceae is known for its medicinal value and considered as a powerful antidote for poisonous snake bites.

Phytochemistry

Alstonia venenata is rich source of indole alkaloids including venoxidine, venoterpine, venserpine, echitoserpidine, 16-epivenenatine and 16-epialstovenine.

Ethnopharmacology

Psychopharmacological investigations of the 4-methoxyindole alkaloids have been reported however detailed studies are missing.

Antibacterial activity: The antibacterial effect of the crude leaf and stem bark extracts in solvent systems like hexane, benzene, isopropanol, ethyl acetate, methanol, and water were investigated. The isopropanol and methanol crude leaf extracts and benzene stem bark extract showed significant antibacterial activity against pathogens and Bacillus sps.

Antifungal activity: A plant quaternary alkaloid Δ^3-alstovenine inhibited the spore germination of most of the 17 fungi tested at a concentration of 250–1000 mg/liter. Saprophytic and biotrophic fungi were equally sensitive to the alkaloid. Cercospora sp. was the most sensitive as 100% inhibition of spore germination was observed at 250 mg/liter. Alternaria species, Curvularia species and Fusarium udum were not affected even at 1000 mg/liters.

The indole alkaloid venenatine exhibited antifungal activity against some plant pathogenics and saprophytic fungi. Venenatine in an aqueous acetic acid solution inhibited spore germination of all the 10 tested fungi, Fusarium udum, Alternaria brassicicola, Ustilago cynodontis and Aspergillus flavus showed an especially high sensitivity towards this compound, exhibiting germination levels below 10%. The spore germination and colony development of the parasitic fungus Erysiphe pisi, which causes powdery mildew in pea (*Pisum sativum*), on excised leaves of pea was also significantly affected.

Further reading

Bhattacharya SK, Ray AB, and Dutta SC. 1975. Psychopharmacological investigations of the 4-methoxyindole alkaloids of *Alstonia venenata*. Planta. Med. 27, 164–170.

Chatterjee A, Majumder PL, and Das BC. 1969. Structure of venserpine: a new alkaloid *Alstonia venenata* R. Br. Chem. India. 27, 1388–1389.

Chatterjee A, Roy DJ, and Mukhopadhyay S. 1981. 16-Epivenenatine and 16-epialstovenine, new stereomers from *Alstonia venenata*. Phytochemistry. 20, 981–1985.

Majumder PL, and Dinda BN. 1974. Echitoserpidine: A new alkaloid of the fruits of *Alstonia venenata*. Phytochemistry. 3, 645–648.

Ray AB, and Chatterjee A. 1968. Venoterpine—a new monoterpenoid alkaloid from the fruits of *Alstonia venenata* R. Br. Tetrahed. Lett. 23, 2763–2766.

Singh SK, Sarma BK, Srivastava JS, Singh UP, and Ray AB. 1999, Antifungal activity of Δ^3-alstovenine, a plant alkaloid isolated from *Alstonia venenata*. Folia. Microbiol. 23, 160–163.

Singh UP, Sharma BK, Mishra PK, and Ray AB. 2000. Antifungal activity of venenatine, an indole alkaloid isolated from *Alstonia venenata*. Folia. Microbiol. 45, 173–6.

Williams SJ, and Thankamani V. 2007. Antibacterial Activity of the Crude Leaf and Stem Bark Extracts of *Alstonia venenata* R. Br. *Icfai*. J. Life Sci. 1, 71–77.

Chapter 19

Ethnopharmacology of *Holoptelea integrifolia* Planch.

19.1 Introduction

The plant is found throughout India at the height of 2000 feet.

In the Ayurvedic system of medicine, the plant is reputed as an anti-obesity agent. In addition, it is considered to be carminative, cholagouge, alterative and anti-diuretic. The leaves and stem bark of *H. integrifolia* are used for skin diseases, obesity, and cancer. The leaves decoction is used in the management of diarrhea.

19.2 Phytochemistry

2α, 3α-dihydroxyolean-12-en-28-oic acid has been reported from heartwood of *H. integrifolia*. The seeds yield 37.4% yellow colored oil. The oil from the seeds of *H. integrifolia* is reported to contain the following acids: lauric 0.2%, myristic 3.5%, palmitic 35.1%, stearic 4.5%, arachidic 1.1%, behenic 0.4%, hexadecenoic 1.9% and oleic 53.3%. Recently, 1, 4-Naphthalenedione has been isolated from the plant.

19.3 Pharmacological investigations

1. ***Anti-inflammatory activity***: The objective of the present study was to investigate the anti-inflammatory property of the aqueous extract of the leaves of *H. integrifolia*. The hind-paw edema was produced in rats by the sub-planter injection of carageenan. The aqueous extract of *H. integrifolia* at dose (250 and 500 mg/kg p+.o) was given to observe % inhibition of paw edema which were comparable with indomethacin

(10 mg/kg p.o) used as a reference drug. The extract administered orally at doses of 250 and 500 mg/kg p.o produced a significant ($P < 0.05$) dose dependent inhibition of edema formation.

2. ***Anti-nausea activity:*** In a rat model, anti-emetic activity of ethanolic extract of leaves of *H. integrifolia* on cisplatin-induced nausea was investigated. Rats react to emetic/nausea-producing stimuli, such as cisplatin, with altered feeding habits, manifested by pica or increased consumption of kaolin. The authors measured pica in rats to quantify cisplatin-induced nausea, and to evaluate the anti nausea effect of pretreatment with ethanolic extract of *H. integrifolia,* which was given orally. Cisplatin at 3 mg/kg (i.p) induced significant pica accompanied by reduced food intake, suggesting the presence of nausea. Hence, this cisplatin dose was selected for testing the anti nausea activity of *H. integrifolia*. Cisplatin-induced pica decreased significantly when animals were pretreated with *H. integrifolia* at doses of 250 mg/kg p.o and 500 mg/kg p.o ($P<0.01$). *H. integrifolia* pretreatment decreased cisplatin induced kaolin intake in the rat model of simulated nausea. The findings suggest that *H. integrifolia* and its active constituent(s) may offer a therapeutic role in chemotherapy-induced emesis.

3. ***Antimicrobial, antioxidant and wound-healing activity:*** A. The methanolic extracts of *H. integrifolia* leaves and stem bark were studied for its wound-healing potential. Since wound healing is severely hampered by microbial infection and reactive oxygen species (ROS), this study was undertaken to evaluate antimicrobial and antioxidant activity apart from wound-healing activity. The antimicrobial property of the *H. integrifolia* was studied against six bacterial and five fungal strains using the agar well diffusion method and minimum microbicidal concentration and minimum inhibitory concentration were determined for each strain. Methanolic extract of stem bark showed a larger zone of inhibition (11.3–20.4 mm) than methanolic extract of leaves (9.6–14.9 mm).

The anti-oxidant activity was evaluated by DPPH free radical scavenging activity using HPLC method. The IC_{50} values obtained for methanolic extract of stem bark (TPC: 78.53 ± 1.26 mg/g) and methanolic extract of leaves (TPC: 57.71 ± 1.45 mg/g) were 37.66 ± 0.48 and 50.36 ± 0.59 µg/well, respectively. In excision wound model, more than 90% wound healing was recorded in treated groups by 14 d of post surgery, where as only 62.99% was observed in the control group. In incision model, higher breaking strengths and higher hydroxyproline content in treated groups suggested higher collagen re-deposition than the control group. Further, histopathology studies conformed wound-healing activity of *H. integrifolia*.

B. In another study, the ethanolic crude extract of stem bark of *H. integrifolia* was screened for its antioxidant activity using α-tocopherol as standard antioxidant. The free radical scavenging potential of the extract was evaluated by two different antioxidant methods; ferric thiocyanate and thiobarbituric acid method. The ethanol extract was found to exhibit good antioxidant property.
C. The inhibitory effect of a phytochemical, 1, 4-naphthalenedione, isolated from *H. integrifolia* on β-lactamase is reported here. This compound was found to have a synergistic effect with the antibiotic amoxicillin against a resistant strain of *Staphylococcus aureus*. The enzyme was purified from the organism and incubated with the compound. An assay showed that the compound can inhibit the enzymatic activity of β-lactamase. Modeling and molecular docking studies indicated that the compound can fit into the active site of α-lactamase. Hence, the compound can serve as a potential lead compound for the development of effective β-lactamase inhibitor that can be used against β-lactam-resistant microbial strains.

4. ***Antidiarrheal activity:*** The ethanolic extract of leaves of *H. integrifolia* was studied for its antidiarrheal properties in experimental diarrhea, induced by castor oil and magnesium sulfate in mice. At the doses of 250 and 500 mg/kg per oral, the ethanolic extract showed significant and dose-dependent antidiarrheal activity in both models. The extracts also significantly reduced the intestinal transit in charcoal meal test when compared to atropine sulate (5 mg/kg; i.m.). The results showed that the ethanolic extract of leaves of *H. integrifolia* have a significant antidiarrheal activity and supports its traditional uses in herbal medicine.

Further reading

Badami RC. 2006. The component acids of *Holoptelea integrifolia* (Urticaceae) seed oil. J. Sci. Food Agricul. 13, 297–9.

Bambhole VB, and Jiddewar GG. 1985. Anti-obesity effect of *Iris versicolor* and *Holoptelea integrifolia* in rats. Sachitra. Ayurved. 37, 557–561.

Graham JG, Quinn ML Fabricant DS, and Farnsworth NR. 2000. Plants used against cancer-an extension of the work of Jonathan Hartwell. J. Ethnopharmacol. 73, 347–377.

Hussain A, Vimani OP, Poplis P, Misra LN. Gupta MM and Srivastva GN. 1992. Dictionary of Indian medicinal plants. Lucknow: Central Institute of Medicinal and Aromatic Plants.

Misra G, Bhatnagar SC, and Nigam SK. 1975. 2 Alpha, 3 alpha-dihydroxyolean-12-en-28-oic acid from *Holoptelea integrifolia* heartwood. Planta. Med. 27, 290–7.

Saraswathy A, Devi S Nandini, and Ramasamy D. 2008. Antioxidant, heavy metals and elemental analysis of *Holoptelea integrifolia* Planch. Indian. J. Pharma. Sci. 70, 232–234.

Sharma S, Kale R, Mante A and Biyani K. 2008. Kailash Ethanolic leaf extract of *Holoptelea integrifolia*, Planch. decreases cisplatin-induced pica in rats. Pharmacog. Mag. 4, 293–7.

Sharma S, Lakshmi KS, Patidar A, Chaudhary A and Dhaker S. 2009. Studies on anti-inflammatory effect of aqueous extract of leaves of Holoptelea integrifolia Planch. in rats. Indian. J. Pharmacol. 41, 87–88.

Sharma S, Lakshmi KS, and Rajesh T. 2009. Evaluation of antidiarrhoeal potentials of ethanolic extract of leaves of *Holoptelea integrifolia* in mice model. Int. J. Pharmaceut. Tech. Res. 1, 832–36.

Srinivas Reddy B, Kiran Kumar Reddy R, Naidu VG, Madhusudhana K, Agwane SB, Ramakrishna S and Diwan PV. 2008. Evaluation of antimicrobial, antioxidant and wound-healing potentials of *Holoptelea integrifolia*. J. Ethnopharmacol. 115, 249–256.

Vinod V, Shijina R, Dileep KV and Sadasivan C. 2010. Inhibition of Beta-Lactamase by 1,4-Naphthalenedione from the Plant Holoptelea integrifolia. App. Biochem. Biotech.160, 1752–1759.

Chapter 20

Ethnopharmacology of *Terminalia belerica* (Gaertn.) Roxb

20.1 Distribution and botany

Terminalia belerica (Gaertn.) Roxb. (Family Combertaceae), is popularly known as Belleric myrobalan. It is found growing wild throughout the Indian subcontinent, Sri Lanka and SE Asia, up to 1200 meters in elevation, in a wide variety of ecologies.

It is a large deciduous tree with a buttressed trunk, a thick brownish-gray bark with shallow longitudinal fissures, attaining a height of between 20 and 30 meters. The leaves are crowded around the ends of the branches, alternately arranged, margins entire, elliptic to elliptic-obovate, rounded tip or subacute, midrib prominent, pubescent when young and becoming glabrous with maturity. The flowers are pale greenish yellow with an offensive odor, borne in axillary spikes longer than the petioles but shorter than the leaves. The fruits are ovoid drupes, gray in color, containing a kernel within.

20.2 Traditional medicinal use

In Ayurveda, the drug is classified as an expectorant. It isan integral part of Ayurvedic laxative formulation, Triphala. *T. belerica* is used in the treatment of the common cold, pharyngitis and constipation. Unripe fruit is a mild laxative and ripe fruit is an astringent. *T. belerica* seeds are used as an aphrodisiac. Oil expressed from the seed pulp is used in leucoderma and alopecia. Modern investigations have proved the laxative activity of the oil.

20.3 Phytochemistry: The fruit contains 21.4% tannin, β-sitosterol, belleric acid, gallic acid ellagic acid, methyl gallate, chebulagic acid and carbohydrates. Fruit-pulp yields 37.7%, dark-yellow colored, fixed oil. Chief constituents of the oil are oleic, palmtitic, stearic and linoleic acid. The bark contains 1.4–7% tannins.

Saponin glycosides bellericoside and bellericanin have been reported. Recently lignans, including termilignan (Fig. 20.1), thannilignan (Fig. 20.2), 7-hydroxy-3',4'-(methylenedioxy) flavone and anolignan B (Fig. 20.3), have been isolated from the fruit . Olean-3β,22β-diol-12-en-28β-D-glucopyranoside-oic acid, new triterpenoid glycoside, has been isolated and characterized from the bark.

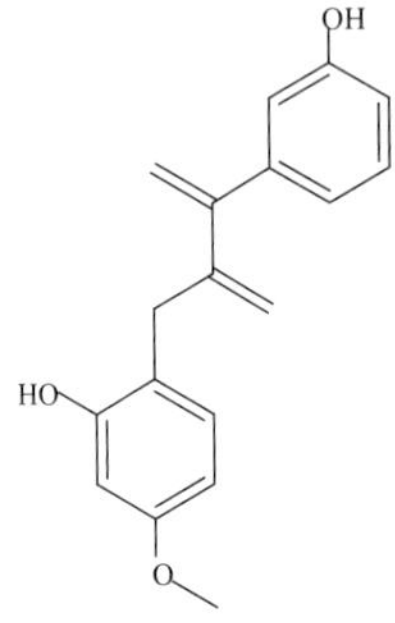

Fig. 20.1 Structure of Termilignan.

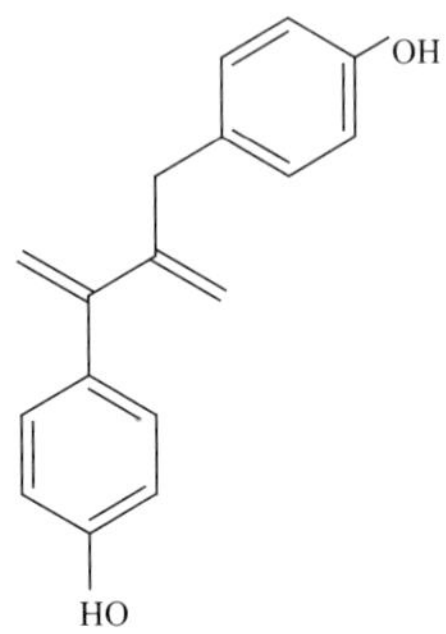

Fig. 20.2 Structure of Thannilignan.

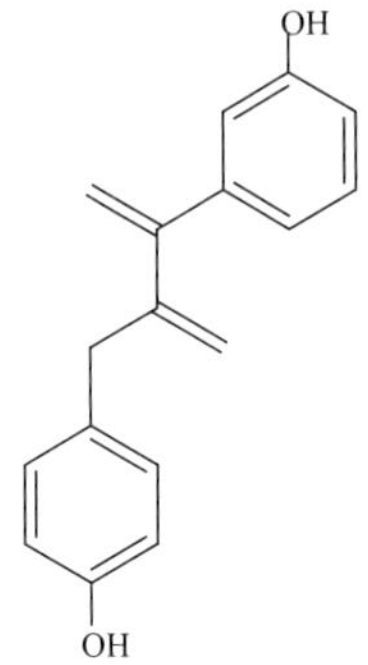

Fig. 20.3 Structure of Anolignan B.

20.4 Physicochemical and nutritional profile: Saran and Singh examined the competent glycerides of the seed oil. The oil was reported to contain oleic (43.21%), linoleic (28.99%), palmitic (11.80%) and stearic acid (16.00%). Singh and Kumar re-examined the seeds in 1946 and reported palmitooleolinoelin (35.24%), stearooleolinoelin (43.51%), palmitodiolein (1.07%) stearodiolein (1.95%) and dioleolinolein (9.54%) and triolein (8.69%) as competent glycerides.

Seed kernel oil of *T belerica* has been reported to yield 12.28% oil on dry basis. Moisture, ash and crude fiber contents of the seed kernel have been found to be 8.43, 2.54, and 8.78%, respectively. The refractive index, co-efficient of viscosity, specific gravity, and energy of activation of the oil were found to be 1.28, 403.6 millipoise at 30Â°C, 0.93 and 6.97 k.cal/mole respectively.

Iodine value, acid value, peroxide value, saponification value, saponification equivalent, ester value, unsaponifiable matter, acetyl value, Reichert-Meissel value, Polenske value, free fatty acids as oleic acid and cholesterol content of the oil are 107, 3.69, 3.14, 189.24, 296.44, 185.55, 1.24%, 3.78, 0.719, 0.945, 0.87% and 26.59 mg per 100 g oil, respectively.

The oil contains 17.70% myristic acid, 21.6% palmitic acid, 45.67% oleic acid and 14.93% stearic acid. The kernel contains 22.57 and 8.38% total lipid and protein respectively. It also contained 0.19mg, 0.45mg, 0.79g and 1.1mg of vitamin B_1, B_2, C and A respectively per 100g of kernel. Ca, Mg, K, Na, P, Fe, Mn, Zn and Cu were found to be 0.3, 0.02, 0.2, 0.2, 0.01%;

23, 1, 12 and 12 ppm respectively in kernel and 0.12, 0.05, 1.15, 0.18, 0.45%; 204, 4, 54 and 50 ppm respectively in oil.

20.5 Ethnopharmacology

Antimalarial and antifungal: An animal study reported antimalarial and antifungal activities of lignans, isolated from *T. belerica*.

Antibacterial: Fruits of *T. belerica* were extracted with petroleum ether, chloroform, acetone, alcohol and water and efficacy of extracts against *Salmonella typhi* and *Salmonella typhimurium* was evaluated. Alcoholic and water extracts of *T. belerica* showed significant anti-salmonella activity and MIC was 12.5 mg/ml against *S. typhimurium*. Results showed that aqueous extract of *T. belerica* was bactericidal at high concentrations whereas low concentrations showed bacteriostatic property. *In vitro* cellular toxicity studies showed no cytotoxicity associated with *T. belerica* extracts.

Anti-diabetic and anti-oxidant activity: Effect of continuous administration of dried 75% methanolic extract of fruits of *T.belerica* suspended in water, was studied in alloxan induced hyperglycemia and antioxidant defense mechanism in rats. *T. belerica* prevented alloxan-induced hyperglycemia significantly from the 6th day of administration and there was 54% reduction on 12th day. Oxidative stress produced by alloxan was found to be significantly lowered by the administration of *T. belerica* extract. Decreased glutathione level produced by alloxan was increased by the administration of the extract in blood and liver.

Antimutagenic: Two polyphenolic fractions, isolated from *T. belerica* were significantly effective against mutagenic effects in *Salmonella typhimurium*. Interaction of the polyphenols with S9 proteins may be the probable cause of the inhibitory effect.

Cholagouge: A study reported bile stimulating activity of *T.belerica* extract.

Hepatoprotective: 1. The effects of an orally administered fraction from the fruits of *T. belerica* in experimental liver injury induced by carbon tetrachloride were studied. Hexobarbitone-induced sleep, zoxazolamine-induced paralysis, transaminases, bilirubin, total protein in serum and microsomal lipid peroxidation and triglycerides in liver were used as indices of liver injury. Pre- and post-treatment with fraction from the fruits of *T. belerica* reduced, in a dose- dependent manner, the elevated levels of serum tranaminases and bilirubin in rats.

2. Another study investigated the protective effect of *T. belerica* fruit extract and gallic acid, at different doses against carbon tetrachloride intoxication. Toxicant caused significant increase in the activities

of serum transaminases and serum alkaline phosphatase. Hepatic lipid peroxidation level increased significantly whereas significant depletion was observed in reduced glutathione level after carbon tetrachloride administration. Treatment with *T. belerica* extract (200, 400 and 800mg/kg, p.o.) and gallic acid (50, 100 and 200mg/kg, p.o.) showed dose-dependent, recovery in all biochemical parameters. Gallic acid was found to be most effective against carbon tetrachloride induced liver and kidney damage.

3. Gallic acid isolated from fraction of the fruits of *T. belerica* was evaluated for its hepatoprotective activity against carbon tetrachloride-induced physiological and biochemical alterations in the liver. The main parameters studied were hexobarbitone-induced sleep, zoxazolamine induced paralysis, serum levels of transaminases and bilirubin. The hepatic markers assessed were lipid peroxidation, drug metabolizing enzymes, glucose-6-phosphatase and triglycerides. Administration of gallic acid led to significant reversal of majority of the altered parameters.

Hypolipidemic activity: A study reported hypolipidemic action of *T. belerica* in animal experiments. The plant also decreased the liver lipids and heart lipids in the drug-treated animals.

Clinical studies: In an open clinical trial of 93 patients suffering from various respiratory conditions *Bibhitaki* was found to have anti-asthmatic, anti-spasmodic, expectorant and anti-tussive activities. However, the design of the study seems to be convoluted.

A study was carried out to evaluate the efficacy of *T. belerica* in 25 patients with a history of diarrhea .The patients received tablets with 150 mgs of the bioactive fraction three times a day. Eleven patients responded to the treatment and required around twelve tablets for recovery; seven patients with *Entamoeba histolytica* cysts, and *Escherichia coli* became negative at the end of the treatment. No side effects were observed during the study.

Toxicity: The seed kernels are reported to be narcotic. Long-term studies have indicated non-toxicity of the *T.belerica* extracts.

References and further reading

Anand KK, Singh B,Saxena AK, Chandan BK, and Gupta VN. 1994. Hepatoprotective studies of a fraction from the fruits of *Terminalia belerica* Roxb. on experimental liver injury in rodents. Phytother. Res. 8, 287–292.

Anand KK, Singh B, Saxena AK, Chandan BK,Gupta VN, and Bhardwaj V. 1997. 3, 4, 5–Trihydroxy benzoic acid (gallic acid), the hepatoprotective principle in the fruits of *Terminalia belerica*-bioassay guided activity. Pharmaco.l Res. 36, 315–21.

Bawa KS, and Kumar A. 1946. Chemcial Examination of fruits of *Terminalia belerica*.Part 11.The component glycerides of fatty oil. J. Pharma. 1, 16.

Chahal, HS, and SS, Agrawal. 2007. Safety evaluation of *Gymnema sylvestre* and *Terminalia belerica*. Int. J. Green. Pharm. 1, 30–32.

Dash B. 1991. Materia Medica of Ayurveda. B. Jain Publishers, New Delhi.

Dhar HL, Miglani BD, and Dhawan D. 1969. Studies on purgative action of an oil obtained from *Terminalia belerica*. Indian. J. Med. Res. 57, 103–5.

Iyengar MA, and Dwivedi S. 1989. *Terminalia belerica* (bahera)—a review. Indian. Drugs. 26, 635–663.

Jadon A, Bhadauria M, and Shukla S. 2007. Protective effect of *Terminalia belerica* Roxb. and gallic acid against carbon tetrachloride induced damage in albino rats. J. Ethnopharmacol. 109, 214–8.

Kirtikar KR, and Basu BD. 1935. Indian Medicinal Plants. 2nd Ed. Vol. 1–4. Reprint. Periodical Experts, Delhi. pp. 48–49.

Madani A, and Jian AK. 2008. Anti-Salmonella activity of *Terminalia belerica*: *in vitro* and *in vivo* studies. Indian. J. Exp. Biol. 46, 817–21.

Molla MTH, Alam MT, and Islam MAU. 2007. Physico-Chemical and Nutritional Studies of *Terminalia belerica* Roxb. Seed Oil and Seed Kernel. J. Bio. Sci. 15, 117–126.

Nadkarni KM. 1954. The Indian Materia Medica, with Ayurvedic, Unani and Home Remedies. Revised and enlarged by AK, Nadkarni. Reprint. Bombay Popular Prakashan PVP. Bombay. p. 34.

Nagraj M. 1995. Floral morphology of *Terminalia belerica* Roxb. Curr. Sci. 24, 194.

Padam SK, Grover IS and Singh M. 1996. Antimutagenic effects of polyphenols isolated from *Terminalia belerica* myroblan in *Salmonella typhimurium*. Indian J Exp Biol., 34, 98–102.

Patnail T, Dey RK, and Gouda P. 2002. Isolation of Triterpenoid Glycoside from Bark of *Terminalia arjuna* using Chromatographic Technique and Investigation of Pharmacological Behavior upon Muscle Tissues. Elec. J. Chem. 4, 474–479.

Patwardhan B, Bhutani KK, Patki PS, Dange SV, Gore DV, Borole DI, Shirolkar RB, and Paranjpe PV. 1990. Clinical evaluation of *Terminalia belerica* in diarrhea. Ancient Sci. Life. 10, 94–97.

Row LR, and Murthy PS. 1970. Chemical examination of *Terminalia belerica* Roxb. Indian. J. Chem. 8, 1047–1048.

Sabu MC, and Kuttan R. 2009. Antidiabetic and antioxidant activity of *Terminalia belerica*. Roxb. Indian. J. Exp. Biol. 47, 270–5.

Shaila HP, Udapa AL, and Udupa SL. 1995. Preventive actions of *Terminalia belerica* in experimentally induced atherosclerosis. Int. J. Cardiol. 49, 101–6.

Siddiqui HH. 1963b. Studies on *Terminalia belerica* Roxb. effect on bile secretion and pharmacodynamic properties. Indian. J. Pharm. 25, 297–302.

Trivedi VP, Nesamany S, and Sharma VK. 1979. A clinical study of the anti-tussive and anti-asthmatic effects of Vibhitakphal Churna (*Terminalia belerica* Roxb.) in the cases of Kasa-Swasa. J. Res. Ayur Sidd. 3, 1–8.

Valsaraj R, Pushpangadan P, Smitt UW, Adsersen A, Christensen SB, Sittie A, Nyman U, Nielsen C and Olsen CE. 1997.New Anti HIV-1, antimalarial and antifungal compounds from *Terminalia belerica*. J. Nat. Prod. 60, 739–42.

Chapter 21

Ethnopharmacology of *Nardostachys jatamansi* DC.

21.1 Introduction

Nardostachys jatamansi DC., is an important plant of the family Valerianceae. It is commonly known as Indian spikenard and is found in the Himalayas. *N. jatamansi* is a perennial herb. Rhizomes occur in short pieces, have a dark gray color and a typical smell. Leaves are sessile and ovate. Flowers are dark–pink in color.

In Ayurveda, *N. jatamansi* is used for nervous headache, excitement, menopausal symptoms, flatulence, epilepsy and intestinal colic. In combination with cold water, the oil is considered to be effective against nausea, stomachache, flatulence, liver problems, jaundice and kidney complaints, insomnia and headache. Externally, the oil is added to a steaming bath to treat inflammation of the uterus. The oils are also used in eye compounds and as poison antidotes. Oil is reported to be useful in the treatment of an atrial flutter.

21.2 Phytochemistry

The roots of the plant contain essential oil, rich in sesquiterpenes and coumarins. Jatamansone or valeranone (Fig. 5.240) is the principal sesquiterpene. Other sesquiterpenes include nardostachone (Fig. 5.241), dihydrojatamansin, jatamansinol, jatamansic acid, jatamansinone, jatamansinol, oroseolol, oroselone, seselin, valeranal, nardostachyin, nardosinone, spirojatamol, jatamol A and B, calarenol, seychellene, seychelane, coumarin: jatamansin or xanthogalin. A new sesquiterpene acid, nardin and new pyranocoumarin: 2′, 2′-dimethyl-3′-methoxy-3′, 4′-dihydropyranocoumarin have been reported. An alkaloid, actinidine has been reported (Fig. 21.1).

21.3 Ethnopharmacology

***Hepatoprotective activity*:** Pretreatment of rats with 800 mg/kg body wt of the 50% ethanolic extract of *N. jatamansi* demonstrated significant hepatoprotective activity against thioacetamide induced hepatotoxicity. Marked reduction in raised levels of serum transaminase and alkaline phosphatase was observed. Pretreatment of the animals with the extract further resulted in an increase in survival of rats intoxicated with LD90 dose of the hepatotoxic drug.

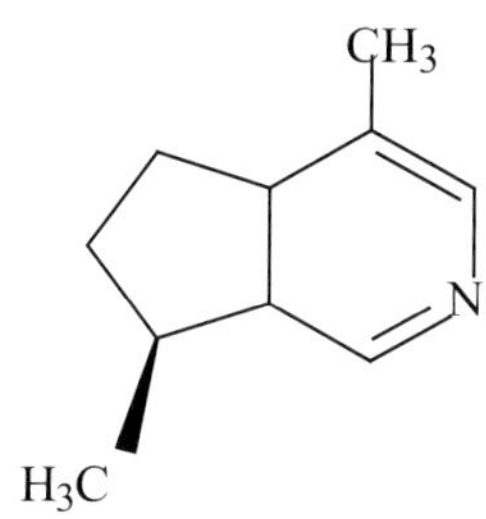

Fig. 21.1 Structure of Actinidine.

Cardio protective and hypolipidemic activity: Rats administered doxorubicin (15 mg/kg, i.p.) showed myocardial damage that was manifested by the elevation of serum marker enzymes (lactate dehydrogenase, creatine phosphokinase, aspartate aminotransaminase and alanine aminotransaminase). The animals showed significant changes in the antioxidant enzymes (superoxide dismutase, catalase, glutathione peroxidase and glutathione-S-transferase) and lipid peroxidation levels. Pretreatment with *N. jatamansi* extract significantly prevented these alterations and restored the enzyme activity and lipid peroxides to near normal levels. Restoration of cellular normality accredits the *N. jatamansi* with a cytoprotective role in doxorubicin-induced cardiac damage.

In a study, ethanolic extract of *N. jatamansi* was studied in Wistar albino rats for cardio protective activity against doxorubicin induced myocardial injury. Doxorubicin is inhibitor of fatty oxidation in the heart and results in cardio toxicity. The rats treated with a single dose of doxorubicin (15 mg/kg) intraperitoneally showed an increase in serum and cardiac lipids (cholesterol, triglycerides, free fatty acids and phospholipids), along with a significant rise in serum low density lipoproteins, very low density lipoproteins and a drop in high density lipoproteins levels, resulting in alteration of serum and cardiac lipid metabolizing enzymes.

Pretreatment with an extract of *N.jatamansi* (500 mg/kg) orally for 7 d to doxorubicin induced rats showed a significant prevention in the lipid status with the activities of the lipid metabolizing enzymes. Histopathological observations were also in correlation with the biochemical parameters. These findings suggest that the protective and hypolipidemic effect of *N. jatamansi* against doxorubicin induced myocardial injury in rats could possibly be mediated through its anti lipid peroxidative properties.

A 50% ethanolic extract of *Curcuma longa* (tuber) and *N. jatamansi* (whole plant) elevated the HDL-cholesterol/total cholesterol ratio in triton-induced hyperlipidemic rats. There also was a reduction in the ratio of total cholesterol/phospholipids.

Nootropic activity: Acetyl cholinesterase inhibitory activity of methanolic and successive water extracts of *N.jatamansi* (rhizome), were investigated for acetyl cholinesterase inhibitory activity *in vitro*. Results indicated that methanolic extracts to be more active than water extracts. The IC (50) values obtained for methanolic and successive water extracts of *N.jatamansi* was 47.21mug/ml. These results partly substantiate the traditional use of *N.jatamansi* for improvement of cognition.

Anticonvulsant activity: Ethanol extract of the roots of *N. jatamansi* was studied for its anticonvulsant activity and neurotoxicity, alone and in combination with phenytoin in rats. The results demonstrated a significant increase in the seizure threshold by *N. jatamansi* root extract against maximal electroshock seizure model as indicated by a decrease in the extension/flexion ratio. However, the extract was ineffective against pentylenetetrazole-induced seizures. *N. jatamansi* root extract also showed minimal neurotoxicity against rotarod test at doses that increased the seizure threshold. Further, pretreatment of rats with phenytoin at a dose of 12.5, 25, 50 and 75 mg/kg in combination with 50mg/kg of *N. jatamansi* root extract resulted in a significant increase in the protective index of phenytoin from 3.63 to 13.18. The dose response studies of phenytoin alone and in combination with *N. jatamansi* extract on the serum levels of phenytoin clearly demonstrated the synergistic action of both the drugs.

Antidepressant activity: A 15-d treatment with an alcoholic root extract of *N. jatamansi* caused an overall increase in the levels of central monoamines and inhibitory amino acids, including a change in the levels of serotonin, 5-hydroxyindole acetic acid, gamma-amino butyric acid, and taurine in rat brain.

Antiparkinson's activity: Rats were treated with 200, 400, and 600 mg/kg body weight of *N. jatamansi* for 3 wk. On day 21, 2 microl of 6-OHDA (12 microg in 0.01% in ascorbic acid-saline) was infused into the right striatum, while the sham-operated group received 2 microl of vehicle. Three weeks after the 6-OHDA injection, the rats were tested for neurobehavioral activity and were scarified after 6 wk for the estimation of lipid peroxidation. The increase in drug-induced rotations and deficits in locomotor activity and muscular coordination due to 6-OHDA injections were significantly and dose-dependently restored by *N. jatamansi*. A significant decrease in the level of dopamine and its metabolites and an increase in the number of dopaminergic D2 receptors in striatum were observed after 6-OHDA injection, and both were significantly recovered following *N. jatamansi* treatment.

Neuroprotective activity: Pretreatment with an alcoholic extract of *N. jatamansi* dosed at 250 mg/kg of for 15 d protected rats against focal ischemia caused by middle cerebral artery occlusion. The protective effect may be associated with improving glutathione content, inhibiting lipid peroxidation, and activity on the Na+/K+ ATPase and catalase enzyme systems.

***Nootropic activity*:** 1. Acetyl cholinesterase inhibitory activity of methanolic and successive water extracts of *N.jatamansi* (rhizome), were investigated for acetyl cholinesterase inhibitory activity *in vitro*. Results indicated that methanolic extracts to be more active than water extracts. The IC (50) values obtained for methanolic and successive water extracts of *N.jatamansi* was 47.21mug/ml. These results partly substantiate the traditional use of *N.jatamansi* for improvement of cognition.

2. The elevated plus maze and the passive avoidance paradigm were employed to evaluate learning and memory parameters. Three doses (50, 100, and 200 mg/kg, p.o.) of an ethanolic extract of *N. jatamansi* were administered for 8 successive days to both young and aged mice. The 200 mg/kg dose of *N. jatmansi* ethanolic extract significantly improved learning and memory in young mice and also reversed the amnesia induced by diazepam (1 mg/kg, i.p.) and scopolamine (0.4 mg/kg, i.p.).

Furthermore, it also reversed aging-induced amnesia due to natural aging of mice. As scopolamine-induced amnesia was reversed, it is possible that the memory improvement may be because of facilitation of cholinergic transmission in the brain. Hence, *N. jatamansi* might prove to be a useful memory restorative agent in the treatment of dementia seen in elderly persons. The underlying mechanism of action can be attributed to its antioxidant property.

Antifungal activity: *N. jatamansi* essential oil demonstrated fungistatic activity against *Aspergillus flavus, Aspergillus niger* and *Fusarium oxysporum*.

21.4 Jatamansone (valeranone): Animal studies done on jatamansone have reported antioestrogenic, antiarrhythmic, antihypertensive, anticonvulsant, sedative and tranquilizing activities. In animal experiments, a limiting effect upon convulsant thresholds and a reduction of motor coordination ability traceable to the sequiterpene ketone, valeranone contained in the drug have been demonstrated. Jatamansone is anticonvulsive in effect without exhibiting neuroleptic characteristics.

Some experiments, typical for tranquillizers jatamansone prolonged barbiturate hypnosis, impaired rotarod performance, potentiate the body-temperature lowering activity of reserpine. Jatamansone exhibited anticonvulsant activity against electric shock and gastro protective actions. In toxicological studies on rats and mice an oral LD50 of greater than 3160 mg/kg was found, which suggests the possibility of a therapeutically useful dose ratio.

Preliminary clinical studies with jatamansone reported reduced incidence of aggressiveness, restlessness, stubbornness and insomnia. In a study conducted on hyperkinetic children, jatamasnone, D-amphetamine and chlorpromazine were compared for efficacy. Jatamasnone and amphetamine significantly improved behavior but amphetamine was better in reducing aggressiveness and restlessness. Children with mental retardation showed little response to any of the drugs. Jatamansone has fewer side effects than D-amphetamine and chlorpromazine.

21.5 Jatamansin: The coumarin: jatamansin is efficacious in internal treatment of varicose veins.

Further reading

Agarwal SS, Sharma RC, and Arora RB. 1973. Antioestrogenic activity of jatamansone semicarbazone. Indian. J. Exper. Biol. 11, 583.

Ahmad M, Yousuf S, Khan Badruzzaman, Hoda Md N. Ahmad MA, Ishrat T, Agarwal AK, and Islam F. 2006. Attenuation by *Nardostachys jatamansi* of 6-hydroxydopamine-induced parkinsonism in rats: behavioral, neurochemical, and immunohistochemical studies. Pharmacol. Biochem. Behav. 83, 150–60.

Ali S, Ansari KA, Jafry, and MA, Kabeer G. 2000. *Nardostachys jatamansi* protects against liver damage induced by thioacetamide in rats. J. Ethnopharmacol. 71, 359–363.

Arora RB, and Arora CK. 1963. Hypotensive and tranquillizing activity of jatamansone (valeranone). Pharmacology of Oriental Plants. Academic Press, London/New York. 7, 51–60.

Arora RB, Singh M, and Arora CK. 1962. Tranquillizing activity of jatamansone. Life Sci. 6, 225.

Arora RB, Arora CK, Sha MJ, and Shet UK. 1967. Animal species variation in hypotensive activity of jatamansone with a report in the clinical trial of this drug. Ind. J. Med. Sci. 21, 455–460.

Arora RB, Sharma PL, and Kapila K. 1985. Antiarrhythmic and anticonvulsant action of jatamansone. Ind. J. Med. Res. 46, 782–791.

Bagchi A, Oshima Y, Hikino H. 1990. Spirojatomol, a new skeletal sesquiterpenoid of *Nardostachys jatamansi* roots. Tetrahedron. 46, 1523–1530.

Bagchi A, Oshima Y, Hikino H. 1991. Jatamols A and B: sesquiterpenoids of *Nardostachys jatamansi* roots. Planta. Med. 57, 282–283.

Biswas K. 1963. Pharmacology of Jatamansin. Active substance of the plant *Nardostachys jatamansi*. Prensa. Med. Argent. 50, 1021–5.

Chatterjee B, Basak U, Datta J, Banerji A, Neuman, and T. Prange. 2005. Studies on the Chemical Constituents of *Nardostachys jatamansi* DC (Valerianaceae). Cheminform. 36, 17.

Dixit VP, Jain P, and Joshi SC. 1998. Hypolipidaemic effects of *Curcuma longa* L. and *Nardostachys jatamansi* DC. in triton-induced hyperlipidaemic rats. Indian. J. Physiol. Pharmacol. 32, 299–304.

Goecht R. 1961. Internal treatment of varicose veins with jatamnsin. Med. Klin. 11, 1370–1.

Gupta PD, and Vermani V. 1968. Clinical trial of jatamansone (syn: Valeranone) in hyperkinetic behaviour disorders. Neurol. India. 6, 168–73.

Harigaya Y, Chatterjee A, Basak B, Saha M, Dutta U, Mukhopadhyay C, Banerji J, and Konda Y. 2000. Structure and stereochemistry of nardostachysin, a new terpenoid ester constituent of the rhizomes of *Nardostachys jatamansi*. J. Nat. Prod. 63, 531–1533.

Hoerster H, Ruecker G, and Tautges J. 1997.Valeranone content in the roots of *Nardostachys jatamansi* and *Valeriana officinalis*. Phytochemistry. 1, 1070–1071.

Joshi H, and Parle M. 2006. *Nardostachys jatamansi* improves learning and memory in mice. J. Med. Food 9, 113–8.

Mishra D, Chaturvedi RV, and Tripathi SC. 1995. The fungitoxic effect of the essential of the herb *Nardostachys jatamansi DC*. Trop. Agricul. 72, 48–52.

Prabhu V, Karanth KS, and Rao A. 1994. Effects of *Nardostachys jatamansi* on biogenic-amines and inhibitory amino-acids in the rat-brain. Planta. Med. 60, 114–117.

Rao VS, Rao A, and Karanth KS. 2005. Anticonvulsant and neurotoxicity profile of *Nardostachys jatamansi* in rats. J. Ethnopharmacol 102, 351–6.

Rucker G, Tautges J, Sieck A. Wenzl H, and Graf E. 1978a. *Nardostachys jatamansi*: a chemical, pharmacological and clinical appraisal. Indian. Med. Res. 28, 17.

Rucker G, Tautges J. Sleck A, Wenzl H, and Graf E. 1978b. Isolation and pharmacological activity of the sesquiterpene valeranone from *Nardostachys jatamansi* DC (in German). *Arzneimittelforschung*. 28, 7–13.

Rücker G, Paknikar SK, Mayer R, Breitmaier E, Will G, and Wiehl L. 1993. Revised structure and stereochemistry of jatamansic acid. Phytochemistry. 33, 141–143.

Salim S, Ahmad M, Zafar KS, Ahmad AS, and Islam F. 2003. Protective effect of *Nardostachys jatamansi in rat cerebral ischemia*. Behav Pharmacol. Biochem. 74, 481–486.

Sastry SD, Maheswari ML, Chakravarti KK, Bhatacharya SC. 1967. Terpenoids-CV1: the structure of calarenol. Tetrahedron. 23, 1997–2000.

Subashini R, Yogeeta S, Gnanapragasam A, and Devaki T. 2006. Protective effect of *Nardostachys jatamansi* on oxidative injury and cellular abnormalities during doxorubicin-induced cardiac damage in rats. J. Pharm. Pharmacol. 58, 257–62.

Subashini R, Ragavendran B, Gnanapragasam A, Yogeeta SK, and Devaki T. 2007. Biochemical study on the protective potential of *Nardostachys jatamansi* extract on lipid profile and lipid metabolizing enzymes in doxorubicin intoxicated rats. Pharmazie. 62, 382–7.

Vinutha JP. 2007. Acetyl cholinesterase inhibitory activity of methanolic and successive water extracts of *Nardostachys jatamansi*. Indian. J. Pharmacol. 23, 127–131.

Zinzius J. 1961.Jatamansin-a new therapeutic agent in venous stasis and related diseases. Dtsch. Med. J. 20, 423–4.

Chapter 22

Ethnopharmacology of *Eclipta alba* Linn.

22.1 Introduction

Eclipta alba (L.) Hassk. (Asteraceae) grows commonly in moist areas as a weed all over the world. In many parts of India, it is grown commercially as a medicinal crop. It is an annual, erect or prostate entirely pubescent herb, often rooting at nodes with opposite, sessile, usually oblong, 2.5–7.5 cm long leaves with appressed hairs. Floral heads 6–8 mm in diameter, solitary, white; achene compressed and narrowly winged. The aerial parts of the plant are used in medicine.

22.2 Traditional medicinal uses

India: According to Ayurvedic philosophy *E.alba* is bitter; alterative and anthelmintic. It is useful in inflammations, hernia, eye diseases, bronchitis, asthma, leucoderma, anemia, heart and skin diseases, night blindness, syphilis etc. It is reported to be beneficial for the complexion, hair, eyes, and teeth. Expressed juice of *E. alba* mixed with goat's milk is used in frontal sinusitis and nasal cattarh in children. Bhringraj taila and Bhringrajadi churana are official preparations.

In the Unani system, the juice of *E. alba* is used in 'Hab Miskeen Nawaz' alongwith aconite, *Croton tiglium*, 'triphala', *Piper nigrum, Piper longum, Zingiber officinale*, and minerals like mercury, sulfur, arsenic, borax etc. for various types of pains in the body. It is also a constituent of 'Roghan Amla Khas' for applying on hair, and of Ma'jun Murrawah-ul-arwah.

Korea: The plant is used as an antidote for snake bites.

The Philippines: A decoction of the dried plant is used for heamoptysis and heamtemesis. For dysentery and heamturia urine, a decoction of the dried herb or tincture is used.

Medicated tea or tinctures are used as household remedies for sprains, furuncle and dermatitis; the tea or tincture is excellent.

Nepal: The plant juice, mixed with an aromatic (essential oil), is used in the treatment of catarrhal problems and jaundice. The leaves are used in the treatment of scorpion stings.

22.3 Phytochemistry

Alkaloids: Alkaloids including ecliptine and nicotine have been reported. Bio-active steroidal alkaloids, verazine, 20-epi-3-dehydroxy-3-oxo-5, 6-dihydro-4, 5-dehydroverazine, ecliptalbine, (20R)-4ß-hydroxyverazine, 4ß-hydroxyverazine, (20R)-25ß-hydroxyverazine and 25ß-hydroxyverazine have been identified from the methanolic extract.

Coumarins: The dried leaves of *E.alba* have been reported to contain wedelolactone (Fig. 22.1), a complex coumarin and its derivatives dimethylewedelolactone-7-glucoside and nor-wedelolactone. Demethylwedelolactone, isodemethylwedelolactone, and strychnolactone have been reported by percolation and hot extraction of the whole plant of *E. alba*.

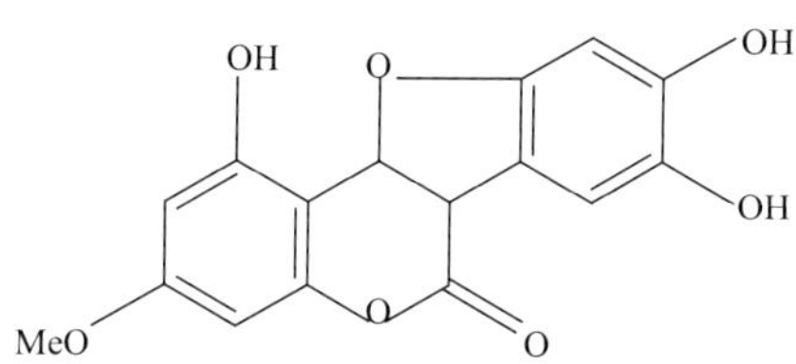

Fig. 22.1 Structure of Wedelolactone.

Hydrocarbons: Ddithienylacetylene ester, ecliptal or α-terthienyl aldehyde, α-terthienyl-methanol and α-formylterthienyl.

Triterpenes: Ecliptasaponin C and D, new triterpenoid glucosides, have been isolated from the whole plant of *E. alba.* A new triterpene saponin, eclalbatin, together with α-amyrin (Fig. 5.245), β-amyrin (Fig. 5.245), ursolic acid (Fig. 5.245), oleanolic acid (Fig. 5.245), and wedelic acid have been isolated. From the whole parts of six new oleanane triterpene glycosides, eclalbasaponins I–VI have been isolated.

Thiopenes: Polyacetylenic thiopenes 5′-senecioyloxymethylene-2-(4-isovaleroxybut-3-ynyl) dithiophene, 5′-tigloyloxymethylene-2-(4-isovaleroxybut-3-ynyl) dithiophene have been reported form the plant. The roots contain polyacetylene substituted thiophenes).

Sterols: The aerial parts of the plant have been reported to contain phytosterol; β glucoside of phytosterol, daucosterol and stigmasterol-

3-O-glucoside. The whole plant contains stigmasterol (Fig. 5.247) and β-sitosterol (Fig. 5.247).

Flavonoids: Apigenin (Fig. 5.112), luteolin (Fig. 15.12) and luteolin-7-glucoside (Fig. 15.9).

Miscellaneous: Nonacosanol, stearic acid, lacceroic acid and 3,4-dihydoxy benzoic acid .

22.4 Ethnopharmacology

Anti aggressive effect: The present study investigated the ability of 100 and 200mg/kg of aqueous extract of *E. alba* to circumvent aggression. Foot shock induced aggression and the water competition test were utilized as models for screening of anti aggressive activity. *E. alba* significantly minimized dominance ($p<0.05$) which is correlated to the level of aggression particularly with 200mg/kg in the water competition test. A tangible behavioral submission was observed with 100 and 200mg/kg and of *E. alba* in the foot shock induced test.

Analgesic effect: The present experimental research investigated the analgesic activity of the total ethanol extract of *E. alba,* and isolated alkaloids in albino mice using the tail clip method, the tail flick method and the acetic acid induced writhing response. The ethanol extract and the total alkaloids produce significant analgesic activity in all the different models of analgesia used. However, the total alkaloidal fraction was the most efficacious in all the models tested.

Anti-inflammatory effect: The methanolic extract administered by the oral route at a concentration of 100 and 200 mgkg-1 showed the significant dose dependent anti-inflammatory activity in carrageenin and egg white induced hind paw oedema in rats. Anti-inflammatory activity of the tested extract was comparable with that of the standard drug indomethacin (10 mgkg-1) and cyproheptadine (8 mgkg-1).

Antibacterial effect: The antimicrobial activity of wedelolactone was evaluated using minimum inhibitory concentration and agar well diffusion method. The compound exhibited good activity against *Staphylococcus epidermidis* and *Salmonella typhimurium.* The MIC test showed the growth inhibition of *S. epidermidis* at a concentration of 15.0 μg/ml, ZOI 10.24 mm and of *S. typhimurium* at a concentration of 25.0 μg/ml, ZOI 9.16 mm.

Antifungal effect: 25beta-hydroxyverazine showed good activity against *Candida albicans* (Abdel Kader et al., 1998). The *in vitro* antifungal activity of the whole plant of *E.alba* extract was investigated against Candida tropicalis, Rhodotorula glutinis and Candida albicans. The extract showed high degree of activity against all tested fungi. The inhibitory effects of extracts are very similar to those of standard antibiotics used.

Antimalarial effect: The anti-malarial activity of *Eclipta alba* leaves extract was evaluated against Plasmodium berghei ANKA strain in mice. A standard inoculum of 1 x 10(6) infected erythrocytes was used. The methanolic leaf extract (250–750 mg/kg) produced a dose-dependant chemosupression or schizontocidal effect during early and established infection and high mean survival time values particularly in the group administered 750 mg/kg/day of extract.

Antihyperglycemic effect: Oral administration of leaf suspension of *E. alba* (2 and 4 g/kg body weight) for 60 d resulted in significant reduction in blood glucose (from 372.0 ± 33.2 to 117.0 ± 22.8), glycosylated hemoglobin HbA1c, a decrease in the activities of glucose-6 phosphatase and fructose1,6-bisphosphatase, and an increase in the activity of liver hexokinase. *E. alba* at dose of 2 g/kg body weight exhibited better sugar reduction than 4 g/kg body weight.

Hepatoprotective effect: 1. Ethanol/water extract significantly counteracted CCl_4-induced inhibition of the hepatic microsomal drug metabolizing enzymes. The loss of hepatic lysosomal acid phosphatase and alkaline phosphatase by CCl_4 was significantly restored by ethanol/water extract.

2. In another study, the EtOAc part of alcoholic extraction exhibited significant hepatoprotective activity against CCl_4-induced liver injury in rats.

3. In yet another study, treatment with 50% ethanol extract of *E.alba* (100 & 250mg/100g body weight) was found to protect the mice from hepatotoxic action of paracetamol as evidenced by significant reduction in the elevated serum transaminase levels. Histopathological studies showed marked reduction in fatty degeneration and centrizonal necrosis in animals receiving different doses of *E.alba* along with paracetamol as compared to the control group.

Hypolipdimeic effect: 1. The total alcoholic extract of *E. prostrata* exhibited a dose-dependent activity in albino rats when compared to standard drugs. The activity was assessed by studying the lipid profiles of serum, liver and heart of the control and drug-treated animals.

2. Charles River Sprague-Dawley CD rats were fed experimental diets supplemented with 0 mg (control), 25 mg (E25), 50 mg (E50), or 100 mg (E100) of a freeze-dried butanol fraction of *E prostrata* per kilogram of diet for 6 wk. Serum triacylglyceride and total cholesterol levels were significantly lower in the E50 and E100 groups by 9.8% to 19.0% and by 10.7% to 13.4%, respectively, and low-density lipoprotein–cholesterol levels were significantly reduced in the same groups by 10.3% to 13.0% compared with the untreated control group.

Neuropharmacological effects: 1. The aqueous, hydroalcoholic extracts and hydrolyzed fraction of the aqueous extract of *E. alba* was subjected to neuropharmacological activities. in rats. The findings indicated nootropic activity of the aqueous extract (300 mg/kg, p.o.) and its hydrolyzed fraction (30 mg/kg, p.o.). The aqueous extract and the hydrolyzed fraction exhibited gastro protective effect and normalized the white blood cell count in the milk induced leukocytosis challenge model.

2. The suspension of *E. alba* containing 100 and 200 mg/kg was administered to rats to evaluate Transfer Latency on an elevated plus maze. Mice were placed at the center of open field apparatus to assess spatial habitual learning, observed for 20 min for rearing and time spent during rearing using varied doses for 30 min, 24 h and 96 and 144 h. The results revealed significant improvement of retrieval memory.

Hair growth promoting effects: The study was aimed to investigate the efficacy of methanol extract of *E. alba* as hair growth promoter. Pigmented C57/BL6 mice, preselected for their telogen phase of hair growth were used. The extract was applied topically to assess telogen to anagen transition. The methanol extract of whole plant when tested for hair growth promoting potential, exhibited dose dependent activity in C57BL6 mice.

Effect on proteolytic and hemorrhagic activities: 1. Wedelolactone and demethylwedelolactone, isolated from *E.alba* demonstrated significant trypsin inhibitory effects.

2. The partially purified ethyl acetate extract (PEE) of *E. prostrata* (containing 47% of wedelolactone) and wedelolactone demonstrated strong antiproteolytic and antihemorrhagic activity against Malayan Pit Viper venom in a dose-dependent manner. The extract, at 5 mg/ml, inhibited proteolytic activity of 100 µg of the venom and hemorrhagic activity of 3 minimum hemorrhagic doses to 95% and 65% respectively. At the same concentration, wedelolactone neutralized the proteolytic activity at around 76% and, at doses of 0.25–1.0 mg/ml, offered protection against hemorrhagic activity of the venom in the range 3–3.5%.

Effect on osteoblast differentiation: Flavonoid, diosmetin (Fig. 22.2), and isoflavonoids, 3′hydroxybiochanin A (Fig. 22.3), and 3′*O*-methylorobol (Fig. 22.4), isolated from the methanol extract of *E. prostrata* significantly increased osteoblast differentiation as assessed by the alkaline phosphatase activity.

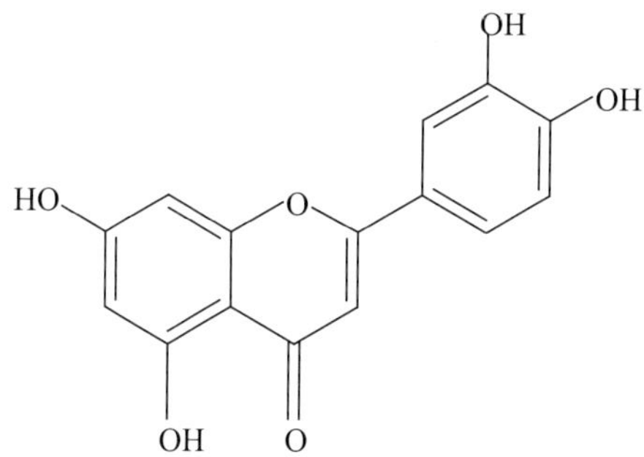

Fig. 22.2 Structure of Diosmetin.

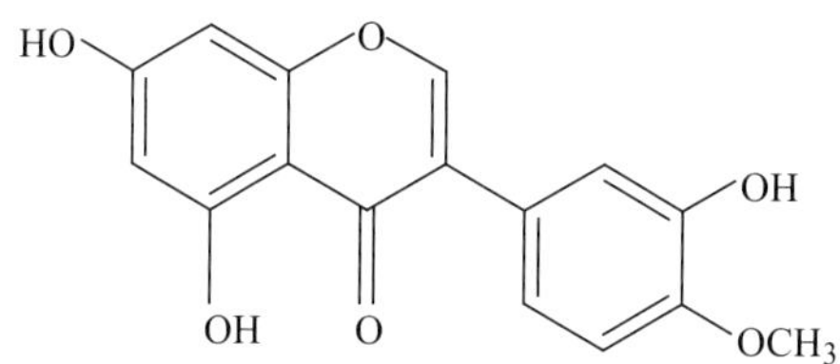

Fig. 22.3 Structure of 3-hydroxybioachanin A.

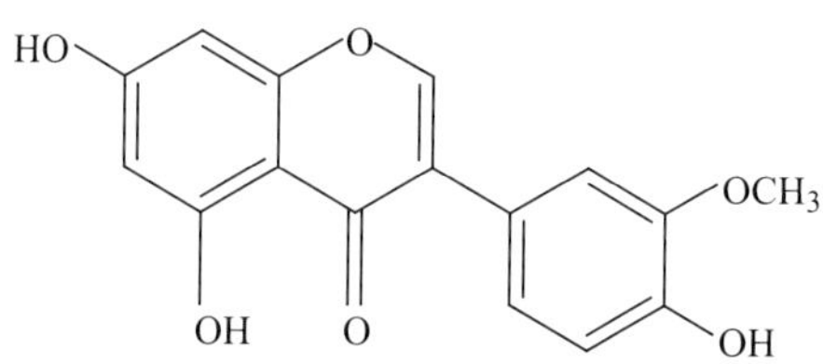

Fig. 22.4 Structure of 3′-*O*-methyllorobol.

22.5 Clinical studies: Two studies reported efficacy of *E.alba* in the treatment of infective hepatitis in adults and jaundice in children, respectively. A clinical study reported diuretic, hypotensive, and hypocholesterolemic properties of *E. alba,* which helps in the alleviating oxidative stress-induced complications in hypertensive patients.

References and further reading

Abdel Kader MS, Bahler BD, Malone S, Werkhoven MC, van Troon F, Wisse JH, Bursuker I. Neddermann KM, Mamber SW, and Kingston DG. 1998. DNA-damaging steroidal alkaloids from *Eclipta alba* from the Suriname rainforest. J. Nat. Prod. 61, 1202–8.

Ananthi J, Prakasam A, and Pugalendi KV. 2003. Antihyperglycemic Activity of *Eclipta alba* Leaf on Alloxan-induced Diabetic Rats. Yale. J. Biol. Med. 1, 97–102.

Anonymous. 1952. The Wealth of India, Raw Materials. 3rd Ed. Council of Scientific and Industrial Research, CSIR. New Delhi, India. pp. 127–128.

Anonymous. 1987. Medicinal Plants in the Republic of Korea. World Health Organization, Manila. p. 33.

Anonymous. 1993. Medicinal Plants of Nepal. Dept. of Medicinal Plants Nepal. p. 45.

Arunachalam G, Subramanian N. Pazhani GP, and Ravichandran V. 2009. Anti-inflammatory activity of methanolic extract of *Eclipta prostrata* L. (Astearaceae). Afr. J. Pharm. Pharmacol. 3, 97–100.

Banji O, Banji D, Annamalai AR, and Manavalan R. 2007. Investigation on the effect of *Eclipta alba* on animal models of learning and memory. Indian. J. Physiol. Pharmacol. 51, 274–8.

Bapna S, Asdule S, Shirshat MS, Jadhav S, Patil LS, and Deskmukh RA. 2007. Anti-malarial activity of *Eclipta alba* against Plasmodium berghei infection in mice. J. Commun. Dis. 9, 91–4.

Bhargava K, Krishnaswamy N, and Seshadri T. 1970. Isolation of demethyl wedelolactone and its glucoside from *Eclipta alba*. Ind. J. Chem. 8, 664–665.

Chopra RN, Nayar SL, and Chopra IC. 1954. Glossary of Indian Medicinal Plants. Council of Scientific and Industrial Research, New Delhi, India. p. 104.

Dalal S, Rana S, Sastry K, and Kataria S. 2009. Wedelolactone as an Antibacterial Agent extracted from *Eclipta alba*. Internet. J. Microbiol. 7, 1.

Dan NV, and Nhu DT. 1989. Medicinal Plants in Vietnam. World Health Organization. pp. 12–14.

Das B, and Chakravarty AK. 1991. Ecliptal, a new terthienyl aldehyde from *Eclipta alba*. Ind. J. Chem. 30, 1052–1053.

Datta K, Singh AT, Mukherjee A, Bhat B, Ramesh B, and Burman AC. 2009. *Eclipta alba* extract with potential for hair growth promoting activity. J. Ethnopharmacol. 124, 450–6.

Dixit S, and Achar M. 1979. Bhringaraja (*Eclipta alba* Linn.) In the treatment of infective hepatitis. Curr. Med. Prac. 23, 237–242.

Dixit S, and Achar M. 1981. Study of bhringaraja (*Eclipta alba*) therapy in jaundice in children. J. Sci. Res. Ind. Med. 2, 96–100.

Han Y, Xia C, Cheng X, Xiang R, Liu H, Yan Q, and Xu D. 1998. Preliminary studies on chemical constituents and pharmacological action of *Eclipta prostrata* L. Zhongguo Zhong Yao Za Zhi. 32, 680–2, 703.

Jain S, and Singh P. 1988. A dithienylacetylene ester from *Eclipta erecta* Linn. Ind. J. Chem. 27, 99–100.

Khargava KK, and Seshadri TR. 1974. Chemistry of medicinal plants: *Eclipta alba* and *Wedelia calandulaceae*. J. Res. Ind. Med. 9, 915.

Kim DI, Lee SH, Choi JH, Lillehoj HS, Yu MH, and Lee GS. 2008. The butanol fraction of *Eclipta prostrata* (Linn.) effectively reduces serum lipid levels and improves antioxidant activities in CD rats. Nut. Res. 28, 550–554.

Kumari SC, Govindasamy S, and Sukumar E. 2006. Lipid lowering activity of *Eclipta prostrata* in experimental hyperlipidemia. J. Ethnopharmacol. 105, 332–335.

Lee MK, Ha NR, Yang H, Sung SH, and Kim YC. 2008. Stimulatory constituents of *Eclipta prostata* on mouse osteoblast differentiation. Phytoter. Res. 3, 145–148.

Lobo O, Banji D, Annamalai AR, and Manavalan R. 2008. Evaluation of antiaggressive activity of *Eclipta alba* in experimental animals. Pak. J. Pharma. Sci. 21, 195–199.

Pal S, and Narasimham N. 1943. The alkaloid in *Eclipta alba* (Hassk). J. Ind. Chem. Soc. 20, 181.

Pithayanukul P, Lapett B, Bavovada R, Pakmanee N, and Suttisri R. 2007. Inhibition of proteolytic and hemorrhagic activities by ethyl acetate extract of *Eclipta prostrata* against Malayan Pit Viper Venom. Pharm. Biol. 45, 282–288.

Rangineni V, Sharada D, and Saxena S. 2007. Diuretic, Hypotensive, and Hypocholesterolemic Effects of *Eclipta alba* in Mild Hypertensive Subjects: A Pilot Study. J. Med. Food 10, 143–148.

Samiulla S, Mundkinajeddu D, Shivanna Y, Arun C, Keerthi M, Prashanth D, Amit A, and Venkataraman BV. 2003. Trypsin inhibitory effect of wedelolactone and demethylwedelolactone. Phytoter. Res. 17, 420–421.

Sawant S, Issac JC, and Narayanan M. 2004. Analgesic studies on total alkaloids and alcohol extracts of *Eclipta alba* (Linn.) Hassk. Phytother. Res. 10, 111–112.

Saxena AK, Singh B, and Anand KK. 1993. Hepatoprotective effects of *Eclipta alba* on subcellular levels in rats. J. Ethnopharmacol. 40, 155–161.

Sikroria BC, Srivastava SJ, and Niranjan GS. 1982. Phytochemical studies on *Eclipta alba*. J. Ind. Chem. Soc. 59, 905.

Singh P. 1988. Naturally-occuring thiophene derivatives from Eclipta species. Bioact. Mol. 7, 179–186.

Tabassum N, and Agrawal SS. 2004. Hepatoprotective activity of *Eclipta alba* Hassk. Against paracetamol induced hepatocellular damage in mice. JK – Pract. 11, 278–280.

Thakur VD, and Mengi SA. 2005. Neuropharmacological profile of *Eclipta alba* (Linn.) Hassk. J. Ethnopharmacol. 102, 23–31.

Upadhyay RK, Pandey MB, Jha RN, and Pandey VB. 2001. Eclalbatin, a triterpene saponin from *Eclipta alba*. J. Asian. Nat. Prod. Res. 3, 213–7.

Venkatesan S, and Ravi R. 2004. Antifungal activity of *Eclipta alba*. Ind. J. Pharmacol. 66, 97–98.

Wagner H, Geyer B, Yoshinobu K, amd Govind SR. 1986. Coumestans as the Main Active Principles of the Liver Drugs *Eclipta alba* and *Wedelia calendulacea*. Planta. Med. 5, 370–2.

Yahara S, Ning D, and Teshihire N. 2006. Six new oleanane triterpene glycosides, eclalbasaponins I-VI isolated from *Eclipta alba* (L.) Hassk. Chem. Pharm. Bull. 42: 1336–1338.

Zhang JS, and Guo QM. 2001. Studies on the chemical constituents of *Eclipta prostrata* (L). Yao-Xue-Xue-Bao. 36, 34–7.

Zhang M, Chen YY, Di XH, and Liu M. 1997. Isolation and identification of ecliptasaponin D from *Eclipta alba* (L.) Hassk. Yao Xue Xue Bao. 32, 633–4.

Chapter 23

Ethnopharmacology of *Cassia siamea* Lam.

23.1 Introduction

Cassia siamea (Lam.) H.S.Irwin & Barneby, Fabaceae is a native of Sri Lanka and is widely cultivated in Southeast Asia. It is traditionally used to treat insomnia, diabetes, fever, hypertension and constipation. *C. siamea* is used in Southwest Nigeria as a remedy for febrile illness.

23.2 Phytochemistry

Barakol (Fig. 23.1), a dioaxaphenalne derivative is a biologically active constituent extracted from the leaves and flowers of *C. siamea*.

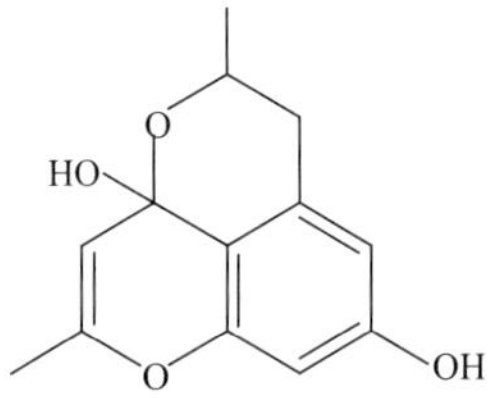

Fig. 23.1 Structure of Barakol.

The trunk and root bark contain alkaloids cassiamine A, B and C. Alkaloids including siaminine A (Fig. 23.2), B and C, have been isolated from the alcoholic extract of the leaves. Two alkaloids, cassiarine A and B have been isolated from the leaves of *C. siamea*.

Three alkaloids, cassiarins C–E (Figs. 23.3 to 23.5), and a new chromone, 10, 11-dihydroanhydrobarakol (Fig. 23.6) having moderate antiplasmodial activity against *Plasmodium falciparum* 3D7, were isolated from the flowers of *C. siamea*.

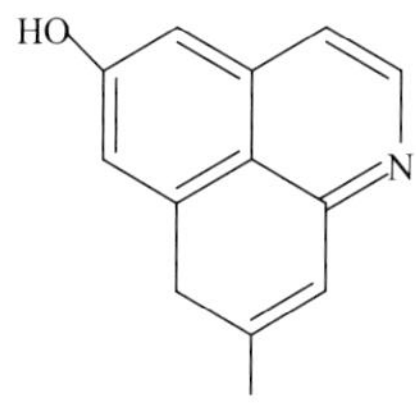

Fig. 23.2 Structure of Siaminine A.

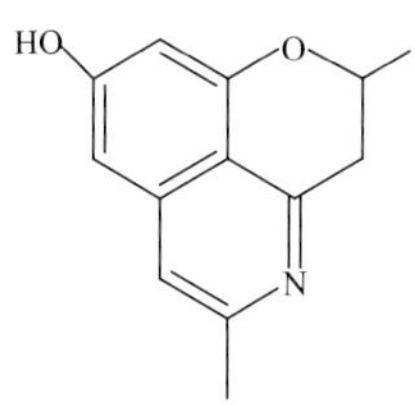

Fig. 23.3 Structure of Cassiarine C.

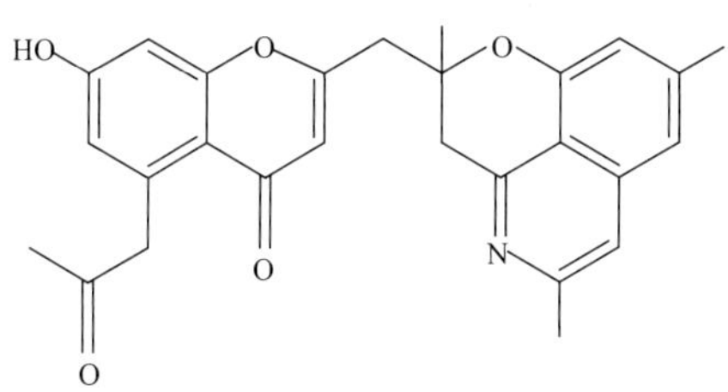

Fig. 23.4 Structure of Cassiarine D.

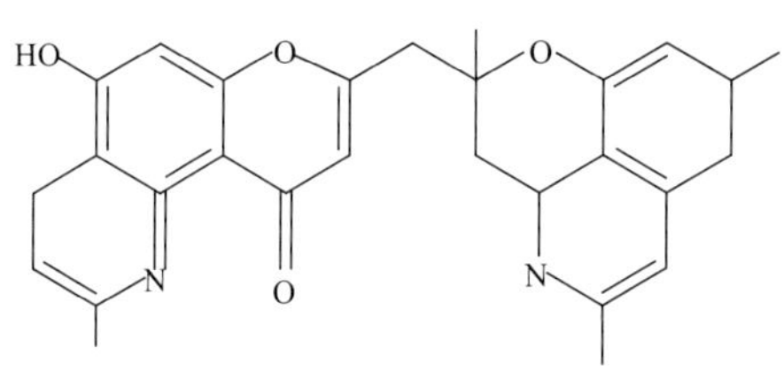

Fig. 23.5 Structure of Cassiarine E.

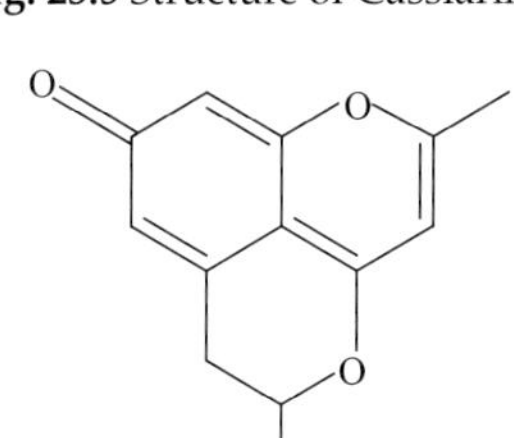

Fig. 23.6 Structure of 10,11-dihydroanhydrobarakol.

New bischromone, chrobisiamone A (Fig. 2.1) with an antiplasmodial activity has been isolated from the leaves.

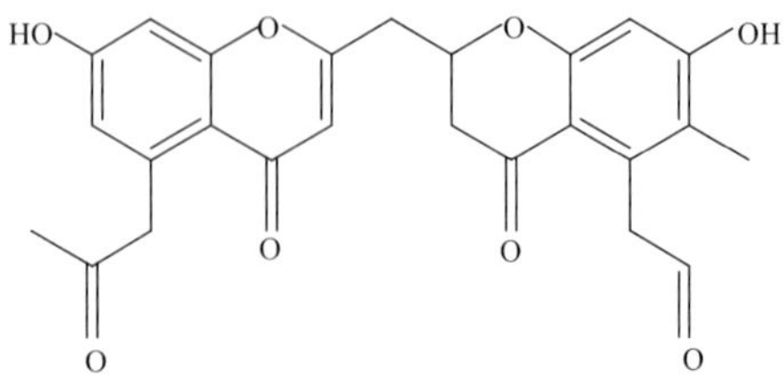

Fig. 23.7 Structure of Chrobisiamone A.

Siameanin, a bianthraquinone has been reported. Chrysophanol and physcion (Fig. 23.8) have been reported from the trunk bark.

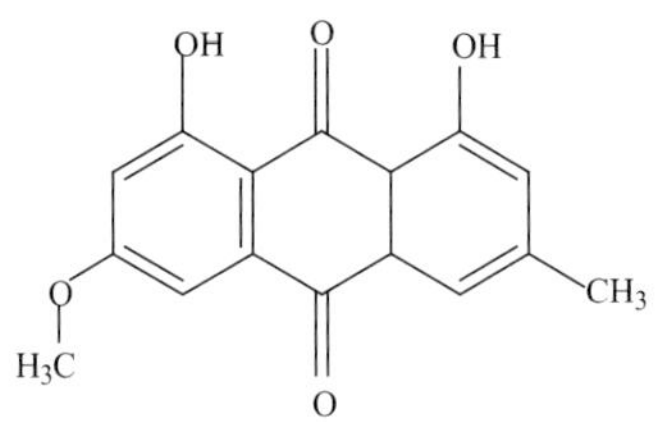

Fig. 23.8 Structure of Physcion.

From heartwood of *Cassia siamea*, a new bianthraquinone, 4,4′-bis(1,3-dihydroxy-2-methyl-6, 8-dimethoxy anthraquinone), along with 1,1′-bis(4,5-dihydroxy-2-methyl anthraquinone), chrysophanol and emodin (Fig. 2.4) have been isolated. New triterpenoid glycoside, 19α, 24-dihydroxyurs-12-ene-28-oic acid- 3-*O*-β-D-xylopyra-noside and anthraquinones have been reported.

Chrysophanol, chrysophanol-1-O-beta-D-glucopyranoside and 1-[(beta-D-glucopyranosyl-(1–>6)-0-beta-D-glucopyranosyl) oxy]-8-hydroxyl-3-methy-9,10-anthraquinone from the stem of *C. siamea* [14]. A new 10-hydroxyl anthrone glycoside, 1, 8, 10-trihydroxyl-1–O–β–D–glucopyrano-syl-3-methyl-10-C (S)–β–D-glucopyranosyl-anthrone-9 (Figure 23.9) has been isolated from the stem of *C. siamea*.

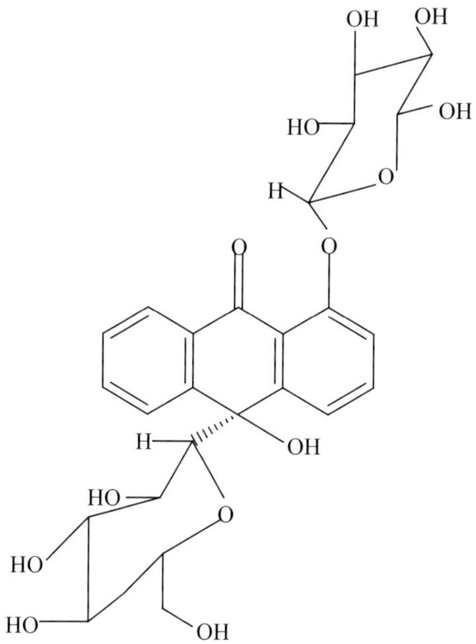

Fig. 23.9 Structure of 1, 8, 10-trihydroxyl-1–O–β–D–glucopyrano-syl-3-methyl-10- C (S)–β –D- glucopyranosyl-anthrone-9.

Triterpenes including lupenone and lupeol have been reported.

Seed oil is reported to be a minor source of vernolic acid and cyclopropenoid fatty acids.

23.3 Ethnopharmacology

Purgative effects: The present study investigated the purgative effects of barakol on the longitudinal smooth muscle contractions of the rat ileum. The extract increased the force of spontaneous muscle contractions in a concentration-dependent manner (EC50 = 0.3 mM). Saxitoxin (0.3 mM) abolished the stimulatory effects of barakol. In addition, atropine (10 mM), partially inhibited barakol-induced smooth muscle contractions suggestive of involvement of cholinergic nerves. The motor effects of barakol were further examined in muscle strips treated with catecholamines to suppress spontaneous contractile activity and decrease muscle tone. Norepinephrine or dopamine (10 mM) decreased the amplitude of spontaneous contractions by 72% and 18%, respectively. The findings suggested the possible role of barakol in gut motility disorders.

Anxiolytic effects: The behavioral effects of an extract of *Cassia siamea,* and barakol, on an elevated plus-maze compared with diazepam were investigated. An aqueous extract of *C. siamea* (1, 6, and 12 g/kg body wt., orally) produced a small increase in the percentage of the open : total number of arm entries and time, time spent on the end of the open arms, total number of arm entries, and number of rears/min. Barakol [10 mg/kg, intraperitoneally] significantly increased all the behavioral parameters in a manner similar to diazepam (I mg/kg, IP, 30 or 60 min before testing), except that barakol and not diazepam increased both the number of rears and total arm entries. Barakol at 25 and 50 mg/kg increased the percentage of the open: total number of arm entries and time and number of rears. The results are suggestive of anxiolytic properties of barakol similar to diazepam. However barakol differed from diazepam in that it increases exploratory and locomotor behavior.

CNS inhibitory effects: Barakol reduced spontaneous locomotor activity, increased the number of sleeping animals and prolonged the thiopental-induced sleeping time, demonstrating a sedative effect. As regards interactions between barakol and standard convulsants, only a high dose (100 mg/kg, i.p.) of barakol slightly prolonged the latency of clonic convulsion induced by picrotoxin. Barakol (25–100 mg/kg, i.p.) suppressed methamphetamine (1 mg/kg, i.p.)-induced hyper-locomotor activity in a dose-dependent manner, indicating an effect on the dopaminergic system. The findings are suggestive of the inhibitory effect of barakol on dopamine release.

Antioxidant effects: 1. The antioxidant activity of alcoholic extract of *C. siamea* rich in polyphenols was investigated. At a concentration of 250 µg/ml, 96% of DPPH radicals and at 500 µg/ml, 42.7, 32.7 and 64.5% of O2–, H2O2 and NO respectively could be scavenged by *C. siamea* flower extract. The extract also inhibited OH radical induced oxidation of protein (BSA) and LPO in murine hepatic microsomes. *C. siamea* flower extract also exhibited a significant antioxidant activity in acute oxidative tissue injury animal model constituted by CCl4 induced hepatotoxicity. The extract also protected against histopathological changes produced by CCl_4.

2. The antioxidant potential of hexane, chloroform, ethyl acetate, methanol, and water extract of bark and leaves of *C. siamea* was explored by superoxide anion radical scavenging assay. The different extracts showed significant inhibition of superoxide radicals in a dose-dependant manner. Among all the bark extracts of *C. siamea*, the methanol extract showed the maximum inhibition of 60.5% at 800 µg/ml concentration and water extract also showed strong antioxidant potential of 51.3% at 1000 µg/ml concentration. The various leaf extracts of *C. siamea* showed moderate antioxidant potential of 25–50% at 1000 µg/ml. This preliminary study indicates the antioxidant activity of the bark and leaves of *C. siamea*.

Antiplasmodial effects: The ethnobotanical use of *C. siamea* as a febrifuge led to the bioassay-guided fractionation of stem bark of the plant extract, using the parasite lactate dehydrogenase assay and multi-resistant strain of *Plasmodium falciparum* (K1) for assessing the *in vitro* antimalarial activity. Emodin and lupeol were isolated from the ethyl acetate fraction. Both compounds were found to be the active principles responsible for the antiplasmodial property with IC_{50} values of 5 µg/mL, respectively. Three alkaloids, cassiarins C–E, and a new chromone, 10,11-dihydroanhydrobarakol, isolated from flowers of *C. siamea* showed moderate antiplasmodial activity against *Plasmodium falciparum* 3D7.

Analgesic and anti-inflammatory effects

Analgesic, anti-inflammatory and antipyretic activities of the petroleum ether, chloroform, ethanol and water extracts of *C. siamea* stem-bark were assessed in rats with the hot plate test, paw pressure and carrageenan induced paw oedema, respectively. Cytotoxicity was assessed against KB and Vero cells. At the doses used (100, 200, and 400 mg/kg) ethanol and water extracts showed significant and dose-dependent analgesic and anti-inflammatory effects. None of the extracts had cytotoxic activity on KB and Vero cell lines and the most active extracts (ethanol and water extracts) had no acute toxicity.

Cardio protective effects

Barakol, isolated from leaves of *C. siamea,* was investigated on aconitine-induced cardiac arrhyihmias in anesthetized rats. The bolus intravenous injection of aconitine induced ventricular fibrillation and ventricular tachycardia. Pretreatment with barakol (10 mg/kg i.v.) reduced the incidence of aconitine-induced ventricular fibrillation and ventricular tachycardia, as well as mortality. The findings suggest that the mechanisms of the protective effects of barakol on aconitine-induced cardiac toxicity were related to the prevention of intracellular Na + accumulation.

Toxicity

Acute: Acute toxicity study of the ethanol extract of *C. siamea* in rats was conducted following intraperitoneal administration of graded doses. LD50 of *C. siamea* extract was found to be 9600 mg kg^{-1} body weight. Mortality occurred in rats treated with high doses of 4000, 5000, 6000, 7000, 8000 and 16000 mg kg^{-1} and appears to be dose dependent.

Subacute: The LD_{50} of barakol after oral administration was 2.33 g/kg. In an acute hepatotoxicity study, single-oral administration of various doses of barakol (60, 100 and 200 mg/kg) were given to rats and the animals were scarified at 24 h after the administration. Liver biopsy revealed no sign of liver damage when compared to those from a paracetamol treated positive control group.

In a subacute toxicity test, barakol at doses of 60, 120 and 240 mg/kg were orally administered daily for a period of 4 wk. A half of 240 mg/kg group, called a recovery group, was maintained for a further 2 wk without barakol administration. No mortality was observed in the controlled and barakol-treated animals. From the histological examination, the barakol-treated group showed only fatty changes in the liver. From blood chemistry determination, bilirubin was increased ($p<0.05$) in a dose-dependent manner and the value returned to normal values within 2 wk. The findings suggested regular monitoring of serum bilirubin with clinical use of barakol .

Chronic: A six-month toxicity study of *C. siamea* powdered leaves (containing 0.17% barakol) was performed in five groups of 15 Wistar rats of each sex. The control group received 5 ml of distilled water/kg BW/day orally, while three of the four experimental groups were given *C. siamea* powdered leaves at the doses of 20, 200 and 2,000 mg/kg BW/day. For the recovery study, the fourth treatment group received the powdered leaves at the dose of 2,000 mg/kg BW/day for 6 mon and was later kept without *C. siamea* administration for 14 more days before being scarified. The levels of bilirubin of the animals receiving the powdered leaf at the doses of 200 and 2,000 mg/kg BW/day were significantly higher than those of their controls. The hepatic damage was more pronounced in male rats. The results showed that long-term consumption of the leaf from *C. siamea* is linked to dose-dependent hepatotoxic effect in rats even at the therapeutic dose.

Further reading

Ajaiyeoba EO, Ashidi JS, Okpako LC, Houghton PJ, and Wright CW. 2001. Antiplasmodial compounds from *Cassia siamea* stem bark extract. Phytother. Res. 22, 254–255.

Chatterjee A, and Bhattacharjee SR. 1965. The structure of siamenain, a new dianthraquinone form *Cassia siamea* Lam. J. Nat. Inst. Sci. 31, 141–150.

Chavalittumron P, Chivapa S, Rattanajarasro S, Suyasootchare B, Soonthornchareonno N, and Punyamun S. 2003. Chronic toxicity of Cassia siamea Lam. leaves in rats. Monograph.

Chen FS, Ito H, Okabe F, Momose Y, Yamamura S, Ridititid W, Chaichantipyuth C, Leelasangaluk V and Tongroach P. 1999. Effects of Barakol on Aconitine-induced Cardiac Arrhythmias. Pharm. Biol. 37, 105–108.

Deachapunya C, Thongsaard W, and Poonyachoti S. 2005. Barakol suppresses norepinephrine-induced inhibition of spontaneous longitudinal smooth muscle contractions in isolated rat small intestine. J. Ethnopharmacol. 101, 227–232.

El-Sayyad SM, Ross SA, and Sayed HM. 1984. New Isoquinolone Alkaloids from the Leaves of *Cassia siamea*. Nat. Prod. 47, 708–710.

Hassanali-Walji A, King TJ, and Wallwork SC. 1969. Barakol, a novel dioxaphenalene derivative from *Cassia siamea*. J. Chem. Soc. Chem. Commun. 2, 678.

Kaur P, and Arora S. 2009. Superoxide anion radical scavenging activity of *Cassia siamea* and *Cassia javanica*. Med. Chem. Res. DOI 10. 1007/s00044-009-9274-9.

Kaura G, Alamb MS, J Zoobi, J Kaleem, and Athara M. 2006. Evaluation of antioxidant activity of *Cassia siamea* flowers. J. Ethnopharmacol. 108, 340–348.

Kinghorn AD, and Balandrin ME. 1992. Human Medicinal Agents from Plants. 2nd Ed. American Chemical Society, Washington. p. 62.

Koyama J, Morita I, Tagahara K, and Aqil M. 2001. Bianthraquinones from *Cassia siamea*. Phytochemistry. 56, 849–851.

Krishan Rao BV, and Murthy MN. 1988. *Cassia siamea* seed oil: a minor source of vernolic acid and cyclopropenoid fatty acids. J. Am. Chem. Soc. 2, 952–954.

Lü TS, Yi YH, Mao SL, Zhou DZ, Xu QZ, Tong HF and Zhang SY. 2001.Studies on the anthraquinones of *Cassia siamea*. Yao Xue Xue Bao. 36, 547–8.

Lu TS, Yi Yh, Zhang Zg, Zhang Zo, and Hua N. 2002. A New 10-Hydroxyl Anthrone Glycoside from *Cassia siamea* Lam. Chinese. Chem. Lett. 13, 731–733.

Monrudee S, Chaichantipyuth C, Murakami M, Michihisa T, Kinzo M, and Hiroshi W. 2003. CNS inhibitory effects of barakol, a constituent of *Cassia siamia* Lamk. J. Ethnopharmacol. 85, 175–176.

Morita H, Oshimi S, Hirasawa Y, Koyama K, Honda T, Ekasari W, Indrayanto G, Zaini NC. 2007. Cassiarins A and B, novel antiplasmodial alkaloids from *Cassia siamea*. Org. Lett. 9, 3691–3.

Ntanduo GFN, Banzouzi JT, Mbatchi B, Elion-Itou RDG, Etou-Ossibi GF, Ramos S, and et al., 2009. Analgesic and anti-inflammatory effects of *Cassia siamea* Lam. stem bark extracts. J. Ethnopharmacol. doi:10.1016/j. jep.2009.09.040.

Nsonde Ntandou GF, Banzouzi JT, Mbatchi B, Elion-Itou RD, Etou-Ossibi AW, Ramos S, Benoit-Vical F, Abena AA and Ouamba JM. Analgesic and anti-inflammatory effects of *Cassia siamea* Lam. stem bark extracts. J. Ethnopharmacol. 127, 108–111.

Oshimi S, Tomizawa Y, Hirasawa Y, Honda T, Ekasari W, Widyawaruyanti A, Rudyanto M, Indrayanto G, Zaini N and Morita H. 2008. Chrobisiamone A, a new bischromone from *Cassia siamea* and a biomimetic transformation of 5-acetonyl-7-hydroxy-2-methylchromone into cassiarin A. Bioorg. Med. Chem. Lett. 18, 3761–3763.

Oshimi S, Deguchi J, Hirasawa Y, Ekasari W, Widyawaruyanti A and Wahyuni TS. 2009. Cassiarins C–E, Antiplasmodial Alkaloids from the Flowers of *Cassia siamea*. J. Nat. Prod. 72, 1899–1901.

Pumpaisalchai W, Kaewvichit S, Siriaunkgul S, Taesothikul T, Niwatananun W, and Sanichwankul K. Toxicity of barakol: hepatotoxicity and subacute toxicity. Vol. 5: Quality, Efficacy, Safety, Processing and Trade in Medicinal and Aromatic Plants.

ISHS Acta Horticulturae 679: III WOCMAP Congress on Medicinal and Aromatic Plants.

Rai PP. 1977. Anthraquinones in *Cassia siamea*. Curr. Sci. 46, 814.

Roana MK, and Sharma M. 1979. Medicinal Plants of India. Ist Ed. Indian Council of Medical Research, New Delhi. pp. 45–46.

Singh HJ, and Agrawal C. 1994. New Triterpenoid Glycoside and Anthraquinones from *Cassia siamea*. Pharma. Biol. 32, 65–68.

Singh V, Singh J, and Sharma JP. 1992. Anthraquinones from heartwood of *Cassia siamea*. Phytochemistry. 31, 2176–2177.

Thongasaard W, Deachapunya C, Pongsakoen S, Boyd EA, Bennett GW, and Marsdern CA. 1996. Barakol: A potential anxiolytic extracted from *Cassia siamea*. Pharmacol. Biochem. Behav. 53, 753–758.

Wiam IM, Jacks TW, and Zongoma YA. 2005. Acute Toxicity and Phytochemical Studies of *Cassia siamea* Extract in Rats. Pak. J. Biol. Sci. 8, 586–589.

Chapter 24

Ethnopharmacology of *Tylophora asthmatica* Wight & Arn.

24.1 Introduction

T. asthmatica Wight & Arn. syn *Tylophora indica* Merr., is a perennial plant native to the South and East India. It belongs to the family Asclepidaceae and is commonly known as Indian Ipecacuanha. In Ayurveda, *T. asthmatica* is known as *arkaparni* or *antamool* [which is based on morphology of the roots]. The drug is official in Bengal pharmacopoeia.

24.2 Traditional uses

In Traditional Indian Medicine (Ayurveda), *T. asthmatica* is used in treatment of asthma, dermatitis and rheumatism. The plant has been described as bronchodilator, emetic, expectorant and diaphoretic.

24.3 Phytochemistry

T, asthmatica contains 0.2–0.3% of alkaloids having phenanthroindalizidine and furoquinoline framework. Tylophorine (Fig. 5.15) and tylophornine are important alkaloids encountered and the percentage is not affected by seasonal variations.

The phenolic alkaloid tylophorinidine (Fig. 24.1) isolated from *T. asthmatica,* along with minor alkaloids, *d*-septicine (Fig. 24.2) and *d*-isotylocrebrine have been isolated from the plant.

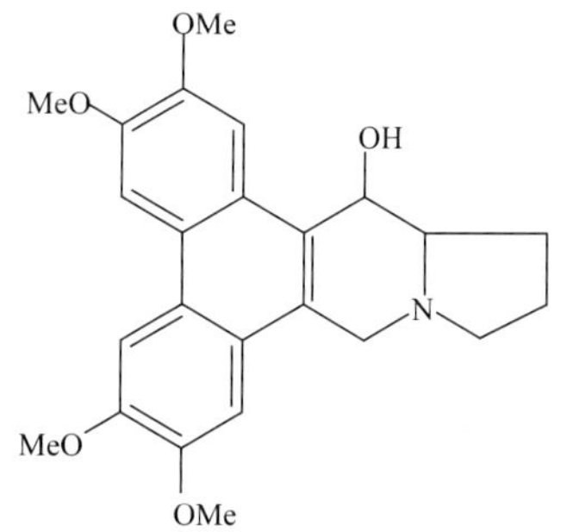

Fig. 24.1 Structure of Tylophorinidine.

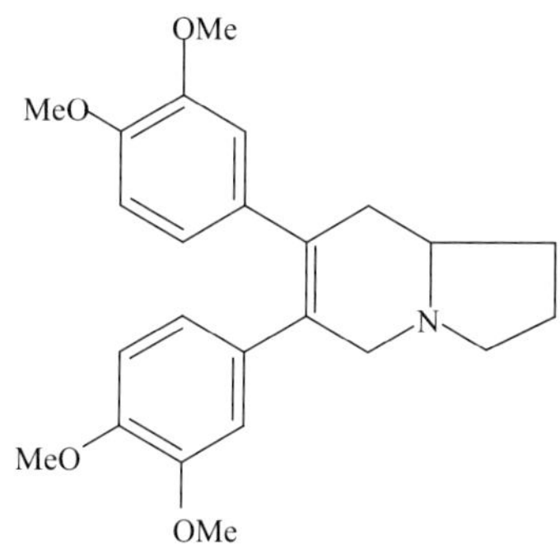

Fig. 24.2 Structure of Septicine.

Tylophorinicine, a minor alkaloid isolated from the roots of *T.asthmatica* has been characterized as 14-hydroxytylophorine. New alkaloids, desmethyltylophorine, and desmethyltylophorinine have been reported. Other alkaloids including, tyloindane (Fig. 24.3) and tyloindicine A-G (Figs. 24.4 and 24.5) have been characterized from the plant.

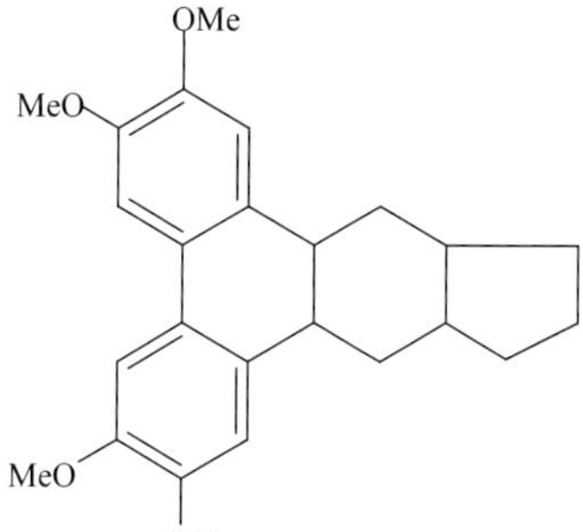

Fig. 24.3 Structure of Tyloindane.

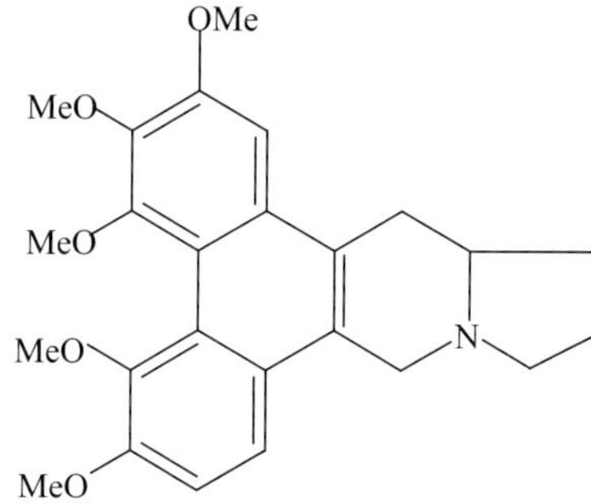

Fig. 24.4 Structure of Tloindicine A.

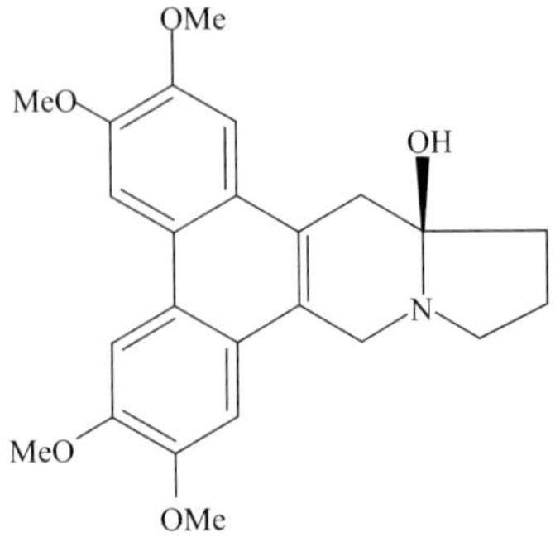

Fig. 24.5 Structure of Tyloindicine G.

The extract of *T. asthmatica,* marketed by pharmaceutical companies is standardized to contain 0.1% of the total alkaloids.

24.4 Ethnoharmacology

Anti asthmatic and anti allergic effects: The total alkaloids of *T. asthmatica* were tested for mast cell stabilizing effect in comparison with disodium cromoglycate by challenging against three different mast cell degranulators, diazoxide, carbachol and polymixin B, *in vitro*. Both tylophora alkaloids and disodium cromoglycate prevented in similar concentrations, the mast cell degranulation occurring with diazoxide alone. Tylophora alkaloids may have similar mechanism of action as disodium cromoglycate through cyclic AMP.

The effect of the alcoholic extract of *T. asthmatica* on weight of the adrenal glands and its functional activities and pituitary adrenal axis was studied on normal, unilaterally adrenalectomised, dexamethasone treated and hypophysectomised male albino rats. The extracts showed stimulation of adrenals as indicated by increase in weight and decrease in cholesterol and vitamin C. It was concluded that probably, *T. asthmatica* acts by direct stimulation of adrenal cortex.

Cytotoxic effect: A study reported cytotoxicity of alkaloids derived from *T. asthmatica* by Ali et al., (2001). Tylophora alkaloids induced apoptosis in K562 cells with characteristic apoptotic features like nuclear condensation, apoptotic body formation, flipping of membrane phosphatidylserine, activation of caspase 3 and release of mitochondrial cytochrome c. These studies suggest that the Tylophora alkaloids, in addition to their antiproliferative effects also induce apoptosis in erythroleukemic cells. These observations imply that Tylophora alkaloids could be useful molecules for their antiproliferative activity and for induction of apoptosis in tumor cells.

Four tylophorine analogs, designated DCB-3500, DCB-3501, DCB-3502, and DCB-3503. All four tylophorine analogs exerted potent growth-inhibitory effects against HepG2, a human hepatocellular carcinoma cell line, and KB, a human nasopharyngeal carcinoma cell line. HepG2 cells. Unlike conventional antitumor drugs, 3 micro M DCB-3503 did not cause DNA breaks or apoptosis in HepG2 cells. Tylophorine analogs had an inhibitory effect on cyclic AMP response elements, activator protein-1 sites, or nuclear factor-kappaB binding site-mediated transcriptions.

Hepatoprotective effect: The methanolic extract of *T. asthmatica* leaves was screened for hepatoprotective activity in carbon tetrachloride induced hepatotoxicity in albino rats. Hepatoprotective activity of methanolic extract at a dose of 200 mg/kg and 300 mg/kg body weight, i.p., was compared with Silymarin (25 mg/kg, i.p.) treated animals. *Tylophora indica* leaves (200 and 300 mg/kg) exhibited significant reduction in serum hepatic enzymes when compared to rats treated with carbon tetrachloride alone. Furthermore, histopathological studies were also done to support the study.

Immunosuppressive effect: Crude extract of the leaves of T. *asthmatica* inhibited delayed hypersensitivity reaction to sheep red blood cells in rats when the alkaloid mixture was administered before and after immunization with these cells. The alkaloid mixture also inhibited contact sensitivity to dinitro-fluorobenzene in mice when given prior to or after contact sensitization. Lymphocytes taken from contact sensitized mice, when treated with tylophora alkaloid *in vitro* and transferred into naive syngeneic hosts, could suppress the transfer of delayed type hypersensitivity (DTH) response.

Imuunomodulator effect: Con A induced proliferation of splenocytes was used as a model system to study the effect of the alkaloids on cellular immune responses. The alkaloid mixture was found to inhibit proliferation of splenocytes at higher concentrations and augment the same at lower concentrations. Both macrophages and T cells were found to be vulnerable to tylophora alkaloids. Inhibition of proliferation at higher concentrations of the alkaloid is due to inhibition of IL-2 production and activation of macrophages, which have a cytostatic effect.

Clinical studies: One such study randomly assigned 110 bronchial asthma patients to receive one *Tylophora asthmatica* leaf (150 mg of the leaf by weight) or a comparable placebo to be chewed and swallowed daily in the early morning for 6 d. At the end of 1 wk, 62% of the patients consuming the tylophora reported experiencing moderate to complete relief of their asthma symptoms compared to 28% in the placebo group. In a follow-up study, the alcoholic extract of crude tylophora leaves in 1 gram of glucose had comparable effects to that of chewing the crude leaf, with 56% of the patients reporting moderate to complete improvement in asthmatic symptoms compared to 32% in the placebo group.

In a double-blind placebo-controlled crossover study of 195 individuals with asthma, participants showed significant improvement when given 40 mg of a *T.asthmatica* alcohol extract daily for 6 d as compared to the placebo. Surprisingly, the difference was even more marked months after use of the herb had been stopped. Similar long-lasting results were seen in two double-blind placebo-controlled studies involving over 200 individuals with asthma.

In another clinical trial, 30 patients with a diagnosis of bronchial asthma for at least 2 yr were assigned at random to one of two treatment groups consisting of 15 individuals each. One group received either 350 mg of tylophora leaf powder or placebo daily in the first week. In comparison, a second group of asthmatics were given a similar amount of the leaf for 7 d followed by an anti-asthmatic drug combination. Overall, the results of the study showed that the amount of oxygen in the lung increased in those using the leaf, but decreased in those using the placebo.

However, the designs of most of these studies are a bit convoluted, and various pieces of information are missing from the reports, making it difficult to evaluate the validity of these trials. A higher quality double-blind study that enrolled 135 individuals found no benefit from *T. asthmatica* in asthma. In another double-blind placebo-controlled study, *Tylophora asthmatica* produced a significant reduction in sneezing and nasal obstruction, and the improvement noted in ventilatory capacity lasted for nearly 10 d.

Toxicology: One report described dermatitis as toxic effect of the alkaloids from *T.asthmatica.* The alkaloids tylophorine and tylophorinine, are described as vesicant but no formal skin testing appears to have been reported in the literature. Administration of pure alkaloid of *T. asthamatica*, suspended in peanut oil and given in single doses (12–100 mg/kg) by gavage, to male rats caused inactivity, respiratory distress, salivation, nasal discharge and diarrhea. The oral LD50 value of the alkaloid was 35.32 mg/kg.

References and further reading

Ali M, Ansari SH, and Grever MR. 2001. Cytotoxic alkaloids from *Tylophora indica*. Pharmazie. 56, 188–90.

Bhavan BV. 1992. Selected Medicinal Plants of India. Tata Press, Bombay , India. p. 33.

Dikshith TS, Raizada RB, and Mulchandani NB. 1990. Toxicity of pure alkaloid of *Tylophora asthamatica* in male rat. Indian J. Exper. Biol. 28, 208–212.

Etherington T, Herbert RB, and Jackson FB. 1977. Furoquinoline alkaloids from *tylophora asthmatica*. Phytochemistry. 7, 1125–1126.

Ganguly T. 2001. Inhibition of cellular immune responses by in experimental models. Phytomedicine. 8, 348–355.

Ganguly T, and Khar A. 2002. Induction of apoptosis in a human erythroleukemic cell line K562 by tylophora alkaloids involves release of cytochrome c and activation of caspase 3. Phytomedicine. 9, 288–95.

Ganguly T, Badheka LP, and Sainis KB. 2001. Immunomodulatory effect of *Tylophora indica* on Con A induced lymphoproliferation. Phytomedicine 8, 431–7.

Gao W, Lam W, Zhong S, Kaczmarek C, Baker DC, and Cheng YC. 2004. Novel mode of action of tylophorine analogs as antitumor compounds. Can. Res. 64, 78–688.

Geetha VS, Viswanathan S, and Kameswaran L. 1981. Comparison of total alkaloids of *Tylophora indica* and disodium cromoglycate on mast cell stabilization. Indian. J. Pharmacol. 13, 199–201.

Gore KV, Rao AK, and Guruswamy MN. 1980. Physiological studies with *Tylophora asthmatica* in bronchial asthma. Indian. J. Med. Res. 71,144–148.

Govindachari TR, Pai BR, Ragade IS, Rajappa S, and Viswanathan N. 1961. Chemical examination of *Tylophora asthmatica*—V: Structure of tylophorinine. Tetrahedron. 1961, 14, 288–295.

Govindachari TR, Viswanathan N, and Radhakrishnan J. 1973. Minor alkaloids of *Tylophora asthmatica* Revised structure of tylophorinidine. Tetrahedron. 29, 891–897.

Gupta S, George P, Gupta V, Tandon VR and Sundaram KR. 1979. *Tylophora indica* in bronchial asthma double blind study. Ind J Med Res. 69, 1, 981–989.

Karnick CR. 1975. Phytochemical investigations of some Tylophora species found in India. Planta. Med. 27, 333–6.

Mathew KK, and Shivpuri DN. 1974. Treatment of asthma with alkaloids of *Tylophora indica*: a double-blind study. Asp. Aller. App. Immunol. 7, 166–179.

Mujeeb M, Aeri V, Bagri P, and Khan SA. 2009. Hepatoprotective activity of the methanolic extract of *Tylophora indica* (Burm. f.) Merill. leaves. Int. J. Green Phar. 3, 125–7.

Mulchandani B, and Venkatachalam SR. 1984. Tylophorinicine, a phenanthroindolizidine alkaloid from *Tylophora asthmatica and pergularia pallida*. Phytochemistry. 23, 1206.

Nadkarni K. 1976. Indian Materia Medica. Vol 1. Popular Prakashan, Bombay, India. p. 78.

Rao KV, Wilson RA, and Cummings B. 2006. Alkaloids of tylophora III: New alkaloids of *Tylophora indica* (burm) merrill and *Tylophora dalzellii* hook. F. J. Pharm. Sci. 11, 1725–1726.

Shivpuri DN, Singhal SC, and Parkash D. 1972. Treatment of asthma with an alcoholic extract of *Tylophora indica*: a cross-over, double-blind study. Ann. Aller. 30, 407–12.

Thiruvengadam KV, Haranatii K, Sudarsan S and Sekar TS. 1971. *Tylophora indica* in bronchial asthma: a controlled comparison with a standard anti-asthmatic drug. J. Indian. Med. Assoc. 71, 172–176.

Udupa AL, Udupa SL, and Guruswamy MN. 1991. The possible site of anti-asthmatic action of *Tylophora asthmatica* on pituitary-adrenal axis in albino rats. Planta. Medica. 57, 409–13.

Chapter 25

Colchicine Containing Medicinal Herbs

25.1 Introduction

Discoroides in his book, De Materia Medica, described the use of the extract of genus *Colchicum* for the treatment of gout. In modern clinical practice, colchicine is prescribed in cases of gout and rheumatism not responding to standard medications. Gout is a clinical condition characterized by high-levels of uric acid in the blood resulting in arthritis.

Colchicine containing medicinal herbs is widely used in traditional systems of medicine including Ayurveda and Unani. In Ayurveda, Indian colchicum (*Colchicum autumnale* Linn. is used in anti-gout prescriptions, *suranjana vati* (tablet preparation) and *suranjanavelha* (confection preparation). Colchicine was official in the London Pharmacopoeia until 1639 and was reintroduced in 1778.

Cochicine (Fig. 5.5) is basically a water-soluble alkaloid, initially isolated from corms of Meadow saffron (*Colchicum luteum* Baker). In addition, colchicine has analgesic and anti-inflammatory activity which makes it ideal for anti-rheumatic therapy.

Colchicine has been also studied as an anticancer agent. Colchicine inhibits polymerization of microtubules by binding to tubulin. In addition to gout, colchicine is used in the treatment of psoriasis, rheumatoid arthritis, Becket's syndrome, penile condylomata acuminata, cardiomyopathy, familial Mediterranean fever, secondary amlydosis and scleroderma. Colchicine is toxic and a typical poisoning case mimics arsenic toxicity. Colchicine is known to cause leucopenia, dermatitis and alopecia and reported to be teratogenic in animal studies.

25.2 Medicinal herbs and colchicine

Table 25.1 Chief medicinal herbs containing colchicine.

S.No	Name of the herb	% of colchicine
1.	*Colchicum autumnale* Linn.	(0.1 to 0.8% (fresh flowers) 1.8% (dried flowers), (0.2 to 0.8%) seeds and corm (0.4 to 0.6%).
2.	*Colchicum luteum* Baker.	0.25% (corm}and 0.4% (seeds)
3.	*Gloriosa superba* Linn.	1.3% (tubers)
4.	*Iphigenia indica* (L.) Kunth	0.6-0.8 %
5.	*Sandersonia aurantiaca* Hook. [38]	4.7%

Colchicum autumnale Linn., commonly known as Autumn Crocus or Meadow Saffron, is native to Central and southeastern Europe and Africa. It has been introduced to Canada and the U.S. *Colchicum autumnale* is a small herbaceous perennial plant 10 to 40 cm high, flowering typically in the autumn after the leaves have disappeared. Leaves are lanceolate, dark green, shiny. They appear in the spring, and then die back before the flowers appear. Flowers: pink, purple or white flowers in groups of 1 to 6 are produced from an underground bulb.

All parts of the plant contain toxins. The greatest concentration of toxins is found in the seeds and the bulb. Colchicine is present in the flowers (0.1 to 0.8% in fresh flowers; up to 1.8% in dried flowers), in the seeds (0.2 to 0.8%) in the bulb (0.4 to 0.6%). The leaves contain very low amounts of colchicine. The other toxins present, which are closely related to colchicine, includes desacetylmethylcolchicine, desacetylthiocolchicine, colchicoside, demethyl desacetylcolchicine.

In traditional medicine, *C. autumnale* is classified as an alterative, aphrodisiac, carminative and laxative and used in gout, rheumatism and disorders of the liver and spleen.

Colchicum luteum Baker., commonly known as Yellow Autumn Crocus, Bitter Colchicum and Golden Collyrium, is found in East Asia-China to the Himalayas. It is an annual herb with stem-base below the ground, Flowers golden yellow, 1–2, on a very short stalk or scape among leaf-sheaths; fruits 2.5–3.8 cm long capsules with long recurved beaks, having numerous seeds; seeds brownish-white, globose or irregularly globular, 2–3 mm in diameter.

The herb contains alkaloids, including, colchicine, colchiceine, demecoicine or colchamine, cornigerine, lutiene, luteidine and collutine N-oxide. Corm contains 0.25% and seeds contain 0.4% of colchicine. Colchicine is readily soluble in water and decomposes into colchiceine. The corm also contains gallic acid, tannic acid, gum and starch. The poisoning of the plant is not unusual and comes through the confusion of the tuber with an onion or preparation of the leaves as a salad. Poisoning has been reported even by the milk of goats and sheep who have eaten the plant.

Gloriosa superba Linn., commonly known as Glory Lily, is another source of the alkaloid. It is a beautiful, herbaceous, tall, glabrous, branching leaf tip climber, about 1–3 m tall. Leaves alternate, opposite or tenately whorled, lanceolate, strongly nerved, nerves parallel, leaf-tip ending in tendril like spiral. Flowers large, showy, axillary, solitary on long pedicels, and nodding.

The plant contains alkaloids: colchicine corresponding to 1.3% (10), superbine and gloriosine. Superbine is toxic and allied to bitter principle of squill. 3-demethyl-N-formyl-N-deacetyl-b-lumicolchicine, 3-demethyl-g-lumicolchicine and 3-demethyl colchicines have been reported. Recently a glycoside (3-O-demethylcolchicine-3-O-alpha-D-glucopyranoside) has been isolated from the seeds.

Four fractions of *Gloriosa superba* L., i.e. hexane fraction, dichloromethane fraction 1, dichloromethane fraction 2, and methanol fraction, were investigated for colchicine-like activity using a mosquito cytogenetic assay. The results revealed that the latter three fractions yielded promisingly high colchicine-like activity, whereas the hexane fraction yielded very low activity compared with 1% colchicine in a 0.85% sodium chloride solution.

The clinical manifestations of colchicine toxicity include gastroenteritis, acute renal failure, cardio toxicity and hematological abnormalities In one case, a young woman ingested 125 g of tubers containing 0.3% colchicine which is equivalent of 350 mg of colchicine. Within 2 h the patient was vomiting and became unconscious on the next day. Acute ascending polyneuropathy and dermatitis have been reported following ingestion of tubers of the plant. Massive generalized alopecia has been reported. Anti-spermatogenic activity of extracts has been reported.

In traditional medicine, *G. superba* is classified as acrid, anthelmintic, laxative and abortifaceint. It is used in the treatment of chronic ulcers, leprosy, inflammation, hemorrhoids, skin diseases and intestinal worms. *G. rothchildiana* O Brien, *G.planti*, *G. lutea*, *G. casuariana* and *G. vuchuria* are also known to contain colchicine.

Iphigenia indica (L.) Kunth is found in the Himalayas, Nepal, India, Sri Lanka, Myanmar, Malaysia and Australia. It is an herb 20–40 cm high; stem simple or sparsely branched. Leaves mostly crowded along steam, linear or gradually tapering, flat or channeled, 5–20 cm long, 1.5–4.5 mm wide, glabrous. Inflorescence 1–4 flowered; pedicles erect, usually 5–20 mm long at anthesis; bracts 1-several per flower. It contains colchicine corresponding to 0.6–0.8%.

The seeds of *Iphigenia stellata* Blatt., indigenous to Maharastra (India) and is reported to contain the highest amount of colchicine. In addition, the plant contains colchicoside, 3-demethylcolchicine, 3-demethyl-N-deacetylcolchicine, multifloramine and kreysiginine.

Merendera persica sensu Hook. (*Colchicum aitchisonii* (Hook. f.) E. Nasir., *Merendera aitchisonii* Hook. f.) is found throughout Afghanistan and Iran. It is known as sweet suranjana. The corm is ovoid; sheath dark reddish-brown, neck long. Leaves 2–6, appearing with the flowers, linear, acute, 2–5 cm long at the flowering stage, upto 12 cm at the fruiting stage. It is reported to contain colchicine (24). Various species of the genus *Androcymbium* inlcuding *A. gramineum, A, gramineum, A. hierrense, A. palaestinum, A. psammophilum, A. rechingerii,* and *A.wyssianum* are reported to contain colchicine.

25.3 Ginkgo biloba Linn. and colchicine

Ginkgo biloba Linn. (Ginkgoaceae), has been used as peripheral vasodilator in traditional medicine. It is used in the treatment of tinnitus, dementia and intermittent claudication. Researchers identified a compound similar to colchicine in pooled placental blood. While tracing the origin of the compound, they examined blood samples of 24 women . Only five women showed the presence of colchicine and were consuming some herbal product (these were identified as Ginkgo and Echinacea products from local retail outlets). The ginkgo-based herbal products were found to contain significant amounts of colchicine. The authors of the study warned that consumption of ginkgo supplements by pregnant women may pose dangers to developing fetuses.

The findings about ginkgo were strongly criticized by several herbal experts and scientists. Some experts were of the view that the investigators misidentified a non-toxic ginkgo compound as colchicine. The ginkgo compound reportedly has a structure similar to that of colchicine and the researchers were not able to differentiate between the two. Further, a comprehensive review of the scientific data available on ginkgo never revealed the presence of colchicine in *Ginkgo biloba*.

References and further reading

Aitchison IET. 1887. Some plants of Afghanistan and their medicinal products. Am. J. Pharm. 59, 1.

Angunawela RM, and Fernando HA. 1971. Acute ascending polyneuropathy and dermatitis following poisoning by tubers of *Gloriosa superba*. Ceylon. Med. J. 16, 233–235.

Chommadov B, and Yusupov MK. 1994. Alkaloids of *Colchicum luteum*. Chem. Nat. Comp. 21, 770–773.

Clewer HWB, Green SJ, Tutin F. 1915. The constituents of *Gloriosa superba*. J. Chem. Soc. Trans. 107, 835–846.

Clewer HWB, and et al., 1915. The constituents of *Gloriosa superba*. J. Chem. Soc. Trans. 107, 835–846.

Dixit VP, Joshi S and Kumar A. Possible anti-spermatogenic activity of *Gloriosa superba* (E to H extract) in male gerbil (Meriones hurrianae Jerdon)—A preliminary study. Comp. Physiol. Ecol. 8, 117–122.

Dobelis I. 1989. Magic and Medicine of Plants. Reader's Digest Books. Pleasantville, NY p. 17.

Dunuwille R, and et al., 1968. Toxic principles of *Gloriosa superba*. Ceylon. J. Med. Sci. 17, 1–6.

Ellenhorn, MJ, Schonwald, S, Ordog, G. and Wasserberger, J. 1996.Toxic principles of *Gloriosa superba*. Ceylon. J. Med. Sci., 1968, 17, 1–6.

Ellington E, Bastida J, Viladomat F, Simanek V and Codin C. 2003. Occurrence of colchicine derivatives in plants of the genus Androcymbium. Biochem. System. Ecol. 31, 715–722.

Gooneratne BW. 1996. Massive generalized alopecia after poisoning by *Gloriosa superba*. Br. Med. J. 1, 023–1024.

Gürer MA, Keskin N, Gülekon A, Karel L, Aksakal B and Baysal V. 1977. Colchicine in Behçet's disease. Lancet. 310, 1037.

Javed M, Khan JA and Siddiqui MMH. 2005. Effect of *Colchicum luteum* Baker in the management of rheumatoid arthritis. Indian. J. Trad. Know. 4, 424–428.

Li W, Fitzloff JF, Farnsworth NR and Fong HH. 2002. Evaluation of commercial *Ginkgo biloba* dietary supplements for the presence of colchicine by high-performance liquid chromatography. Phytomedicine. 9, 442–446.

Mehra PN, and Khoshoo TN. 1951. Observations on some colchicine-containing plants. J. Pharm. Pharmacol. 3, 486–496.

Mendis S. 1989. Colchicine cardiotoxicity following ingestion of *Gloriosa superba* tubers. Postgrad. Med. J. 65, 752–755.

Mizushima. Y, and et al., 1977. Colchicine in Behçet's disease. Lancet. 310, 1037.

Nagaratnam N, De Silva DPKM and De Silva, N. 1973. Colchicine poisoning following ingestion of *Gloriosa superba* tubers. Trop. Geogr. Med. 25, 15–17.

Petty HR, and et al., 2001. Identification of colchicine in placental blood from patients using herbal medicines. Chem. Res. Toxicol. 14, 1254–8.

Sarin YK, and et al., 1974. Colchicine from the seeds of *Gloriosa superba*. Curr. Sci. 43, 87.

Singh NP, and Karthikeyan S. 2000. Flora of Maharashtra State: Dicotyledones, Vol. I (Ranunculaceae to Rhizophoraceae). Botanical Survey of India. Calcutta. 898, 84.

Srivastava UC, and Chandra V. 1975. *Gloriosa superba* Linn. (kalihari). An important colchicine producing plant J. Res. Ind. Med. 10, 92–95.

Stapczynski JS, and et al., 1981. Colchicine overdose: report of two cases and review of the literature. Ann. Emer. Med. 10, 364–369.

Tewari PV, and et al., 1972b. Experimental studies on the ecbolic properties of *Gloriosa superba* Linn. (kalihari). J. Res. Ind. Med. 7, 27–38.

Tewari PV, and et al. 1976. Preliminary pharmacological studies on *Gloriosa superba* Linn. (kalihari). J. Res. Ind. Med. 11, 34–39.

Trease GE, and Evans WC. 1966. A Textbook of Pharmacognosy. 9th Ed. Baillière, Tindall and Cassell. London. p. 902.

von Krogh G. 1978. Topical treatment of penile condylomata acuminata with podophyllin, podophyllotoxin and colchicine. A comparative study. Acta. Dermato-Venereologica. 58, 163–168.

Zachariae H, and et al., Colchicine in generalized pustular psoriasis: clinical response and antibody-dependent cytotoxicity by monocytes and neutrophils. Arch. Dermatol. Res. 1982, 274, 327–333.

Chapter 26

Lakshmana-Ayurvedic Drug of Controversial Origin

26.1 Introduction

Lakshmana is an important medicinal plant of Ayurveda, the ancient system of healing. Lakshmana is a type of *kantkāri,* a medicinal plant widely used in Ayurveda in the treatment of respiratory diseases. Lakshmana has been described as a white variety of kantkari, making it a possible representative of the Natural Order Solanaceae. Kantkari is an ingredient of dashmula, the Ayurvedic anti-inflammatory. The drug is of great interest as it has been mentioned as a possible treatment of female infertility. Bhāvamishra, an ancient Ayurvedic physician, also mentions kantkari for promoting conception in females.

26.2 Lakshmana in ancient Ayurvedic texts

Bhāvprakash Nighantu

Actions: Pungent, laxative, appetizer, light and hot in potency.

Therapeutics: Cough, asthma, fever, chronic rhinitis, myalgia, worm infestation and heart ailments. It pacifies *vāta* and *kapha*.

Raj Nighantu

Actions: Pungent, laxative, appetizer, light and hot in potency. It pacifies *vāta* and *kapha*.

Therapeutics: Loss of appetite and eye-ailments.

Dhanwantri Nighantu

The author has described *lakśmanā* as a variety of *brahatī.*

Actions and therapeutics: Bitter, pacifies *Vāta* and *Kapha* and cures indigestion and cough.

26.3 Medicinal plants of Solanaceae in Ayurveda

Several medicinal plants of the Natural Order of Solanaceae have found application in *Āyurvedic* formulations. *Kantkārī* (*Solanum xanthocarpum* Schrad et Wendl., *Solanum surattense* Burm.f., *Solanum virginianum* L.), *kākāmacī* (*Solanum nigrum* L.), *brahatī* (*Solanum indicum* L.) and *Solanum trilobatum* L. are some important plants.

26.4 Kantkāri (*Solanum xanthocarpum* Schrad et Wendl.)

English name: Yellow-berried-night shade. It is distributed in India, Sri Lanka and Pakistan. It is prickly, a much-branched herb, usually spreading or diffused; young branches are densely covered with minute star-shaped hair, prickles are yellow, shining about 1.5 cm long. Leaves are upto 10 cm long, their midribs and other leaves with sharp, yellow prickles. Flowers are purple, about 2 cm long, a few together in small branches, opposite to leaves. Fruit are 1.5–2.0 cm, round yellow or pale with green veins.

Phytochemistry: It contains alkaloids (scopolamine, solanidine and solasonine), ß-sitosterol and steroid saponin (disogenin).

Actions: Antitussive, bronchodilator, bitter, carminative and anodyne.

Therapeutics: *S. xanthocarpum* is primarily used in the treatment of chronic bronchitis and bronchial asthma. Taken with honey, tulsi (Ocimum sanctum), datura (Datura metal), and black pepper makes it effective in cases of bronchial asthma. Expressed juice of the berries is used in sore throat. Flowers and fruits are used to resolve burning sensation of the feet. Leaves are used to relieve pain locally.

Ethnopharmacology: Anti-nociceptive, antispermatogenic and hypotensive. Fruits and shoots have been reported to be antibacterial.

Clinical studies: Clinical efficacy of *S. xanthocarpum* was studied in bronchial asthma in a pilot study. *S. xanthocarpum* demonstrated anti-asthmatic effect in terms of various parameters of pulmonary function. However, the effect was less when compared to standard bronchodilators.

26.5 Possible representatives of lakshmana

26.5.1 *Ipomoea muricata* (L.) Jacq.(*Convolvulaceae*)

Common name: Purple moonflower. The seeds of *Ipomoea muricata* are largely imported into Bombay, from Iran, under the name of *tukm-i-nil*.

Distribution: Native to eastern India and Bangladesh.

Botany: Perennial vining climber to 30 feet. It is a rare climber, sporting unusual aerial rootless and white, funnel-shaped blossoms in the second year.

Phytochemistry: Work done in the Philippines has demonstrated the presence of indolizidine alkaloids in the seeds. Two resin glycosides and muricatins VII and VIII have been isolated from the seeds

Actions: Aphrodisiac.

Therapeutics: The juice of this plant is used to destroy bedbugs, and the seeds are said to be identical in their medicinal properties with those of the official plant. *Ipomoea muricata* (L.) Jacq, locally known as 'Tonkin', has been used for generations by the Dominicans in the Philippines for medicinal purposes. The seeds, stems and leaves are said to be effective in treating several types of skin ailments such as chronic and gangrenous wounds, cuts and blisters due to burns.

Ethnopharmacology: Analgesic, antiseptic, antimicrobial and antifungal.

26.5.2 *Cynoglossum lanceolatum* Forssk.(Boraginaceae)

Mishra in his work on rare Ayurvedic drugs has indicated *C. lanceolatum* as a possible candidate for lakshmana.

Distribution: Common throughout parts of Africa and Asia. It is also distributed in Madagascar.

Botany: Annual or biennial herb, the taproot 1-8 mm in diam.; stems erect, to *c.* 1 m tall, with sparse to moderate, appressed to spreading pubescence. Basal leaves in an evident rosette or smaller plants apparently immediately erect and lacking a basal rosette. Inflorescences terminal, once to several times dichotomously branched cymes, the branches strigillose; flowers on pedicels 1–7 mm long, bisexual; sepals narrowly ovate. Fruits 4.5–5.5 mm broad; nutlets ovoid, 2–3 mm broad.

Phytochemistry: Pyrrolizidine alkaloids: cynaustraline and cynaustine.

26.5.3 *Solanum ferox* Linn. (Solanaceae)

Foundation for Revitalization of Local Health Traditions (FRLHT) has mentioned *S. ferox* as a possible candidate for lakshmana.

Syn: *S. lasiocarpum* Dunal, *S. zeilanicum* Blanco.

Common name: Tarambulo (The Philippines).

Distribution: The Philippines, North East India, Malaysia and South China.

Botany: *S. ferox* is a small weed, suberect, prickly, hairy herb 0.5 to 1.5 meters in height the leaves are ovate, 15 to 20 cm long, 12 to 23 cm wide, lobed at the margins, and densely covered with stiff wooly hairs above and prickly spines on the nerves beneath; the lobes are triangular, and 2.5 to 4 cm deep. The flowers are borne on lateral racemes. Fruit is yellow, rounded, 2.5 to 3.5 cm in diameter, densely covered with needle- like hairs, and man-seeded.

Phytochemistry: Seeds contain fatty acids.

Therapeutics: In the Philippines, leaves of the plant are used as cataplasma for indolent swellings. The decoction is used in syphilis. Seeds are useful in toothache.

26.6 Further work

Proper identification of ancient drug *lakṣmanā* is a debatable topic. The drug has been mentioned as a cure for female infertility in ancient texts. Further, it is considered to be a type of *kantkārī*. Disogenin has been reported from various Solanum species like *S. xanthocarpum* Schrad et Wendl. and *Solanum khasianum* C.B. Clarke. Fruits of these species are in high demand for production of progesterones of natural origin. These are prized drugs for curing conditions like infertility and habitual abortions (coherence with the ancient texts). *S. khasianum* has white flowers. Ancient texts have not mentoined the detailed morphology of lakshmana, but presence of white flowers have been mentoined. Work on *S. khasianum* as a possible representative of lakshmana is warranted.

References and further reading

Chhote L, and Chunekar KC. 1985. Study of lakshmana in Samhitas, Sachitra. Ayurved. 37, 10, 601–605.

Garg, SK, and Gupta DR. 2006. Chemical Examination of the Seed Fat of *Solanum ferox* L. Fette, Seifen, Anstrichimittel. 68, 449–450.

Govindan S, Viswanathan S, Vijayasekaran V, and Alagappan R. 1999. A pilot study on the clinical efficacy of *Solanum xanthocarpum* and *Solanum trilobatum* in bronchial asthma. J. Ethnopharmacol. 66, 205–10.

Heble MR, and Narayanaswami S. 1968. Diosgenin and ß-Sitosterol: Isolation from *Solanum xanthocarpum* Tissue Cultures. Science. 161, 1145.

Karnick CR. 1976b. On the correct identity of the plants termed as "lakshmana"; a comparative, botanical, chemical, pharmacological, and Ayurvedic confirmation. Nat. Med. Gaz. 15, 1–9.

Sharma N, Sharma AK, and Zafar R. 1975. Indole alkaloids in the callus culture of *Ipomoea muricata* Linn. Indian Journal of Pharmaceutical Sciences. 52, 111–2.

Verma R. 1975. Lakshmana. Dhanvantri. 49, 320–323.

Yadev CL, and Chunekar KC. 1984. The Wonder Ayurvedic lakshmana for progeny. A historical appraisal. Ind. J. His. Sci. 19, 272–278.

Ysrael MC, Waterman P, and Nonato MG. 1997. Identification of phenylpropanoids, phenylpropanoid glucosides and a phenylethanol diglucoside from *Ipomoea muricata* Jacq. Convolvulaceae. *Acta-Manilana* (Philippines). 5, 13–20.

Chapter 27

Antidiabetic potential of *Pterocarpus marsupium*

27.1 Introduction

P. marsupium, or the Indian Kino Tree is a medium to large, deciduous tree that can grow up to 30 meters tall. It is native to India, Nepal, and Sri Lanka. In India it is found in parts of the Western Ghats in the Karnataka-Kerala region. *Pterocarpus marsupium* is one of the most highly valued trees of Eastern herbal medicine, renowned for its blood sugar-lowering effect. It is commonly known as vijaysar and was recommended as early as 1000 BC by Sushruta for the treatment of diabetes. In traditional practice, diabetic patients were given water which had been allowed to sit overnight in a goblet made from *P. marsupium* heartwood.

27.2 Phytochemistry: *P. marsupium* is a rich source of polyphenolic compounds including homopterocarpin (Fig. 27.1) and pterocarpin (Fig. 27.2).

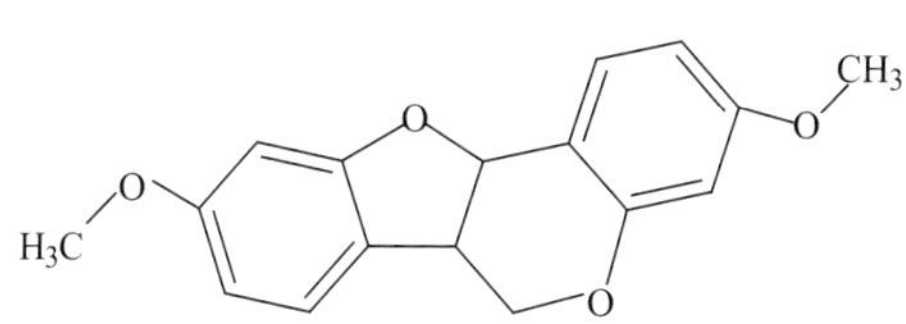

Fig. 27.1 Structure of Homopterocarpin.

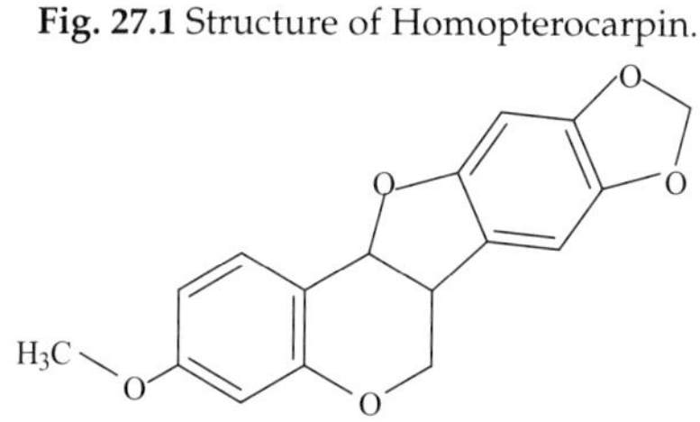

Fig. 27.2 Structure of Pterocarpin.

The other key compounds include the diaryl propane derivative, propterol (Fig. 27.3); the stilbene, pterostilbene (Fig. 27.4); the hydrochalcone, pterosupin (Fig. 27.5); the benzofuranone, marsupsin (Fig. 27.6); the flavanoid, liquiritigenin and the catechin, (-)-epicatechin. Propterol B has been characterized from a heartwood extract.

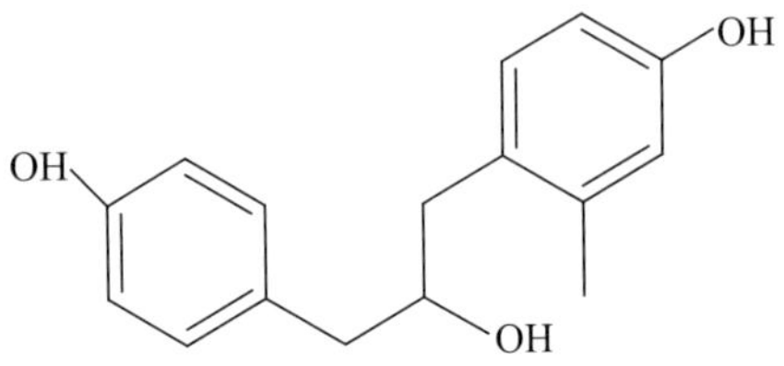

Fig. 27.3 Structure of Propterol.

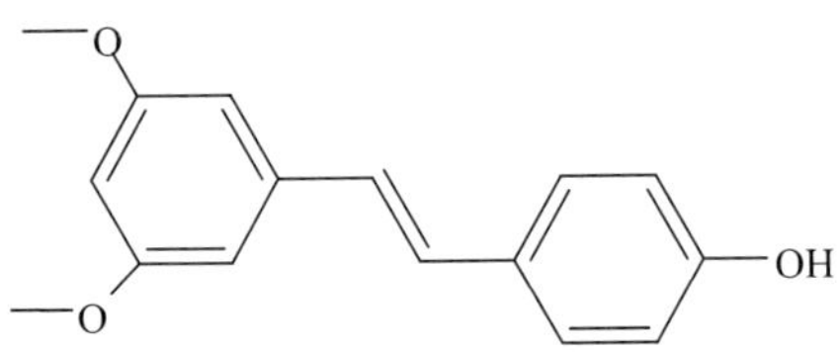

Fig. 27.4 Structure of Pterostilbene.

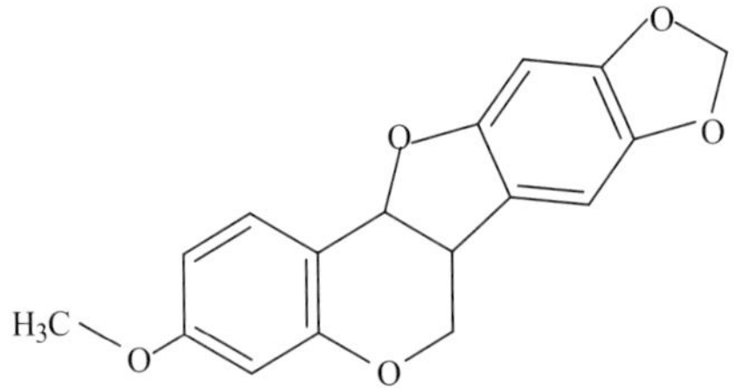

Fig. 27.5 Structure of Pterosupin.

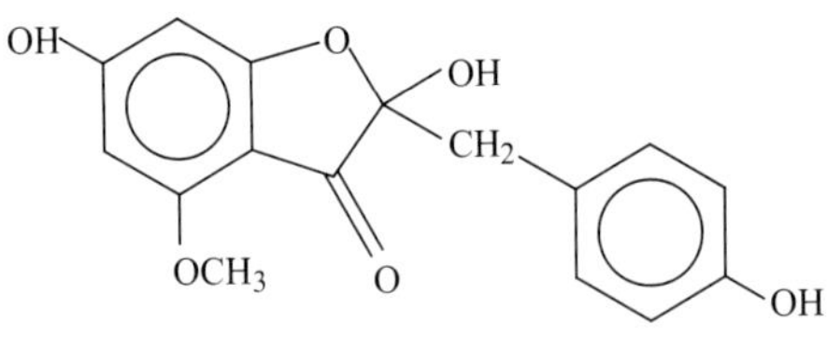

Fig. 27.6 Structure of Marsupsin.

Propterol, pterostilbene, pterosupin and marsupsin are the main components of heartwood, while (-)-epicatechin is found in the bark. The phenolics, marsupsin, pterosupin, and, pterostilbene have been identified as the blood sugar lowering components of heartwood, while (-)-epicatechin has been identified as the blood sugar lowering compound in the bark.

From the petrol extract of *P. marsupium,* a new sesquiterpene alcohol; elin-4(15)-en-1β,11-diol, β-eudesmol, erythrodiol-3-monoacetate and pterostilbene have been reported.

A novel isoflavonoid glycol, marsupol has been characterized from heartwood. 2-hydroxy-2-benzylcoumaranone, carpusin, has been an extractive of heartwood. Pterocarposide, an isoaurone *C*-glucoside has been reported from aqueous extract of heartwood.

27.3 Ethnopharmacology

1. A flavonoid fraction extracted from the bark of *P. marsupium* was studied for the hypoglycemic activity on normal and alloxanized albino rats. The flavonoid fraction did not show a consistent effect on normal blood sugar levels but it effectively reversed the alloxan-induced changes in the blood sugar level and the beta-cell population in the pancreas. The flavonoid fraction also showed a protective effect when it was given prior to alloxan administration.
2. Glucose levels in rats with hyperglycemia induced by streptozotocin were determined after i.p. administration of marsupsin, pterosupin, and pterostilbene. Marsupsin and pterostilbene significantly lowered the blood glucose level of hyperglycemic rats, and the effect was comparable to that of metformin.
3. Protective effects of the methanolic extract of *P. marsupium* heartwood were evaluated on NIDDM-induced rat gastric ulceration and mucosal offensive and defensive factors. NIDDM was produced in 5-d-old rat puppies by administering streptozotocin (70 mg/kg, i.p). The animals showing blood glucose level > 140 mg/dl after 12 wk of STZ administration were considered as NIDDM positive rats. *P. marsupium* (750 mg/kg) decreased the blood sugar level both in normal and NIDDM rats.
4. A study investigated the influence of methanolic extract of *P. marsupium* and isolated 7-O-α-l-rhamnopyranosyl oxy-4′-methoxy-5-hydroxy isoflavone on a battery of cellular targets Glut-4, PPARγ and PI3 kinase. The significant glucose uptake showed by *P. marsupium* crude and pure was comparable with insulin and rosiglitazone. Elevation of Glut-4 and PPARγ gene expression in parallel with glucose uptake supported the *in vitro* glucose uptake activity of *P. marsupium* methanolic extract and isoflavone. PI3 kinase plays an important role in glucose transport and activated by *Pterocarpus marsupium* methanolic extract but not the isolated pure isoflavone.

5. The antidiabetic activity of various subfractions of the alcohol extract of the bark of *P. marsupium* was evaluated in alloxan-induced diabetic rats. The results indicated that *P. marsupium* can control the diabetes related metabolic alterations apart from controlling the glucose levels. Among the fractions tested the butanol subfraction was found to be more active in comparison with other subfractions.

27.4 Clinical trials

1. A small clinical trial was carried out with 10 patients, who were given 200 ml of water, which was stored in a glass overnight, twice a day for one month, taken after lunch, whereas water stored for the whole day was drunk after dinner. There was an encouraging reduction of blood sugar from the second week of treatment and this hypo-glycemic activity continued as long as heartwood was given.
2. In clinical trials with human subjects, *P. marsupium* has shown remarkable anti-diabetic action. In two clinical trials, diabetic patients were given no treatment other than administration of an extract of *P. marsupium*. The first study evaluated both newly-diagnozed and untreated type 2 diabetics. Ninety seven patients were given varying doses of an extract of *P. marsupium* ranging from 2 to 4 gms a day over a period of 12 wk. At the end of the study period, parameters were evaluated for all the patients. The results showed that 67% of the patients were able to reduce and maintain glucose levels by using various amounts of *P. marsupium* extract. Of this group, 73% showed stabilized glucose levels at a daily dose of 2 gms of Pterocarpus extract, whereas 16% required 3 gms a day and 10% stabilized at 4 gms per day.
3. In another study using human subjects, 22 diabetics, mostly with type 2 diabetes, ranging in age from 29 to 70 yr were given a decoction of either 2 or 4 ounces three times daily made from 36 or 72 gms of dry bark of Pterocarpus marsupium respectively for 7 d.

Four parameters were monitored during this study: 1) fasting blood sugar, 2) glucose tolerance, 3) urine sugar content and 4) diabetic symptoms. The subjects were separated into two groups; Group A with 10 participants and Group B with 12. Group A received a decoction of 2 ounces 3 times a day, while Group B received 4 ounces 3 times a day.

Among the subjects in Group A, 3 out of 10 patients showed improvement in only one area of testing; glucose tolerance. In contrast, 9 of the 12 patients in Group B experienced benefits in all tested areas. These patients showed significant improvement in glucose tolerance and glucose uremia, and also a decrease in fasting blood sugar and amelioration of some diabetic symptoms. No undesirable side effects were noted during the course of the study.

Further reading

Adinarayana D, and Syamasundar KV. 1982. A new sesquiterpene alcohol from *Pterocarpus marsupium*. Phytochemistry. 21, 1083–1085.

Ahmad F, Khalid P, Khan MM, Chaaubey M, Rastogi AK, Jalil R, and Kidwai. 1991. Hypoglycemic activity of *Pterocarpus marsupium* wood. J. Ethnopharmacol. 35, 71–75.

Anandharajan R, Pathmanathan K, Shankernarayanan NP, Vishwakarma RA, and Balakrishnan A. 2005. Upregulation of Glut-4 and PPARγ by an isoflavone from *Pterocarpus marsupium* on L6 myotubes: a possible mechanism of action. J. Ethnopharmacol. 97, 253–260.

Chakravarthy BK, Gupta SJ, SS Gambhir SS, and Gode KD. 1980. Pancreatic beta cell regeneration—a novel antidiabetic mechanism of *Pterocarpus marsupium* Roxb. Ind. J. Pharmacol. 12, 123–127.

Dhanabal SP, Kokate K, Ramanathan M, Kumar EP, and Suresh B. 2006. Hypoglycaemic activity of *Pterocarpus marsupium* Rox. Phytother. Res. 20, 4–8.

Handa SS, Singh R, Maurya R, Satti NK, Suri KA, and Suri OP. 2000. Pterocarposide, an isoaurone C-glucoside from *Pterocarpus marsupium*. Tetrahedron. Lett. 41, 1579–1581.

Indian Council of Medical Research. 1998. Flexible dose open trail of Vijayasar in cases of newly-diagnosed non insulin dependent diabetes mellitus. Indian. J. Med. Res. 108, 24–9.

Jain SC, Sharma SK, Kumar R, Rajwanshi VK, and Babu BR. 1997. A homoisoflavanone from *Pterocarpus marsupium*. Phytochemistry 44, 765–766.

Joshi MC, Dorababu M, Prabha T, Kumar MM, and Goel RK. 2004. Effects of *Pterocarpus marsupium* on NIDDM-induced rat gastric ulceration and mucosal offensive and defensive factors. Ind. J. Pharmacol. 36, 296–302.

Manickam M, Ramanathan M, Farboodniay MA, Jahromi J, Chansouria PN, and Ray AB. 1997. Antihyperglycemic Activity of Phenolics from *Pterocarpus marsupium*. J. Nat. Prod. 60, 609–610.

Pandey MC, and Sharma PV. 1975. Hypoglycaemic effect of bark of *Pterocarpus marsupium* Roxb. Med, Sur. 2, 21–23.

Subba Rao AV, and Mathew J. 1982. Marsupol: A novel isoflavonoid glycol from *Pterocarpus marsupium*. Phytochemistr. 21, 1837–1838.

Subba Rao AV, and Mathew J. 1983. Carpusin: a novel 2-hydroxy-2-benzylcoumaranone from *Pterocarpus marsupium*. Phytochemistry. 22, 794–795.

Subba Rao AV, and Mathew J. 1984. Propterol B, a further 1,3-diarylpropan-2-ol from *Pterocarpus marsupium*. Phytochemistry. 23, 1814–1815.

Index

A

B

C

I

Q

Color Plate Section

Chapter 1

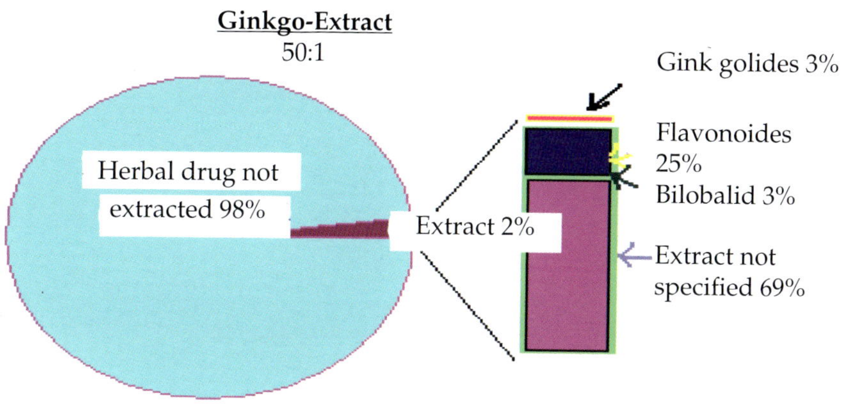

Fig. 1.3 Specification for Ginkgo extract.

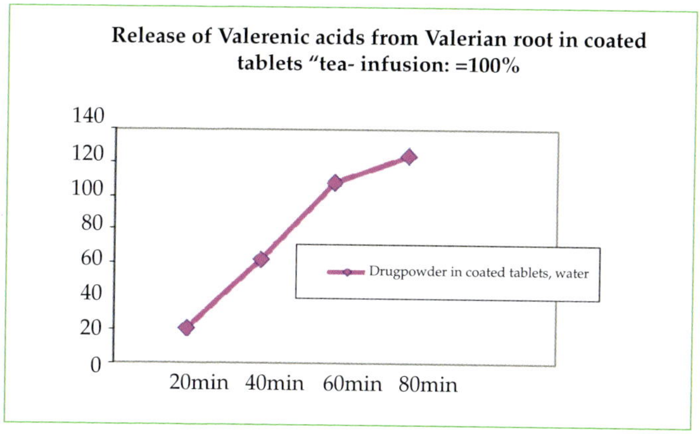

Fig. 1.4

Release of Silymarin % of the specified content

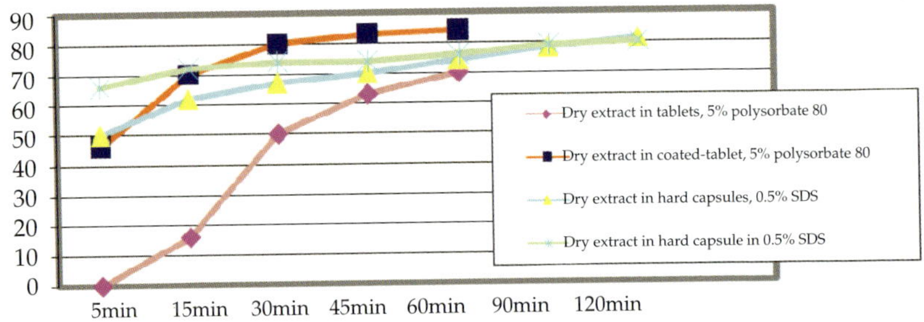

Fig. 1.5
Source: Keller (2001).

Bioavailability/Bioequivalence Problems/Question

Bearberry extract (Arctostaphylos uva-ursl, folium) active constituents: Arbutin, indication: cystitis

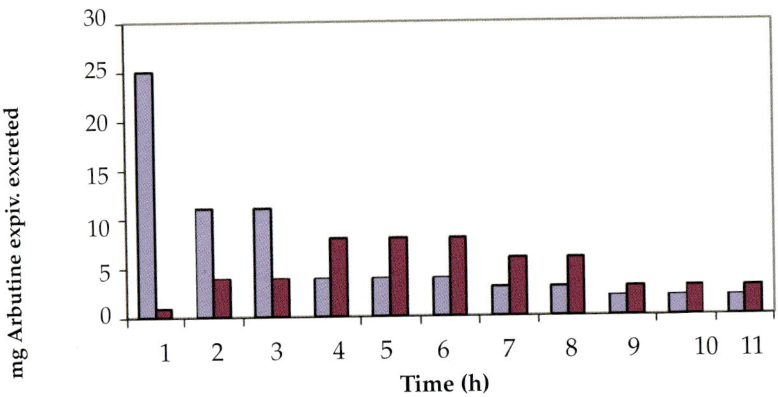

Fig. 1.6
Source: Keller (2001).

Chapter 2

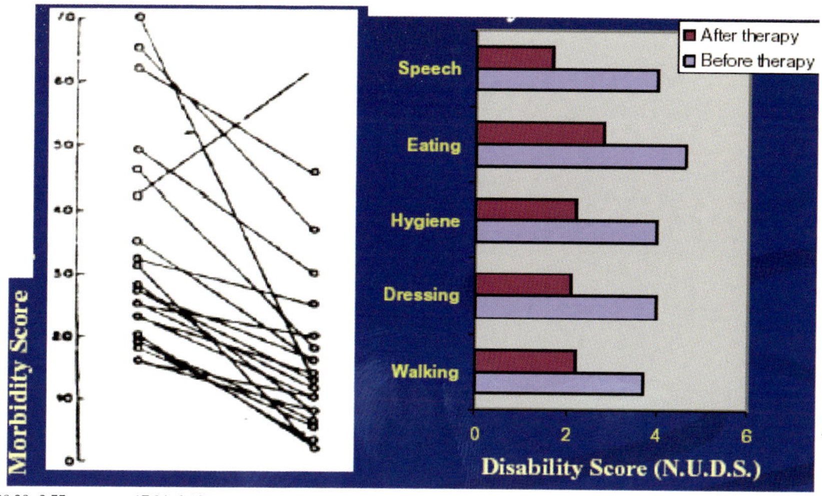

38.30±3.55 17.84±3.12
Mean ±S.E

Fig. 2.3 Clinical activity of *M. pruriens* in Parkinson's disease.

Chapter 6

BOTANICAL US Mass Market, 1998 (US Million)

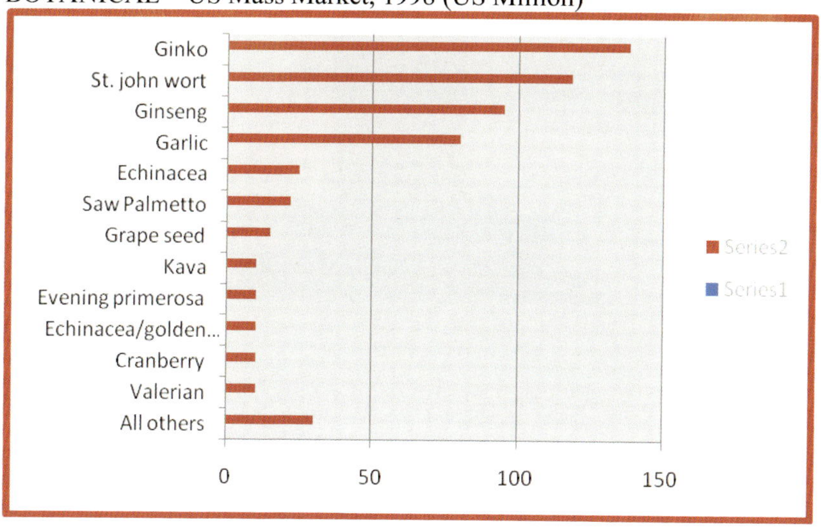

Fig. 6.18 Best selling herbs in US market in 1998.

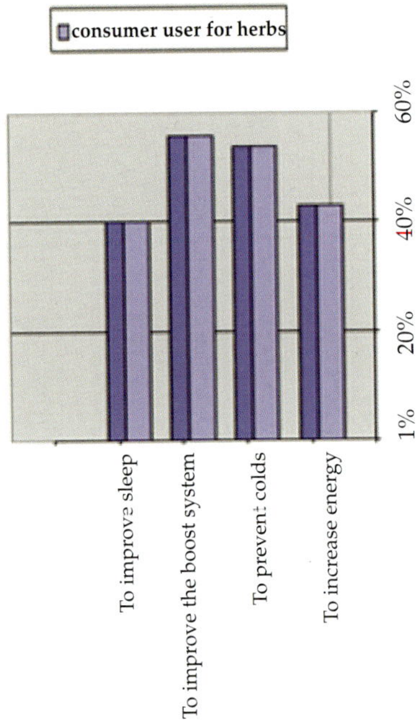

Fig. 6.19 Consumer perception of benefits of herbs.

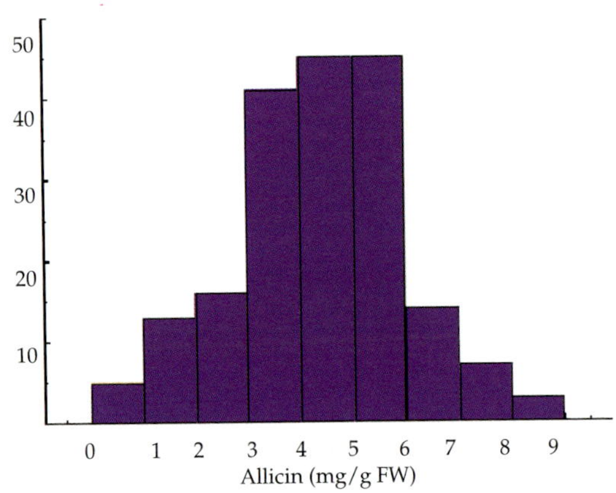

Fig. 6.21 Frequency distribution of allicin levels in 200 garlic samples. *Source*: Adapted with permission from Rural Industries Research and Development Corporation, Victoria.